The Kaggle Book
Second Edition

Master data science competitions with machine learning, GenAI, and LLMs

Luca Massaron

Bojan Tunguz

Konrad Banachewicz

‹packt›

Packt and this book are not officially connected with Kaggle. This book is an effort from the Kaggle community of experts to help more developers.

The Kaggle Book
Second Edition

Portfolio Director: Sunith Shetty

Relationship Lead: Tushar Gupta

Project Manager: Shashank Desai

Content Engineer: Tiksha Abhimanyu Lad

Technical Editor: Seemanjay Ameriya

Copy Editor: Safis Editing

Indexer: Rekha Nair

Proofreader: Tiksha Abhimanyu Lad

Production Designer: Ganesh Bhadwalkar

Growth Lead: Merlyn M Shelley

First published: April 2022

Second edition: December 2025

Production reference: 1151225

Published by Packt Publishing Ltd.

Grosvenor House

11 St Paul's Square

Birmingham

B3 1RB, UK.

ISBN 978-1-83508-320-8

www.packtpub.com

Foreword

I had a background in econometrics but became interested in machine learning techniques, initially as an alternative approach to solving forecasting problems. As I started discovering my interest, I found the field intimidating to enter: I didn't know the techniques or terminology and didn't have the credentials that would allow me to break in.

It was always my dream that Kaggle would allow people like me the opportunity to break into this powerful new field. Perhaps the thing I'm proudest of is the extent to which Kaggle has made data science and machine learning more accessible. We've had many Kagglers go from newbies to top machine learners, being hired at places such as NVIDIA, Google, and OpenAI, and starting companies such as DataRobot.

Luca, Bojan, and Konrad's book helps make Kaggle even more accessible. It offers a guide to both how Kaggle works and many of the key learnings that they have taken out of their time on the site. Collectively, they've been members of Kaggle for over 20 years, entered 330 competitions, made over 2,000 posts to Kaggle forums, and shared over 100 notebooks and 50 datasets. They are all top-ranked users and well-respected members of the Kaggle community.

Those who complete this book should expect to be able to engage confidently on Kaggle—and engaging confidently on Kaggle has many rewards.

Firstly, it's a powerful way to stay on top of the most pragmatic developments in machine learning. Machine learning is moving very quickly. In 2019, over 300 peer-reviewed machine learning papers were published per day. This volume of publishing makes it impossible to be on top of the literature. Kaggle ends up being a very valuable way to filter what developments matter on real-world problems—and Kaggle is useful for more than keeping up with the academic literature. Many of the tools that have become standard in the industry have spread via Kaggle. For example, XGBoost in 2014 and Keras in 2015 both spread through the community before making their way into industry.

Secondly, Kaggle offers users a way to "learn by doing." I've heard active Kagglers talk about competing regularly as "weight training" for machine learning. The variety of use cases and problems they tackle on Kaggle makes them well prepared when they encounter similar problems in industry. Because of competition deadlines, Kaggle trains the muscle of iterating quickly. There's probably no better way to learn than to attempt a problem and then see how top performers tackled the same problem (it's typical for winners to share their approaches after the competition).

So, for those of you who are reading this book and are new to Kaggle, I hope it helps make Kaggle less intimidating. For those who have been on Kaggle for a while and are looking to level up, I hope this book from three of Kaggle's strongest and most respected members helps you get more out of your time on the site.

Anthony Goldbloom

Founder of Kaggle

A Word from Some Kaggle Grandmasters

The Kaggle Book distills what really wins: structured problem framing, reproducible pipelines, and an honest treatment of feature engineering and validation. The first edition is the closest thing to a field manual I recommend to data teams—useful whether you're aiming for gold medals or production-grade models.

- Fahrettin Firat Gonen, PhD

I missed the first edition, but I won't miss the second. It's Data Science, Machine learning and AI from the perspective and experience of three Kaggle Competition GrandMasters. Any plans to climb the Kaggle rankings? This is the book.

- Marília Prata, retired Dental Doctor and Kaggle Legacy Grandmaster (mpwolke)

A practical and comprehensive guide by Kaggle pioneers, paving the way to Grandmaster level.

- Shotaro Ishihara, Senior Research Scientist at Japanese Media Company

I remember reading The Kaggle book when it was published, and I think that many parts are still relevant nowadays. I believe that the chapters about the metrics and the validation setup are the most important ones. Whether you participate in an ML competition or work on a project at your job, it is crucial to set up the validation approach. You need to be able to evaluate your approach and measure the improvements from the incremental experiments. The book does a great job at describing this and provides enough links to the materials for further study.

- Andrey Lukyanenko, Kaggle GDE. MLE @ Meta

Participating in Kaggle competitions has been an invaluable step in my journey to mastering data science and machine learning topics - and it has had a significant impact on my career. Luca, Bojan and Konrad are among the most knowledgeable and respected Kaggle Grandmasters in the community. Starting from foundational elements like proper validation patterns and scoring metrics, to more advanced topics such as stacking, the authors demonstrate a deep understanding of machine learning and Kaggle's inner workings, providing valuable insights to both the beginner and the experienced data scientist.

- Alberto Danese, Head of Data Science & Advanced Analytics at Nexi, Kaggle Competitions Grandmaster.

The Kaggle Book not only offers a detailed guide to tackle and participate in Kaggle competitions, but its insights and learnings can easily be applied to real-world industry problems.

- Parul Pandey, ML Consultant, Prev H2O.ai & Weights & Biases

As a Kaggle Grandmaster with over a decade of competition experience, I found The Kaggle Book to be an invaluable resource that I wish I had when starting out. The practical competition strategies and technical approaches shared here compress years of trial-and-error into actionable insights that will accelerate any data scientist's journey from beginner to medalist.

- Dmitry Larko, 3x Kaggle Competition GrandMaster

One of Kaggle's greatest gifts to the community is the opportunity to learn from the very best. The Kaggle Book distils this wisdom into a treasure trove of winning strategies and expert advice for machine learning practitioners.

- Martin (Head or Tails), Staff Data Scientist at Crunchbase

Contributors

About the authors

Luca Massaron is a data scientist with over a decade of experience in transforming data into high-impact, innovative artifacts, solving real-world problems, and generating value for businesses and stakeholders. He is the author of numerous bestselling books on AI, machine learning, and algorithms. Luca is also a 3x Kaggle Grandmaster who reached number 7 in the worldwide user rankings for his performance in data science competitions. Additionally, he is recognized as a **Google Developer Expert (GDE)** in AI, Kaggle, and the cloud.

My warmest thanks go to my family, Yukiko and Amelia, for their support and loving patience as I prepared this new book in a long series.

Bojan Tunguz is the founder and CEO of TabulAI, a start-up focused on applying machine learning and AI to structured-data problems. Before founding TabulAI, he worked at three other machine learning start-ups and most recently at NVIDIA. He holds a PhD in theoretical physics from the University of Illinois and has taught as a professor at three liberal arts colleges.

Konrad Banachewicz holds a PhD in statistics from Vrije Universiteit Amsterdam. His academic work focused on extreme dependency modeling in credit risk. In addition to his research activities, he was a tutor and supervised master's students. He transitioned from classical statistics to data mining and machine learning before "data science" became a buzzword.

Over the next decade, he tackled quantitative analysis problems in various financial institutions, becoming an expert in the full life cycle of a data product. His work spanned high-frequency trading to credit risk, predicting potato prices, and analyzing anomalies in the performance of large-scale industrial equipment. He is a believer in knowledge sharing and also competes on Kaggle.

I would like to thank my brother for being a fixed point in a chaotic world and continuing to provide inspiration and motivation. Dzięki, Braciszku.

About the reviewers

As a data science solutions architect, **Andrey Kostenko** specializes in the end-to-end development of data-driven solutions leveraging both predictive modeling and generative AI. With a PhD in mathematics and statistics and over a decade of diverse data science and machine learning experience, he translates technical vision into execution, mentoring teams on best practices for model development and deployment at scale. His extensive expertise across AWS, Azure, and GCP is key to operationalizing these models, delivering seamlessly integrated solutions that drive tangible business value.

Laura Fink is a Senior Data Scientist at H2O.ai, where she works on autonomous agentic systems and contributes to h2oGPTe, a platform that achieves top-tier results on GAIA, a benchmark that evaluates AI agents on diverse, real-world tasks. Her current work focuses on building robust autonomous agent systems that address complex challenges and demonstrate adaptive, autonomous decision-making inspired by biological organization. With a background in physics and as a Kaggle Notebooks Grandmaster, she enjoys exploring data puzzles, uncovering hidden insights, and sharing learning experiences. She combines curiosity, technical skill, and creativity to solve challenging problems and reveal what data can tell us about the real world.

Join our community on Discord

Join our community's Discord space for discussions with the authors and other readers:

`https://packt.link/kaggle`

Table of Contents

Preface   **xxi**

Free Benefits with Your Book xxvii

Part 1: Your Kaggle Launchpad: Mastering the Essentials   1

Chapter 1: Introducing Kaggle and Other Data Science Competitions   3

Free Benefits with Your Book • 4

The rise of data science competition platforms 4

The Kaggle competition platform • 7

Other competition platforms • 11

Introducing Kaggle competitions 13

Stages of a competition • 13

Types of competitions and examples • 17

Submission and leaderboard dynamics 24

Explaining the Common Task Framework paradigm • 24

Understanding what can go wrong in a competition • 26

Computational resources 28

Kaggle Notebooks • 30

Teaming and networking 32

Performance tiers and rankings 36

Criticism and opportunities 39

Summary 41

Chapter 2: Organizing Data with Datasets   **43**

Setting up a dataset 43

Gathering the data 49

Working with datasets 55

Using Kaggle datasets in Google Colab 56

Legal caveats 60

Summary 60

Chapter 3: Working and Learning with Kaggle Notebooks   **63**

Technical requirements 63

Setting up a notebook 65

Running your notebook 74

Saving notebooks to GitHub 76

Setting a notebook as a utility script 79

Getting the most out of notebooks' resources 80

Upgrading to Google Cloud Platform (GCP) 81

Going one step beyond 83

Kaggle Learn courses 89

Summary 93

Chapter 4: Kaggle Models   **95**

Selecting a Kaggle model 96

   Task • 96

   Data Type • 97

   Framework • 98

   Language • 99

   License • 100

   Fine Tunable • 101

   Size • 102

Using Kaggle Models 103

Uploading your model to Kaggle Models 108

Summary 110

Chapter 5: Leveraging Discussion Forums   113

How Kaggle forums work 114

Sample approaches to discussions 120

Discussions on common challenges • 120

Information leakage and overfitting • 122

Netiquette 125

Summary 125

Part 2: Elevating Your Game:
Advanced Techniques for Competitive Success   127

Chapter 6: Competition Tasks and Metrics   129

Evaluation metrics and objective functions 130

Basic types of tasks 132

Regression • 132

Classification • 133

Ordinal • 134

The Meta Kaggle dataset 135

Handling never-before-seen metrics 138

Metrics for regression (standard and ordinal) 142

Mean squared error (MSE) and R squared • 143

RMSE • 144

RMSLE • 145

MAE • 146

Metrics for classification (label prediction and probability) 148

Accuracy • 148

Precision and recall • 151

The F1 score • 153

Log loss • 154

ROC-AUC • 154

Matthews correlation coefficient (MCC) • 156

Metrics for multi-class classification 157

Metrics for object detection and segmentation problems 164

IoU • 167

Dice • 168

Metrics for multi-label classification and recommendation problems 169

Optimizing evaluation metrics 171

Custom metrics and custom objective functions • 172

Post-processing your predictions • 175

Probabilistic adjustments of the predictions • 178

Summary 183

Chapter 7: Designing Good Validation   185

Snooping on the leaderboard 186

The importance of validation in competitions 189

Bias and variance • 193

Trying different splitting strategies 196

The basic train-test split • 197

Probabilistic evaluation methods • 199

k-fold cross-validation • 199

Understanding how k-fold cross-validation works • 200

K-fold variations • 202

Sequential cross-validation in time series • 205

Validation in financial time series • 207

Nested cross-validation • 210

Producing OOF predictions • 212

Subsampling • 213

The bootstrap • 213

Tuning your model validation system 218

Using adversarial validation 221

Example implementation • 223

Handling different distributions of training and test data • 225

Handling data leakage 230

Feature leakage • 231

Leakage at the example level • 232

Handling data leakage in Kaggle • 233

Summary 234

Chapter 8: Modeling for Tabular Competitions 237

The Tabular Playground Series 238

Setting a random state for reproducibility 245

The importance of EDA 247

Performing EDA in Kaggle • 248

Dimensionality reduction with t-SNE and UMAP • 250

Reducing the size of your data 252

Speeding up data processing 254

Applying feature engineering 256

Easily derived features • 257

Meta-features based on rows and columns • 260

Target encoding • 262

Using feature importance to evaluate your work • 267

Pseudo-labeling 269

Denoising with autoencoders 271

AutoML for tabular competitions 276

Neural networks for tabular competitions 277

Summary 284

Chapter 9: Hyperparameter Optimization 287

Basic optimization techniques 288

Grid search • 289

Random search • 291

Halving search • 293

Key parameters and how to use them 296

Linear models • 296

Support vector machines • 297

Random forests and extremely randomized trees • 298

Gradient tree boosting • 300

LightGBM • 300

XGBoost • 302

CatBoost • 304

HistGradientBoosting • 306

Bayesian optimization 309

Using scikit-optimize • 311

Customizing a Bayesian optimization search • 317

Extending Bayesian optimization to neural architecture search • 325

Creating lighter and faster models with KerasTuner • 333

The TPE approach in Optuna • 342

Using Weights & Biases 352

Tracking experiments • 352

Versioning with artifacts • 356

Implementing Sweeps optimization • 356

Summary 359

Chapter 10: Ensembling with Blending and Stacking Solutions — 361

A brief introduction to ensemble algorithms 362

Averaging models into an ensemble 366

Majority voting • 369

Averaging of model predictions • 372

Weighted averages • 373

Averaging in your cross-validation strategy • 375

Correcting averaging for ROC-AUC evaluations • 376

Blending models using a meta-model 377

Best practices for blending • 378

Stacking models together 383

Performing stacking • 385

Stacking variations • 388

Creating complex stacking and blending solutions 389

Summary 394

Chapter 11: Modeling for Computer Vision 395

Augmentation strategies 396

Keras built-in augmentations • 401

The ImageDataGenerator approach • 401

Preprocessing layers • 404

A deeper dive: the albumentations package • 406

Image classification 409

Object detection 419

Semantic segmentation 433

Exploring a capstone case study: CZII – CryoET Object Identification competition • 452

The key characteristics of the dataset • 453

Modeling implications • 453

Talking about the data format • 453

Evaluation strategy and challenges • 454

Data exploration overview • 456

A first baseline solution, the 3D U-Net segmentation approach • 461

Starting from a Kaggle baseline • 463

Examining the second-place solution: an ensemble of lightweight 3D models • 465

How to perform ensembling and inference • 467

Looking at other top solutions and insights • 468

Overall trends emerged in the competition • 470

Summary 471

Chapter 12: Modeling for NLP   473

Sentiment analysis 473

Open domain Q&A 483

Toxic comments classification 499

Text classification with TF-IDF and logistic regression • 500

Importing the necessary libraries • 500

Defining the target labels and loading the dataset • 501

Word-level TF-IDF vectorization • 501

Character-level TF-IDF vectorization • 502

Combining word and character features • 503

Training the logistic regression model and cross-validation • 503

Calculating the total cross-validation score • 504

Saving the predictions to a CSV file • 504

Text preprocessing and cleanup • 504

Importing libraries, loading files, and setting up global variables • 505

Loading pretrained dictionaries • 507

Setting preprocessing parameters • 507

Defining contraction patterns • 508

Splitting toxic words • 509

Tokenizing with TweetTokenizer • 509

URL replacement • 510

Normalizing by dictionary • 510

Loading a spaCy model • 511

The main normalization function • 511

Reading and normalizing data • 512

Saving the processed data • 513

Text classification with RNNs 514

Imports and environment setup • 514

Loading preprocessed data • 515

Loading embeddings • 516

Splitting the datasets • 516

Building the Keras model • 516

Training and averaging multiple seeds • 517

Creating a submission file • 519

Text classification with DistilBERT 519

Setting up the environment and dependencies • 520

Loading and preparing the training data • 520

Creating a custom Dataset class for multi-label classification • 521

Splitting the data into training and validation sets • 522

Initializing the tokenizer and creating data loaders • 523

Defining the model architecture • 524

Preparing the model and optimizer for training • 524

Training loop • 525

Preparing and processing the test data • 526

Inference on the test data • 526

Formatting and saving the predictions • 527

Text classification with AutoTrain 528

Setting up the environment and dependencies • 528

Setting up AutoTrain parameters • 529

Initializing and creating the Autotrain project • 530

Loading a pretrained model and tokenizer • 531

Preparing the test data for inference • 531

Creating a custom Dataset class • 531

Running predictions with the trainer • 532

Text classification with LLM embeddings and logistic regression 533

OpenAI embeddings • 533

Initializing the OpenAI client • 534

Defining a helper function for embeddings • 534

Loading and cleaning the data • 535

Handling specific data anomalies • 535

Generating embeddings for the data • 535

Converting embeddings to NumPy arrays • 536

Saving the embeddings for later use • 536

NVIDIA embeddings • 536

Defining a function to get embeddings • 537

Setting the stage: Data and dependencies • 537

Cross-validation and training • 538

Iterating over each target • 538

Making predictions and recording performance • 539

Preparing the submission • 539

Text augmentation strategies 540

Basic techniques • 540

Text augmentation with back-and-forth translation • 546

nlpaug • 548

Summary 551

Chapter 13: Generative AI in Kaggle Competitions   **553**

Understanding generative AI and LLMs 554

The working of LLMs • 555

Unlocking global communication with Gemma: fine-tuning LLMs for new languages .. 560

Competition format and data • 560

Top solutions overview • 560

Fine-tuning Gemma in practice: example • 561

Key techniques used in fine-tuning • 562

LLM prompt recovery • 564

Competition overview • 564

Third-place solution (team prompt = "don't say anything") • 567

Origins of the mean prompt • 569

Quantitative outcome and qualitative lessons • 572

AI assistants for data tasks with Gemma 573

Competition overview • 573

Top solution: "PyGEM" – a Python programming chatbot • 575

Summary 578

Chapter 14: Simulation and Optimization Competitions 581

Technical requirements 582

Working with Connect X 582

Rock, Paper, Scissors 587

Santa competition 2020 590

A few other Kaggle game agent competitions 594

FIDE and the Google Efficient Chess AI Challenge 596

The 4th place solution: enhancing Stockfish with a small neural network • 598

Choosing the base engine • 598

Optimizing memory usage • 598

Enhancing the evaluation function with a neural network • 599

Code example and repository • 601

Broader implications • 602

Summary 604

Part 3: Kaggle for Your Career: Building Your Profile and Finding Opportunities 607

Chapter 15: Creating Your Portfolio of Projects and Ideas 609

Building your portfolio with Kaggle 610

The golden rules of a good portfolio • 615

Leveraging notebooks • 617

Leveraging discussions • 619

Leveraging datasets • 620

Arranging your online presence beyond Kaggle 625

Blogs and publications • 626

GitHub • 629

Making an online demo • 630

Writing a paper on arXiv • 630

Monitoring ongoing competitions 631

Summary 632

Chapter 16: Finding New Professional Opportunities   635

Building connections with other competition data scientists 636

Participating in Kaggle Days and other Kaggle meetups 649

Getting spotted and other job opportunities 649

How Kaggle can help • 650

The STAR approach • 651

Summary (and some parting words) 653

Chapter 17: Unlock Your Exclusive Benefits   655

Other Books You May Enjoy   661

Index   665

Preface

Having competed on Kaggle for so many years, we have experienced highs and lows in many competitions. While on this long journey, we often found ourselves refocusing our efforts on different activities related to Kaggle. Over time, we devoted ourselves not only to competitions but also to creating content and code based on the demands of the data science market and our own professional aspirations.

After a certain point in our journey, we started to feel that our combined experience and still-burning passion for competitions could really help other participants who have just begun or who would like to get inspired and make the decision to start, by giving them access to the essential expertise they need to begin their own journey in data science competitions.

We then decided to work on a book on Kaggle with a purpose:

To offer, in a single place, the best tips for being competitive and approaching most of the problems you may find when participating in Kaggle as well as other data science competitions

To offer enough suggestions to allow anyone to reach at least the Expert level in any Kaggle discipline: Competitions, Datasets, Notebooks, or Discussions

To provide tips on how to get the most out of Kaggle and leverage this experience for professional growth in data science

To gather the most significant perspectives on the experience of participating in competitions in a single source by interviewing Kaggle Masters and Grandmasters and listening to their stories and suggestions

In short, we have written a book that demonstrates how to participate in competitions successfully and take advantage of all the opportunities Kaggle offers.

We present here the second edition of this book, with updated chapters and content. It aims to provide even more help in an evolving landscape where, alongside the established tabular data, time series, computer vision, and NLP competitions, there is a growing number of competitions revolving around AutoML and powerful **Large Language Models (LLMs)**. These newer LLM competitions require skills in fine-tuning models like Llama, Mistral, Gemma, Qwen, and Deep-Seek for specific tasks.

As in the previous edition, this book is also intended as a practical reference to save you time and effort, thanks to its selection of many competition tips and tricks that are hard to learn about and find on the internet or on Kaggle forums.

Nevertheless, the book doesn't limit itself to providing practical help; it also aspires to help you figure out how to boost your career in data science by participating in competitions.

Please note that this book does not teach data science from the ground up. We do not provide detailed explanations of linear regression, random forests, or gradient-boosting functions. Instead, we focus on how to use these methods effectively to achieve the best results in data-related problems. We expect our readers to have a solid foundation in data science topics and at least a basic proficiency in Python usage.

If you are still a data science beginner, you must supplement this book with other data science, machine learning, and deep learning textbooks and train on online courses, such as those offered by Kaggle itself or by **Massive Open Online Courses (MOOCs)** such as edX or Coursera.

If you want to practically strengthen your data science knowledge through hands-on experience, challenge yourself with intriguing data problems, and simultaneously build a network of fellow data scientists as passionate as you are, then this is definitely the book for you.

Let's get started, then!

Who this book is for

At the time of this new edition's completion, there were 122,388 Kaggle novices (users who have just registered on the website) and 67,924 Kaggle contributors (users who have just filled in their profiles and made at least one submission to a competition) enlisted in the rankings of Kaggle competitions. This book has been written for all of them and for anyone else wanting to break the ice and start taking part in competitions on Kaggle and learning from them.

What this book covers

Chapter 1, Introducing Kaggle and Other Data Science Competitions, discusses how competitive programming evolved into data science competitions. It explains why the Kaggle platform is the most popular site for these competitions and gives you an idea about how it works.

Chapter 2, Organizing Data with Datasets, introduces you to Kaggle Datasets, the platform's standard method of data storage. We discuss setting up and gathering data, and utilizing it in your work on Kaggle.

Chapter 3, Working and Learning with Kaggle Notebooks, discusses Kaggle Notebooks, the baseline coding environment. We talk about the basics of notebook usage, how to leverage the GCP environment, and how to use the notebooks to build up your data science portfolio.

Chapter 4, Kaggle Models, introduces the new feature that allows you to upload models in a distinct way from datasets, explaining how to discover, utilize, and share pre-trained models directly on the platform, streamlining workflows, particularly for complex tasks like NLP and computer vision.

Chapter 5, Leveraging Discussion Forums, allows you to familiarize yourself with discussion forums, the primary manner of communication and idea exchange on Kaggle.

Chapter 6, Competition Tasks and Metrics, illustrates how evaluation metrics for certain kinds of problems strongly influence the way you can operate when building your model solution in a data science competition. The chapter also addresses the large variety of metrics available in Kaggle competitions.

Chapter 7, Designing Good Validation, introduces you to the importance of validation in data competitions. It discusses overfitting, shake-ups, leakage, adversarial validation, different kinds of validation strategies, and strategies for your final submissions.

Chapter 8, Modeling for Tabular Competitions, discusses tabular competitions, especially those on Kaggle in recent years, the Tabular Playground Series. Tabular problems are standard practice for most data scientists, and there is a lot that can be learned from Kaggle.

Chapter 9, Hyperparameter Optimization, explores how to extend the cross-validation approach to find the best hyperparameters for your models – in other words, those that can generalize in the best way on the private leaderboard – under pressure and with the scarcity of time and resources that you experience in Kaggle competitions.

Chapter 10, Ensembling with Blending and Stacking Solutions, explains ensembling techniques for multiple models, such as averaging, blending, and stacking. We will provide you with some theory, practice, and code examples that you can use as templates when building your solutions on Kaggle.

Chapter 11, Modeling for Computer Vision, discusses problems related to computer vision, one of the most popular topics in AI in general, and on Kaggle specifically. We demonstrate full pipelines for building solutions to challenges in image classification, object detection, and image segmentation.

Chapter 12, Modeling for NLP, focuses on the frequently encountered types of Kaggle challenges related to natural language processing. We demonstrate how to build an end-to-end solution for widespread problems like answering open-domain questions.

Chapter 13, Generative AI in Kaggle Competitions, explores the rapidly growing field of generative AI competitions on Kaggle and covers strategies for tasks involving LLMs, such as fine-tuning, prompt engineering, and evaluating generated text or images.

Chapter 14, Simulation and Optimization Competitions, provides an overview of simulation competitions, which have gained popularity on Kaggle over recent years.

Chapter 15, Creating Your Portfolio of Projects and Ideas, explores ways to stand out by showcasing your work on Kaggle and other sites appropriately.

Chapter 16, Finding New Professional Opportunities, concludes the overview of how Kaggle can positively affect your career by discussing the best ways to leverage all your Kaggle experience to find new professional opportunities.

To get the most out of this book

The Python code in this book has been designed to be run on a Kaggle notebook, without any installation on a local computer. Therefore, don't worry about what machine you have available or what version of Python packages you should install.

All you need is a computer with access to the internet and a free Kaggle account. In fact, to run the code on a Kaggle notebook (you will find instructions about the procedure in *Chapter 3*), you first need to open an account on Kaggle. If you don't have one yet, just go to www.kaggle.com and follow the instructions on the website.

We link to many resources throughout the book that we think you will find helpful. When you encounter a link, we recommend that you explore it: you will find code available on public Kaggle notebooks that you can reuse or further materials to illustrate concepts and ideas that we have discussed in the book.

Download the example code files

The code bundle for the book is hosted on GitHub at https://github.com/PacktPublishing/ The-Kaggle-Book-2nd-Edition. We also have other code bundles from our rich catalog of books and videos available at https://github.com/PacktPublishing/. Check them out!

Download the color images

We also provide a PDF file that has color images of the screenshots/diagrams used in this book. You can download it here: https://packt.link/gbp/9781835083208.

Conventions used

There are some text conventions used throughout this book.

CodeInText: Indicates code words in text, database table names, folder names, filenames, file extensions, pathnames, dummy URLs, user input, and Twitter handles. For example: "The dataset will be downloaded to the Kaggle folder as a .zip archive – unpack it, and you are good to go."

A block of code is set as follows:

```
from google.colab import drive
drive.mount('/content/gdrive')
```

Any command-line input or output is written as follows:

```
pip install git+git://github.com/AutoViML/AutoViz.git
```

Bold: Indicates a new term, an important word, or words that you see on the screen. For instance, words in menus or dialog boxes appear in the text like this. For example: "The specific limits at the time of writing are **100 GB per private dataset** and a **100 GB total** quota."

Warnings or important notes appear like this.

Tips and tricks appear like this.

Get in touch

Feedback from our readers is always welcome.

General feedback: Email `feedback@packtpub.com` and mention the book's title in the subject of your message. If you have questions about any aspect of this book, please email us at `questions@packtpub.com`.

Errata: Although we have taken every care to ensure the accuracy of our content, mistakes do happen. If you have found a mistake in this book, we would be grateful if you reported this to us. Please visit `http://www.packtpub.com/submit-errata`, click **Submit Errata**, and fill in the form.

Piracy: If you come across any illegal copies of our works in any form on the internet, we would be grateful if you would provide us with the location address or website name. Please contact us at `copyright@packtpub.com` with a link to the material.

If you are interested in becoming an author: If there is a topic that you have expertise in and you are interested in either writing or contributing to a book, please visit `http://authors.packtpub.com`.

Share your thoughts

Once you've read _The Kaggle Book, Second Edition_, we'd love to hear your thoughts! Scan the QR code below to go straight to the Amazon review page for this book and share your feedback.

`https://packt.link/r/183508320X`

Your review is important to us and the tech community and will help us make sure we're delivering excellent quality content.

Free Benefits with Your Book

This book comes with free benefits to support your learning. Activate them now for instant access (see the "*How to Unlock*" section for instructions).

Here's a quick overview of what you can instantly unlock with your purchase:

PDF and ePub Copies   **Next-Gen Web-Based Reader**

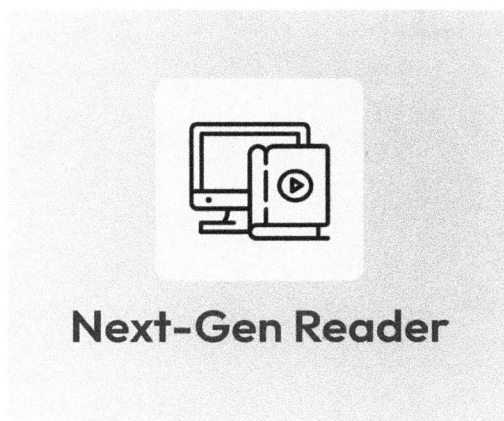

Free PDF and ePub versions

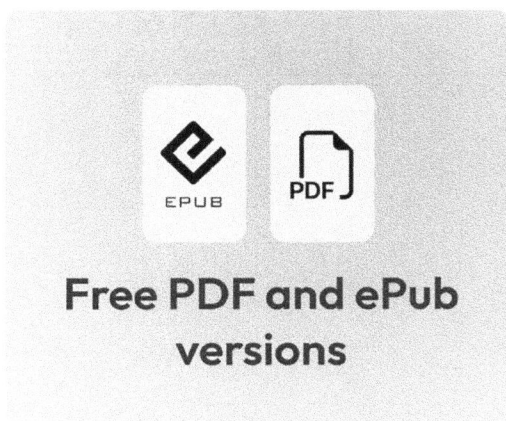

Next-Gen Reader

Access a DRM-free PDF copy of this book to read anywhere, on any device.

Use a DRM-free ePub version with your favorite e-reader.

Multi-device progress sync: Pick up where you left off, on any device.

Highlighting and notetaking: Capture ideas and turn reading into lasting knowledge.

Bookmarking: Save and revisit key sections whenever you need them.

Dark mode: Reduce eye strain by switching to dark or sepia themes.

How to Unlock

Scan the QR code (or go to packtpub.com/unlock). Search for this book by name, confirm the edition, and then follow the steps on the page.

Note: Keep your invoice handy. Purchases made directly from Packt don't require one.

Part 1

Your Kaggle Launchpad: Mastering the Essentials

In this foundational part, you'll embark on your Kaggle journey, demystifying data science competitions and the essential tools Kaggle provides. We'll explore the landscape of competitions, learn how to manage data with Kaggle Datasets, master the interactive environment of Kaggle Notebooks, discover pre-trained Kaggle Models, and tap into the collective wisdom of Discussion forums. By the end of this part, you'll be equipped with the fundamental knowledge and practical skills to confidently start participating, learning, and collaborating on Kaggle.

This part of the book includes the following chapters:

- *Chapter 1, Introducing Kaggle and Other Data Science Competitions*
- *Chapter 2, Organizing Data with Datasets*
- *Chapter 3, Working and Learning with Kaggle Notebooks*
- *Chapter 4, Kaggle Models*
- *Chapter 5, Leveraging Discussion Forums*

1

Introducing Kaggle and Other Data Science Competitions

Data science competitions have long been around, and they have experienced growing success over time, starting from a niche community of passionate competitors, drawing more and more attention, and reaching a much larger audience of millions of data scientists. As longtime competitors on the most popular data science competition platform, Kaggle, we have witnessed and directly experienced all these changes through the years.

Currently, there are plenty of resources available about Kaggle, and generally about data science competitions. You can easily find a large number of meetups, discussion panels, podcasts, interviews, and even online courses explaining how to win such competitions. However, apart from the book you are reading now, you won't find any structured guides about navigating data science competitions and getting the most out of them – not just in score or ranking but also regarding professional experience.

In this book, rather than just packaging up a few hints about how to win or score highly on Kaggle, we intend to present you with a guide on how to compete better and get back the maximum possible from your competition experiences, whether on Kaggle or any other competition platform, particularly from the perspective of your professional life. Additionally, there are interviews with Kaggle Masters and Grandmasters accompanying the contents of the book. We hope they will offer different perspectives and insights on specific aspects of competing on Kaggle and inspire the way you will test yourself and learn to do competitive data science.

By the end of this book, you'll have absorbed the knowledge we drew directly from our own experiences, resources, and learning from competitions, and everything you need to pave the way for yourself to learn and grow, competition after competition.

As a starting point, in this chapter, we will explore how competitive programming evolved into data science competitions, why the Kaggle platform is the most popular site for such competitions, and how it works.

We will cover the following topics:

- The rise of data science competition platforms
- The Common Task Framework paradigm
- The Kaggle platform and some other alternatives
- How a Kaggle competition works: stages, competition types, submission and leaderboard dynamics, computational resources, networking, and more

---

Free Benefits with Your Book

Your purchase includes a free PDF copy of this book along with other exclusive benefits. Check the *Free Benefits with Your Book* section in the Preface to unlock them instantly and maximize your learning experience.

The rise of data science competition platforms

Competitive programming has a long history, starting in the 1970s with the first iterations of the **ICPC**, the **International Collegiate Programming Contest**. In the original ICPC, small teams from universities and companies participated in a competition that required solving a series of problems using a computer program (at the beginning, participants coded in FORTRAN). Teams had to display good teamwork, problem-solving, and programming skills to achieve a good final rank.

The experience of participating in the heat of such a competition and the opportunity to stand in the spotlight for recruiting companies provided the students with ample motivation, making the competition popular for many years. Among ICPC finalists, a few have become renowned: there is *Adam D'Angelo*, the former CTO of Facebook and founder of Quora, *Nikolai Durov*, the co-founder of Telegram Messenger, and *Matei Zaharia*, the creator of Apache Spark. Together with many other professionals, they all share the same experience: having taken part in an ICPC.

After the ICPC, programming competitions flourished, especially after 2000, when remote participation became more feasible, allowing international competitions to run more easily and at a lower cost. The format is similar in most of these competitions: there is a series of problems, and you have to code a solution to solve them. The winners are given a prize but also make themselves known to recruitment companies or simply become famous.

Typically, problems in competitive programming range from combinatorics and number theory to graph theory, algorithmic game theory, computational geometry, string analysis, and data structures. Recently, problems relating to artificial intelligence have successfully emerged, in particular after the launch of the **KDD Cup**, a contest in knowledge discovery and data mining, held by the **Association for Computing Machinery's (ACM's) Special Interest Group (SIG)** during its annual conference (`https://kdd.org/conferences`).

The first KDD Cup, held in 1997, involved a problem with direct marketing for lift curve optimization, and it started a long series of competitions that continues today. The archives contain datasets, instructions, and winners at `https://www.kdd.org/kdd-cup`. To underscore the practical significance of KDD Cups and real-world issues, just have a look at the latest challenges available in 2024 at the time of writing. The first challenge, to be found at `https://www.aicrowd.com/challenges/amazon-kdd-cup-2024-multi-task-online-shopping-challenge-for-llms`, is sponsored by Amazon and it invites participants to design powerful LLMs that can better assist users in navigating online shopping, making it a more intuitive and satisfying experience, similar to a knowledgeable shopping assistant in real life. The second challenge, `https://www.aicrowd.com/challenges/meta-comprehensive-rag-benchmark-kdd-cup-2024`, sponsored by Meta, invites participants to build **Retrieval-Augmented Generation (RAG)** systems to mitigate the hallucinations that may happen in responses from LLM-based models, thus making LLMs more trustworthy in providing information. Finally, the last challenge, `https://www.biendata.xyz/kdd2024/`, involves leveraging academic graph mining and detecting incorrect author assignations, tracing papers' sources, and managing academic questions and answers.

KDD Cups proved quite effective in establishing best practices in data science, with many published papers describing solutions, techniques, and competition dataset sharing. These have been useful for many practitioners for experimentation, education, and benchmarking.

The successful examples of both competitive programming events and the KDD Cup inspired companies (such as Netflix) and entrepreneurs (such as *Anthony Goldbloom*, the founder of Kaggle) to create the first data science competition platforms where companies can host data science challenges that are hard to solve and might benefit from crowdsourcing. In fact, given that no golden rule works for all the problems in data science, many problems require a time-consuming approach that can be summed up as *try all that you can try*.

In the long run, no algorithm can beat all the others on all problems, as stated by the **No Free Lunch** theorem by David Wolpert and William Macready. The theorem reveals what every practitioner eventually discovers through hardship: each machine learning algorithm performs effectively only when its hypothesis space includes the solution. Consequently, as you cannot know beforehand if a machine learning algorithm can best tackle your problem, you have to test it directly on your problem before being assured that you are doing the right thing. There are no theoretical shortcuts or other holy grails of machine learning – only empirical experimentation can tell you what works.

For more details, you can look up the No Free Lunch theorem for a theoretical explanation of this practical truth. Here is a complete article from Analytics India Magazine on the topic: `https://analyticsindiamag.com/what-are-the-no-free-lunch-theorems-in-data-science/`.

Crowdsourcing proves ideal when you lack the workforce and computer power, but you must extensively test algorithms and data transformations to find the best possible combinations. That's why, for instance, governments and companies commonly resort to competitions to advance in certain fields and problems:

- On the government side, we can quote DARPA and its many competitions surrounding self-driving cars, robotic operations, machine translation, speaker identification, fingerprint recognition, information retrieval, OCR, automatic target recognition, and many others.
- On the business side, we can quote a company such as Netflix, which entrusted the outcome of a competition to improve its algorithm for predicting user movie selection.

The Netflix competition was based on the idea of improving existing collaborative filtering. The purpose was simply to predict the potential rating a user would give a film solely based on the ratings they gave other films, without knowing specifically who the user was or what the films were. Since no user description or movie title or description were available (all being replaced with identity codes), the competition required entrants to develop smart ways to use the past ratings available. The grand prize of US $1,000,000 was to be awarded only if the solution could improve the existing Netflix algorithm, Cinematch, above a certain threshold.

The competition ran from 2006 to 2009 and saw victory for a team made up of the fusion of many previous competition teams: a team from Commendo Research & Consulting GmbH, *Andreas Töscher* and *Michael Jahrer*, quite renowned also in Kaggle competitions; two researchers from AT&T Labs; and two others from Yahoo!. Ultimately, winning the competition required so much computational power and the ensembling of different solutions that teams were forced to merge to keep pace.

This situation was also reflected in the actual usage of the solution by Netflix, which preferred not to implement it but simply took the most interesting insight from it to improve its existing Cinematch algorithm. You can read more about it in this Wired article: `https://www.wired.com/2012/04/netflix-prize-costs/`.

At the end of the Netflix competition, what mattered was not the solution per se, which was quickly superseded by the change in the business focus of Netflix from DVDs to online movies. The insights gained from the competition were the real benefit for both the participants, who gained a considerable reputation in collaborative filtering, and the company, which could transfer its improved recommendation knowledge to its new business.

The Kaggle competition platform

Companies other than Netflix have also benefitted from data science competitions. The list is long, but we can quote a few examples where the company running the competition reported a clear benefit. For instance:

- The insurance company Allstate was able to improve its actuarial models built by their experts, thanks to a competition involving hundreds of data scientists (`https://www.kaggle.com/c/ClaimPredictionChallenge`).

- In another well-documented example, General Electric was able to improve by 40% on the industry-standard performance (measured by the root mean squared error metric) for predicting arrival times of airline flights, thanks to a similar competition (`https://www.kaggle.com/c/flight`).

- The Kaggle competition, Vesuvius Challenge - Ink Detection (`https://www.kaggle.com/competitions/vesuvius-challenge-ink-detection/`) has been part of the efforts to read ancient scrolls carbonized by the eruption of Mount Vesuvius in 79 AD. The effort has finally led to reading one of the scrolls in its entirety (`https://scrollprize.org/grandprize`).

To this day, the Kaggle competition platform has held hundreds of competitions, and these are just a few examples of companies that have used them successfully. Let's take a step back from specific competitions for a moment and talk about the story of the Kaggle company, which is the common thread throughout this book.

Kaggle took its first steps in February 2010, thanks to *Anthony Goldbloom*, an Australian economist with a degree in economics and econometrics. After working at Australia's Department of the Treasury and the Research Department at the Reserve Bank of Australia, Goldbloom interned in London at The Economist, the international weekly newspaper on current affairs, international business, politics, and technology.

At The Economist, he had the occasion to write an article about big data, which inspired his idea to build a competition platform that could crowdsource the best analytical experts to solve interesting machine learning problems (https://www.smh.com.au/technology/from-bondi-to-the-big-bucks-the-28yearold-whos-making-data-science-a-sport-20111104-1myq1.html). Since the crowdsourcing dynamics played a relevant part in the business idea for this platform, he derived the name *Kaggle*, which recalls by rhyme the term *gaggle*, a flock of geese, the goose also being the symbol of the platform.

After moving to Silicon Valley in the USA, his Kaggle start-up received $11.25 million in Series A funding from a round led by Khosla Ventures and Index Ventures, two renowned venture capital firms. The first competitions were rolled out, the community grew, and some of the initial competitors came to be quite prominent, such as *Jeremy Howard*, the Australian data scientist and entrepreneur, who, after winning a couple of competitions on Kaggle, became the President and Chief Scientist of the company.

Jeremy Howard left his position as President in December 2013 and established a new start-up, **fast.ai** (www.fast.ai), offering machine learning courses and a deep learning library for coders.

At the time, there were some other prominent **Kagglers** (the name indicating frequent participants in competitions held by Kaggle), such as *Jeremy Achin* and *Thomas de Godoy*. After reaching the top 20 global rankings on the platform, they promptly decided to retire and founded their own company, DataRobot. Soon after, they started hiring their employees from among the best participants in the Kaggle competitions in order to instil the best machine learning knowledge and practices into the software they were developing. Today, DataRobot is one of the leading companies in developing AutoML solutions (software for automatic machine learning).

The Kaggle competitions claimed more and more attention from a growing audience. Even Geoffrey Hinton, the "godfather" of deep learning, participated in (and won) a Kaggle competition hosted by Merck in 2012 (https://www.kaggle.com/c/MerckActivity/overview/winners). Kaggle was also the platform where François Chollet launched his deep learning package Keras during the Otto Group Product Classification Challenge (https://www.kaggle.com/c/otto-group-product-classification-challenge/discussion/13632) and Tianqi Chen launched XGBoost, a speedier and more accurate version of gradient boosting machines, in the Higgs Boson Machine Learning Challenge (https://www.kaggle.com/c/higgs-boson/discussion/10335).

Besides Keras, François Chollet has also provided the most useful and insightful perspective on how to win a Kaggle competition in one of his answers on the Quora website: https://www.quora.com/Why-has-Keras-been-so-successful-lately-at-Kaggle-competitions.

Fast iterations of multiple attempts, guided by empirical (more than theoretical) evidence, are, in truth, all that you need. We don't think there are many more secrets to winning a Kaggle competition than he pointed out in his answer.

Notably, François Chollet also hosted his own competition on Kaggle (`https://www.kaggle.com/c/abstraction-and-reasoning-challenge/`), widely recognized as the world's first general AI competition.

Competition after competition, the community revolving around Kaggle grew to touch one million in 2017, the same year as, during her keynote at Google Next, *Fei-Fei Li*, Chief Scientist at Google, announced that Google Alphabet would acquire Kaggle. Since then, Kaggle has been part of Google and kept being active and growing. Although data science competitions are Kaggle's most influential and impactful aspect, recent statistics show that most users primarily engage with the platform in other ways. These include downloading public datasets (reinforcing Kaggle's role as a major data hub), creating public Notebooks in Python or R, and expanding their knowledge through Kaggle's microcourses. Such trends continue strongly today, and Kaggle can now claim almost twenty-three million users, a number that is constantly growing, which you can check using this helpful and regularly updated notebook by Bojan Tunguz: `https://www.kaggle.com/code/tunguz/unique-kaggle-users`.

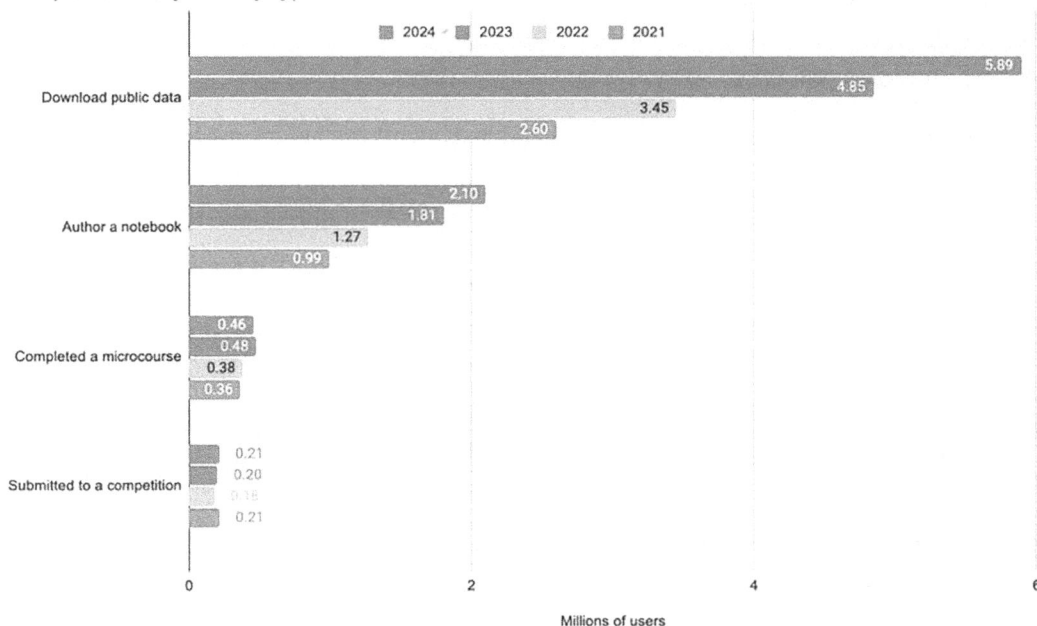

Figure 1.1: A bar chart showing how users used Kaggle in years from 2021 to 2024

In 2022, after 12 years, Anthony Goldbloom left Kaggle (`https://www.kaggle.com/discussions/general/329411`) and joined AIX Ventures, a venture capital fund that invests in artificial intelligence start-ups, as an Investment Partner. He has also founded a new company, Sumble, with Ben Hamner, the previous co-founder and CTO of Kaggle.

Kaggle's CEO (until recently, on January 2025, when he resigned and leave the lead to Nate Keating - see: `https://www.kaggle.com/discussions/general/555988`) has become D. Sculley has become D. Sculley, a former Director of Engineering at Google Brain, with long experience working with large-scale machine learning systems and leading research teams. In a recent interview during a podcast (`https://www.youtube.com/watch?v=1aajTQvZJ94`), Sculley mentioned how Kaggle might further evolve into an even more comprehensive resource for machine learning professionals, researchers, and learners, with a focus on practical applications and inclusivity. There are a few notable points about future directions in the discussion. First, Kaggle will continue to be novice-friendly, offering everyone on Kaggle pursuing their machine learning path an opportunity to learn, test themselves, and connect to other learners. In addition, some Kaggle competitions may, in the future, steer towards more production-grade machine learning (MLOps) solutions. This would mean going beyond focusing solely on accuracy metrics and evaluating machine learning models more often on dimensions like model efficiency and scalability. Another notable aspect that Sculley mentioned is publishing more results and insights from Kaggle (by the way, there are already 50,000 papers around citing Kaggle in some way), especially as meta-analyses. An example of something similar could be the 2023 Kaggle AI Report on competitions (`https://www.kaggle.com/competitions/2023-kaggle-ai-report`).

Through the years, Kaggle has offered many of its participants even more opportunities, such as:

- Creating their own company
- Launching machine learning software and packages
- Getting interviews in magazines (`https://www.wired.com/story/solve-these-tough-data-problems-and-watch-job-offers-roll-in/`)
- Writing machine learning books (`https://twitter.com/antgoldbloom/status/745662719588589568`)
- Pursuing their dream career: AI enterprises such as Nvidia and H2O.ai assemble teams comprised of Kaggle Grandmaster talent

Most importantly, we need to add to the list the fact that, by competing on Kaggle, you learn more about the skills and technical aspects involved in data science modeling.

Other competition platforms

Though this book focuses on competitions on Kaggle, we must remember that many data competitions are held on private or other competition platforms. In truth, most of the information in this book will also work for other competitions since they essentially all operate under similar principles, and the benefits for the participants are more or less the same.

Although many other platforms are localized in specific countries or are specialized only for certain kinds of competitions, such as those sponsored by academic institutions, for completeness, we will briefly introduce some of them, at least those we have some experience and knowledge of:

- **DrivenData** (`https://www.drivendata.org/competitions/`) is a crowdsourcing competition platform devoted to social challenges (see `https://www.drivendata.co/blog/intro-to-machine-learning-social-impact/`). The company is a social enterprise aiming to bring data science solutions to organizations tackling the world's biggest challenges, thanks to data scientists building algorithms for social good. For instance, as you can read in this article, `https://www.engadget.com/facebook-ai-hate-speech-covid-19-160037191.html`, Facebook has chosen DrivenData for its competition on building models against hate speech and misinformation. According to The State of Competitive Machine Learning 2023 (`https://mlcontests.com/state-of-competitive-machine-learning-2023`), DrivenData is the second most popular competition platform by number of registered users.

- **Numerai** (`https://numer.ai/`) is an AI-powered, crowdsourced hedge fund in San Francisco. It hosts a weekly tournament where you can submit your predictions on hedge fund obfuscated data and earn prizes in the company's cryptocurrency, **Numeraire**.

- **EvalAI** (`https://eval.ai/`) is an open-source platform designed for hosting and participating in AI and machine learning challenges using a standardized environment for benchmarking AI models and fostering collaboration and competition within the research community.

- **CrowdANALYTIX** (`https://www.crowdanalytix.com/community`) is a bit less active now, but this platform used to host quite a few challenging competitions a short while ago. The community blog is quite an interesting place for getting an idea of what challenges you can find on this platform: `https://www.crowdanalytix.com/jq/communityBlog/listBlog.html`.

- **Signate** (`https://signate.jp/competitions`) is a Japanese data science competition platform. It is quite rich in contests and offers a ranking system similar to Kaggle (`https://signate.jp/users/rankings`).

- **Zindi** (`https://zindi.africa/competitions`) is a data science competition platform from Africa. It hosts competitions to solve Africa's most pressing social, economic, and environmental problems.

- **Alibaba Cloud** (`https://www.alibabacloud.com/campaign/tianchi-competitions`) is a Chinese cloud computing and AI provider that has launched the **Tianchi Academic competitions** (`https://tianchi.aliyun.com/`), partnering with academic conferences such as SIGKDD, IJCAI-PRICAI, and CVPR and featuring challenges such as image-based 3D shape retrieval, 3D object reconstruction, and instance segmentation.

- **Analytics Vidhya** (`https://datahack.analyticsvidhya.com/`) is the largest Indian community for data science, offering a platform for data science hackathons.

- **CodaLab** (`https://codalab.lri.fr/`) is a French-based data science competition platform created as a joint venture between Microsoft and Stanford University in 2013. They feature a free cloud-based notebook called **Worksheets** (`https://worksheets.codalab.org/`) for knowledge sharing and reproducible modeling.

- **AIcrowd** (`https://www.aicrowd.com/`) was originally launched in 2016 as crowdAI.org. It was relaunched as AIcrowd in 2018 and it is backed by the Digital Epidemiology Lab at **École Polytechnique Fédérale de Lausanne** (**EPFL**), one of the Swiss Federal Institutes of Technology. The platform has already run numerous successful challenges, including official NeurIPS challenges in 2017 and 2018 (and it is running two challenges of the KDD Cup 2024).

Other minor platforms are InnoCentive (`https://www.innocentive.com/`), Grand-Challenge (`https://grand-challenge.org/`) for biomedical imaging, the Chinese company DataFountain (`https://www.datafountain.cn/competitions`), and the list could go on. Recently, Hugging Face, the French-American open-source platform for developing and sharing machine learning models, datasets, and applications for **natural language processing** (**NLP**), has experimented with launching a competition platform that is progressively being improved (`https://huggingface.co/docs/competitions`).

In our brief review, we described some of the most important data science competition platforms, aside from Kaggle. To have a more complete and up-to-date overview of the competition landscape, which is vibrant and diverse, you can actually consult a few sources online. You can always find an extensive list of ongoing major competitions at the Russian community Open Data Science (`https://ods.ai/competitions`) and occasionally even discover new competition platforms. In addition, you can see an overview of the running competitions on platforms and in an independent format at the `mlcontests.com` website, together with the current costs for renting GPUs.

The website is constantly updated, and it is an easy way to glance at what's going on with data science competitions across different platforms. More importantly, ML Contests publishes a yearly *State of Competitive Machine Learning* report (you can read the 2023 report here: `https://mlcontests.com/state-of-competitive-machine-learning-2023`), which is an incredibly rich source for everything related to machine learning competitions of any kind.

Kaggle is always the best platform where you can find the most exciting competitions and obtain the widest recognition for your competition efforts. However, picking up a challenge outside of it makes sense, and we recommend it as a strategy when you find a competition matching your personal and professional interests. There are many alternatives and opportunities besides Kaggle, which means that if you consider more competition platforms alongside Kaggle, you can more easily find a competition that might interest you because of its specialization or data. In addition, you can expect less competitive pressure during these challenges (and consequently a better ranking or even winning something) since they are less known and advertised. However, also expect less sharing among participants, since no other competition platform has reached the same richness of sharing and networking opportunities as Kaggle.

Introducing Kaggle competitions

At this point, we need to delve more deeply into how Kaggle works. In this section, we will discuss the various types of Kaggle competitions, and you'll get a flavor of what it means to compete on Kaggle. However, there are different paths to explore on Kaggle beyond just competitions. These include datasets, notebooks, and discussions, which are alternative and parallel competitive paths that will be discussed in the next chapters. Kaggle can also be seen as a learning platform and a hub for data and models, too.

Later in the book, we'll come back to discuss many of these aspects of Kaggle in much more detail, with more suggestions and strategies.

Stages of a competition

A competition on Kaggle is arranged into different steps. By looking at each of them, you can better understand how a data science competition works and what to expect.

When a competition is launched, there are usually some posts on social media, for instance, on the Kaggle Twitter profile, `https://twitter.com/kaggle`, that announce it, and a new tab will appear in the Kaggle section about **Active Competitions** on the **Competitions** page (`https://www.kaggle.com/competitions`).

You'll be taken to its page if you click on a particular competition's tab. At a glance, you can check if the competition will have prizes (and if it awards points and medals, a secondary consequence of participating in a competition), how many teams are currently involved, and how much time is still left for you to work on a solution:

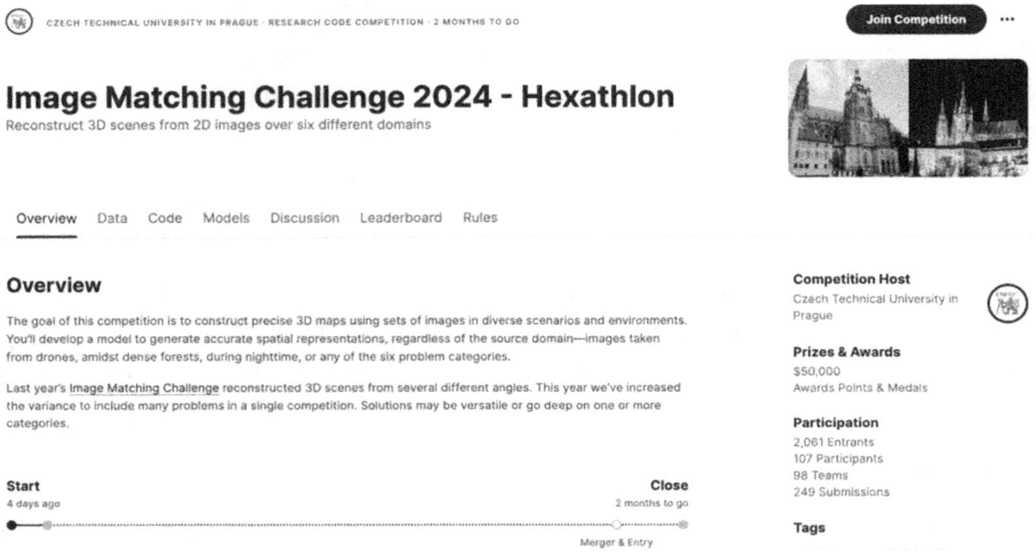

Figure 1.2: A competition's page on Kaggle

There, you can explore the **Overview** menu first, which provides information about:

- The topic of the competition
- Its evaluation metric (that your models will be evaluated against)
- The timeline of the competition
- The prizes
- The legal or competition requirements

Usually, the timeline is a bit overlooked, but it should be one of the first things you check; it doesn't tell you simply when the competition starts and ends, but it will provide you with the **rule acceptance deadline**, which is usually referred to as **Entry** and ranges from seven days to two weeks before the competition closes. The rule acceptance deadline marks the last day you can join the competition (by accepting its rules). There is also the **team merger deadline,** often referred to as **Merger**: you can arrange to combine your team with another competitor's at any point before that deadline, but it won't be possible after that. Often these two deadlines can be found combined together, where you have **Merger & Entry** falling on the same day.

The **Rules** menu is also quite often overlooked (with people just jumping to **Data**), but it is crucial to check it because it can tell you about the requirements of the competition. Among the key information you can get from the rules, there is:

- Your eligibility for a prize
- Whether you can use external data to improve your score
- How many submissions (tests of your solution) a day you get
- How many final solutions you can choose

Once you have accepted the rules, you can download any data from the **Data** menu or start working on Kaggle Notebooks (online, cloud-based notebooks) from the **Code** menu, reusing code others have made available or creating your own code from scratch.

If you decide to download the data, also consider that you have a **Kaggle API** that can help you run downloads and submissions in an almost automated way. It is an important tool for running your models on your local computer or cloud instance. You can find more details about the API at `https://www.kaggle.com/docs/api`, and you can get the code from GitHub at `https://github.com/Kaggle/kaggle-api`.

If you examine the Kaggle GitHub repo closely, you can also find all the Docker images (`https://github.com/Kaggle/docker-python` and `https://github.com/Kaggle/docker-rstats`) they use for their online notebooks, Kaggle Notebooks:

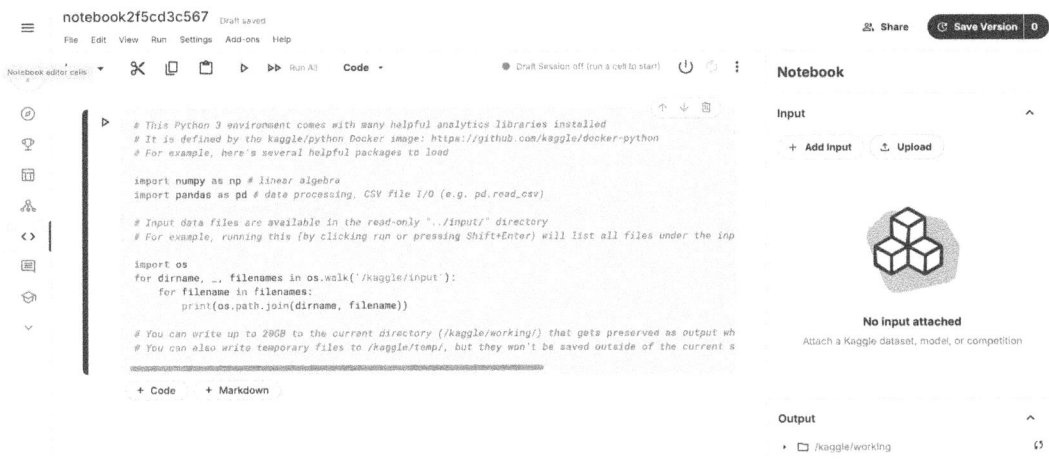

Figure 1.3: A Kaggle Notebook ready to be coded

At this point, as you develop your solution, it is our warm suggestion not to continue in solitude but to contact other competitors through the **Discussion** forum, where you can ask and answer questions specific to the competition.

You will often find useful hints about specific problems with the data or even ideas to help improve your solution. Many successful Kagglers have reported finding ideas on the forums that have helped them perform better and, more importantly, learn more about modeling in data science.

Once your solution is ready, you can submit it to the Kaggle evaluation engine in adherence to the competition's specifications. Some competitions will accept a CSV file as a solution. Others will require you to code and produce results in a Kaggle Notebook. You can keep submitting solutions throughout the competition.

Every time you submit a solution, soon after, the leaderboard will provide you with a score and a position among the competitors (the wait time varies depending on the computations necessary for the score evaluation). That position is only roughly indicative because it reflects your model's performance on the part of the test set, called the **public test set**, since your performance is made public during the competition for everyone to know.

Before the competition closes, each competitor can choose the number (usually two) of their solutions for the final evaluation.

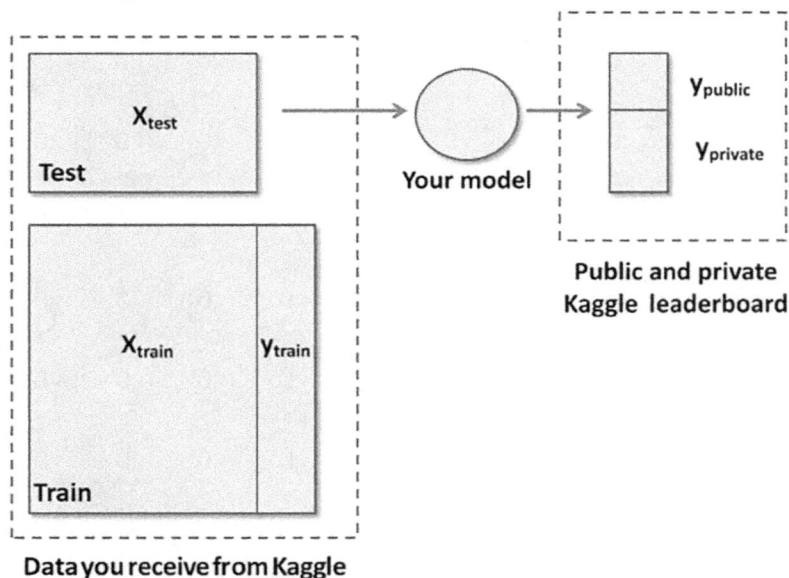

Figure 1.4: A diagram demonstrating how data turns into scores for the public and private leaderboard

Only when the competition closes, based on the models the contestants have decided to be scored, is their score on another part of the test set, called the **private test set**, revealed. This new leaderboard, the private leaderboard, constitutes the final, effective scores for the competition, but it is still not official and definitive in its rankings. In fact, the Kaggle team will take some time to check that everything is correct and that all contestants have respected the competition rules.

After a while (and sometimes after some changes in the rankings due to disqualifications), the private leaderboard will become official and definitive, the winners will be declared, and many participants will unveil their strategies, solutions, and code on the competition discussion forum. At this point, it is up to you to check the other solutions and try to improve your own. We strongly recommend that you do so since this is another important source of learning in Kaggle.

Types of competitions and examples

Kaggle competitions are categorized based on **competition categories**, and each category has a different implication regarding how to compete and what to expect. The type of data, the problem's difficulty, awarded prizes, and competition dynamics are quite diverse inside the categories. Therefore, it is important to understand beforehand what each implies.

Here are the official categories that you can use to filter out the different competitions:

- Featured
- Annuals (actually part of the Featured competitions)
- Getting Started
- Research
- Community
- Playground
- Simulations
- Analytics

There are also some categories that were present in the past but are now deprecated. However, they are worth mentioning because they can help you understand the potential of data science competitions:

- Masters
- Recruitment

Featured is the most common type of competition, involving a business-related problem from a sponsor company and a prize for the top performers. The winners will grant a non-exclusive license of their work to the sponsor company; they will have to prepare a detailed report of their solution and sometimes even participate in meetings with the sponsor company.

There are examples of Featured competitions every time you visit Kaggle. Currently, many of them are problems relating to applying deep learning methods to unstructured data like text, images, videos, or sound. In the past, tabular data competitions were commonly seen, that is, competitions based on problems relating to structured data that can be found in a database.

By leveraging examples using ensembles of decision trees, deep learning, or gradient boosting methods with clever feature engineering, solutions derived from Kaggle could indeed improve how a tabular data problem is solved, defining a new benchmark or approach. Clear examples of approaches that emerged from Kaggle competitions are denoising autoencoders (`https://www.kaggle.com/c/porto-seguro-safe-driver-prediction/discussion/44629`) and using adversarial validation to detect distribution shifts (`https://www.kaggle.com/code/konradb/adversarial-validation-and-other-scary-terms/`).

Unfortunately, nowadays, these competitions are run less often because a crowdsourced solution won't often be much better than what a good team of data scientists or even AutoML software can do. Given the spread of better software and good practices, the increase in result quality obtainable from competition is indeed marginal. However, a good deep learning solution could still make a big difference in the unstructured data world. For instance, pre-trained networks such as BERT brought about double-digit increases in previous standards for many well-known NLP task benchmarks.

Masters have not shown up for a long time now, but they were private, invite-only competitions. The purpose was to create competitions only for experts (generally competitors ranked as Masters or Grandmasters, based on Kaggle medal rankings), based on their rankings on Kaggle.

Annuals are competitions that always appear during a certain period of the year. Among the Annuals, we have the *Santa Claus* competitions (usually based on an algorithmic optimization problem) and the *March Machine Learning Mania* competition, which has run yearly since 2014 during the US College Basketball Tournaments.

Research competitions imply a research or science purpose instead of a business one, sometimes for serving the public good. That's why these competitions do not always offer prizes. In addition, these competitions sometimes require the winning participants to release their solution as open-source.

Recent examples of research competitions are:

- Image Matching Challenge 2024 – Hexathlon (`https://www.kaggle.com/competitions/image-matching-challenge-2024`) where you reconstruct 3D scenes from 2D images over six different domains and it is sponsored by the Czech Technical University in Prague
- HMS – Harmful Brain Activity Classification (`https://www.kaggle.com/competitions/hms-harmful-brain-activity-classification`), sponsored by Harvard Medical School, requires you to classify seizures and other patterns of harmful brain activity in critically ill patients
- NeurIPS 2023 – Machine Unlearning (`https://www.kaggle.com/competitions/neurips-2023-machine-unlearning`), sponsored by Google, requires you to erase the influence of requested samples from a model without hurting its accuracy

Sponsors that want to test the ability of potential job candidates hold **Recruitment** competitions, which are another kind of competition that has stopped being held. These competitions were limited to teams of one and offered the best-placed competitors an interview with the sponsor as a prize. The competitors had to upload their CVs at the end of the competition if they wanted to be considered for being contacted.

Examples of Recruitment competitions are:

- *The Facebook Recruiting Competition* (`https://www.kaggle.com/c/FacebookRecruiting`); Facebook has held a few of this kind of competition
- The *Yelp Recruiting Competition* (`https://www.kaggle.com/c/yelp-recruiting`)

Getting Started competitions do not offer any prizes but friendly and easy problems for beginners to get accustomed to Kaggle principles and dynamics. They are usually semi-permanent competitions whose leaderboard is refreshed from time to time. Suppose you are looking for a tutorial in machine learning. In that case, these competitions are the right places to start because you can find a highly collaborative environment, and there are many Kaggle Notebooks that show you how to process the data and create different types of machine learning models.

Famous ongoing **Getting Started** competitions are:

- *Digit Recognizer* (`https://www.kaggle.com/c/digit-recognizer`)
- *Titanic – Machine Learning from Disaster* (`https://www.kaggle.com/c/titanic`)
- *House Prices – Advanced Regression Techniques* (`https://www.kaggle.com/c/house-prices-advanced-regression-techniques`)

Playground competitions are a little bit more difficult than the Getting Started ones. Still, they are also meant for competitors to learn and test their abilities without the pressure of a fully-fledged Featured competition (though in Playground competitions, sometimes the heat of the competition may also turn relatively high). The usual prizes for such competitions are just swag (an acronym for "Stuff We All Get," such as a cup, a t-shirt, or socks branded by Kaggle; see `https://www.kaggle.com/general/68961`) or a bit of money.

One famous Playground competition is the original *Dogs vs. Cats* competition (`https://www.kaggle.com/c/dogs-vs-cats`), where the task is to create an algorithm to distinguish dogs from cats.

Simulation competitions became quite popular a few years ago, featuring a variety of challenges ranging from simple games like rock-paper-scissors (`https://www.kaggle.com/competitions/rock-paper-scissors`) to complex space simulations such as Halite (`https://www.kaggle.com/competitions/halite`). Currently, Lux AI is the main simulation competition running, which is also part of the NeurIPS 2023 competition. The goal of the Lux AI competition is to develop AI agents capable of analyzing opponents, optimizing resource gathering and allocation, and outperforming others (`https://www.kaggle.com/competitions/lux-ai-season-2-neurips-stage-2`).

We should mention **Analytics** competitions, where the evaluation is qualitative, and participants are required to provide ideas, drafts of solutions, PowerPoint slides, charts, and so on, and to **Community** (previously known as InClass) competitions, which academic institutions as well as Kagglers hold. You can read about the launch of the Community competitions at `https://www.kaggle.com/product-feedback/294337`, and you can get tips about running one of your own at `https://www.kaggle.com/c/about/host` and `https://www.kaggle.com/community-competitions-setup-guide`.

We spoke to *Parul Pandey*, Kaggle Notebooks Grandmaster, Datasets Master, and principal data scientist at H2O.ai, about her experience with Analytics competitions and more.

Interview: Parul Pandey

`https://www.kaggle.com/parulpandey`

What's your favorite kind of competition and why? In terms of techniques and solving approaches, what is your specialty on Kaggle?

I really enjoy the data analytics competitions, which require you to analyze the data and provide a comprehensive analysis report at the end. These include the **Data Science for Good** competitions (**DS4G**), sports analytics competitions (NFL etc.), and the general survey challenges. Unlike the traditional competitions, these competitions don't have a leaderboard to track your performance compared to others, nor do you get any medals or points.

On the other hand, these competitions demand end-to-end solutions touching on multi-faceted aspects of data science like data cleaning, data mining, visualizations, and conveying insights. Such problems provide a way to mimic real-life scenarios and provide your insights and viewpoints. There may not be a single best answer to solve the problem, but it gives you a chance to deliberate and weigh up potential solutions, and imbibe them into your solution.

How do you approach a Kaggle competition? How different is this approach to what you do in your day-to-day work?

My first step is always to analyze the data as part of **EDA** (short for **exploratory data analysis**). It is something that I also follow as part of my work routine. Typically, I explore the data to look for potential red flags like inconsistencies in data, missing values, outliers, etc., which might pose problems later. The next step is to create a good and reliable cross-validation strategy. Then I read the discussion forums and look at some of the Notebooks shared by people. It generally acts as a good starting point, and then I can incorporate things in this workflow from my past experiences. It is also essential to track the model performance.

For an Analytics competition, however, I like to break down the problem into multiple steps. For instance, the first part could be related to understanding the problem, which may require a few days. After that, I like to explore the data, followed by creating a basic baseline solution. Then I continue enhancing this solution by adding a piece at a time. It might be akin to adding Lego bricks one part at a time to create that final masterpiece.

Tell us about a particularly challenging competition you entered, and what insights you used to tackle the task.

As I mentioned, I mostly like to compete in Analytics competitions, even though occasionally I also try my hand in the regular ones too. I'd like to point out a very intriguing Data Science for Good competition titled Environmental Insights Explorer (`https://www.kaggle.com/c/ds4g-environmental-insights-explorer`). The task was to use remote sensing techniques to understand environmental emissions instead of calculating emissions factors from current methodologies.

What really struck me was the use case. Our planet is grappling with climate change issues, and this competition touched on this very aspect. While researching for my competition, I was amazed to find the amount of progress being made in this field of satellite imagery and it gave me a chance to understand and dive more deeply into the topic. It gave me a chance to understand how satellites like Landsat, Modis, and Sentinel worked, and how they make the satellite data available. This was a great competition to learn about a field I knew very little about before the competition.

In your experience, what do inexperienced Kagglers often overlook? What do you know now that you wish you'd known when you first started?

I will cite some of the mistakes that I made in my initial years on Kaggle.

Firstly, most of the newbies think of Kaggle as a competitions-only platform. If you love competitions, there are plenty here, but Kaggle also has something for people with other specialties. You can write code and share it with others, indulge in healthy discussions, and network. Curate and share good datasets with the community. I initially only used Kaggle for downloading datasets, and it was only a couple of years ago that I actually became active. Now when I look back, I couldn't have been more wrong. A lot of people get intimidated by competitions. You can first get comfortable with the platform and then slowly start participating in the competitions.

Another important thing that I would like to mention is that many people work in isolation, lose motivation, and quit. Teaming up on Kaggle has many unseen advantages. It teaches you to work in a team, learn from the experiences, and work towards a common goal in a limited time frame.

Do you use other competition platforms? How do they compare to Kaggle?

While most of my current time is spent on Kaggle, in the past I have used Zindi, a data science competition platform focused on African use cases. It's a great place to access datasets focused on Africa. Kaggle is a versatile platform, but there is a shortage of problem statements from different parts of the world. Of late, we have seen some diversified problems too, like the recently held chaii competition — an NLP competition focusing on Indian languages. I believe similar competitions concentrating on different countries will be helpful for the research and the general data science community as well.

Cross-sectional to the taxonomy of Kaggle competitions, you must also consider that competitions may have different formats. The usual format is the so-called **simple format**, where you provide a solution, and it is evaluated as we previously described. More sophisticated, the **two-stage competition** splits the contest into two parts, and the final dataset is released only after the first part has finished and only to the participants of the first part. The two-stage competition format has emerged to limit the chance of some competitors cheating and infringing the rules since the evaluation is done on a completely untried test set that is available for a short time only. Contrary to the original Kaggle competition format, in this case, competitors have a much shorter amount of time and much fewer submissions to figure out any useful patterns from the test set. A variation of the two-stage competition occurs when the test data for the competition is collected in the future. An example is *CAFA 5 Protein Function Prediction* (`https://www.kaggle.com/competitions/cafa-5-protein-function-prediction`) or *JPX Tokyo Stock Exchange Prediction* (`https://www.kaggle.com/competitions/jpx-tokyo-stock-exchange-prediction`). After the final submission deadline, there are periodic updates to the leaderboard to reflect how the submissions perform with the collected data. There can be a few interim updates before the final evaluation, adding excitement, as the leaderboard changes radically in respect of the initial settings.

For the same reason, the **Code** competitions have recently appeared, where all submissions are made from a Kaggle Notebook, and any direct upload of submissions is disabled.

For Kagglers at different stages of their competition careers, there are no restrictions at all in taking on any kind of competition. However, we have some suggestions against or in favor of the format or type of competition depending on your level of experience in data science and your computational resources:

- For complete beginners, the **Getting Started** or the **Playground** competitions are good places to begin since you can easily get more confident about how Kaggle works without facing high competitive pressure. That being said, many beginners have successfully started with Featured and Research competitions, because being under pressure helped them to learn faster. Our suggestion is, therefore, to decide based on your learning style. Some Kagglers need to learn by exploring and collaborating (and the Getting Started or the Playground competitions are ideal for that). Others need the heat of a fast-paced competition to find their motivation.

- For **Featured** and **Research** competitions, also take into account that these competitions are often about fringe applications of AI and machine learning and, consequently, you often need a solid background or the willingness to study all the relevant research in the field of application of the competition.

Finally, keep in mind that most competitions require you to have access to **computational resources** that are often not available to most data scientists in the workplace. This can turn into growing expenses if you use a cloud platform other than Kaggle. **Code** competitions and competitions with time or resource limitations strive to put all the participants on the same resource level at inference time, but you still can benefit from training your models on better computation resources.

Submission and leaderboard dynamics

The way Kaggle works seems simple: the test set is hidden from participants; you fit your model; if your model is the best in predicting the test set, then you score highly, and you possibly win. Unfortunately, this description renders the inner workings of Kaggle competitions in an overly simplistic way. It doesn't consider that there are dynamics regarding the direct and indirect interactions of competitors or the nuances of the problem you are facing and its training and test set.

Explaining the Common Task Framework paradigm

A more comprehensive description of how Kaggle works is actually given by Professor *David Donoho*, professor of statistics at Stanford University (https://statistics.stanford.edu/people/david-donoho), in his paper *50 Years of Data Science*. It first appeared in the *Journal of Computational and Graphical Statistics* and was subsequently posted in the *MIT Computer Science and Artificial Intelligence Laboratory* (see http://courses.csail.mit.edu/18.337/2015/docs/50YearsDataScience.pdf).

Professor Donoho does not refer to Kaggle specifically but to all data science competition platforms. Quoting computational linguist *Mark Liberman*, he refers to data science competitions and platforms as part of a **Common Task Framework (CTF)** paradigm that has been silently and steadily progressing data science in many fields over the last decades. He states that a CTF can work incredibly well at improving the solution of a problem in data science from an empirical point of view, quoting the Netflix competition and many DARPA competitions as successful examples. The CTF paradigm has contributed to reshaping the best-in-class solutions for problems in many fields.

A CTF is composed of **ingredients** and a **secret sauce**. The ingredients are simple:

- A publicly available dataset and a related prediction task
- A set of competitors who share the common task of producing the best prediction for the task
- A system for scoring the predictions by the participants fairly and objectively, without providing hints about the solution that are too specific (or limiting them, at least)

The system works best if the task is well defined and the data is of good quality. In the long run, the performance of solutions improves by small gains until it reaches an asymptote. The process can be sped up by allowing a certain amount of sharing among participants (as happens on Kaggle through discussions and sharing Kaggle Notebooks and extra data provided by the datasets found in the Datasets section). According to the CTF paradigm, competitive pressure in competition suffices to produce always-improving solutions. When the competitive pressure is paired with some degree of sharing among participants, the improvement happens even faster – hence why Kaggle introduced many incentives for sharing.

This is because the **secret sauce** in the CTF paradigm is the competition itself, which, within the framework of a practical problem whose empirical performance has to be improved, always leads to the emergence of new benchmarks, new data and modeling solutions, and in general to an improved application of machine learning to the problem posed by the competition. A competition can, therefore, provide a new way to solve a prediction problem, new ways of feature engineering, and new algorithmic or modeling solutions. For instance, deep learning did not simply emerge from academic research. Still, it first gained a tremendous boost because of successful competitions that signaled its efficacy. We have already mentioned, for instance, the Merck competition, won by *Geoffrey Hinton's* team using one of the first deep learning solutions: https://www.kaggle.com/c/MerckActivity/overview/winners.

Coupled with the open software movement, which allows everyone access to powerful analytical tools (such as scikit-learn, TensorFlow, and PyTorch), the CTF paradigm brings about even better results because all competitors are on the same level at the start. On the other hand, the reliance on a solution to a competition on specialized or improved hardware can limit achievable results because it can prevent competitors without access to such resources from properly participating and contributing directly to the solution or indirectly by exercising competitive pressure on the other participants. Understandably, this is why Kaggle started offering free cloud services to participants of its competitions, Kaggle Notebooks, which we will introduce in the **Computational resources** section. It can flatten some differences in hardware-intense competitions (as most deep learning ones are) and increase the overall competitive pressure.

Understanding what can go wrong in a competition

Given our previous description of the CTF paradigm, you may be tempted to imagine that all a competition needs is to be set up on a proper platform and good results, such as positive involvement for participants and outstanding models for the sponsor company, will automatically come in. However, some things can go wrong and instead lead to a disappointing result in a competition, both for the participants and the institution running it:

- Leakage from the data
- Probing from the leaderboard (the scoring system)
- Overfitting and consequent leaderboard shake-up
- Private sharing

You have **leakage** from data when part of the solution can be retraced in the data itself. For instance, certain variables could be posterior to the target variable, so they reveal something about it. This happens in fraud detection when you use variables that are updated after fraud happens or in sales forecasting when you process information relating to the effective distribution of a product (more distribution implies more requests for the product, hence more sales).

Another issue could be that the training and test examples are ordered predictably or that the values of the identifiers of the examples hint at the solution. Examples are, for instance, when the identifier is based on the ordering of the target or the identifier value is correlated with the flow of time, and time affects the probability of the target.

Such solution leakage, sometimes named **golden features** by competitors (because getting a hint of such nuances in the data can turn into gold prizes for the participants), invariably leads to a solution that is not reusable. This also implies a sub-optimal result for the sponsor, but they at least can learn something about leaking features that can affect solutions to their problem.

Another problem is the possibility of **probing a solution** from the leaderboard. In this situation, you can take advantage of the evaluation metrics shown to you and snoop the solution by repeated submission trials on the leaderboard. Again, in this case, the solution is completely unusable under different circumstances. A clear example of this happened in the competition *Don't Overfit II*. The winning participant, *Zachary Mayers*, submitted every individual variable as a single submission, gaining information about the possible weight of each variable that allowed him to estimate the correct coefficients for his model (you can read Zach's detailed solution here: `https://www.kaggle.com/c/dont-overfit-ii/discussion/91766`). Generally, time series problems, or other problems where there are systematic shifts in the test data, may be seriously affected by probing since they can help competitors to successfully define some kind of *post-processing* (like multiplying their predictions by a constant) that is most suitable for scoring highly on the specific test set.

Another form of leaderboard snooping (getting a hint about the test set and overfitting it) happens when participants rely more on the feedback from the public leaderboard than their own tests. Sometimes, this turns into a complete failure of the competition, causing a wild shake-up – a complete and unpredictable reshuffling of the positions on the final leaderboard. In such a case, the winning solutions may not turn out to be optimal for the problem or even just dictated by chance. This has led to the diffusion of techniques analyzing the potential gap between the training and public test sets. This kind of analysis, called **adversarial testing**, can provide insights into how much to rely on the leaderboard and whether there are features that are so different between the training and test set that it would be better to avoid them altogether.

For an example, you can look at this Notebook by *Bojan Tunguz*: `https://www.kaggle.com/tunguz/adversarial-ieee`.

Another kind of defense against leaderboard overfitting is choosing safe strategies to avoid submitting solutions that are based too much on the leaderboard results. For instance, since (typically) two solutions are allowed to be chosen by each participant for final evaluation, a good strategy is to submit the best-performing one based on the leaderboard and the best-performing one based on your own cross-validation tests.

To avoid problems with leaderboard probing and overfitting, Kaggle has recently introduced different innovations based on Code competitions, where the evaluation is split into two distinct stages, as we previously discussed, with participants being completely blind to the actual test data so they are forced to consider their own local validation tests more.

Finally, another possible distortion of a competition is due to **private sharing** (sharing ideas and solutions in a closed circle of participants) and other illicit moves such as playing through multiple accounts or playing in multiple teams and stealing ideas. All such actions create an information asymmetry between participants that can be favorable to a few and detrimental to most. Again, the resulting solution may be affected because sharing has been imperfect during the competition, and fewer teams can exercise full competitive pressure. Moreover, if these situations become evident to participants (for instance, see `https://www.kaggle.com/c/ashrae-energy-prediction/discussion/122503`), it can lead to distrust and less involvement in the competition or subsequent competitions.

Computational resources

Some competitions pose limitations in order to render feasible solutions available to production or otherwise useful for the sponsors. This typically happens in code competitions. For instance, the *Vesuvius Challenge - Ink Detection* competition (`https://www.kaggle.com/competitions/vesuvius-challenge-ink-detection`) had strict limits on execution time, reproducibility, and documentation requirements, and some limits to the data you could use for solutions. Other competitions, instead, for instance, *The Learning Agency Lab - PII Data Detection* (`https://www.kaggle.com/competitions/pii-detection-removal-from-educational-data`), feature specific tracks focused on efficiency called **Efficiency Tracks** where a prize is given based on the speed of prediction, given enough predictive performance of the model or based on an efficiency score, which comprises both aspects in a single formula. Further examples of competitions with an efficiency track are:

- Severstal: Steel Defect Detection (`https://www.kaggle.com/c/severstal-steel-defect-detection`)
- Feedback Prize - Predicting Effective Arguments (`https://www.kaggle.com/c/feedback-prize-effectiveness`)
- Feedback Prize - English Language Learning (`https://www.kaggle.com/c/feedback-prize-english-language-learning`)
- Learning Equality - Curriculum Recommendations (`https://www.kaggle.com/c/learning-equality-curriculum-recommendations`)
- Predict Student Performance from Game Play (`https://www.kaggle.com/c/predict-student-performance-from-game-play`)

In such competitions, for example, participants may be required to operate within constraints such as limited memory resources or utilizing only CPUs. Usually, this is achieved using techniques such as:

- In classical machine learning, by trading more complex ensemble methods with an ensemble of a few models or even single models
- Downsizing your deep learning model, for instance, by **knowledge distillation** techniques where you train a smaller model (the student model) on the basis of the outputs of a larger one (the teacher model)
- Converting a model to **ONNX format**, a more efficient way to store and utilize a deep learning model
- **Pruning** a neural network by reducing its layers, or specific weights, such as filters or channels, thus reducing the size of the model
- Reducing the precision of a neural network by **quantization**, for instance, from 32-bit floating point values to 8-bit unsigned integers, thus reducing the network size and time to compute through it

You can read a complete report on the efficiency tracks and how they contributed to testing the efficacy of the various methods on the field by reading the notebook Towards Green AI (`https://www.kaggle.com/code/iamleonie/towards-green-ai/notebook`) by Leonie (`https://www.kaggle.com/iamleonie`).

Notebook-only (previously known as Kernel-Only) competitions, which require both training and inference to be executed on Kaggle Notebooks (preventing participants from submitting pre-trained models or external code), do not pose a problem for the resources you have to use. This is because Kaggle will provide you with all the resources you need (and this is also intended as a way to put all participants on the same start line for a better competition result).

More disparities arise when you have competitions that only limit the use of Notebooks to inference time, a type of competition that has now become the norm (**Code-only competitions**). In these cases, you can train your models on your own machine, and the only limit is then, at test time, the time complexity due to the number and complexity of models you produce. Since most competitions at the moment require deep learning solutions, you have to be aware that you will need specialized hardware, such as GPUs, in order to achieve a competitive result.

Even in some of the now-rare **tabular competitions**, you'll soon realize that you need a robust machine with quite some processors and a lot of memory to easily apply feature engineering to data, run experiments, and build models quickly.

Standards change rapidly, so it is difficult to specify standard hardware that you should have to compete at least in the same league as other teams. We can get hints about the current standard by looking at what other competitors use, either as their own machine or as a machine on the cloud.

For instance, HP launched a program that awarded an HP Z4 or Z8 to a few selected Kaggle participants in exchange for brand visibility. For instance, a Z8 machine has up to 72 cores, 3 TB of memory, 48 TB of storage (a good share by solid storage hard drive standards), and usually dual NVIDIA RTX as the GPU. We understand that this may be a bit out of reach for many; even renting a similar machine for a short time on a cloud instance such as Google's GCP or Amazon's AWS is out of the discussion, given the expenses for even moderate usage.

The cloud costs for each competition naturally depend on the amount of data to process and the number and type of models you build. Free credit giveaways in Kaggle competitions for the GCP and AWS cloud platforms usually range from US $200 to US $500.

Our suggestion, as you start your journey to climb to the top rankings of Kaggle participants, is, therefore to go with the machines provided free by Kaggle, Kaggle Notebooks (previously known as Kaggle Kernels).

Kaggle Notebooks

Kaggle Notebooks are versioned computational environments based on Docker containers running in cloud machines that allow you to write and execute scripts and notebooks in R and Python. Kaggle Notebooks:

- Are integrated into the **Kaggle environment** (you can make submissions from them and keep track of what submission refers to what Notebook)
- Come with most data science packages pre-installed
- Allow some customization (you can download files and install further packages)

The basic Kaggle Notebook is just CPU-based, but you can have versions boosted by an NVIDIA Tesla P100 or a TPU v3-8. TPUs are hardware accelerators specialized for deep learning tasks.

Though bound by a usage number and time quota limit, Kaggle Notebooks give you access to the computational workhorse to build your baseline solutions on Kaggle competitions:

Notebook type	CPU cores	Memory	Number of notebooks that can be run at a time	Weekly quota
CPU	4	30 GB	10	Unlimited
GPU	4	29 GB	2	30 hours
TPU	96	330 GB	2	20 hours

Table 1.1: Kaggle Notebook Types and Usage Limits

Also, the **GPU notebooks** may come in different configurations:

- P100 GPU: 1 NVIDIA Tesla P100 GPU
- T4 x2 GPU: 2 NVIDIA Tesla T4 GPUs

Besides the total runtime, CPU and GPU notebooks can run for a maximum of 12 hours per session before stopping (TPU notebooks for just 9 hours), meaning you won't get any results from the run apart from what you have saved on disk. You have a 20 GB disk saving allowance to store your models and results, plus an additional scratchpad disk that can exceed 20 GB for temporary usage during script running (as of April 28th, 2024, please refer to https://www.kaggle.com/docs/notebooks to verify the presently provided resources and limitations).

In certain cases, the GPU-enhanced machine provided by Kaggle Notebooks may not be enough. For instance, the recent *Deepfake Detection Challenge* (https://www.kaggle.com/c/deepfake-detection-challenge) required the processing of data consisting of around 500 GB of videos. That is especially challenging because of the 30-hour time limit of weekly usage and because of the fact that you cannot have more than two machines with GPUs running at the same time. Even if you can double your machine time by changing your code to leverage the usage of TPUs instead of GPUs (which you can find some guidance for easily achieving here: https://www.kaggle.com/docs/tpu), that may still not prove enough for fast experimentation in a data-heavy competition such as the Deepfake Detection Challenge.

For this reason, in , *Working and Learning with Kaggle Notebooks*, we will provide you with tips for successfully coping with these limitations to produce decent results without buying a heavy-performing machine. We will also show you how to integrate Kaggle Notebooks with GCP or, alternatively, in , *Organizing Data with Datasets*, how to move all your work into another cloud-based solution, Google Colab.

Teaming and networking

While computational power plays its part, only human expertise and ability can make a real difference in a Kaggle competition. For a competition to be handled successfully, it sometimes requires the collaborative efforts of a team of contestants. Apart from Recruitment competitions, where the sponsor may require individual participation to evaluate each participant's abilities better, there is typically no restriction against forming teams. Usually, teams consist of a maximum of five contestants.

Teaming has its own advantages because it can multiply efforts to find a better solution. A team can spend more time on the problem together, and different skills can be of great help; not all data scientists will have the same skills or the same level of skill when it comes to different models and data manipulation.

However, teaming is not all positive. Coordinating different individuals and efforts toward a common goal may prove not so easy, and some suboptimal situations may arise. A common problem is when some of the participants are not involved or are simply idle, but no doubt, the worst is when someone infringes the rules of the competition – to the detriment of everyone since the whole team could be disqualified – or even spies on the team to give an advantage to another team, as we mentioned earlier.

Despite any negatives, teaming up in a Kaggle competition is a great opportunity to get to know other data scientists better, collaborate for a purpose, and achieve more since Kaggle rules somehow reward teams over lone competitors by awarding more points in total. In fact, it's true that being part of a team typically results in earning fewer points for the same position in a competition. However, the cumulative points earned by a team usually surpass those that an individual would have earned alone. For instance, in a competition with 1,000 Kagglers participating, an individual ranking 10th would receive 10,706 points. Meanwhile, a team of two would earn 7,570 points each (totaling 15,140 points), and a team of three would earn 6,181 points each (totaling 18,543 points). Competing alone or as part of a team presents a clear trade-off: opting to compete individually may result in more points gained, while joining a team involves giving up a portion of potential points to increase the likelihood of achieving a better position on the leaderboard.

Teaming up is not the only possibility for networking in Kaggle, though it is certainly more profitable and interesting for the participants. You can also network with others through discussions on the forums or by sharing datasets and notebooks during competitions. All these opportunities on the platform can help you get to know other data scientists and be recognized in the community.

There are also many occasions to network with other Kagglers outside the Kaggle platform. First of all, a few Slack channels can be helpful. For instance, **KaggleNoobs** (`https://www.kaggle.com/getting-started/20577`) is a channel opened up in 2016 that features many discussions about Kaggle competitions. They have a supportive community that can help you if you have some specific problem with code or models.

Quite a few other channels are devoted to exchanging opinions about Kaggle competitions and data science-related topics. Some channels are organized on a regional or national basis, for instance, the Japanese channel **Kaggler-ja** (`http://kaggler-ja-wiki.herokuapp.com/`) or the Russian community **Open Data Science Network** (`https://ods.ai/`), created in 2015, which later opened also to non-Russian speaking participants. The Open Data Science Network doesn't offer simply a Slack channel but also courses on winning competitions, events, and reporting on active competitions on all known data science platforms (see `https://ods.ai/competitions`).

Aside from Slack channels, several local meetups themed around Kaggle in general or around specific competitions have sprung up, some just temporarily, others in a more established form. A meetup focused on Kaggle competitions, usually built around a presentation from a competitor who wants to share their experience or suggestions, is the best way to meet other Kagglers in person, exchange opinions, and build alliances for participating in data science contests together.

In this league, mention should be made of **Kaggle Days**, an initiative founded by *Maria Parysz* and *Paweł Jankiewicz* and active from 2018 to 2022. Although Kaggle Days is presently inactive, it is noteworthy because of its significant impact and effectiveness within the Kaggle community. Over the years, the Kaggle Days organization arranged a few events in major locations worldwide, including Warsaw, Paris, Dubai, San Francisco, and Tokyo, intending to bring together a conference of Kaggle experts. It also created a network of local meetups in various countries, many of which remain active today (such as **Kaggle Days Milan** organized by Alberto Danese, a Kaggle Grandmaster: `https://www.youtube.com/watch?v=5pkNnXLbFqs&list=PLCoiw8pILkTfTav1tSjrYA-LKADYeH1CP`). We had the opportunity to catch up with Paweł, a Competitions Grandmaster, about his experiences with Kaggle.

Interview: Paweł Jankiewicz

```
https://www.kaggle.com/paweljankiewicz
```

What's your favorite kind of competition and why? In terms of techniques and solving approaches, what is your specialty on Kaggle?

Code competitions are my favorite type of competition because working in a limited environment forces you to think about different kinds of budgets: time, CPU, memory. Too many times in previous competitions I needed to utilize even up to 3-4 strong virtual machines. I didn't like that in order to win I had to utilize such resources, because it makes it a very uneven competition.

How do you approach a Kaggle competition? How different is this approach to what you do in your day-to-day work?

I approach every competition a little bit differently. I tend to always build a framework for each competition that allows me to create as many experiments as possible. For example, in one competition where we needed to create a deep learning convolutional neural network, I created a way to configure neural networks by specifying them in the format C4-MP4-C3-MP3 (where each letter stands for a different layer). It was many years ago, so the configuration of neural networks is probably now done by selecting the backbone model. But the rule still applies. You should create a framework that allows you to change the most sensitive parts of the pipeline quickly.

Day-to-day work has some overlap with Kaggle competitions in terms of modeling approach and proper validation. What Kaggle competitions taught me is the importance of validation, data leakage prevention, etc. For example, if data leaks happen in so many competitions, when people who prepare them are the best in the field, you can ask yourself what percentage of production models have data leaks in training; personally, I think 80%+ of production models are probably not validated correctly, but don't quote me on that.

Another important difference in day-to-day work is that no one really tells you how to define the modeling problem. For instance:

- Should the metric you report or optimize be RMSE, RMSLE, SMAPE, or MAPE?
- If the problem is time-based, how can you split the data to evaluate the model as realistically as possible?

And these are not the only important things for the business. You also must be able to communicate your choices and why you made them.

Tell us about a particularly challenging competition you entered, and what insights you used to tackle the task.

The most challenging and interesting was the Mercari Price Prediction Code competition. It was very different from any other competition because it was limited to 1 hour of computation time and only 4 cores with 16 GB of memory. Overcoming these limitations was the most exciting part of the challenge. My takeaway from this competition was to believe more in networks for tabular data. Before merging with my teammate Konstantin Lopukhin (`https://www.kaggle.com/lopuhin`), I had a bunch of complicated models including neural networks, but also some other boosting algorithms. After merging, it turned out that Konstantin was using only one architecture which was very optimized (number of epochs, learning rate). Another aspect of this competition that was quite unique was that it wasn't enough to just average solutions from the team. We had to reorganize our workflow so that we had a single coherent solution and not something quickly put together. It took us three weeks to combine our solutions together.

In your experience, what do inexperienced Kagglers often overlook? What do you know now that you wish you'd known when you first started?

Software engineering skills are probably underestimated a lot. Every competition and problem is slightly different and needs some framework to streamline the solution (look at `https://github.com/bestfitting/instance_level_recognition` and how well their code is organized). Good code organization helps you to iterate faster and eventually try more things.

What's the most important thing someone should keep in mind or do when they're entering a competition?

The most important thing is to have fun.

Performance tiers and rankings

Apart from monetary prizes and other material items, such as cups, t-shirts, hoodies, and stickers, Kaggle offers many immaterial awards. Kagglers spend a considerable amount of time and effort during competitions (not to mention developing the skills they use to compete, which are, in truth, quite rare in the general population). The monetary prizes usually cover the efforts of the top few Kagglers, if not just the one in the top spot, leaving the rest with an astonishing number of hours spent voluntarily with little return. In the long term, participating in competitions with no tangible results may lead to disaffection and disinterest, lowering the competitive intensity.

Hence, Kaggle has found a way to reward competitors with an honor system based on medals and points. The idea is that the more medals and points you have, the more relevant your skills are, leaving you open to opportunities in your job search or any other relevant activity based on your reputation.

First, there is a **general leaderboard** that combines all the leaderboards of the individual competitions (`https://www.kaggle.com/rankings`). Based on the position they attain in each competition, Kagglers are awarded some points that, all summed together, provide their ranking on the general leaderboard. At first glance, the formula for scoring points in a competition may look a bit complex:

$$\left[\frac{100000}{\sqrt{N_{teammates}}}\right] * [Rank^{-0.75}] * [log_{10}(1 + log_{10}(N_{teams}))] * [e^{-t/500}]$$

There is even a website (`https://kagglepoints.com/`) to help you compute the points you may gain in a competition!

Nevertheless, in reality, it is simply based on a few ingredients:

- Your rank in a competition
- Your team size
- The popularity of the competition
- How old the competition is

Intuitively, ranking highly in popular competitions brings many points. Less intuitively, the size of your team matters in a non-linear way. That's due to the inverse square root part of the formula, since the proportion of points you have to give up grows with the number of people involved.

It is still quite favorable if your team is relatively small (two to three people, maximum) due to the advantage in wits and computational power brought about by collaboration.

Another point to keep in mind is that points decay with time. The decay is not linear, but as a rule of thumb, keep in mind that, after a year, very little is left of the points you gained. Therefore, glory on the general leaderboard of Kaggle is temporary unless you keep participating in competitions with similar results to before. As a consolation, on your profile, you'll always keep the highest rank you ever reached.

Also, the points you get from notebooks and datasets are subject to decay based on the simpler formulation:

$$e^\wedge(t/500)$$

The only key element here to consider for the decay dynamics is the time, t, which is the number of days since the notebook or dataset was created, expressed in days. Points decay over time according to the above formula. Each upvote on a created notebook or dataset is initially worth 1 point, and it decays from the day it was assigned (see https://www.kaggle.com/progression/code and https://www.kaggle.com/progression/datasets).

More long-lasting than points is the medal system, which covers all three areas of Competitions in Kaggle. Based on your results, you will be awarded medals for your performance on Competitions, Notebooks, and Datasets. Previously, there were also discussions among Grandmaster paths, but they were ruled out due to the impact of large language models, voting rings, and spamming links on the discussion dynamics. In Competitions, medals are awarded based on your position on the leaderboard. In the other two areas, medals are awarded based on the upvotes of other competitors, which can lead to some suboptimal situations, as upvotes are a less objective metric and also depend on popularity. The Kaggle team implemented a mitigation to prevent such unfair situations, as votes must be upvoted by users in the Expert tier or higher to count toward medals (note that self-votes are not counted).

The more medals you get, the higher the ranks of Kaggle mastery you can enter. The ranks are Expert, Master, and Grandmaster. The page at https://www.kaggle.com/progression explains everything about how to get medals, and how many and what kinds are needed to access the different ranks.

Keep in mind that these ranks and honors are always relative and that they may change over time. Some years ago, in fact, the scoring system and the ranks were quite different and have been changed. The resulting system was first modified on May 13th, 2015, and then again on July 8[th,] 2025 (read the announcement on the new Kaggle progression system here: https://www.kaggle.com/discussions/product-announcements/588704). However, the rules and requirements are likely to change again in the future to maintain the rarity and value of higher achievements and prevent attempts to game the system to obtain a title from Kaggle.

Recently, Kaggle introduced an additional system of achievements called Awards and Badges (you can read the details here: `https://www.kaggle.com/discussions/product-feedback/536045`). The new Kaggle Awards system offers a new way to recognize Kagglers' outstanding contributions beyond the existing progression metrics. This includes celebrating achievements from unique events, special programs, and one-off awards. It also highlights successes in analytics competitions and acknowledges the efforts of competition hosts. You can browse the list of Awards here: `https://www.kaggle.com/discussions/general/536047`.

Figure 1.5: A set of Kaggle Awards

Besides Awards, Badges are given for trying out different features across Kaggle and for your activity across the site. They are somewhat more lightweight recognition and are not intended as credentials; they just measure your level of engagement with the platform.

Figure 1.6: A set of Kaggle Badges

As part of the gamification approach on the platform, Badges are seen by many Kagglers as funny and collectible additions. You can browse all the available Badges here: `https://www.kaggle.com/rankings/awards/`.

Criticism and opportunities

Kaggle has drawn quite a few criticisms since it appeared. Participation in data science competitions is still a subject of debate today, with many different opinions out there, both positive and negative.

On the side of negative criticism:

- Kaggle provides a false perception of what machine learning really is since it is just focused on leaderboard dynamics.

- Kaggle is just a game of hyperparameter optimization and ensembling many models just to scrape a slightly higher score on a single unique evaluation metric (while in reality overfitting the test set), which is quite distant from the real objectives of any data science project in industry or academia.

- Kaggle is filled with inexperienced enthusiasts who are ready to try anything under the sun to get a score and a spotlight in hopes of being spotted by recruiters.

- As a further consequence, competition solutions are too complicated and often too tailored to a specific test set, rendering them a waste of time to implement because they won't work on different data. Some Kaggle solutions even do not produce "useful" models, that is models that should work for the task they were trained on, as discussed in this post by Lauren Oakden-Rayner, a medical AI researcher and radiologist: `https://laurenoakdenrayner.com/2019/09/19/ai-competitions-dont-produce-useful-models/`.

Many perceive Kaggle, like many other data science competition platforms, to be far from what data science is in reality. The point the critics raise is that business problems do not come from anywhere, and you seldom already have a well-prepared dataset to start with since you usually build it along the way based on refining business specifications and the understanding of the problem at hand. Moreover, many critics emphasize that Kagglers don't learn or excel at creating **production-ready** models since a winning solution cannot be constrained by resource limits or considerations about technical debt (though this is not always true for all competitions).

All such criticism is related, in the end, to how Kaggle's standings can be compared to other kinds of experience in the eyes of an employer, especially relative to data science education and work experience. One persistent myth is that Kaggle competitions won't help get you a job or a better job in data science and that they do not put you on another plane compared to data scientists who do not participate.

Our perspective is that it is a misleading belief that Kaggle rankings do not hold an inherent value beyond the Kaggle community. For instance, in a job search, Kaggle can provide you with some very useful competencies in modeling data and problems and effective model testing. It can also expose you to many techniques and data/business problems beyond your actual experience and comfort zone. Still, it cannot supplement you with everything you need to place yourself successfully as a data scientist in a company, nor can it replace the dynamics of properly managing and developing useful data science projects through competition alone.

You can use Kaggle for learning (there is also a section on the website, Courses, devoted to just learning) and for differentiating yourself from other candidates in a job search; however, how this will be considered varies considerably from company to company. Regardless, what you learn on Kaggle will invariably prove useful throughout your career and provide a hedge when solving complex and unusual problems with data modeling; by participating in Kaggle competitions, you build strong competencies in modeling and validating. You also network with other data scientists, which can get you a reference for a job more easily and provide you with another way to handle complex problems beyond your skills because you will have access to other people's competencies and opinions.

Hence, we believe that Kaggle functions more indirectly to help you in your career as a data scientist in various ways. Of course, sometimes Kaggle will help you to be contacted directly as a job candidate based on your successes. Still, Kaggle will often provide you with the intellectual skills and experience you need to succeed, first as a candidate and then as a practitioner.

In fact, after playing with data and models on Kaggle for a while, you'll have had the chance to see enough different datasets, problems, and ways to deal with them under time pressure that, when faced with similar problems in real settings, you'll be skilled in finding solutions quickly and effectively.

This latter opportunity for a skill upgrade is why we were motivated to write this book in the first place and what this book is actually about. You won't find a guide solely on how to win or score highly in Kaggle competitions, but you will find a guide about how to compete better on Kaggle and get the most back from your competition experiences.

Use Kaggle and other competition platforms smartly. Kaggle is not a passepartout – being first in a competition won't assure you a highly paid job or glory beyond the Kaggle community, and the top model in a Kaggle competition won't turn into the most suitable model for successfully handling a specific task or problem.

However, consistently participating in competitions is a card to be played smartly to show interest and passion in your data science job search and to improve specific skills that can differentiate you as a data scientist and not make you obsolete in front of AutoML solutions.

If you follow us through this book, we will show you how.

Summary

In this starting chapter, we first discussed how data science competition platforms have risen and how they actually work, both for competitors and the institutions that run them, referring in particular to the convincing CTF paradigm discussed by Professor David Donoho.

We illustrated how Kaggle works without forgetting to mention other notable competition platforms and how it could be helpful for you to take on challenges outside Kaggle as well. Regarding Kaggle, we detailed how the different stages of a competition work, how competitions differ, and what resources the Kaggle platform can offer you.

In the following chapters, we will explore Kaggle in more detail, starting with how to work with datasets.

Join our book's Discord space

Join our community's Discord space for discussions with the authors and other readers:

`https://packt.link/kaggle`

2

Organizing Data with Datasets

In his story, *The Adventure of the Copper Beeches*, Arthur Conan Doyle has Sherlock Holmes shout "Data! Data! Data! I cannot make bricks without clay." This mindset, which served the most famous detective in literature so well, should be adopted by every data scientist. For that reason, we begin the more technical part of this book with a chapter dedicated to data: specifically, in the Kaggle context, leveraging the power of the Kaggle Datasets functionality for our purposes.

In this chapter, we will cover the following topics:

- Setting up a dataset
- Gathering the data
- Working with datasets
- Using Kaggle datasets in Google Colab
- Legal caveats

Setting up a dataset

In principle, any data you can use (subject to limitations; see the *Legal caveats* section later on in this chapter) can be uploaded to Kaggle. The specific limits at the time of writing this book are **200 GB per private dataset** and an overall 200 GB maximum for private datasets, which, if exceeded, requires you either to make some of your datasets public or delete unused ones. Keep in mind that the size limit per single dataset is calculated uncompressed; uploading compressed versions speeds up the transfer but does not help against the limits. You can check the most recent documentation for the datasets at this link: `https://www.kaggle.com/docs/datasets`. There is also an additional technical specification that requires you to limit to 50 top-level files in your root; if you have more, you have to create a directory structure to hold them.

Kaggle promotes itself as the *"home of open data science"* and the impressive collection of datasets available from the site certainly lends some credence to that claim. The topics range from oil prices to anime recommendations, and newsworthy subjects make their way to Kaggle with rather remarkable speed. As an example, when the emails of Anthony Fauci were released under the Freedom of Information Act in May 2021 (`https://www.washingtonpost.com/politics/interactive/2021/tony-fauci-emails/`), they were uploaded as a Kaggle dataset a mere 48 hours later.

Before uploading the data for your project into a dataset, make sure to check the existing content. For several popular applications (image classification, NLP, and financial time series), there is a chance the dataset you want to use might have already been stored on Kaggle. For the sake of this introduction, let us assume the kind of data you will be using in your project is not already there, so you need to create a new dataset:

1. When you head to the menu with three lines on the left-hand side (technically called an hamburger menu) and click on **Datasets**, you will be redirected to the **Datasets** page:

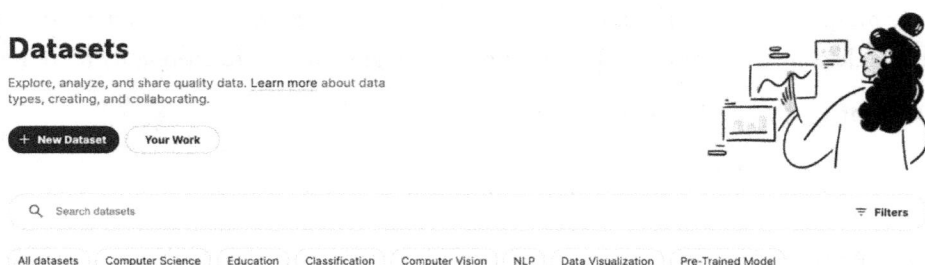

Figure 2.1: The Datasets page

2. When you click on **+ New Dataset**, you will be prompted for the basics: uploading the actual data and giving it a title. The icons on the left-hand side correspond to the different sources you can utilize for your dataset. We describe them in the order in which they are shown on the page: first, *uploading a file from the local drive*. This is a self-explanatory option:

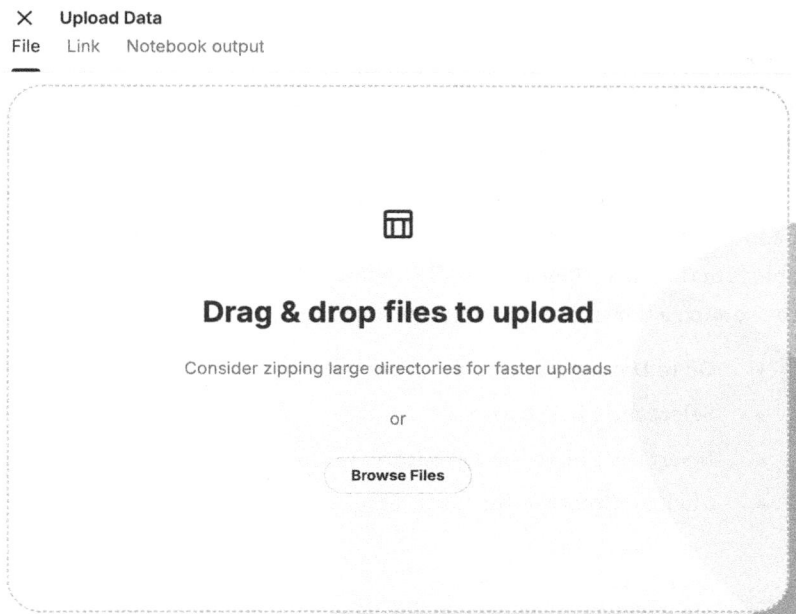

Figure 2.2: Creating a dataset by uploading a file from the local drive

3. Next, we can create a dataset from a **remote URL**, pointing to a specific file:

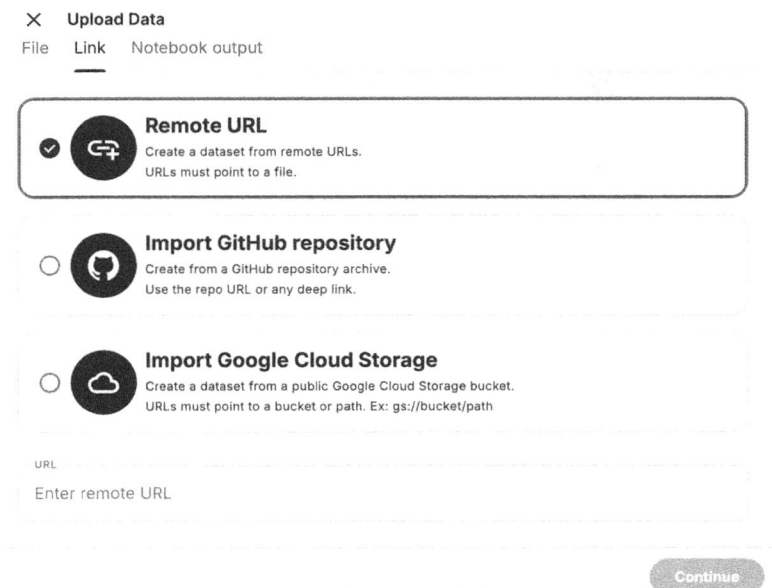

Figure 2.3: Creating a dataset by uploading a file from a remote URL

As stated in the technical documentation for Kaggle Datasets, you can upload any file you can think of. However, the system explicitly supports, by means of previews and data exploration, only CSV files, JSON files, SQLite format databases, common archive formats like ZIP or 7-Zip, and BigQuery datasets.

4. We can create a dataset by importing a public GitHub repository: this feature is particularly handy when it comes to experimental libraries. While frequently offering hitherto unavailable functionality, they are usually not included in the Kaggle environment. If you want to use such a library in your code, you can import it as a dataset, as demonstrated below:

    - Go to **Datasets** and click **+ New Dataset**.
    - Select the GitHub icon.
    - Insert the link to the repository, as well as the title for the dataset.
    - Click on **Create** at the bottom right.

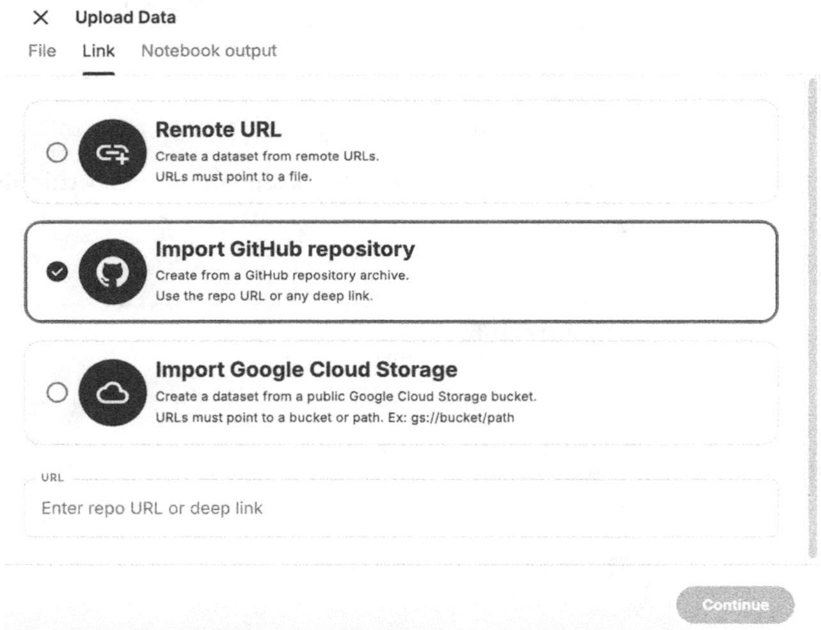

Figure 2.4: Creating a dataset by uploading a file from a GitHub repository

5. Another option for creating a dataset is using **output files from an existing notebook:**

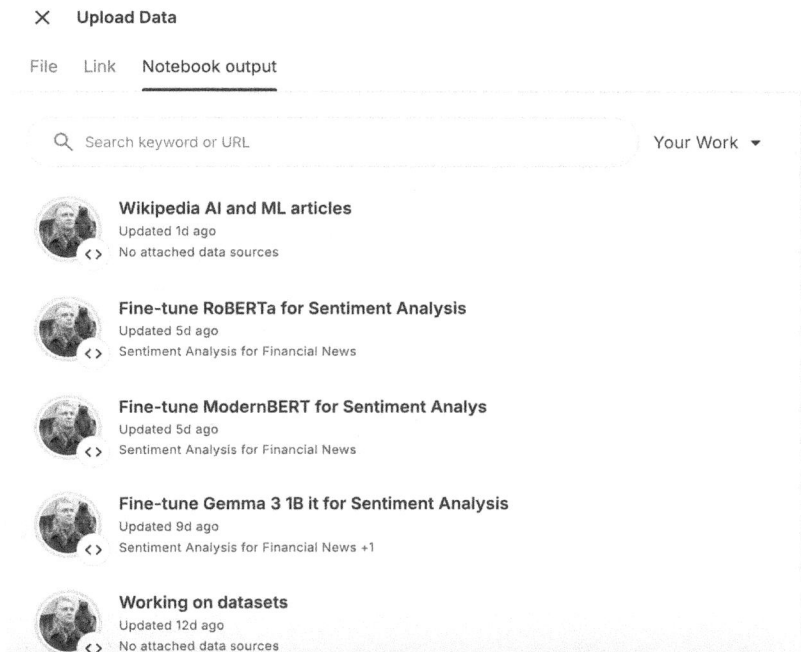

Figure 2.5: Creating a dataset by uploading the output of a Notebook

6. Finally, we can **import a Google Cloud Storage file:**

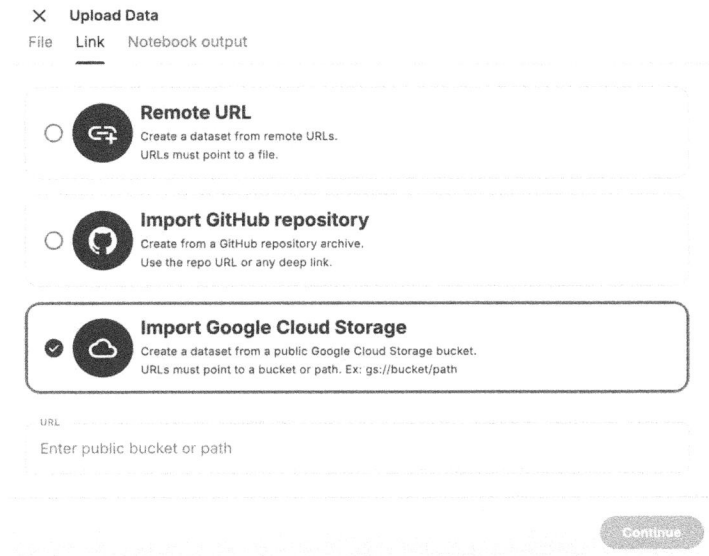

Figure 2.6: Creating a dataset from a Google Cloud Storage file

Irrespective of the input method you chose, next to the **Create** button, there is another one marked **Private**. By default, any dataset you create is private: only you, its creator, can view and edit it. It is probably a good idea to leave this setting at default at the dataset creation stage and only make it public (available to either a select list of contributors, or everyone) at a later stage.

Keep in mind that Kaggle is a popular platform, and many people upload their datasets – including private ones – so try to think of a non-generic title. This will increase the chance of your dataset being noticed.

7. Once you have completed all the steps and clicked **Create**, voilà! Your first dataset is ready. You can then head to the **Data Card**:

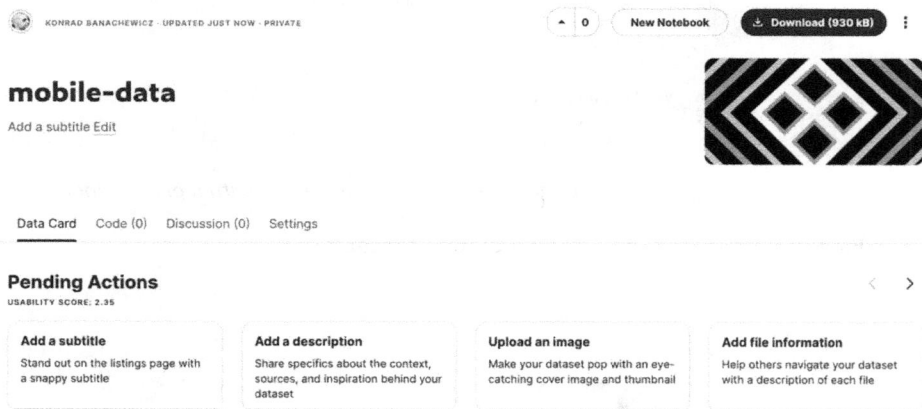

Figure 2.7: The Data Card

The preceding screenshot demonstrates the different information you can provide about your dataset; the more you provide, the higher the **usability index**. This index is a synthetic measure summarizing how well your dataset is described. Datasets with higher usability indices appear higher in the search results. For each dataset, the usability index is based on several factors:

- **Self-explanatory ones:** Subtitle, tags, description, and image
- **Metadata/format:** File information column descriptors, license, and file format
- **Data maintenance:** Data provenance, expected update frequency, and notebook published

In principle, you do not handle all the aspects listed above; your newly created dataset is perfectly usable without them (and if it is a private one, you probably do not care – after all, you know what is in it). However, community etiquette would suggest filling out the information for the datasets you make public; the more you specify, the more usable the data will be to others.

Gathering the data

Apart from legal aspects, there is no real limit on the kind of content you can store in the datasets (tabular data, images, or text). If it fits within the size requirements, you can store it. This includes data harvested from other sources; **posts** by hashtag or topic are among the popular datasets at the time of writing:

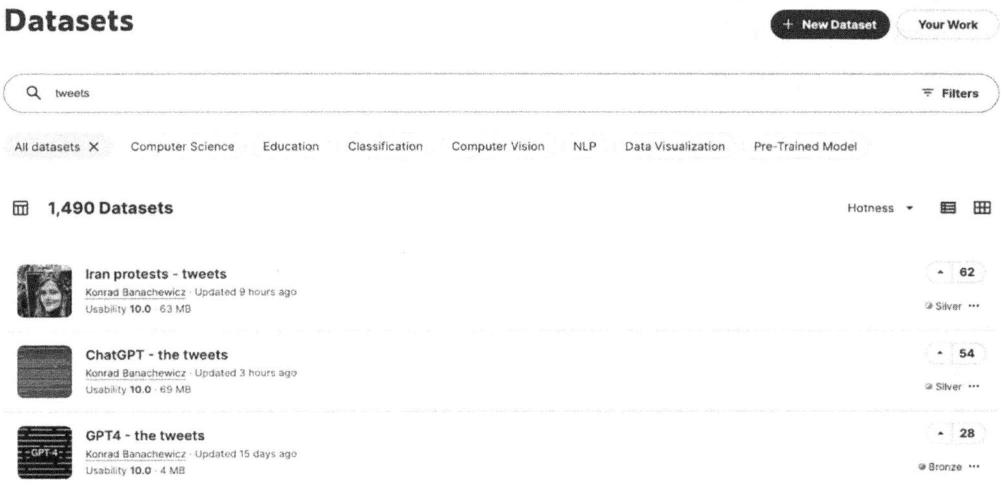

Figure 2.8: Aggregated X posts are a popular category

Discussion of the different frameworks for harvesting data from social media (X, Reddit, and so on) is outside the scope of this book, but a useful starting point is provided by the work of Gabriel Preda (`https://www.kaggle.com/gpreda`):

- Gathering posts from X: `https://github.com/gabrielpreda/covid-19-tweets/blob/master/covid-19-tweets.ipynb`

- Gathering Reddit posts: `https://github.com/gabrielpreda/reddit_extract_content/blob/main/reddit_birds_are_not_real.py`

Other ideas for gathering original data to upload to Kaggle Datasets include setting up an online survey, scraping data freely available from the web, and collecting data from a mobile app or a recording instrument.

To brainstorm some further ideas on data collection and publication on Kaggle, we spoke to *Andrew Maranhão* (aka Larxel), Datasets Grandmaster) and senior data scientist at the Hospital Albert Einstein in São Paulo. He previously managed to reach the number 1 position in Datasets and we found it interesting to discuss with him his rise to dataset success, his tips for creating datasets, and his general experiences on Kaggle.

Interview: Andrew Maranhão

https://www.kaggle.com/andrewmvd

What's your favorite kind of competition and why? In terms of techniques and solving approaches, what is your specialty on Kaggle?

Medical imaging is usually my favorite. It speaks to my purpose and job. Among medical competitions, NLP is language-bound, and tabular data varies widely among hospitals, but imaging is mostly the same, so any advancement in this context can bring about benefits for many countries across the world, and I love this impact potential. I also have a liking for NLP and tabular data, but I suppose this is pretty standard.

Tell us about a particularly challenging competition you entered, and what insights you used to tackle the task.

In a tuberculosis detection in X-ray images competition, we had around 1,000 images, which is a pretty small number for capturing all the manifestations of the disease. I came up with two ideas to offset this:

- Pre-train on external data of pneumonia detection (~20k images), as pneumonia can be mistaken for tuberculosis.
- Pre-train on the multilabel classification of lung abnormalities (~600k images) and use grad-CAM with a simple SSD to generate bounding box annotations of classification labels.

In the end, a simple blend of these two achieved 22% more compared to the result that the second-place team had. It happened at a medical convention, with about 100 teams participating.

You have become a Dataset Grandmaster and achieved the number 1 rank in Datasets. How do you choose topics and find, gather, and publish data for your datasets on Kaggle?

This is a big question; I'll try to break it down piece by piece.

1. *Set yourself a purpose.*

    The first thing that I have in mind when choosing a topic is the reason I am doing this in the first place.

    When there is a deeper reason underneath, great datasets just come off as a result, not as a goal in itself. Fei Fei Li, the head of the lab that created ImageNet, revealed in a TED talk that she wanted to create a world where machines would be able to reason and appreciate the world with their vision in the same way her children did.

    Having a purpose in mind will make it more likely that you'll engage and improve over time, and will also differentiate you and your datasets. You can certainly live off tabular data on everyday topics, though I find that unlikely to leave a lasting impact.

2. *A great dataset is the embodiment of a great question.*

    If we look at the greatest datasets in current literature, such as ImageNet and others, we can see some common themes:

    - It is a daring, relevant question with great potential for all of us (scientific or real-world application)
    - The data was well-collected, controlled for quality, and well-documented
    - It has an adequate amount of data and diversity for our current hardware
    - It has an active community that continuously improves the data and/or builds upon that question.

    As I mentioned before, I feel that asking questions is a primary role of a data scientist and is likely to become even more prominent as automated machine and deep learning solutions advance. This is where datasets can certainly exercise something unique to your skill set.

3. *Create your process for success, rather than only pursuing success for the sake of success.*

    Quality far overshadows quantity; you only need 15 datasets to become a Grandmaster and the flagship datasets of AI are few and well-made.

I have thrown away as many datasets as I have published. It takes time, and it is not a one-and-done type of thing as many people treat it – datasets have a maintenance and continuous improvement side to them.

One thing that is very often overlooked is supporting the community that gathers around your data. Notebooks and datasets are mutual efforts, so supporting those who take the time to analyze your data goes a long way for your dataset too. Analyzing their bottlenecks and choices can give directions as to what preprocessing steps could be done and provided, and also the clarity of your documentation.

All in all, the process that I recommend starts with setting your purpose, breaking it down into objectives and topics, formulating questions to fulfill these topics, surveying possible sources of data, selecting and gathering, preprocessing, documenting, publishing, maintaining and supporting, and finally, improvement actions.

For instance, let's say that you would like to increase social welfare; you break it down into an objective, say, racial equity. From there, you analyze topics related to the objective and find the Black Lives Matter movement. From here, you formulate the question: how can I make sense of the millions of voices talking about it?

This narrows down your data type to NLP, which you can gather data for from news articles, YouTube comments, and X posts (which you choose, as it seems more representative of your question and feasible). You preprocess the data, remove identifiers, and document the collection process and dataset purpose.

With that done, you publish it, and a few Kagglers attempt topic modeling but struggle to do so because some posts on X contain many foreign languages that create encoding problems. You support them by giving them advice and highlighting their work and decide to go back and narrow the posts down to English, to fix this for good.

Their analysis reveals the demands, motivations, and fears relating to the movement. With their efforts, it was possible to break down millions of posts into a set of recommendations that may improve racial equity in society.

4. *Doing a good job is all that is in your control.*

Ultimately, it is other people that turn you into a Grandmaster, and votes don't always translate into effort or impact. In one of my datasets, about Cyberpunk 2077, I worked on it for about 40 hours total and, to this day, it is still one of my least upvoted datasets. But it doesn't matter. I put in the effort, I tried, and I learned what I could – that's what is in my control, and next week I'll do it again no matter what. Do your best and keep going..

Are there any particular tools or libraries that you would recommend using for data analysis/ machine learning?

Strangely enough, I both recommend and unrecommend libraries. LightGBM is a great tabular ML library with a fantastic ratio of performance to compute time; CatBoost can sometimes outperform it, but it comes at the cost of increased compute time, during which you could be having and testing new ideas. Optuna is great for hyperparameter tuning, Streamlit for frontends, Gradio for minimum viable products (MVP), FastAPI for microservices, Plotly and Plotly Express for charts, and PyTorch and its derivatives for deep learning

While libraries are great, I also suggest that, at some point in your career, you take the time to implement machine learning algorithms yourself. I first heard this advice from Andrew Ng and then from many others of equal caliber. Doing this creates very in-depth knowledge that sheds new light on what your model does and how it responds to tuning, data, noise, and more.

In your experience, what do inexperienced Kagglers often overlook? What do you know now that you wish you'd known when you first started?

Over the years, the things I wished I realized sooner the most were:

1. Absorbing all the knowledge at the end of a competition
2. Replication of winning solutions in finished competitions

With the pressure of a competition drawing to a close, you can see the leaderboard shaking more than ever before. This makes it less likely that you will take risks and take the time to see things in all their detail. When a competition is over, you don't have that rush and can take as long as you need; you can also replicate the rationale of the winners who made their solutions known.

If you have the discipline, this will do wonders for your data science skills, so the bottom line is: stop when you are done, not when the competition ends. I have also heard this advice from an Andrew Ng keynote, where he recommended replicating papers as one of his best ways to develop yourself as an AI practitioner.

Also, at the end of a competition, you are likely to be exhausted and just want to call it a day. No problem there; just keep in mind that the discussion forum after the competition is done is one of the most knowledge-rich places on planet Earth, primarily because many rationales and the code for winning solutions are made public there. Take the time to read and study what the winners did; don't give in to the desire to move on to something else, as you might miss a great learning opportunity.

Has Kaggle helped you in your career? If so, how?

Kaggle helped my career by providing a wealth of knowledge and experience and also building my portfolio. My first job as a data scientist was largely due to Kaggle and DrivenData competitions. Throughout my career, I studied competition solutions and participated in a few more. Further engagement on datasets and notebooks also proved very fruitful in learning new techniques and asking better questions.

In my opinion, asking great questions is the primary challenge faced by a data scientist. Answering them is surely great as well, although I believe we are not far from a future where automated solutions will be more and more prevalent in modeling. There will always be room for modeling, but I suppose a lot of work will be streamlined in that regard. Asking great questions, however, is far harder to automate – if the question is not good, even the best solution could be meaningless.

Have you ever used something you have done in Kaggle competitions in order to build your portfolio to show to potential employers, and that helped you in getting a job?

Absolutely. I landed my first job as a data scientist in 2017 using Kaggle as proof of knowledge. To this day, it is still a fantastic CV component, as educational backgrounds and degrees are less representative of data science knowledge and experience than a portfolio is.

A portfolio with projects with competitions shows not just added experience but also a willingness to go above and beyond for development, which is arguably more important for long-term success.

Do you use other competition platforms? How do they compare to Kaggle?

I also use DrivenData and AICrowd. The great thing about them is that they allow organizations that don't have the same access to financial resources, such as start-ups and research institutions, to create competitions.

Great competitions come from a combination of great questions and great data, and this can happen regardless of company size. Kaggle has a bigger and more active community, and the hardware they provide coupled with the data and notebook capabilities make it the best option, yet both DrivenData and AICrowd introduce just as interesting challenges and allow for more diversity.

What's the most important thing someone should keep in mind or do when they're entering a competition?

Assuming your primary goal is development, my recommendation is that you pick a competition on a topic that interests you and a task that you haven't done before. Critical sense and competence require depth and diversity. Focusing and giving your best will guarantee depth, and diversity is achieved by doing things you have not done before or have not done in the same way.

Working with datasets

Once you have created a dataset, you probably want to use it in your analysis. There are different methods of going about this, but the principal method is starting a Kaggle notebook where you use your dataset as a primary source. Other approaches are using the data on a Google Colab notebook or downloading it to your local computer or cloud instance. In this chapter, we will discuss the first two options, working on a **Kaggle notebook** and on **Google Colab**. As for working on your Kaggle dataset on Kaggle, you can do this by navigating to the dataset page and then clicking on the more options menu (the three-dots menu) and select the **New Notebook** from the menu:

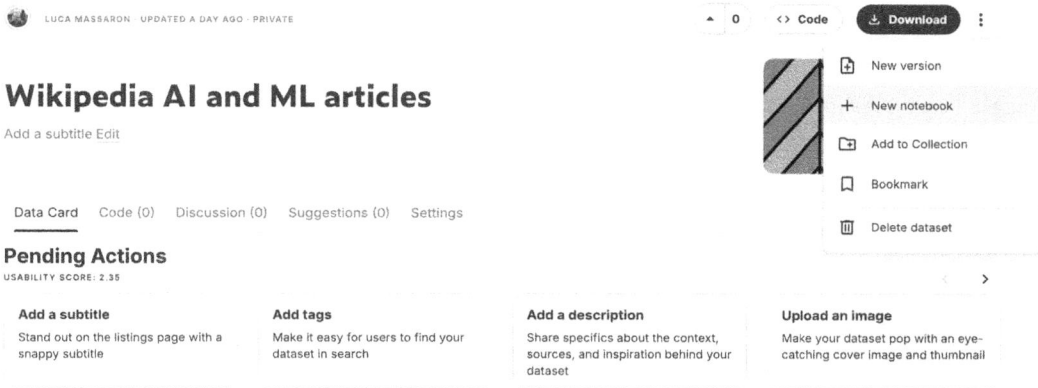

Figure 2.9: Creating a notebook from the dataset page

Once you have done this, you will be redirected to your **Notebook** page:

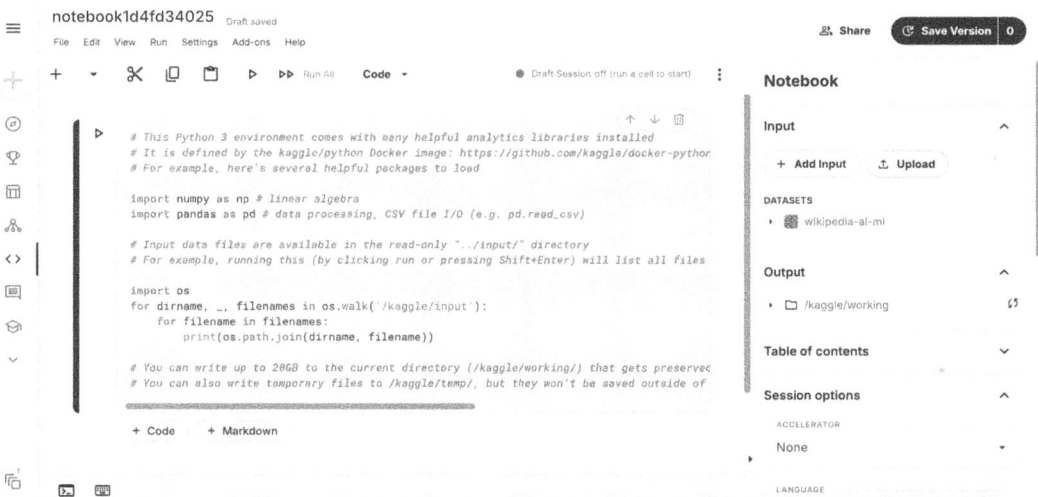

Figure 2.10: Starting a notebook using your dataset

Here are a few pointers around this:

- The **alphanumeric title** is generated automatically; you can edit it by clicking on it.
- On the right-hand side under **Input**, you see the list of data sources attached to your notebook; the dataset I selected can be accessed under **../input/** or from **/kaggle/input/**.
- The opening block (with the imported packages, descriptive comments, and printing the list of available files) is added automatically to a new Python notebook.

With this basic setup, you can start to write a notebook for your analysis and utilize your dataset as a data source. We will discuss notebooks at greater length in .

Using Kaggle datasets in Google Colab

Kaggle notebooks are free to use, but not without limits (more on that in the next chapter, *Working and Learning with Notebooks*), and the first one you are likely to hit is the time limit – 12 hours for a CPU/GPU session, 9 for TPU. A popular alternative is Google Colab, a free Jupyter Notebook environment that runs entirely in the cloud: `https://colab.research.google.com`.

Even once we've moved the computations there, we might still want to have access to the Kaggle datasets, so importing them into Colab is a rather handy feature. The remainder of this section discusses the steps necessary to use Kaggle datasets through Colab.

The first thing we do, assuming we are already registered on Kaggle, is head to the account page to generate the **API token** (an access token containing security credentials for a login session, user identification, privileges, and so on):

1. Go to your account's settings, which can be found at `https://www.kaggle.com/settings/account`, and click on **Create New Token**:

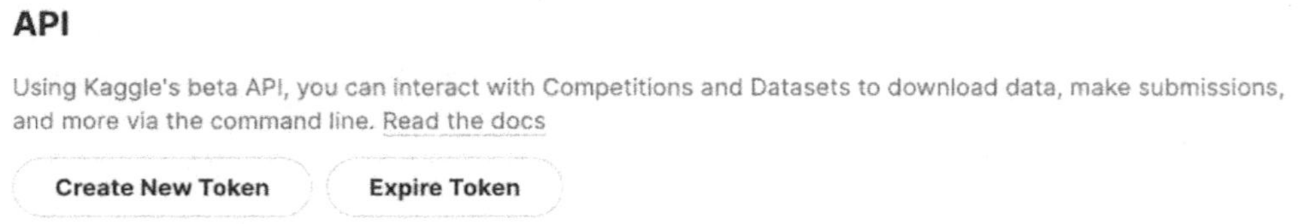

   API

   Using Kaggle's beta API, you can interact with Competitions and Datasets to download data, make submissions, and more via the command line. Read the docs

   (**Create New Token**)   (**Expire Token**)

Figure 2.11: Creating a new API token

A file named **kaggle.json** containing your username and token will be created.

2. The next step is to create a folder named **Kaggle** in your Google Drive and upload the **.json** there:

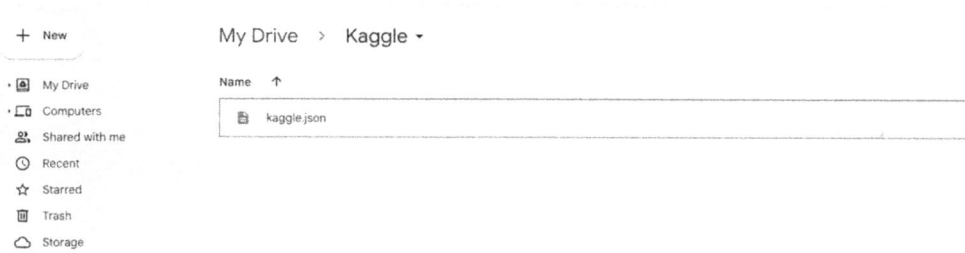

Figure 2.12: Uploading the .json file into Google Drive

3. Once done, you need to create a new Colab notebook and mount your drive by running the following code in the notebook:

```
from google.colab import drive
drive.mount('/content/gdrive')
```

4. Get the authorization code from the URL prompt and provide it in the empty box that appears, then execute the following code to provide the path to the *.json config*:

```
import os

# content/gdrive/My Drive/Kaggle is the path where kaggle.json is
# present in the Google Drive
os.environ['KAGGLE_CONFIG_DIR'] = "/content/gdrive/My Drive/Kaggle"

# change the working directory
%cd /content/gdrive/My Drive/Kaggle

# check the present working directory using the pwd command
```

5. As an alternative to uploading your *kaggle.json* file to Google Drive, you can use the recently added Colab Secrets functionality that provides a secure way to store and manage sensitive API keys on Google Colab. If you decide to use this procedure, instead, after obtaining your Kaggle JSON file, open Google Colab, navigate to the key icon on the left bar, and add two new keys (allowing access to the Colab notebook). The two new keys are `KAGGLE_USERNAME` and `KAGGLE_KEY` and their values can be obtained by opening the kaggle.json file and retrieving the **username** and **key** values, respectively.

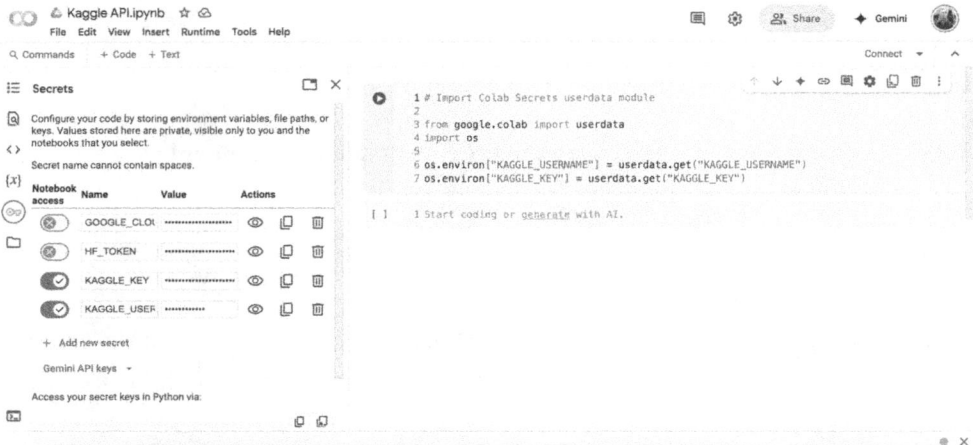

Figure 2.13: Setting Colab Secrets with your Kaggle API token

6. Once done, you can immediately use it with this code snippet:

```python
from google.colab import userdata
import os

os.environ["KAGGLE_USERNAME"] = userdata.get('KAGGLE_USERNAME')
os.environ["KAGGLE_KEY"] = userdata.get('KAGGLE_KEY')
```

Using Colab Secrets has advantages. In fact, you need to set your secrets just once and you can reuse them for multiple Colab notebooks. In addition, you can actually set various secrets, not just of the Kaggle API but also, for instance, for Hugging Face or OpenAI API access.

7. We can download the dataset now. Begin by going to the dataset's page on Kaggle, by clicking on the **code** button you can get some hints about how to download your dataset by kagglehub, as a pandas or Hugging Face dataset, or by mlcroissant (`https://github.com/mlcommons/croissant`):

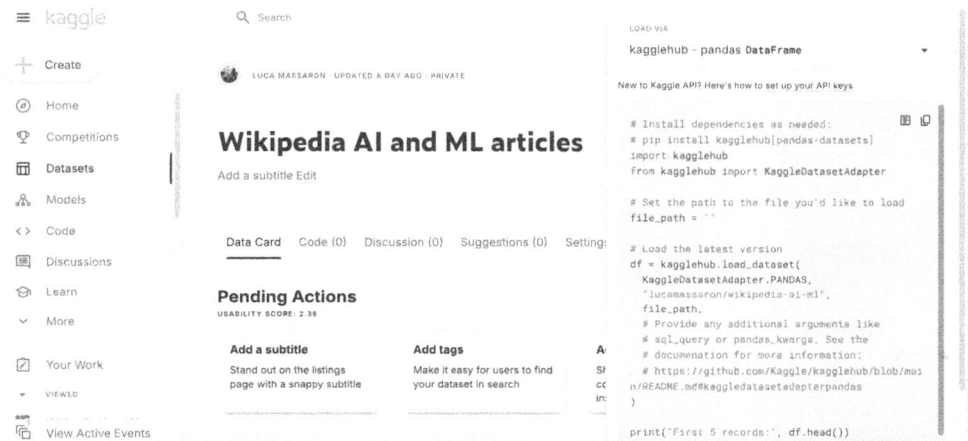

Figure 2.14: Copying the code for downloading a dataset using the kagglehub

8. An alternative way to download a Kaggle dataset could be using the **kagglehub** package, which is also pre-installed on Google Colab:

```
import kagglehub
path = kagglehub.dataset_download(
    "ajaypalsinghlo/world-happiness-report-2021"
)
```

This command will return the path where the files have been downloaded.

9. The dataset will be downloaded to the **Kaggle** folder as a **.zip** archive – unpack it and you are good to go.

As you can see from the list above, using a Kaggle dataset in Colab is a straightforward process – all you need is an API token, and making the switch gives you the possibility of using more GPU hours than what is granted by Kaggle.

Legal caveats

Just because you can put some data on Kaggle does not necessarily mean that you should. An excellent example would be the *People of Tinder dataset*; in 2017, a developer used the Tinder API to scrape the website for semi-private profiles and uploaded the data on Kaggle. After the issue became known, Kaggle ended up taking the dataset down. You can read the full story here: `https://www.forbes.com/sites/janetwburns/2017/05/02/tinder-profiles-have-been-looted-again-this-time-for-teaching-ai-to-genderize-faces/?sh=1afb86b25454`

In general, before you upload anything to Kaggle, ask yourself two questions:

- **Is it allowed from a copyright standpoint?** Remember to always check the licenses. When in doubt, you can always consult `https://opendefinition.org/guide/data/` or contact Kaggle.

- **Are there privacy risks associated with this dataset?** Even though posting certain types of information is not strictly speaking illegal, doing so might be harmful to another person's privacy.

The limitations are commonsensical, so they are not likely to hamper your efforts in that sphere on Kaggle. The same caveats are valid the other way around. Before downloading a dataset, it is really a good practice to check whether its license terms and usage restrictions allow the use you plan. In fact, some datasets may be offered free for educational and non-commercial purposes, but they could be restricted for research and business usage. Standard open-source licenses include:

- CC0 (Creative Commons Zero)
- MIT License
- Apache License2.0

You may err on the side of caution; however, a further good practice would be to review the description and information about the data to look up what the original granted license from the source was. A useful tool for this purpose is the Google Dataset Search engine, a search engine from Google that helps locate online data. Entries from the Google Dataset Search engine may corroborate or cast doubt on the license declared on a Kaggle dataset.

Summary

In this chapter, we introduced Kaggle Datasets, the standardized manner of storing and using data in the platform. We discussed dataset creation, ways of working outside of Kaggle, and the most important functionality: using a dataset in your notebook.

Besides being an essential tool to use in Kaggle competitions, Kaggle datasets also play an important role in helping you handle real-world data since Kaggle provides a vast array of datasets across various domains, such as healthcare, finance, and social sciences. Kagglers may use this data deluge to conduct exploratory data analysis, build ML models, and even derive insights that can influence decision-making in businesses or research. Finally, Kaggle datasets are another great collaboration opportunity because users can share data and work together on projects.

This essential data tool provides a good segue to our next chapter, where we focus our attention on Kaggle Notebooks.

Join our book's Discord space

Join our community's Discord space for discussions with the authors and other readers:

```
https://packt.link/kaggle
```

3

Working and Learning with Kaggle Notebooks

Kaggle notebooks are Jupyter notebooks in the browser that can be run free of charge. Although they seem simple, having a Jupyter notebook completely preconfigured with all the necessary packages, running directly on Kaggle, free of charge, on a CPU, GPU, or even TPU, is a powerful tool for exploring datasets, building models, sharing your work with the Kaggle community and the world, and, of course, successfully submitting to competitions.

In this chapter, you will learn how to set up and work with notebooks, while also learning how to get the most out of them. More specifically, we will cover the following topics:

- Setting up a notebook
- Running your notebook
- Setting a notebook as a utility script
- Getting the best out of notebooks
- Upgrading to **Google Cloud Platform (GCP)**
- One step beyond: Using notebooks in your data science portfolio
- Kaggle Learn courses

Technical requirements

You can execute your experiments from any device with an internet connection, although something bigger than a mobile phone is probably a good idea. The technical specifications of the environment (as of the time of writing) are quoted here from the Kaggle website (`https://www.kaggle.com/docs/notebooks`):

- 12 hours execution time for CPU and GPU sessions, 9 hours for TPU
- 20 gigabytes of auto-saved disk space (*/kaggle/working*)
- Additional scratchpad disk space (outside /kaggle/working) that will not be saved outside of the current session

The following are some other specifications:

- CPU specifications:

    - Four CPU cores
    - 30 gigabytes of RAM

- GPU: P100 specifications:

    - One Nvidia Tesla P100 GPU
    - Four CPU cores
    - 29 gigabytes of RAM

- GPU: T4x2 specifications:

    - Two Nvidia Tesla T4 GPUs
    - Four CPU cores
    - 29 gigabytes of RAM

- TPU VM v3-8 specifications:

    - 96 CPU cores
    - 330 gigabytes of RAM

On the side of the GPU offerings, if you're aiming for raw, single-precision power, the P100 generally takes the lead over the T4. However, when it comes to mixed precision (essential for deep learning workloads where speed and efficiency matter), the T4 excels, bringing a noticeable boost to performance. Plus, with a T4x2 setup (which comprises two GPUs), you're effectively doubling your GPU memory, giving you a substantial advantage for larger models and datasets.

As for TPUs, you will find TPU v3-8s in Kaggle notebooks, one of the most advanced hardware available for accelerating deep learning tasks. Each TPU v3-8 is equipped with four dual-core TPU chips, providing a total of eight TPU cores dedicated to rapid computation.

This kind of hardware particularly shines when handling mixed-precision calculations and intensive matrix operations, making it an ideal choice for modeling tasks that need both speed and precision.

Whether you go for a P100 or a T4, or even decide on a TPU, please keep in mind that you have some quotas on using all these resources. At the moment, quotas are set to 30 hours a week for GPUs and 20 hours for TPUs. Further limitations are that you cannot run more than two batch GPU sessions and a single batch TPU session. As for interactive sessions (when you are running the code as you are editing the notebook), you are constrained to a single session for both GPU and TPU instances.

Having clarified these limitations, without further ado, let us jump into working with Kaggle notebooks. The first thing we need to do is figure out how to set up a notebook.

Setting up a notebook

There are a few methods of creating a notebook on Kaggle:

- **From the homepage**: Select the **+ Create** button. A window will appear with various options, with the first choice being to create a **New Notebook**:

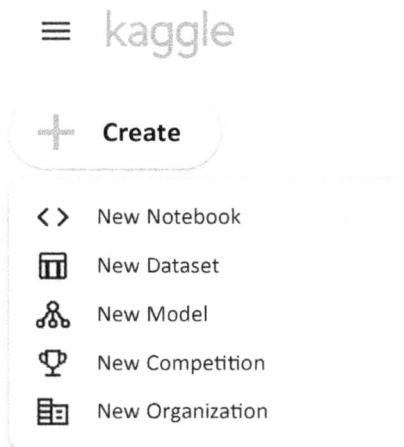

Figure 3.1: Using the + Create button on the main page

- **From the front page** (including the main page of a competition): Navigate to the **Code** section of the menu on the left-hand side of the landing page at https://www.kaggle.com/ and click the **+ New Notebook** button. This is the preferred method if you are planning an experiment that involves uploading your own dataset:

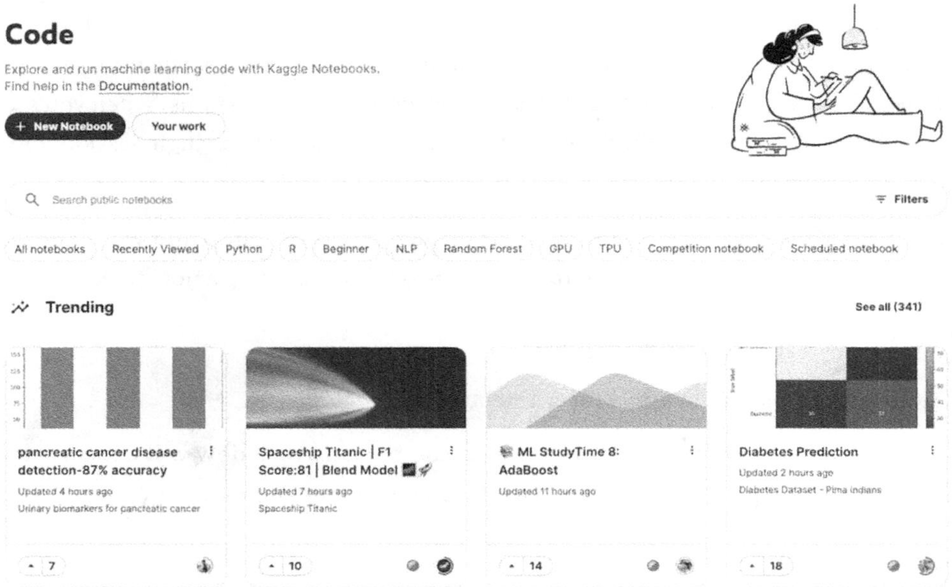

Figure 3.2: Creating a new notebook from the Code page

- **From a dataset:** You can go to the page of the dataset you are interested in and click the **New Notebook** button there, as we saw in the previous chapter:

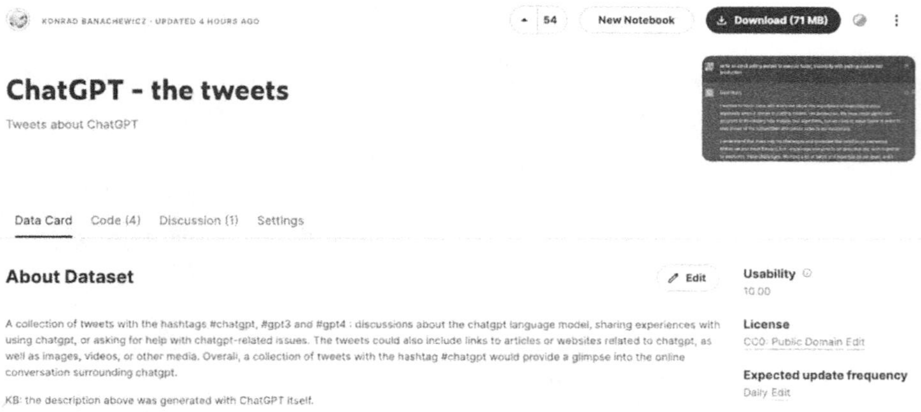

Figure 3.3: Creating a new notebook from the Dataset page

Whichever method you choose, after clicking **New Notebook**, you will be taken to your notebook page:

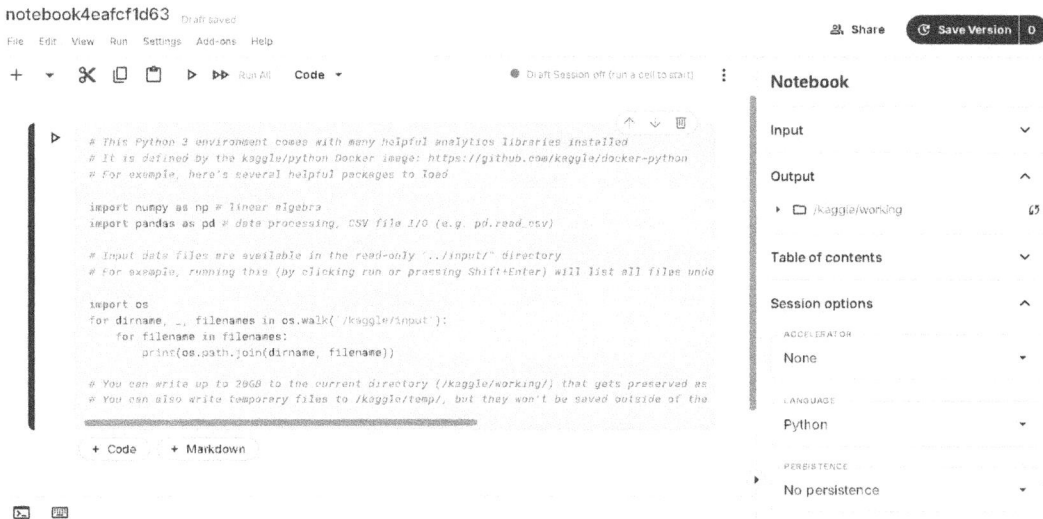

Figure 3.4: The notebook page

On the right-hand side of the new notebook page shown in the preceding screenshot, we have several settings that can be adjusted. Let's zoom into this area and take a closer look:

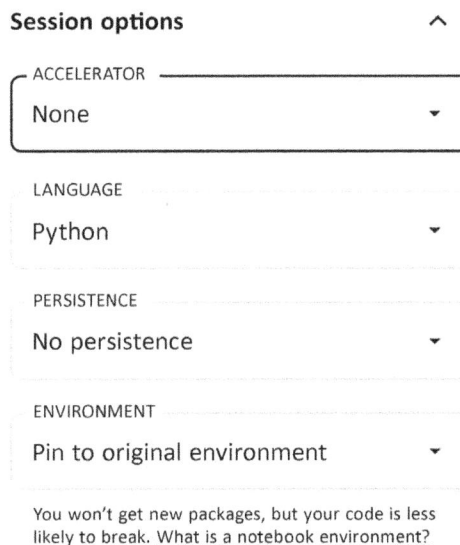

Figure 3.5: Notebook options

Let us discuss the settings briefly:

1. First, there is the **Accelerator**, as shown in the following screenshot. You can choose to run your code on a CPU (accelerator **None**), one of the two GPU options, or on a TPU. Keep in mind that moving from a CPU to (a single) GPU requires only minimal changes to the code and can be handled via system device detection. Migrating your code to a TPU requires more elaborate rewriting, starting with data processing. Switching between a CPU, GPU, and TPU when you are working on your notebook is always possible, but each time you do, the environment is restarted and you will need to run all your code from the beginning.

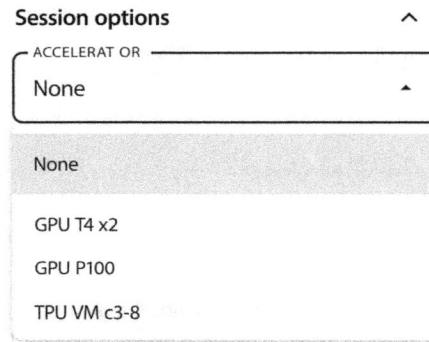

Figure 3.6: Accelerator choice from notebook options

2. Next, we have the coding **Language**. At the time of writing, the Kaggle environment only allows Python and R as available options for coding your notebooks. By default, a new notebook is initialized with the language set to **Python** – if you want to use R instead, click on the dropdown and select **R**.

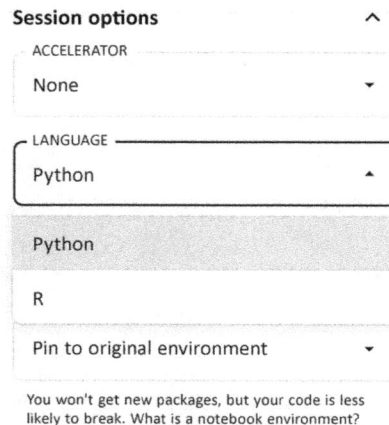

Figure 3.7: Language choice

3. **Persistence** is part of the customization options for a notebook: it allows you to control how much progress is saved between notebook runs. There are four modes of persistence:

- **No persistence**: Every time a notebook is executed, you start with a clean slate
- **Variables only**: When the notebook session ends, variables are saved; they will be restored the next time a notebook is run
- **Files only**: Files in your */kaggle/working* directory will carry over from one run of your notebook to the next
- **Variables and Files**: A combination of the previous two variants

These options are shown in the following screenshot:

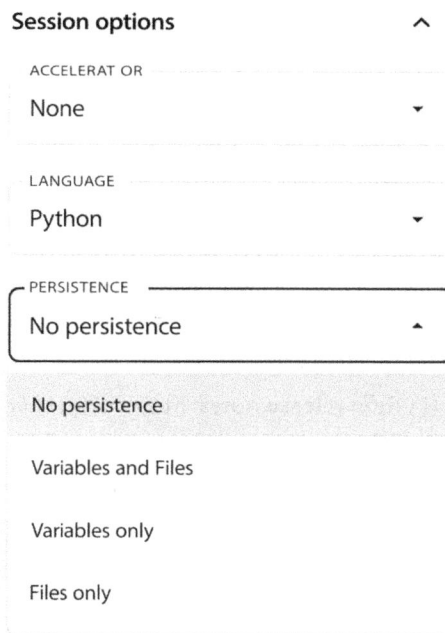

Figure 3.8: Persistence options

The feature is relatively new, so it comes with some caveats (from the product feedback page: https://www.kaggle.com/discussions/product-feedback/355440):

- Persistence must be enabled before the session stops for it to take effect (you can't persist what is already lost).
- All persistence modes are best-effort; there are circumstances where data loss is still possible (e.g., notebook crashes).

- Persistence adds some time to stopping/starting your notebooks (to save/restore your data).

- Persistence can't restore all types of data (e.g., GPU/TPU memory), or data that gets to be too large to fit in your/kaggle/working directory.

- This only affects **interactive notebooks**, not *Save Versions*.

4. Next comes **Environment**: This toggle allows you to decide whether to always use the latest Docker environment (the risky option; it is fast to get updates but dependencies might break with future updates) or pin the notebook to the version of the environment used to create the notebook (the safer choice). The latter option is the default one, and unless you are conducting very active development work, there is no real reason to tinker with it. Typically, a new release of the working environment is available every two weeks and these environments are progressively being aligned with Google Colab (read the following announcement: https://www.kaggle.com/discussions/product-announcements/552460). The idea is to make developing across Kaggle and Colab easier. The great news is that the Rust-based Python package and project manager **uv** (https://github.com/astral-sh/uv) is now pre-installed instead of conda. It can be used in place of pip to install packages at a faster rate using the notebook cell command:

```
!uv pip install --system <package>
```

If needed, you can be notified of every new environment change by subscribing to release notes on GitHub (Python release notes: https://github.com/Kaggle/docker-python/releases, R release notes: https://github.com/Kaggle/docker-rstats/releases) as stated at https://www.kaggle.com/discussions/product-feedback/161327.

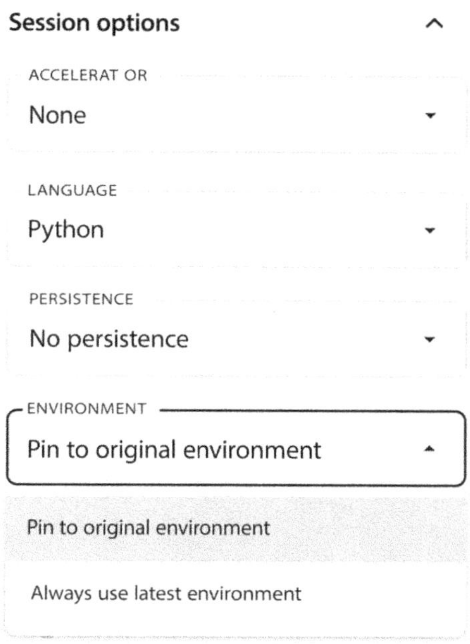

Session options   ∧

ACCELERAT OR

None   ▾

LANGUAGE

Python   ▾

PERSISTENCE

No persistence   ▾

ENVIRONMENT

Pin to original environment   ▴

Pin to original environment

Always use latest environment

Figure 3.9: Environment options

5. Finally, we have the **Internet** toggle, which enables or disables online access. If you are
    connected and need to, for example, install an extra package, the download and installation
    of dependencies will take place automatically in the background. The most common
    situation in which you need to explicitly disable internet access is for submission to a
    competition that explicitly prohibits online access at submission time.

    There is also a switch for connecting your notebook to GitHub and uploading it every time
    you save a new version under your GitHub account (read more: https://www.kaggle.com/
    discussions/product-feedback/295170). This functionality requires you to first create a
    repository on GitHub. Once linked, you can effectively version control your notebook work.

If you need to run a notebook repeatedly (e.g., to download recent data in regular updates), it can be tedious to do that manually every time. This is where **notebook scheduling** comes in: you can have up to four notebooks that are automatically re-run – either daily or with a custom-defined frequency.

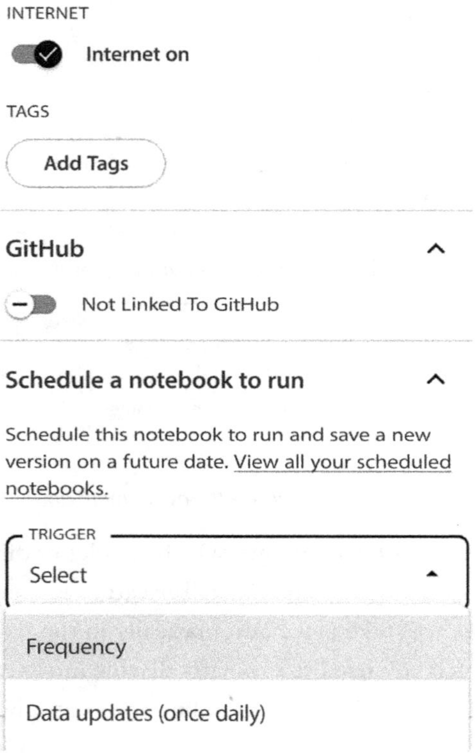

Figure 3.10: Scheduling notebook execution

An important aspect of using notebooks is that you can always take an existing one (created by yourself or another Kaggler) and clone it to modify and adjust to your needs. This can be achieved by clicking the three dots in the upper-right-hand corner of your notebook page and selecting **Copy & edit notebook**. In Kaggle parlance, the process is referred to as **forking**:

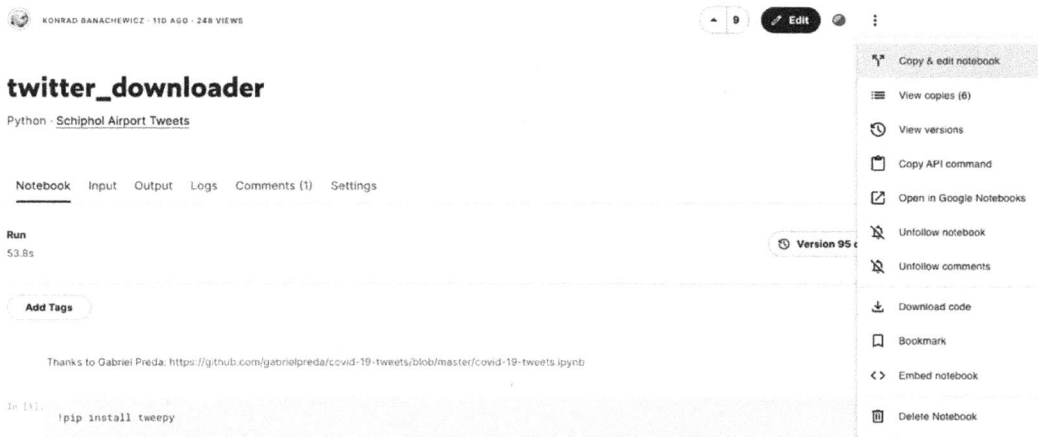

Figure 3.11: Forking an existing notebook

A note on etiquette: If you have participated in a Kaggle competition before, you will probably have noticed that the leaderboard is flooded with forks of forks of well-scoring Notebooks. There is nothing wrong with building on somebody else's work – but if you do, remember to upvote the original author and give explicit credit to the creator of the work referenced.

A notebook you create is private (only visible to you) by default. If you want to make it available to others, you can choose between adding collaborators, so that only the users explicitly added to the list will be able to view or edit the content, or making the notebook public, in which case everybody can see it.

Running your notebook

All the coding is finished, the notebook seems to be working fine, and you are ready to execute. To do that, go to the upper-right corner of your notebook page and click **Save Version**:

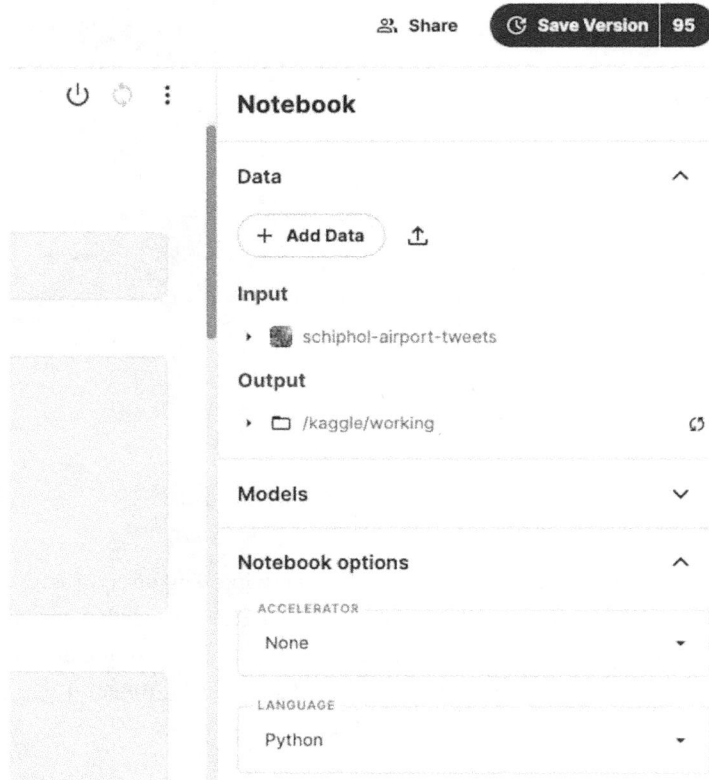

Figure 3.12: Saving your script

Save & Run All is usually used to execute the script, but there is also a **Quick Save** option, which can be used to save an intermediate version of the script before it is ready for submission:

✕   Save version

VERSION NAME

Version 97

10 / 50

VERSION TYPE

✔✔   Save & Run All (Commit)   ▾

save the output

✔✔   Save & Run All (Commit)

✔   Quick Save

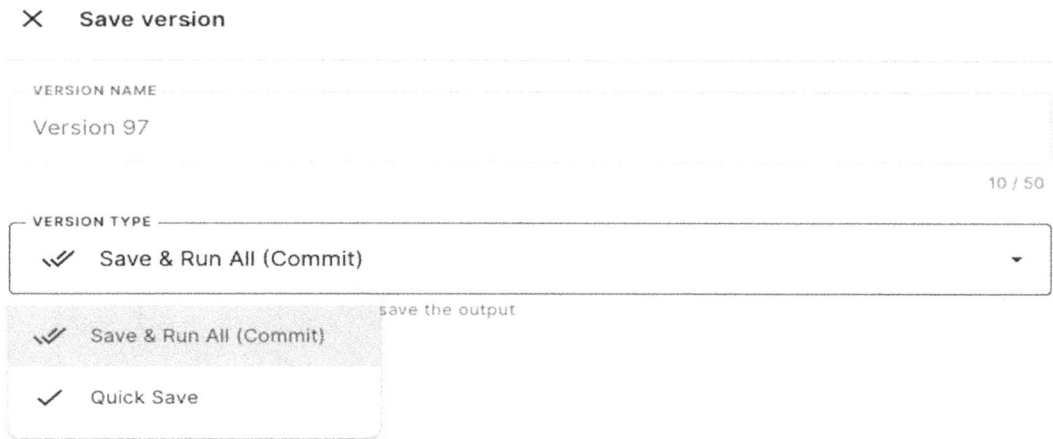

Figure 3.13: Different options for Save version

Once you have launched your script(s), you can head to the lower-left corner and click on **Active Events**:

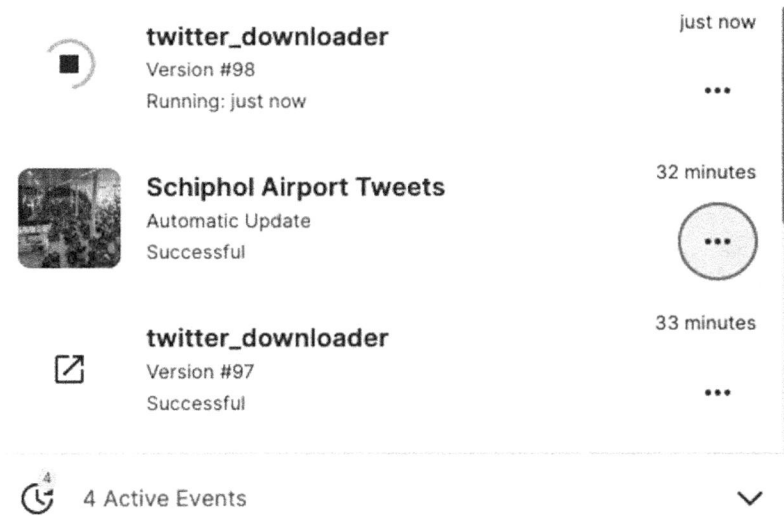

twitter_downloader
Version #98
Running: just now

just now

•••

Schiphol Airport Tweets
Automatic Update
Successful

32 minutes

•••

twitter_downloader
Version #97
Successful

33 minutes

•••

4 Active Events   ⌄

Figure 3.14: Monitoring active events

This allows you to monitor the behavior of your notebooks. Normal execution is associated with the message **Running**; otherwise, it is displayed as **Failed**. Should you decide that you want to kill a running session for whatever reason (for instance, you realize you forgot to use the most recent data), you can do it by clicking on the three dots on the right-hand side of your script entry under **Active Events** and you will receive a popup like the one shown in the figure below:

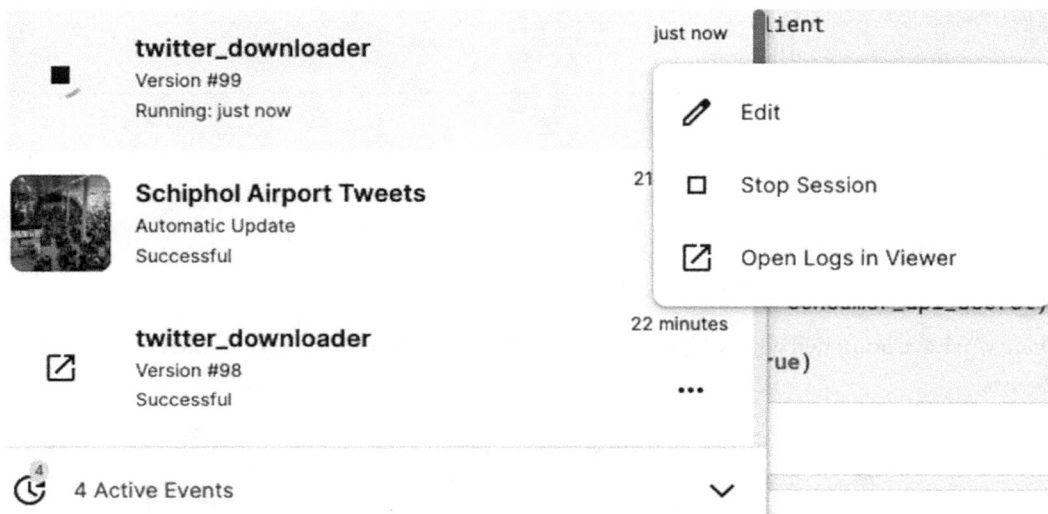

Figure 3.15: Canceling notebook execution

Next, we move to saving the notebooks.

Saving notebooks to GitHub

Version control is a crucial part of any data science workflow – and a Kaggle feature allows you to store your code or your notebook in the version control repository GitHub. You can store your work in both **public** and **private** repositories, having it committed automatically as you save a version of your code. The functionality can prove quite useful for sharing your work with your Kaggle teammates, as well as showcasing your work to the wider public.

In order to enable this feature, you need to edit your notebook and from the **File** menu, choose the **Link to GitHub** option:

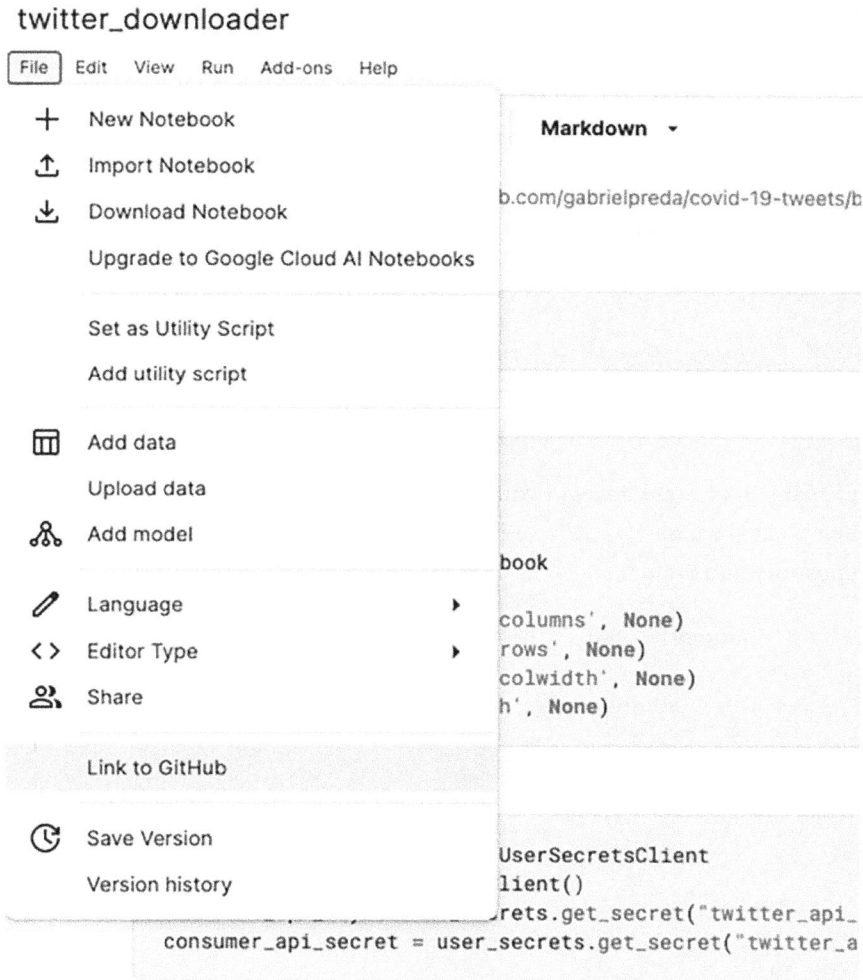

twitter_downloader

| File | Edit | View | Run | Add-ons | Help |

+ New Notebook

⬆ Import Notebook

⬇ Download Notebook

Upgrade to Google Cloud AI Notebooks

Set as Utility Script

Add utility script

▦ Add data

Upload data

⚛ Add model

⟋ Language ▶

<> Editor Type ▶

👥 Share

Link to GitHub

⏱ Save Version

Version history

Markdown ▾

b.com/gabrielpreda/covid-19-tweets/b

book

```
columns', None)
rows', None)
colwidth', None)
h', None)
```

```
UserSecretsClient
lient()
rets.get_secret("twitter_api_
consumer_api_secret = user_secrets.get_secret("twitter_a
```

Figure 3.16: Enabling GitHub integration

After choosing this option, you will have to link your GitHub account to the notebook. You will be explicitly asked for linking permissions only the first time you choose to link. For any subsequent links to new notebooks, the operation will be carried out automatically.

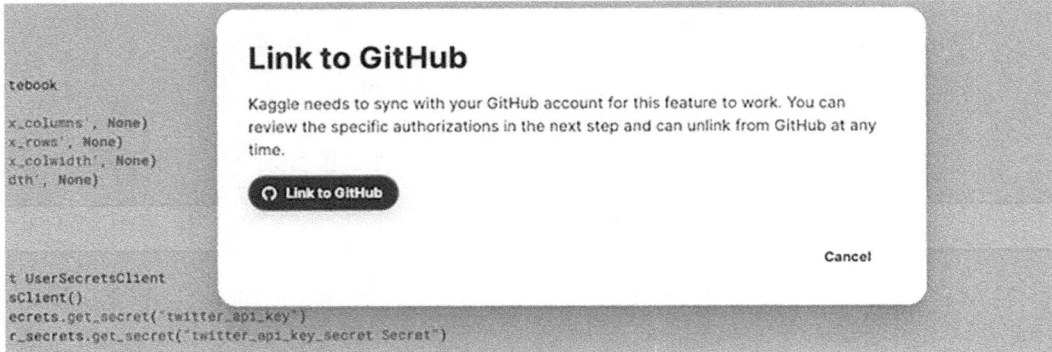

Figure 3.17: Linking to GitHub

Only after linking your notebook will you be allowed to sync your work to a repository of your choice when you save a new version of it. You can at that point upload your notebook to an existing repository and branch:

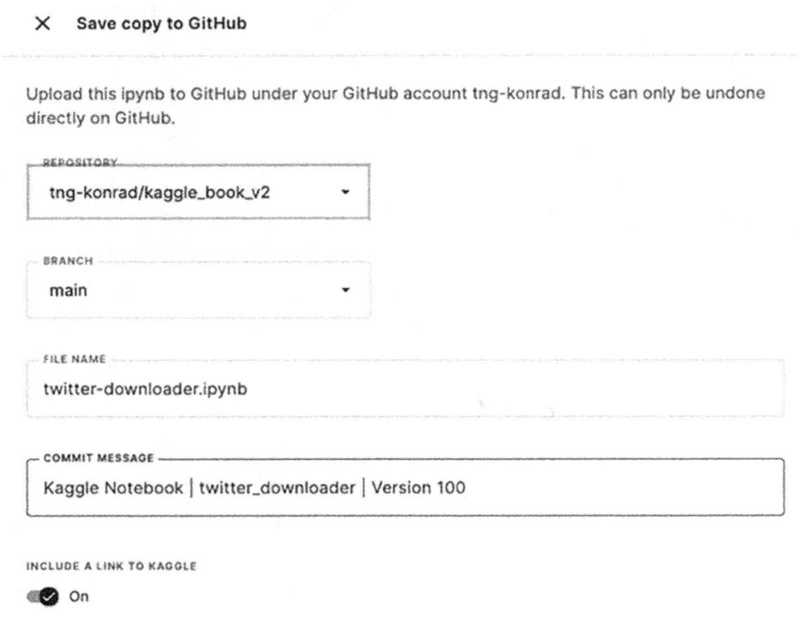

Figure 3.18: Committing your work to GitHub

After deciding on a repository and a branch (thus allowing you to store different developments of your work), you can change the name of the file you are going to push to the repository and modify the **commit message**.

If you decide not to sync a particular notebook on GitHub anymore, all you have to do is go back to the **File** menu and select **Unlink from GitHub**. Finally, if you want Kaggle to stop connecting with your GitHub repository, you can unlink your accounts from either your Kaggle account page under **My linked accounts** or from GitHub's settings pages: `https://github.com/settings/applications`.

Setting a notebook as a utility script

Apart from storing your code on GitHub, reusing your code created in a notebook for other notebooks (a typical situation that often happens in the same competition or across competitions) can be facilitated by using the **utility scripts function** (`https://www.kaggle.com/discussions/product-feedback/91185`).

Creating utility scripts is a straightforward process. These scripts are based on existing notebooks that you've already committed, and they use the latest version of the notebook you've chosen. To convert a notebook into a reusable utility script, you simply select the notebook, open it, and choose the **Set as Utility Script** option from the **File** menu. This action flags the notebook as a utility script, making the process easy and intuitive.

When you need to reuse a utility script in your new notebook, you just select the **Add input** menu item from the **File** menu. Then, on the **Add Input** panel, select the **Utility Scripts** tag (also select the **Your Work** tag to highlight only your scripts) and add the script you want:

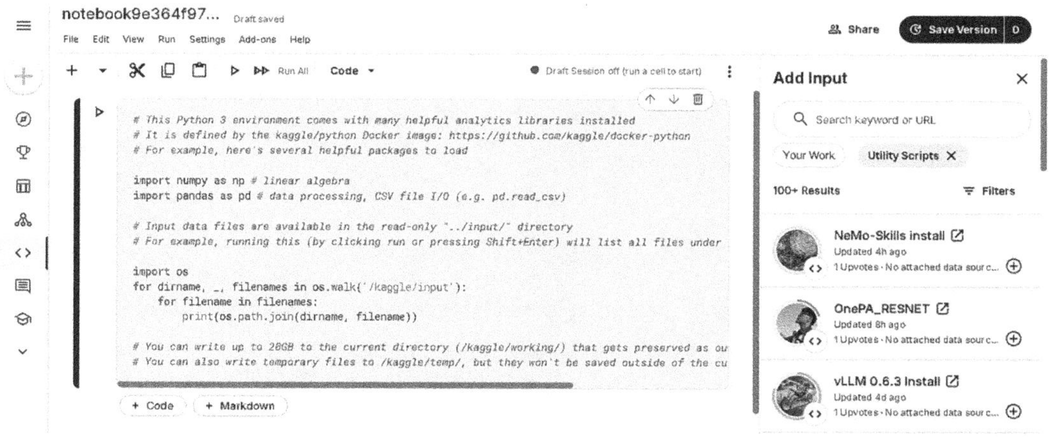

Figure 3.19: Selecting a utility script

At this point, you can import any module or function on the utility script just using the format `import <name_of_script>`. `<name_of_script>` is just the name of the utility script, lowercase and with spaces replaced by underscores (for instance, NeMo-Skills install becomes `nemo_skills_install`). The import will trigger the execution of the script, and all modules and functions present in the script will be available in your notebook's working memory.

Utility scripts are particularly useful for situations when you want to install something from GitHub or elsewhere that is not available on Kaggle notebooks: just create a notebook pulling and installing the packages that you need, set it as a utility script, add it to your notebook, and import it (as an example, you can have a look at this notebook from Darragh: `https://www.kaggle.com/code/darraghdog/nemo-skills-install/notebook`).

Getting the most out of notebooks' resources

Kaggle provides a certain amount of resources free of charge, with the quotas resetting weekly. You can monitor your usage in your own profile:

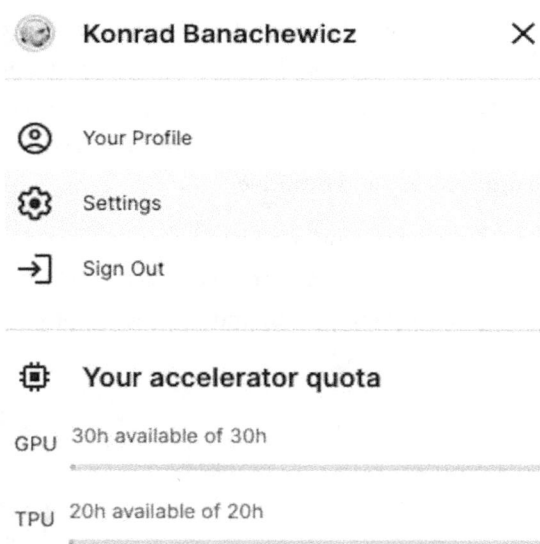

Konrad Banachewicz   ✕

⊙ Your Profile

⚙ Settings

→] Sign Out

▣ **Your accelerator quota**

GPU   30h available of 30h

TPU   20h available of 20h

Figure 3.20: Current status for accelerator quotas

While the amounts might seem large at first glance, this initial impression can be deceptive; it is actually fairly easy to use your quota very quickly. Some practical suggestions that can help you control the usage of the resources are as follows:

- The counter for the quota (measuring how long you have been using your chosen accelerator, GPU, or TPU) starts running the moment you **initialize** your notebook.

- This means that you should always start by checking that the GPU is disabled under the settings (see *Figure 3.6* above). Write the boilerplate first, check your syntax, and enable/disable the GPU for when you add the parts of the code that actually depend on GPU initialization. Remember that the notebook will restart when you change the accelerator.

- It is usually a good idea to run the code from end to end on a small subset of data to get a feel for the execution time. This way, you minimize the risk that your code will crash due to exceeding this limit.

Sometimes the resources provided freely by Kaggle are not sufficient for the task at hand, and you need to move to a beefier machine. If your raw data is over 100 GB, you need to either resize/downsample your images (which is likely to have an adverse impact on your model performance) or train a model in an environment capable of handling high-resolution images. You can set up the whole environment yourself (an example of this setup is the section *Using Kaggle Datasets in Google Colab in*), or you can stay within the framework of notebooks – but swap the underlying machine. This is where **Google Cloud AI notebooks** come in.

Upgrading to Google Cloud Platform (GCP)

The obvious benefits of upgrading to GCP include getting access to more powerful hardware and not being constrained to time or quota limits. In fact, the GPUs provided free by Kaggle are not top of the line in terms of performance, and also, the available RAM can be quite limiting – especially in resource-intensive applications like large NLP models or high-resolution image processing. However, while the improvement in execution time is obvious, leading to faster iteration through the development cycle, it comes at a cost: you need to decide how much you are prepared to spend. For a powerful machine crunching the numbers, time is quite literally money.

To sum up, the decision to use GCP or stick with free resources on Kaggle, such as the free GPU tier, depends on your project's requirements. For resource-intensive models, long training times, complex hyperparameter tuning, or production-level deployments, GCP offers powerful hardware, extended runtime, and even scalability, making it ideal for larger projects and resource-intensive competitions.

On the other hand, for lightweight models, prototyping, or educational experiments, free resources on Kaggle can provide enough power without cost, especially when dealing with smaller datasets or simpler machine learning algorithms. Ultimately, GCP excels when performance and scale are essential, while free resources on Kaggle are optimal for lighter, cost-sensitive tasks.

In order to migrate your notebook to the GCP environment, go to the **Add-ons** menu from the Notebook menu and click on **Upgrade to Google Cloud AI Notebooks**:

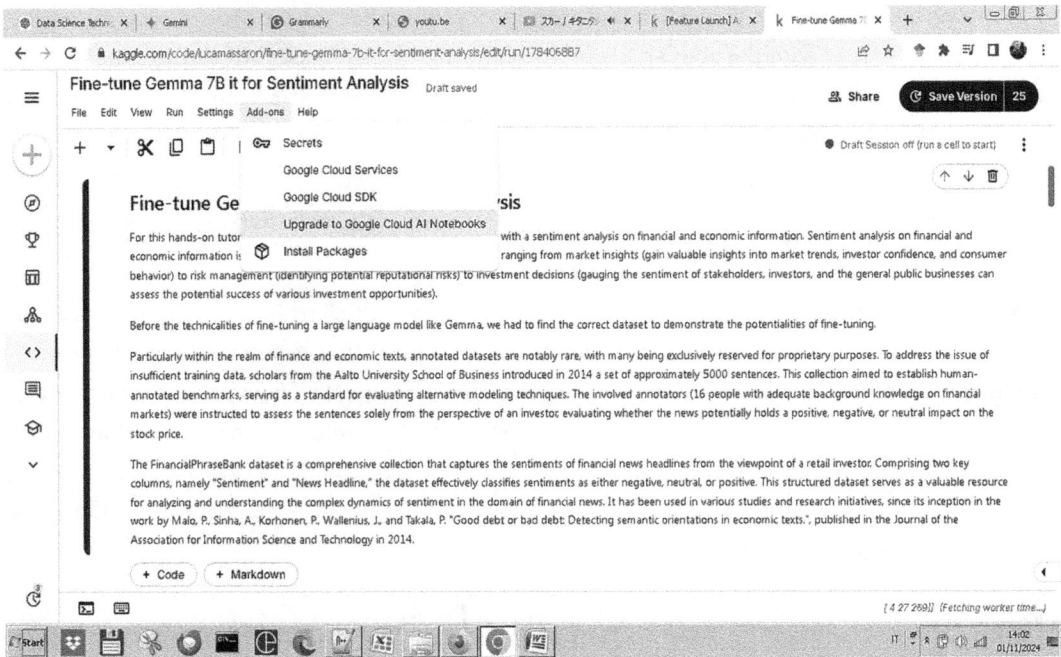

Figure 3.21: Upgrading to Google Cloud AI Notebooks

You will be greeted by the following prompt:

Upgrade to Google Cloud AI Platform Notebooks

☁ Google Cloud

Access more compute power by exporting your notebook and its dependencies to Google Cloud AI Platform Notebooks where you can customize a virtual machine without quotas or runtime limits.

This process is three steps:

1. Setup a billing-enabled Google Cloud Project
2. Setup your notebook instance and optionally customize your machine
3. Run your code without limits

Cancel   Continue

Figure 3.22: Upgrade to Google Cloud AI Platform Notebooks prompt

When you click **Continue**, you will be redirected to the GCP console, where you need to configure your billing options. As a reminder: *GCP is not free*. If it is your first time, you will need to complete a tutorial guiding you through the necessary steps.

> As general suggestions, consider these recommendations based on our personal experience:
>
> - Try prototyping your models on Kaggle first to optimize the code and spot potential bugs. Once the model's in good shape, migrate it to GCP, where it can run faster on a more powerful machine.
> - In GCP, you can customize the machine type to better fit your project requirements, scaling up or down as needed. Start with smaller instances, and only upgrade if needed.
> - Consider using Spot instances for cost-effective, non-critical workloads.
> - For deep learning, leverage GPU-accelerated instances like the A100.
> - Monitor usage and consider setting alerts in the GCP console to avoid unexpected expenses.
> - Tools like the GCP pricing calculator can also help with estimating costs in advance.
> - Use **Google Cloud Storage (GCS)** to store large datasets. You can quickly move data between GCS and Kaggle using the gsutil command, which minimizes data transfer time and allows better control over data updates or modifications.
>
> Using GCP and Kaggle in tandem allows flexibility and helps to achieve more in your competitions.

Going one step beyond

Kaggle notebooks are a fantastic tool for education and participating in competitions but they also serve another useful purpose: as a component of a portfolio you can use to demonstrate your data science skills. There are many potential criteria to consider when building your **data science portfolio** (branding, audience reach, enabling a pitch to your potential employer, and so on) but none of them matter if nobody can find it. Because Kaggle is part of Google, the notebooks are indexed by the most popular search engine in the world – so if someone is looking for a topic related to your code, it will show up in their search results.

Here is a personal example from one of the authors: a few years ago, Konrad wrote a notebook for a competition. The problem he wanted to tackle was *adversarial validation*. For those unfamiliar with the topic: a fairly easy way to see if your training and test sets have a similar distribution is to build a binary classifier trained to tell them apart (the concept is covered in more detail in , *Designing Good Validation*). When writing this chapter, Konrad tried to search for the notebook and, lo and behold, it showed up high up in the search results (notice the fact that Konrad did not mention Kaggle or any personal details like his name in the query):

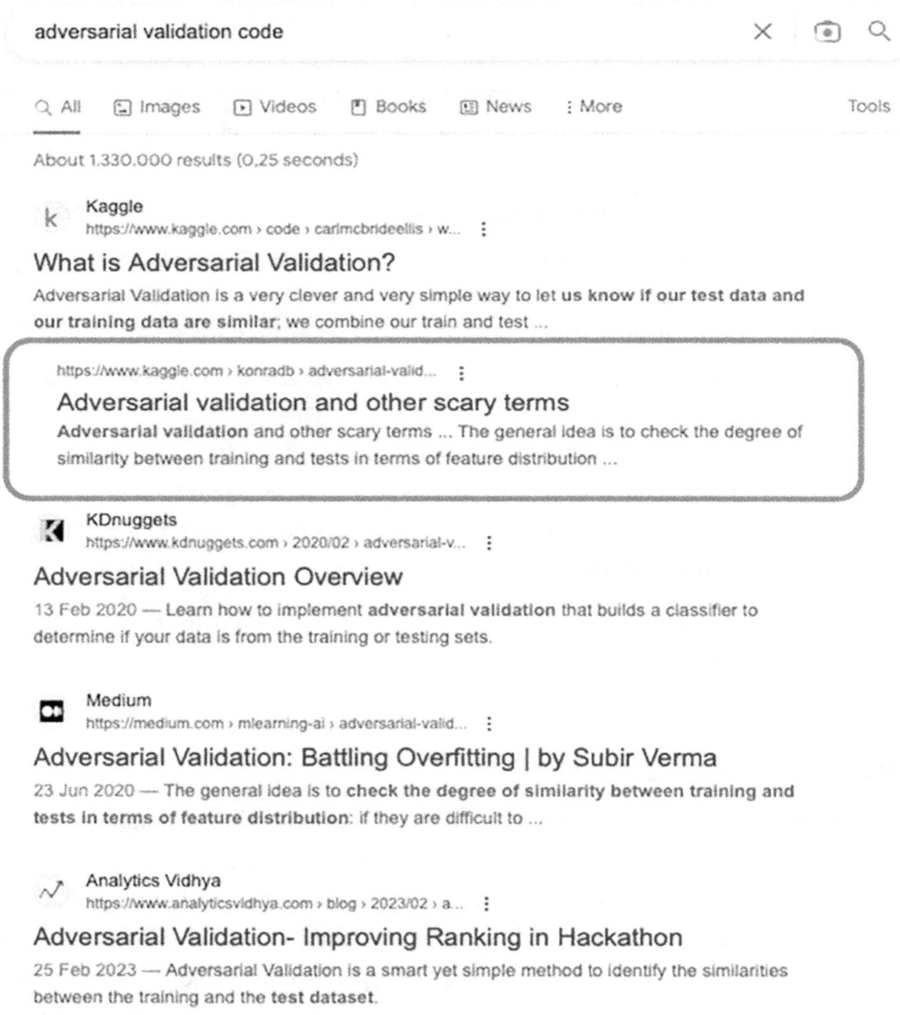

Figure 3.23: A Kaggle notebook showing up in Google results

Moving on to other benefits of using notebooks to demonstrate your skillset: just like competitions, datasets, and discussions, notebooks can be awarded votes/medals and thus position you in the progression system and ranking. You can stay away from the competitions track and become an Expert, Master, or Grandmaster purely by focusing on high-quality code that the community appreciates. The most up-to-date version of the progression requirements can be found at `https://www.kaggle.com/progression`; the following is a snapshot of progression tiers relevant to notebooks:

Expert

You've completed a significant body of work on Kaggle in one or more categories of expertise. Once you've reached the expert tier for a category, you will be entered into the site wide Kaggle Ranking for that category.

Competitions	Datasets	Notebooks	Discussions
☑ ◉ 2 bronze medals	☑ ◉ 3 bronze medals	☑ ◉ 5 bronze medals	☑ ◉ 50 bronze medals

Master

You've demonstrated excellence in one or more categories of expertise on Kaggle to reach this prestigious tier. Masters in the Competitions category are eligible for exclusive Master-Only competitions.

Competitions	Datasets	Notebooks	Discussions
☑ ◉ 1 gold medal	☐ ◉ 1 gold medal	☑ ◉ 10 silver medals	☑ ◉ 50 silver medals
☑ ◉ 2 silver medals	☑ ◉ 4 silver medals		☑ 200 medals in total

Figure 3.24: Tier progression requirements

Progressing in the Notebooks category can be a challenging experience; while easier than Competitions, it is definitely harder than Discussions. The most popular notebooks are those linked to a specific competition: *exploratory data analysis*, end-to-end proof-of-concept solutions, as well as leaderboard chasing. It is an unfortunately common practice that people clone the highest-scoring public notebook, tweak some parameters to boost the score, and release it to wide acclaim (if upvotes can be considered a measure of sentiment). However, the platform remains a vibrant community and the majority of Kagglers appreciate innovative ideas and quality work. We encourage you to continue exploring new approaches and sharing your unique insights. Remember, Kaggle is a place for learning, growth, and collaboration, and your contributions are valued.

In this day and age, it is easy to be skeptical about claims of "community building," but in the case of Kaggle, it happens to actually be true. Their brand recognition in the data science universe is second to none, both among practitioners and among recruiters who actually do their homework. In practice, this means that a (decent enough) Kaggle profile can get you through the door already – which, as we all know, is frequently the hardest step. Remember that your Kaggle profile comes with a count of followers and gives you the possibility of linking other professional networks like LinkedIn or GitHub, hence you can leverage the connections you gain inside the community.

To give you an insider's perspective, we had the pleasure of speaking to *Martin Henze*, aka *Heads or Tails*, a Kaggle Grandmaster in Notebooks and Discussions, and staff data scientist at Crunchbase. Martin is also the author of Notebooks of the Week: *Hidden Gems*, a weekly collection of the very best notebooks that have escaped public notice. You can get notified of new releases of his Hidden Gems by simply following his Kaggle profile or his accounts on X and LinkedIn.

Interview: Martin Henze

`https://www.kaggle.com/headsortails`

What's your favorite kind of competition and why? In terms of techniques and solving approaches, what is your specialty on Kaggle?

For a long time, my focus was on EDA notebooks rather than leaderboard predictions themselves. Most of my experience prior to Kaggle had been with tabular data, and the majority of my EDA notebooks deal with extracting intricate insights from newly launched tabular challenges. I still consider this my specialty on Kaggle, and I have spent a significant amount of time crafting the structure, data visualizations, and storytelling of my notebooks.

How do you approach a Kaggle competition? If you work in data science, how different is this approach to what you do in your day-to-day work?

Even as Kaggle has shifted away from tabular competitions, I strongly believe that the data itself is the most important aspect of any challenge. It is easy to focus too early on model architectures and hyperparameter tuning. But in many competitions, the key to success remains a data-centric approach that is built on detailed knowledge of the dataset and its quirks and peculiarities. This is true for image data, NLP, time series, and any other data structures you can think of. Therefore, I always start with an extensive EDA before building a simple baseline model, a CV framework, and then slowly iterate the complexity of this pipeline.

The main difference compared to my data science day job is probably that the kind of baseline models that most experienced people can build within the first week of a new challenge would be considered sufficient to put into production. In many cases, after those first few days, we're more than 80% on the way to the ultimate winner's solution, in terms of scoring metric. Of course, the fun and the challenge of Kaggle are to find creative ways to get those last few percent of, say, accuracy. But in an industry job, your time is often more efficiently spent tackling a new project instead.

Has Kaggle helped you in your career? If so, how?

Kaggle has shaped and supported my career tremendously. The great experience in the Kaggle community motivated me to transition from academia to industry. Today, I'm working as a data scientist in a tech startup and I'm continuously growing and honing my skills through Kaggle challenges.

In my case, my focus on constructing extensive Kaggle notebooks helped me a lot, since I could easily use those as my portfolio. I don't know how often a hiring manager would actually look at those resources, but I frequently got the impression that my Grandmaster title might have opened more doors than my PhD did. Or maybe it was a combination of the two. In any case, I can much recommend having a portfolio of public notebooks. Moreover, during my job search, I used the strategies I learned on Kaggle for various take-home assignments and they served me well.

In your experience, what do inexperienced Kagglers often overlook? What do you know now that you wish you'd known when you first started?

I think that we are all constantly growing in experience. And we're all wiser now than we were ten years, five years, or even one year ago. With that out of the way, one crucial aspect that is often overlooked is that you want to have a plan for what you're doing, and to execute and document that plan. And that's an entirely understandable mistake to make for new Kagglers, since everything is novel and complex and at least somewhat confusing. I know that Kaggle was confusing for me when I first joined. So many things you can do: forums, datasets, challenges, courses. And the competitions can be downright intimidating: Neuronal Cell Instance Segmentation; Stock Market Volatility Prediction. What even are those things? But the competitions are also the best place to start.

Because when a competition launches, nobody really has a clue about it. Yeah, maybe there is a person who has done their PhD on almost the same topic. But those are rare. Everyone else, we're all pretty much starting from zero. Digging into the data, playing with loss functions, running some simple starter models. When you join a competition at the beginning, you go through that learning curve in an accelerated way, as a member of a community. And you learn alongside others who will provide you with tons of ideas. But you still need a plan.

And that plan is important, because it's easy to just blindly run some experiments and see all that GPU RAM being used and feel good about it. But then you forget which version of your model was doing best, and is there a correlation between local validation and LB? Did I already test this combination of parameters? So write down what you are going to do and then log the results. There are more and more tools that do the logging for you, but this is also easily done through a custom script. Machine learning is still mostly an experimental science, and the key to efficient experiments is to plan them well and to write down all of the results so you can compare and analyze them.

What mistakes have you made in competitions in the past?

I have made lots of mistakes and I hope that I managed to learn from them. Not having a robust cross-validation framework was one of them. Not accounting for differences between train and test. Doing too much EDA and neglecting the model building – that one was probably my signature mistake in my first few competitions. Not doing enough EDA and missing something important – yep, done that too. Not selecting my final two submissions. (That ended up making not much of a difference, but I still won't forget it again.)

The point about mistakes, though, is similar to my earlier point about experiments and having a plan. Mistakes are fine if you learn from them and if they help you grow and evolve. You still want to avoid making easy mistakes that could be avoided by foresight. But in machine learning (and science!), failure is pretty much part of the process. Not everything will always work. And that's fine. But you don't want to keep making the same mistakes over and over again. So the only real mistake is not to learn from your mistakes. This is true for Kaggle competitions and in life.

Are there any particular tools or libraries that you would recommend using for data analysis or machine learning?

I know that we increasingly live in a Python world, but when it comes to tabular wrangling and data visualization I still prefer R and its tidyverse: dplyr, ggplot2, lubridate, etc. The new tidymodels framework is a serious competitor to sklearn. Even if you're a die-hard Python aficionado, it pays to have a look beyond pandas and friends every once in a while. Different tools often lead to different viewpoints and more creativity. In terms of deep learning, I find PyTorch most intuitive with its FastAI interface. And, of course, everyone loves Hugging Face nowadays – and for very good reason.

What's the most important thing someone should keep in mind or do when they're entering a competition?

The most important thing is to remember to have fun and to learn something. So much valuable insight and wisdom is shared both during and after a competition that it would be a shame not to take it in and grow from it. Even if the only thing you care about is winning, you can only accomplish that by learning and experimenting and standing on the shoulders of the giants in this community. But there is so much more to Kaggle than the leaderboards, and once you start contributing and giving back to the community you will grow in a much more holistic way. I guarantee it.

Kaggle Learn courses

A great many things about Kaggle are about acquiring knowledge. Be it the learnings from a competition, datasets you manage to find in the ever-growing repository, or a demonstration of a hitherto unknown model class, there is always something new to find out. An important component of that collection is the courses gathered under the **Kaggle Learn** label: https://www.kaggle.com/learn. These are micro-courses marketed by Kaggle as *"the single fastest way to gain the skills you'll need to do independent data science projects,"* the core unifying theme being a crash course introduction across a variety of topics. Each course is divided into small chapters, followed by coding practice questions. The courses are delivered using notebooks, where portions of the necessary theory and exposition are intermingled with the bits you are expected to code and implement yourself.

The following list includes a short overview of the most relevant ones for those beginning to explore different subfields of data science:

- **Intro to ML/Intermediate ML** (https://www.kaggle.com/learn/intro-to-machine-learning and https://www.kaggle.com/learn/intermediate-machine-learning): These courses are best viewed as a two-parter: the first one introduces different classes of models used in machine learning, followed by a discussion of topics common to different models like under/overfitting or model validation. The second one goes deeper into feature engineering, dealing with missing values and handling categorical variables.

  Useful for people beginning their ML journey.

- **pandas** (https://www.kaggle.com/learn/pandas): This course provides a crash-course introduction to one of the most fundamental tools used in modern data science. You first learn how to create, read, and write data, and then move on to data cleaning (indexing, selecting, combining, grouping, and so on).

Useful for both beginners (pandas functionality can be overwhelming at times) and practitioners alike (as a refresher/reference).

- **Game AI** (`https://www.kaggle.com/learn/intro-to-game-ai-and-reinforcement-learning`): This course is a great wrap-up of the tech-focused part of the curriculum introduced by Kaggle in the learning modules. You will write a game-playing agent, tinker with its performance, and use the minimax algorithm.

A practice-oriented introduction to reinforcement learning.

- **Machine Learning Explainability** (`https://www.kaggle.com/learn/machine-learning-explainability`): Building models is fun, but in the real world, not everybody is a data scientist, so you might find yourself in a position where you need to explain what you have done to others. This is where this mini-course on model explainability comes in: you will learn how to assess how relevant your features are with three different methods: permutation importance, SHAP, and partial dependence plots.

Extremely useful for anybody working with machine learning in a commercial setting, where projects live or die on how well the message is conveyed.

- **AI Ethics** (`https://www.kaggle.com/learn/intro-to-ai-ethics`): This last course is a very interesting addition to the proposition: it discusses the practical tools to guide the moral design of AI systems. You will learn how to identify the bias in AI models, examine the concept of AI fairness, and find out how to increase transparency by communicating ML model information.

Very useful for practitioners, as "responsible AI" is a phrase we are hearing more and more.

Apart from the original content created by Kaggle, there are other learning opportunities available on the platform through user-created notebooks; the reader is encouraged to explore them on their own.

To conclude this chapter, we caught up with *Andrada Vulpe*, a Kaggle notebooks Grandmaster who very much encourages learning from notebooks. Andrada is a Z by HP Global Data Science Ambassador, data scientist at Endava, and dev expert at Weights & Biases. We discussed Andrada's notebook competitions, her career, and more.

Interview: Andrada Vulpe

https://www.kaggle.com/andradaolteanu

What's your favorite kind of competition and why? In terms of techniques and solving approaches, what is your specialty on Kaggle?

I would say my specialty on Kaggle leans more toward data visualization, as it enables me to combine art and creativity with data.

I would not say I have a favorite type of competition, but I would rather say I like to switch it up occasionally and choose whatever I feel is interesting. The beauty of Kaggle is that one can learn multiple areas of data science (computer vision, NLP, exploratory data analysis and statistics, time series, and so on) while also becoming familiar and comfortable with many topics (like sports, the medical field, finance and cryptocurrencies, worldwide events, etc.).

Another great thing is that, for example, if one wants to become more proficient in working with text data, there is almost always a Kaggle competition that requires NLP. Or, if one wants to learn how to preprocess and model audio files, there are competitions that enable that skill as well.

Tell us about a particularly challenging competition you entered, and what insights you used to tackle the task.

The most challenging "competition" I have ever entered is the Kaggle Data Science and Machine Learning Annual Survey. I know this is not a "real" competition – with a leaderboard and heavy-duty machine learning involved – however, for me, it was one of the competitions I "sweated" during and learned the most.

This is a Notebook competition, where the users have to become creative in order to win one of the five prizes Kaggle puts on the table. I have participated in it 2 years in a row. In the first year (2020), it challenged my more "basic" visualization skills and forced me to think outside the box (I took third place); in the second year (2021), I prepared for it for around 4 months by learning D3, in an attempt to get to a whole other level on my data visualization skills (still in review; so far, I have won the "Early Notebook Award" prize). The best insights I can give here are:

- First, do not get lost within the data, and try to create graphs that are as accurate as possible; if necessary, build double verification methods to be sure that what you are representing is clear and concise. Nothing is worse than a beautiful graph that showcases inaccurate insights.
- Try to find inspiration around you: from nature, from movies, from your work. By doing so, you can draw amazing themes and interesting ways to spruce up your visualization.

Has Kaggle helped you in your career? If so, how?

Yes. Tremendously. I believe I owe a big part of where I am now in my career to Kaggle, and for this I am forever grateful. Through Kaggle I became a Z by HP Ambassador; I also discovered Weights & Biases, which is an amazing machine learning experiment platform, and now I am a proud dev expert for them. Last but not least, through this platform I connected with my now lead data scientist at Endava, who recruited me, and I have been working with him since. In short, my position at Endava and the connection I have to two huge companies (HP and Weights & Biases) are a direct result of my activity on the Kaggle platform.

I believe the most overlooked aspect of Kaggle is the community. Kaggle has the biggest pool of people, all gathered in one convenient place, from which one could connect, interact, and learn from.

The best way to leverage this is to take, for example, the first 100 people from each Kaggle section (Competitions, Datasets, Notebooks – and, if you want, Discussions), and follow on Twitter/LinkedIn everybody that has this information shared on their profile. This way, you can start interacting on a regular basis with these amazing people, who are so rich in insights and knowledge.

What mistakes have you made in competitions in the past?

The biggest mistake I have made in competitions in the past is to not participate in them. I believe this is the biggest, most fundamental mistake beginners make when they enter onto the platform.

Out of fear (and I am talking from personal experience), they believe they are not ready, or they just don't know how to start. Fortunately, if you follow a simple system, it will become very easy to enter any competition:

- Enter any competition you like or sounds interesting.

- Explore the description page and the data.

- If you have no idea how to start, no worries! Just enter the "Code" section and look around for notebooks that have a lot of upvotes, or are made by experienced people, like Grand-masters. Start doing a "code along" notebook, where you look at what others have done and "copy" it, researching and trying to improve it yourself. This is, in my opinion, the best way to learn – you never get stuck, and you learn by doing in a specific project.

What's the most important thing someone should keep in mind or do when they're entering a competition?

They should keep in mind that it is OK to fail, as usually it is the best way to learn.

What they should also keep in mind is to always learn from the competition Grandmasters, because they are usually the ones who share and explain machine learning techniques that one may never think of. The best way of learning something is to look at others who "have already made it," so your road to success will not be as bumpy, but rather much more painless, smooth, and quick. Take 2-3 Grandmasters that you really admire and make them your teachers; study their notebooks, code along, and learn as much as possible.

Do you use other competition platforms? How do they compare to Kaggle?

I have never used any other competition platform – simply because I feel like Kaggle has it all.

Summary

In this chapter, we explored the potential of Kaggle notebooks, free-to-use Jupyter notebooks equipped with pre-installed packages and compute options like CPUs, GPUs, and TPUs. These notebooks offer a flexible and powerful way to experiment with data, develop machine learning models, and share insights with the broader Kaggle community. We covered fundamental skills, including setting up and running notebooks, using them as utility scripts, and maximizing their features to enhance your workflow. Additionally, we examined the advantages of upgrading to GCP for larger projects and how to leverage Kaggle notebooks in building a data science portfolio. Finally, we noted the value of Kaggle Learn courses as a resource for expanding your knowledge and skills.

You are now able to create your own notebook, efficiently utilize the available resources, and use the results for competitions or your individual projects.

Next, we will talk about Kaggle models: a repository of pre-trained models that are easy to use in Kaggle competition notebooks.

Get This Book's PDF Version and Exclusive Extras

UNLOCK NOW

Scan the QR code (or go to packtpub.com/unlock). Search for this book by name, confirm the edition, and then follow the steps on the page.

Note: Keep your invoice handy. Purchases made directly from Packt don't require an invoice.

Join our book's Discord space

Join our community's Discord space for discussions with the authors and other readers:

https://packt.link/kaggle

4

Kaggle Models

Pretrained models are the name of the game in machine learning in 2024. With a dedicated hub for models, Kaggle plans to make using pretrained models in competitions easier, leading to a virtuous cycle for the entire community with increased usage of pretrained models and many Kagglers posting their own models (to know more, read this post from D. Sculley: `https://www.kaggle.com/discussions/product-feedback/470613`). Over the last decade, **Kaggle Competitions** have proved to be the place you go to find out what works and what does not – stress testing and validating the boundaries of terra incognita have never been more important (anybody remember capsule networks? They actually never worked on Kaggle!).

Kaggle Models is a repository of TensorFlow and PyTorch pretrained models that are easy to use in Kaggle competition notebooks. It is also the place where you can publish your own models and be guided by a template that helps you define a reference framework (i.e., Keras, TensorFlow, PyTorch, Transformers, and so on), variation, and license. The best way to think of **Kaggle Models** is to imagine it as a sort of specialized subset of **Datasets** that provides a convenient way to discover, use, and share public pretrained models for machine learning.

In this chapter, we will cover the following topics:

- Selecting a Kaggle model
- Using Kaggle Models
- Uploading your model to Kaggle Models

Selecting a Kaggle model

When faced with a problem, you do not always need to reinvent the wheel; sometimes it just suffices to know how to look for an existing solution that fits your needs and solves your problem. In this section, we will just go through the necessary steps to successfully filter out a model from the **Kaggle Models** selection according to your specifications and needs.

Models can be found in the menu left alongside **Datasets** and **Code** – it is the direct link to the **Models** landing page, where you can search and apply a set of filters to choose a model necessary for your application:

Models

Search and discover hundreds of trained, ready-to-deploy machine learning models in one place.

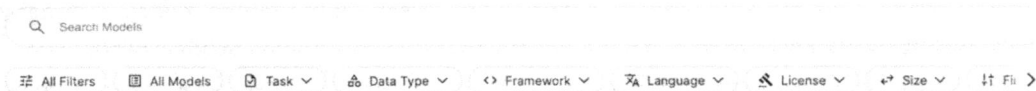

🔍 Search Models

⇄ All Filters   ▤ All Models   🗋 Task ⌄   ⊹ Data Type ⌄   <> Framework ⌄   🗛 Language ⌄   ⚒ License ⌄   ↵ Size ⌄   ↕ Fil ❯

Figure 4.1: Kaggle Models front page

Most of the filters are self-explanatory, but for the sake of completeness, we will review the different options available here to point out the selection filters that can mostly be of advantage to you in your search for the right pretrained model for your problem.

Task

You can choose the type of task you want to focus on – you can have multiple objectives, so it is possible to select, for example, both **Image Classification** and **Image Feature Vector** if you need image classification as your ultimate task, but obtaining the embeddings is important as well:

All Filters   ✕

☑ Task   ⌃

🔍 Search

⌖ Image Classification
🖾 Object Detection
▦ Image Feature Vector   ○ Other

⛁ Data Type   ⌄

‹ › Framework   ⌄

🗛 Language   ⌄

⚒ License   ⌄

↓↑ Fine Tunable   ⌄

↩ Size   ⌄

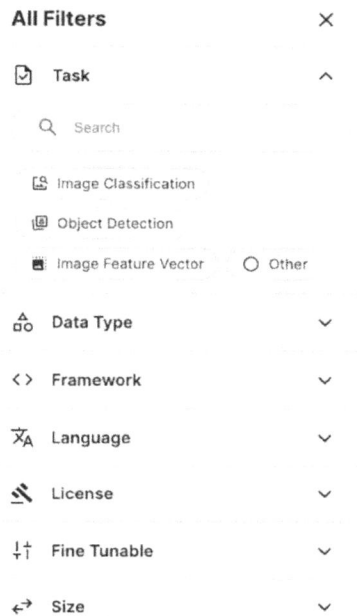

Figure 4.2: Kaggle Models filters: Task

Data Type

Data Type selection refers to the modality you wish to use. Unlike **Task**, with data type or modality selection, you can only pick one option:

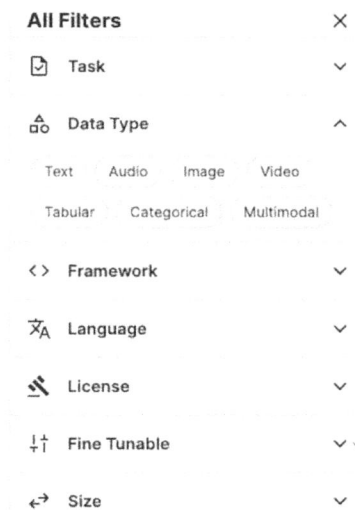

All Filters   ✕

☑ Task   ⌄

⛁ Data Type   ⌃

Text   Audio   Image   Video

Tabular   Categorical   Multimodal

‹ › Framework   ⌄

🗛 Language   ⌄

⚒ License   ⌄

↓↑ Fine Tunable   ⌄

↩ Size   ⌄

Figure 4.3: Kaggle Models filters: type of data the model can handle

Framework

Framework selection offers a broad range of choices, comprising both the most common deep learning frameworks such as TensorFlow, Jax, Keras, and PyTorch to other popular choices such as Transformers (Hugging Face package for language models) and even pickled scikit-learn models. You can make your choice based on the availability, your preferences, or the flexibility that each framework offers to you:

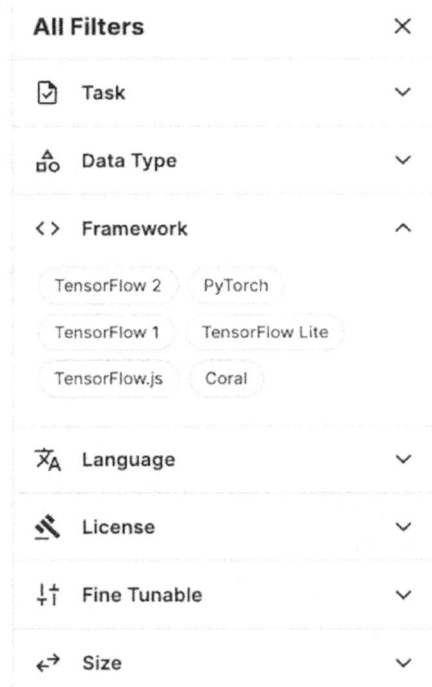

Figure 4.4: Kaggle Models filters: Framework

Language

Language selection matters mostly in the context of NLP-oriented tasks; picking a language yields all models trained on that specific language corpus, as well as multilingual ones:

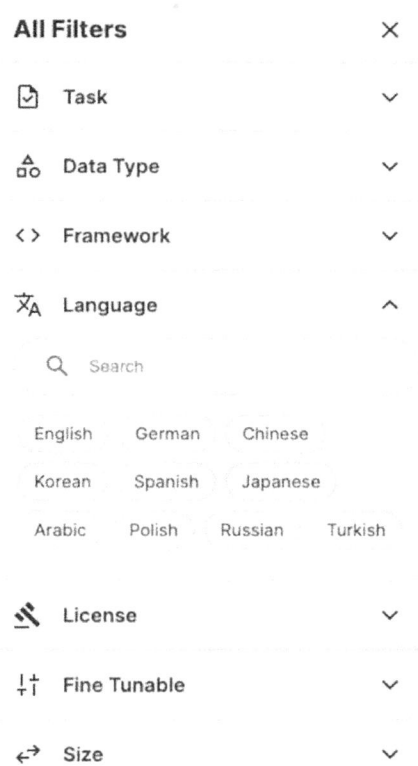

Figure 4.5: Kaggle Models filters: Language

License

There are two main situations where the **License** type might be especially relevant: either you participate in a competition that imposes some extra terms in that aspect, or you are experimenting with a model that can be later used in the business context. Either way, this toggle allows you to subset your selection in a desired fashion:

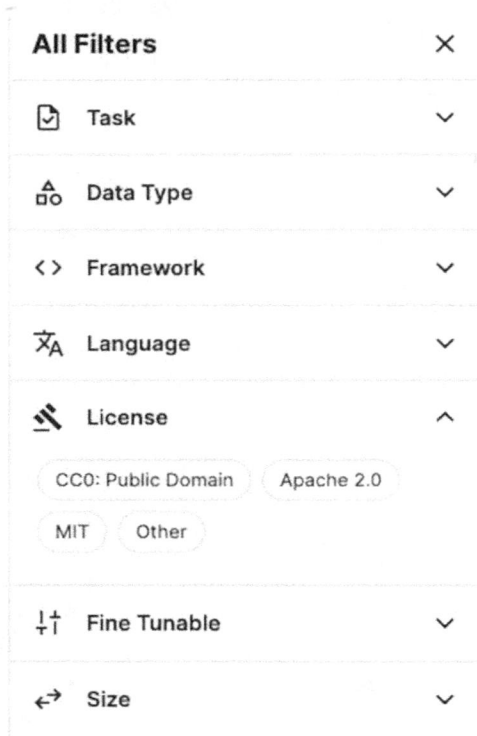

Figure 4.6: Kaggle Models filters: License

Fine Tunable

The need for fine-tuning a model is situation-dependent, but this toggle allows you to customize your selection and filter out foundational models that have already been fine-tuned for a specific task:

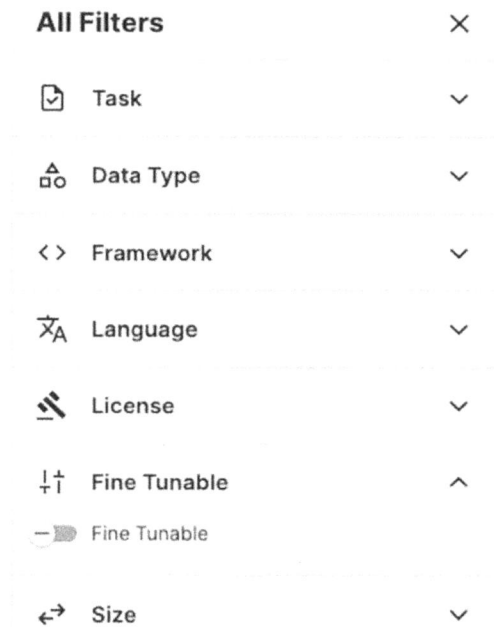

All Filters   ✕

☑ Task   ⌄

▵ Data Type   ⌄
□○

< > Framework   ⌄

文A Language   ⌄

⚒ License   ⌄

↓↑ Fine Tunable   ⌃

— ⫸ Fine Tunable

↩ Size   ⌄

Figure 4.7: Kaggle Models filters: Fine Tunable

Size

As we are constantly reminded across popular culture, size does matter – and the old adage holds true in the context of machine learning models as well since a lot of emerging capabilities are assigned to models based on their size in terms of parameters and, consequently, of memory gigabytes necessary to run them properly. The models are segmented into three size classes, distinguishing small models to medium and large ones weighing over one gigabyte (typically, large language models):

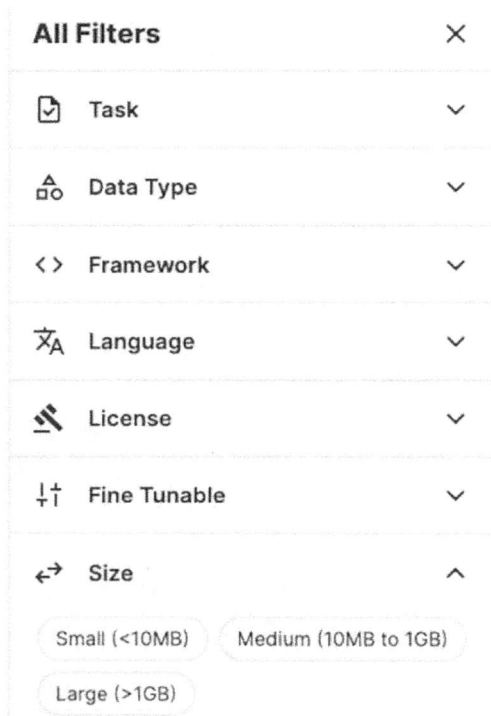

All Filters   ✕

☑ Task   ⌄

△□○ Data Type   ⌄

< > Framework   ⌄

文A Language   ⌄

⚒ License   ⌄

↓↑ Fine Tunable   ⌄

↩→ Size   ⌃

    Small (<10MB)   Medium (10MB to 1GB)

    Large (>1GB)

Figure 4.8: Kaggle Models filters: Size

The filters we have reviewed allow you to restrict the search for a suitable model based on various features, from task, modality, and framework to language, license, and size. Now that you've selected the specifics of the model you want, you can go ahead and start using your chosen model.

Using Kaggle Models

When you click on a model, you will be taken to the details page for that model; an example for **BERT** (https://www.kaggle.com/models/google/bert) is shown in the following screenshot:

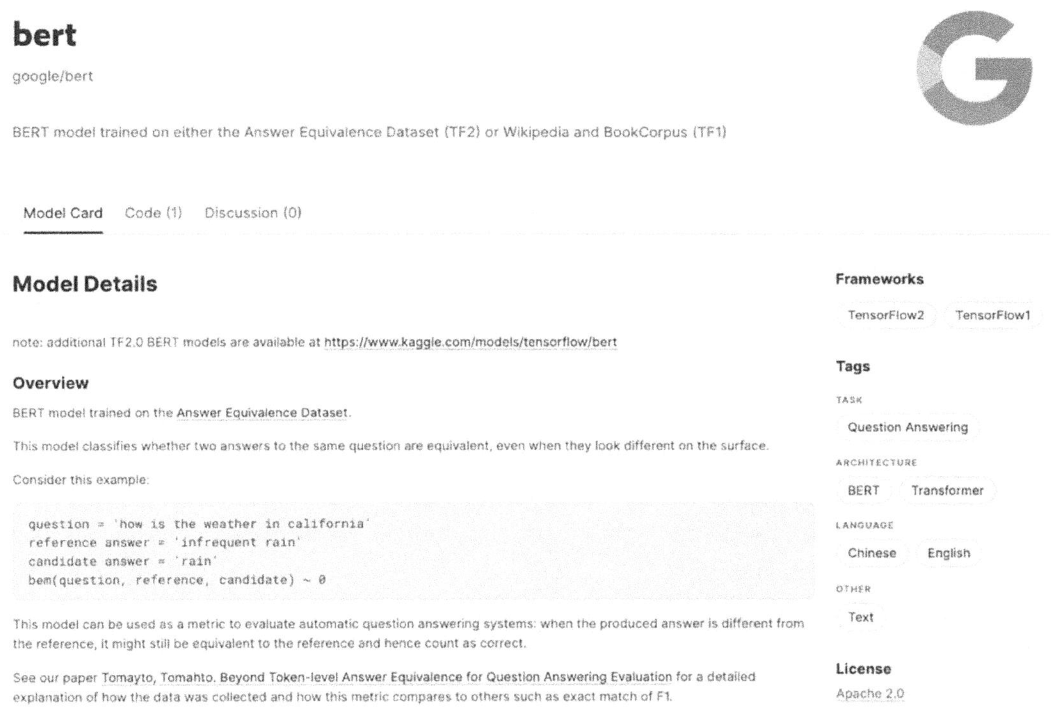

bert

google/bert

BERT model trained on either the Answer Equivalence Dataset (TF2) or Wikipedia and BookCorpus (TF1)

Model Card   Code (1)   Discussion (0)

Model Details

Frameworks

TensorFlow2   TensorFlow1

note: additional TF2.0 BERT models are available at https://www.kaggle.com/models/tensorflow/bert

Tags

Overview

TASK

BERT model trained on the Answer Equivalence Dataset.

Question Answering

This model classifies whether two answers to the same question are equivalent, even when they look different on the surface.

ARCHITECTURE

Consider this example:

BERT   Transformer

```
question = 'how is the weather in california'
reference answer = 'infrequent rain'
candidate answer = 'rain'
bem(question, reference, candidate) ~ 0
```

LANGUAGE

Chinese   English

OTHER

This model can be used as a metric to evaluate automatic question answering systems: when the produced answer is different from the reference, it might still be equivalent to the reference and hence count as correct.

Text

See our paper Tomayto, Tomahto. Beyond Token-level Answer Equivalence for Question Answering Evaluation for a detailed explanation of how the data was collected and how this metric compares to others such as exact match of F1.

License

Apache 2.0

Figure 4.9: Model card for BERT

At a glance, the page can be briefly divided into three parts:

- **Model Details**: Containing the description of the model, the literature references, and some usage examples
- **Model Variations**: Allowing you to download the different variations of the model (in terms of size, framework, version, and task)
- **Metadata**: Referencing collaborators, authors, provenance, and the preferred way to cite the model

The **Model Details** page contains an **Overview** tab with a **Model Card** (metadata and information about how the model was trained, what its acceptable use cases are, any limitations, etc.), alongside **Tags** and **License** information. There are also tabs for notebooks and discussions.

Beyond the overall metadata, the **Model Details** page also organizes all variations (same model with a different number of parameters) and frameworks (compatibility across, e.g., both TensorFlow and PyTorch) for a given model. You can view and use the specific framework (in this case, TensorFlow 2 or TensorFlow 1) and variation (a single option for TensorFlow 2 and multiple models, for instance, on the TensorFlow 1 framework) by selecting the options shown in the following screenshot:

Model Instance

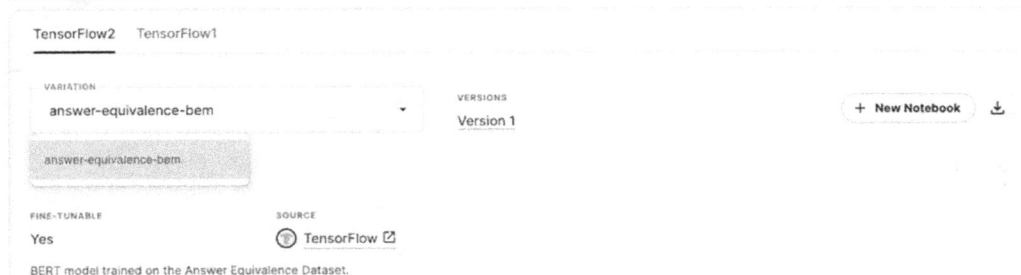

TensorFlow2   TensorFlow1

VARIATION
answer-equivalence-bem   VERSIONS   + New Notebook
                                                          Version 1

FINE-TUNABLE   SOURCE
Yes   TensorFlow

BERT model trained on the Answer Equivalence Dataset.

Figure 4.10: Model Instance for BERT

A useful part of the model page is **Example Use**, which can help you understand the specific nature of the model variation – and if the creator of the model was conscientious, there is likely a code sample to get you started:

Example Use
Overview

BERT model trained on the Answer Equivalence Dataset.

This model classifies whether two answers to the same question are equivalent, even when they look different on the surface.

Consider this example:

```
question = 'how is the weather in california'
reference answer = 'infrequent rain'
candidate answer = 'rain'
bem(question, reference, candidate) ~ 0
```

This model can be used as a metric to evaluate automatic question answering systems: when the produced answer is different from the reference, it might still be equivalent to the reference and hence count as correct.

See our paper Tomayto, Tomahto. Beyond Token-level Answer Equivalence for Question Answering Evaluation for a detailed explanation of how the data was collected and how this metric compares to others such as exact match of F1.

Example use

```
import tensorflow_hub as hub

# Create BERT inputs of the usual form:
input_ids, segment_ids = tokenize_and_convert_to_ids(...)
inputs = {
  'input_ids': input_ids,
  'segment_ids': segment_ids
}
```

Figure 4.11: Example usage on the model page

If you click on **New Notebook,** a new page will open that should look familiar if you have read on notebooks:

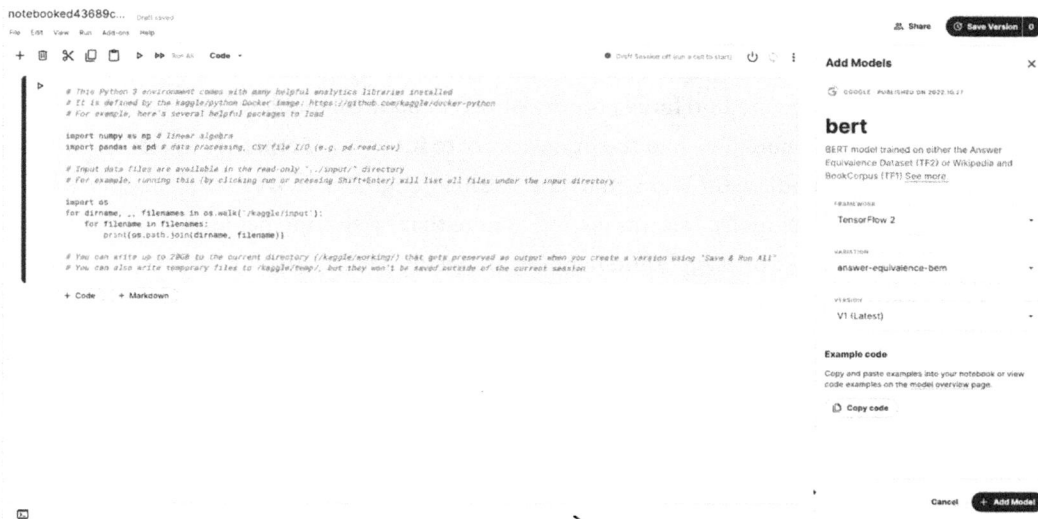

Figure 4.12: New notebook started from a model page

A very convenient option worth pointing out is **Copy code**; it allows you to reuse the example usage from the model page as a starting point in your notebook:

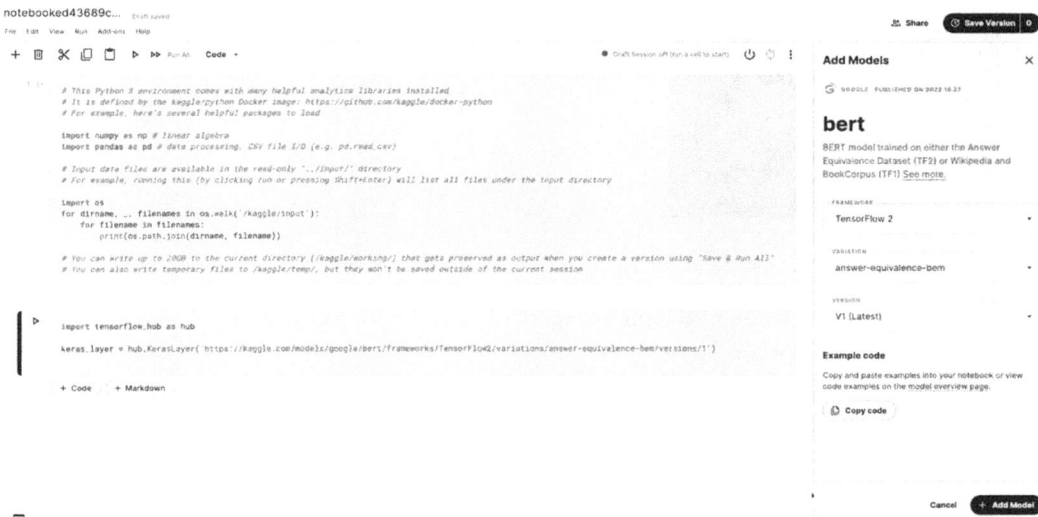

Figure 4.13: Copying the example usage from the model page

Once the model has been added by clicking **Add Model**, the remaining workflow follows the pattern described in .

Selecting a **pretrained model** and using it for a Kaggle competition has become an essential skill for Kagglers nowadays, especially since more and more competitions are based on language models, which are pretrained on large corpora of texts and are capable of generating text. For this new edition of the book, we had the opportunity to interview *Leonie Monigatti*. Leonie is a Kaggle Notebooks Grandmaster and a machine learning engineer at Weaviate, an open source vector database. Professionally, it is interesting to note that by sharing her learning on machine learning and data science on Kaggle and her blog alongside her day-to-day job, she managed to build a portfolio and was able to first transition to a role in developer advocacy, and later, became a machine learning engineer. On her blog, she also shares her insights into everything related to data science and machine learning and, of course, about her usage of models for crafting Kaggle notebooks and participating in analytics competitions.

Interview: Leonie Monigatti

https://www.kaggle.com/iamleonie

What initially attracted you to Kaggle competitions, and how has your involvement evolved over time?

I originally signed up on Kaggle to learn the Python programming language and some basic data science concepts. As a complete beginner, I found that data exploration was a great way to practice working with the pandas library, which led me to share some **exploratory data analysis (EDA)** notebooks publicly and participate in my first analytics competition.

Winning a prize in the 2019 Kaggle Machine Learning & Data Science Survey competition just a few months after signing up on Kaggle and getting positive feedback from the Kaggle community (special shoutout to Marília Prata, who has been continuously motivating Kaggle novices) on my work encouraged me to continue sharing more notebooks publicly.

Once I had the fundamentals of Python down, I continued learning about new data science and machine learning concepts. I documented my learning by creating notebooks around these topics and sharing them for others to learn from. While I mostly only actively participate in analytics competitions, nowadays, I enjoy checking out different competitions on Kaggle to learn new concepts and create learning material for myself and others.

Describe your process for selecting competitions to participate in. What criteria do you consider, and how do you prioritize your time and resources?

I mainly enjoy participating in analytics competitions. As there are only a few a year, I participate in all of them if time permits.

In your opinion, what distinguishes a successful Kaggle competitor from an average one? Are there any specific skills or qualities that contribute to success in data science competitions?

Time commitment. This applies to both regular Kaggle competitions and analytics competitions. While regular Kaggle competitions often require extensive experimentation with different components that may not make it into the final training pipeline, analytics competitions require extensive data analysis, and the majority of plots will not make it into the final report.

In your experience, what are some common pitfalls or oversights that inexperienced Kagglers tend to make? If you could offer one piece of advice to aspiring Kagglers, what would it be?

A common oversight in analytics competitions is underestimating the power of text and storytelling. I see many entries that focus on the code and the plots they output but lack text.

Here are some actionable tips to avoid this common pitfall. Start with a descriptive title (to invite people to click on your entry in the first place), then introduce what your notebook is about (to ease your reader into the topic of your notebook), explain your code and plots in a few words (to highlight what is interesting about this plot), and add some text to your plot directly (e.g., title, axis labels, maybe an arrow with a comment pointing to the most exciting insight).

As you can see, you don't have to be a distinguished writer to communicate your insights effectively—just start using it in general, and it will make a big difference.

Reflecting on your experience, how has participating in Kaggle competitions influenced your learning and professional growth in data science?

Being active in the Kaggle community has directly impacted my personal career path. It has allowed me to pivot from electrical engineering to machine learning.

Although creating Kaggle notebooks and writing blog posts was only a way for me to learn new concepts in the beginning, over time, the collection of notebooks and blog posts became a public portfolio of my technical knowledge. Although having a portfolio used to be more common in creative careers, today, having a public portfolio of your work (e.g., GitHub profile, Kaggle profile, blog, website, etc.) can also be beneficial in technical careers. In my case, I was approached for my current role because of my portfolio. Now, I get to do for a living what I used to do for fun.

Looking ahead, what trends or developments do you anticipate shaping the future of Kaggle competitions and the broader field of data science? How do you envision the role of generative AI evolving in Kaggle competitions and the broader field of data science?

What excites me most is that we're now at a point where we not only get to develop powerful ML models but also build end-to-end pipelines and applications using them, as we've seen in the recent "Google – AI Assistants for Data Tasks with Gemma" analytics competition. However, as is common with analytics competitions, scoring the entries transparently is difficult. I could imagine more competitions like these appearing once robust evaluation methods are established for these end-to-end pipelines.

In the next section, we will complete the overview of **Kaggle Models** by hinting at how you can upload a model of yours onto the platform, a procedure quite reminiscent of **Kaggle Datasets**.

Uploading your model to Kaggle Models

Besides being a model user, by selecting them and applying them properly to your problem, you may also become a model creator and publish your own models on **Kaggle Models**. Doing so is quite straightforward thanks to the various aspects we covered when dealing with their selection and usage.

You start from the Kaggle **Models** page and select the **New Model** button:

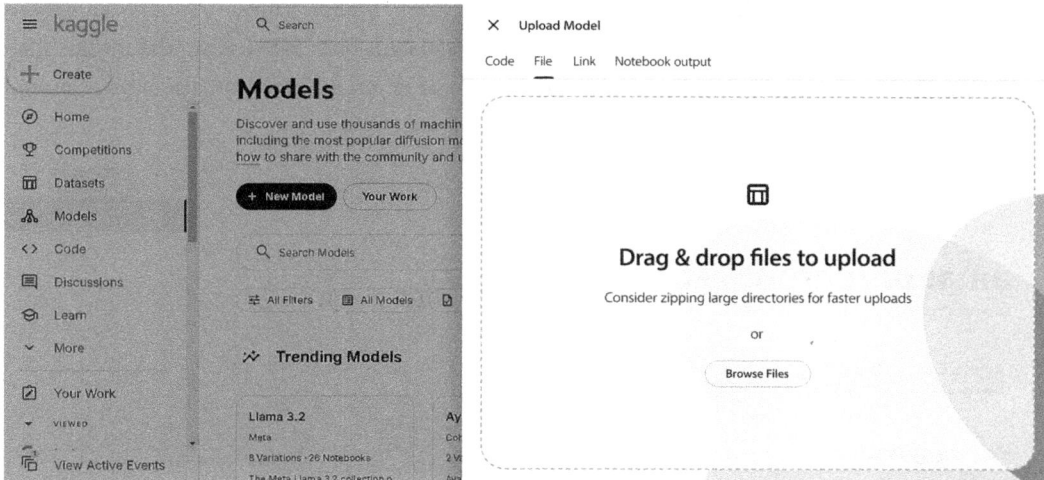

Figure 4.14: Uploading a model on Kaggle Models

You'll get quite similar options to the ones seen with Kaggle Datasets. You can upload a model by code, that is by using KaggleHub (and the page will provide you with a code example), uploading files, providing a link to a remote URL, GitHub repository, or Google Cloud Storage, or, finally, just using the output from a Kaggle notebook.

At this point, you just need to make a few choices about the model name, the variation name you are uploading, the used framework, the license, and whether the model should be a public or private one.

Once the model has been uploaded (a task that may require some time if your model is quite huge), you will be taken to the model's page. There you will be faced with a template that will help you enrich the documentation around your model, which is useful both for sharing the model with the community and for your own sake because having as much information about your model will be handy when you later you will have to use it:

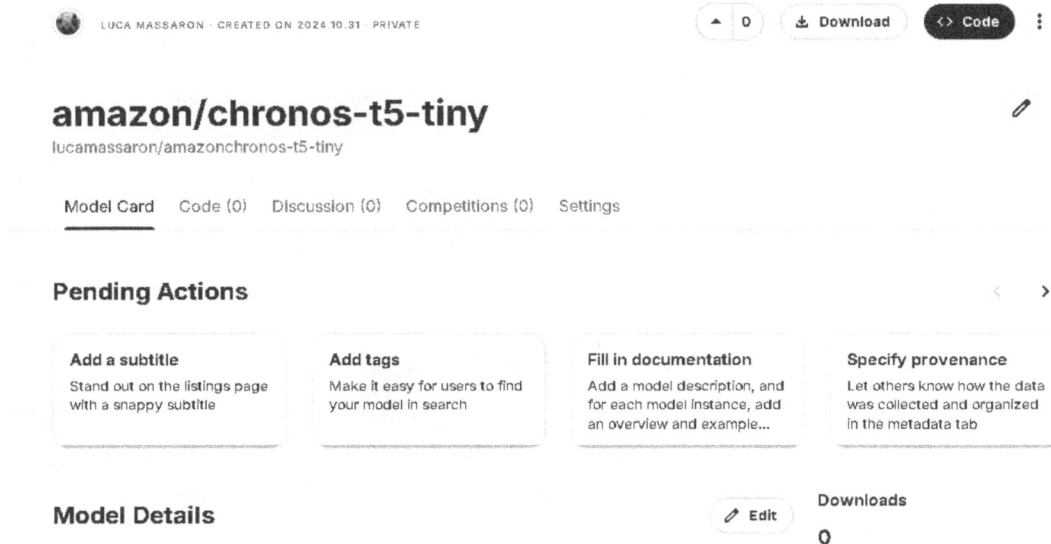

Figure 4.15: A newly uploaded model on Kaggle Models

You will be prompted to insert a subtitle, tags, documentation, and model provenance, and enrich the model with some Kaggle notebook as an example. The procedure is quite similar to what happens on **Kaggle Datasets** and you can refer to for further details. Specifically with **Kaggle Models**, however, you can also expand the model by uploading new variations anytime.

At this point, your model is readily available on Kaggle and you can use it in a competition or for creating new Kaggle notebooks!

Summary

In this chapter, we discussed **Kaggle Models**: a repository of pretrained models that are easy to use in Kaggle competition notebooks and that also allows you to upload your own models to share with the community or reuse for your competitions. As more and more competitions are shifting toward using pretrained models, this repository allows you to separate data (using Kaggle datasets) from models, helping you to find more easily and organize better the tools you need to excel in your projects and competitions.

Kaggle Models also points the direction toward the future of Kaggle itself, since now the data science field is driven more and more by generative models and pretrained models in general.

Coming up next, we will introduce discussion forums, the primary form of exchanging ideas and opinions on Kaggle.

Join our book's Discord space

Join our community's Discord space for discussions with the authors and other readers:

`https://packt.link/kaggle`

5

Leveraging Discussion Forums

Discussion forums are the primary means of information exchange on Kaggle. Whether it's discussing an ongoing competition, a dataset, or a notebook presenting a novel approach, Kagglers talk about things all the time. Kaggle discussion forums are the most direct way for you to learn, collaborate, and grow from your participation. Discussion posts offer knowledge sharing, problem-solving, networking, and inspiration. You can freely browse them in order to pick up ideas for competitions or for your own projects, but only by actively participating can you build strong relationships within the data science community on Kaggle.

In this chapter, we present the Discussion forums: how they are organized, the code of conduct, and how the wealth of information within them can be used. We will cover the following topics:

- How forums work
- Discussion points for an example competition
- Netiquette

How Kaggle forums work

Entering the Discussion forums gives you access to general topics and to all the discussions on the platform, as well as to discussions on each competition. The most direct way to access this is by clicking on **Discussions** in the left-hand side panel on the Kaggle home page:

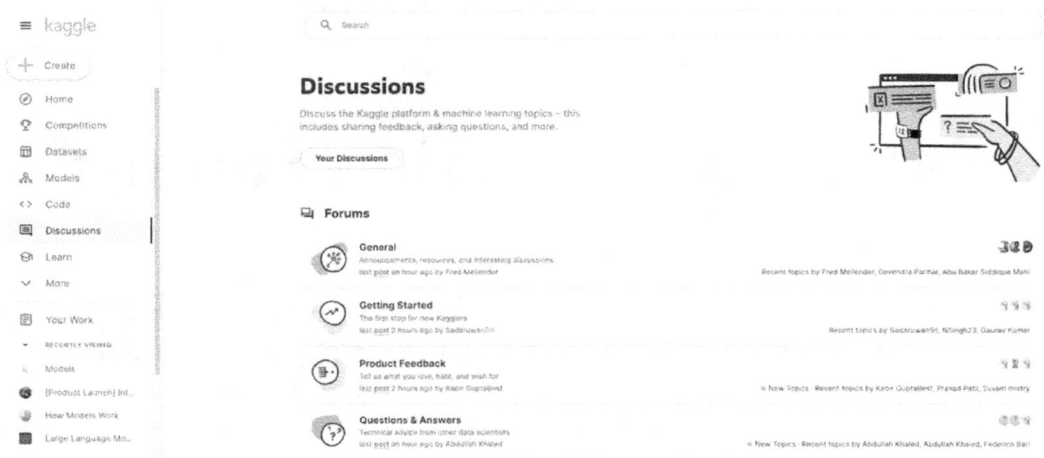

Figure 5.1: Entering the Discussions page from the main menu

The top section contains **Forums**, which has categories of general topics. Perusing those topics is useful whether you are participating in your first competition, have a suggestion to make, or just have a general question.

Below **Forums**, you can find a combined view of discussions across Kaggle: mostly conversations related to competitions (which form the bulk of activity on Kaggle), but also notebooks or notable datasets. By default, they are sorted by Hotness: those with the highest participation and the most activity are shown closer to the top. This section below Forums is where you can find content more relevant to the dynamic nature of the field – a collection of discussions from different subsets of Kaggle, with the ability to filter on specific criteria:

Discussion from across Kaggle

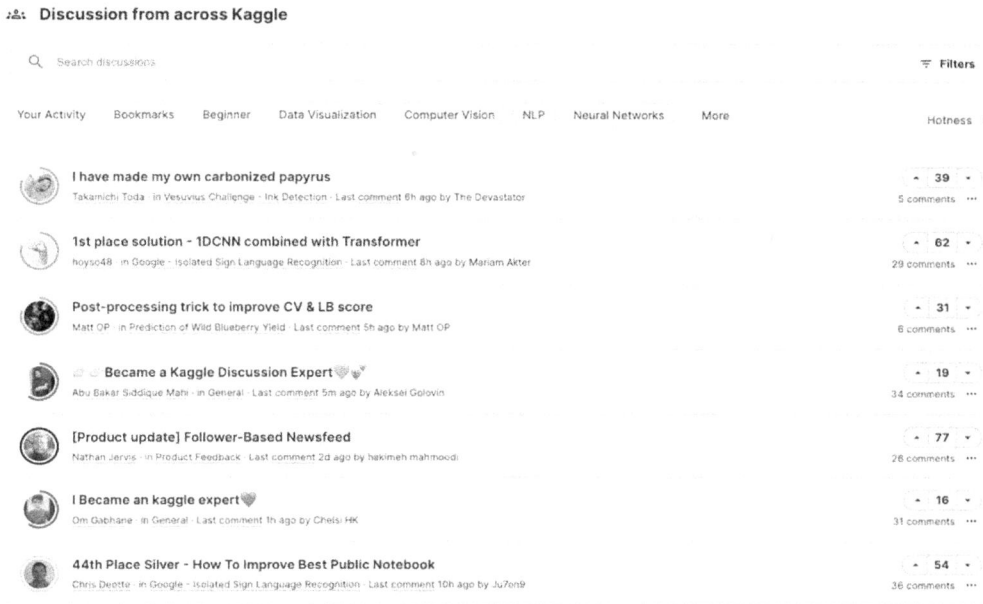

Figure 5.2: Discussions from across Kaggle

Depending on your interest, you can start personalizing the content by using the filters. You can filter by:

- **Recency**: Allows you to control the range of information you are catching up on.

- **My Activity**: Gives an overview of your comments/publications/views across all forums; useful if you are involved in multiple discussions simultaneously.

- **Admin**: Provides a quick overview of announcements from Kaggle admins.

- **Types**: Discussions can take place in the general forums, regarding specific competitions, or around datasets.

- **Tags**: While not present everywhere, several discussions are tagged, and this functionality allows you to filter on these tags.

These categories are shown in the following screenshot:

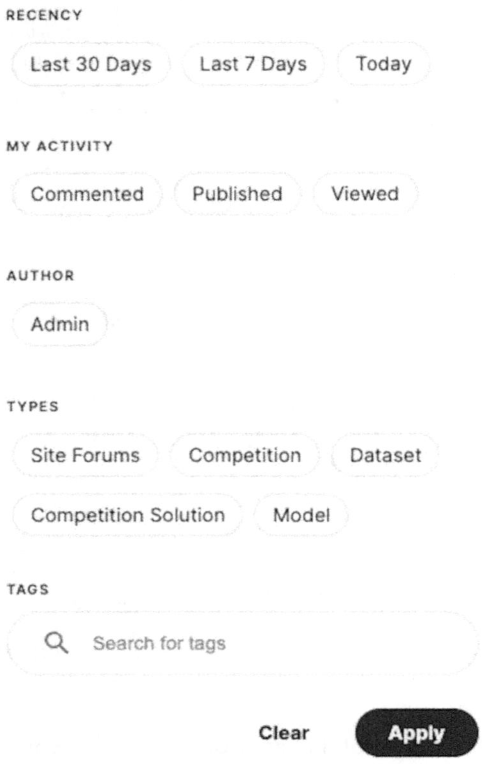

RECENCY

| Last 30 Days | | Last 7 Days | | Today |

MY ACTIVITY

| Commented | | Published | | Viewed |

AUTHOR

| Admin |

TYPES

| Site Forums | | Competition | | Dataset |

| Competition Solution | | Model |

TAGS

| Q   Search for tags |

Clear   **Apply**

Figure 5.3: Available filters for discussions

The next figure shows a sample output of filtering on discussions on the tag **Beginner**. This will help you just see discussions about methods and ideas suitable for a beginner to Kaggle or data science modeling:

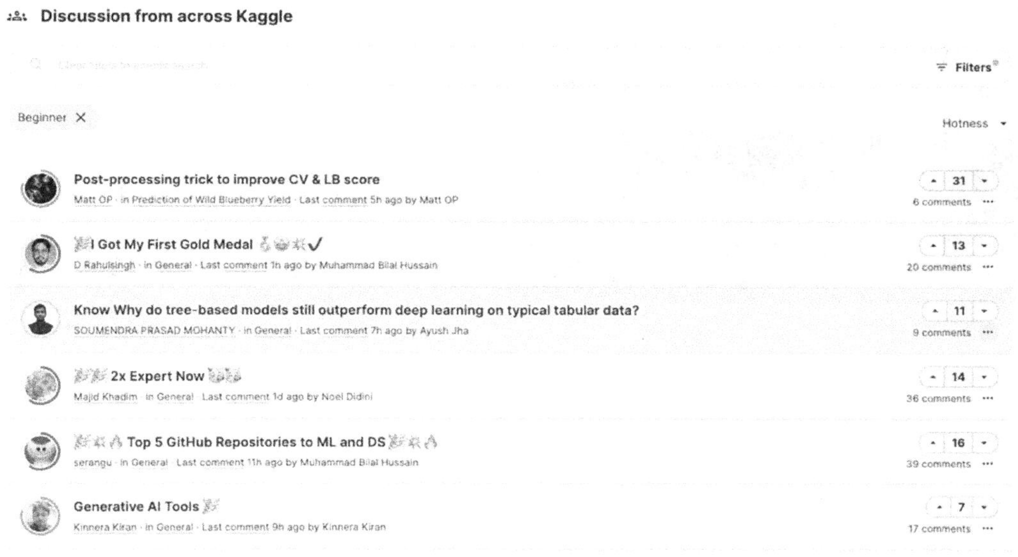

👥 **Discussion from across Kaggle**

Figure 5.4: Filtering discussions to those tagged "Beginner"

As an alternative, you can also focus on a specific topic. You can sort by **Hotness, Recent Comments, Recently Posted, Most Votes,** or **Most Comments:**

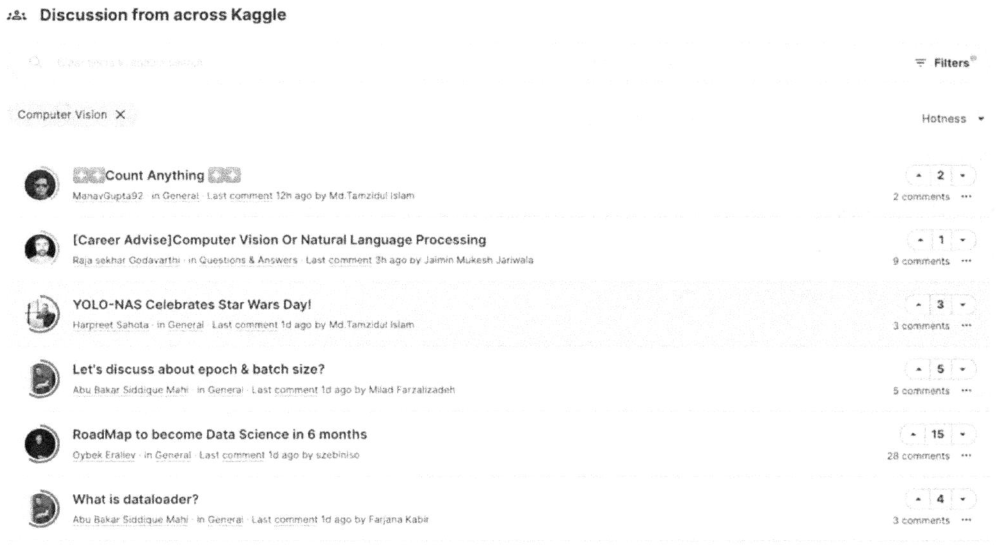

👥 **Discussion from across Kaggle**

Figure 5.5: Sorting options for discussion topics, such as by Hotness or Most Votes

People come to Kaggle for diverse reasons, but competitions remain the primary attraction. Each Kaggle competition has its own dedicated discussion forum, which you can enter by going to the competition page and selecting **Discussion**:

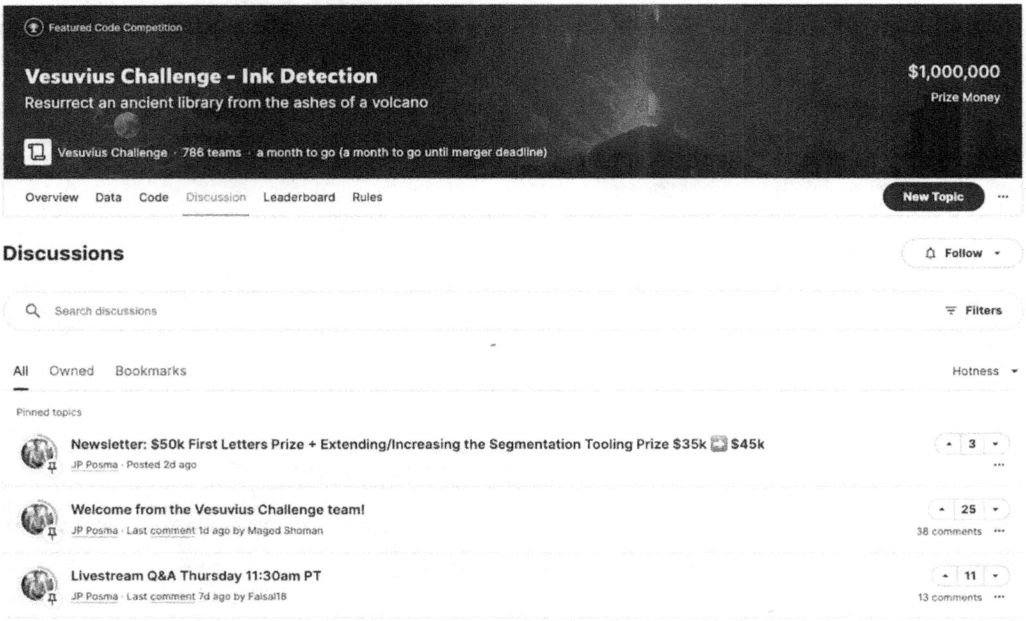

Figure 5.6: Discussion forum for a competition

It was not always the case, but these days virtually all competitions have a FAQ topic pinned at the top of their dedicated discussion forum. Starting there is a good idea for two main reasons:

- It saves you time as the most common queries will have been addressed there.
- You avoid asking redundant or duplicate questions in the remainder of the forum, making everyone's experience better.

Like notebooks, discussion forums have an option for you to bookmark particularly relevant topics for later reference:

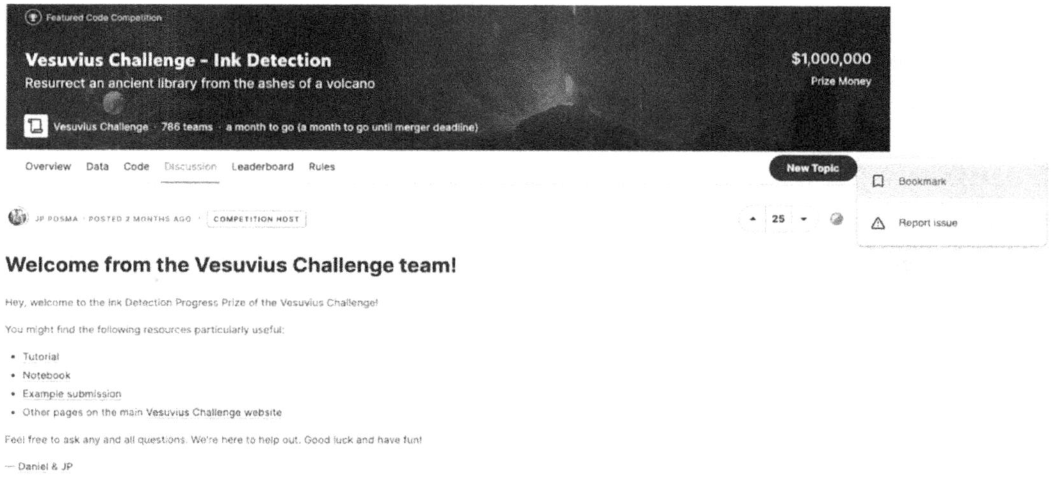

Figure 5.7: Bookmarking a topic in a discussion forum

An overview of all your bookmarked topics can be found on your profile page:

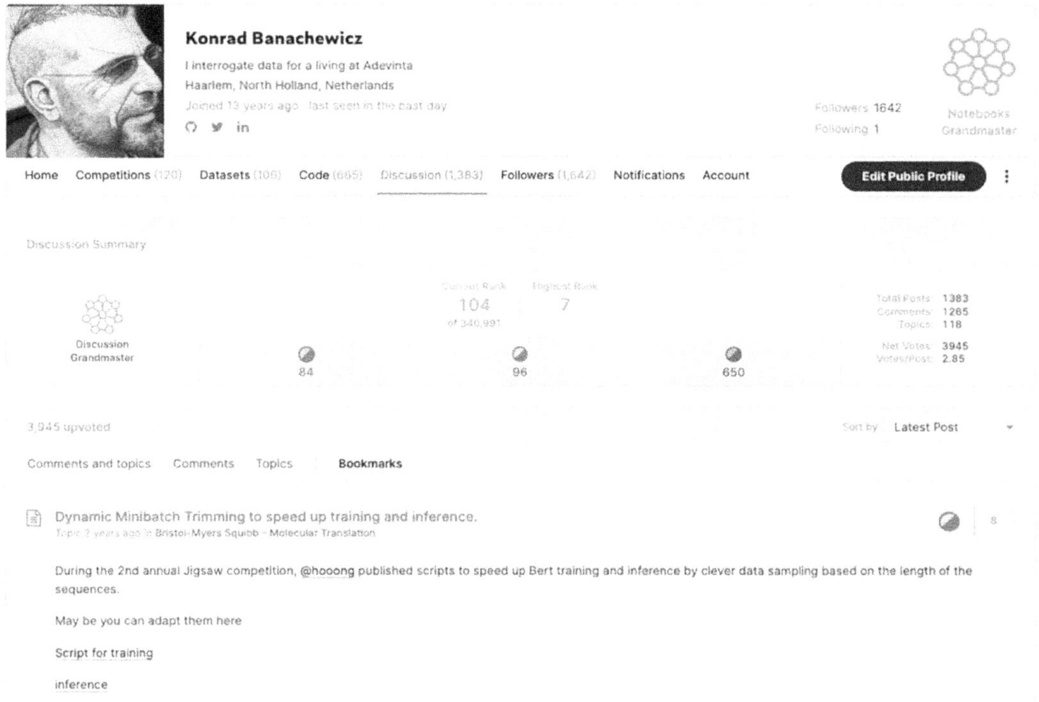

Figure 5.8: Bookmarked topic on a user profile

Having discussed how Kaggle forums work, in the next section, we are going to get an overview of a few sample discussion approaches, demonstrating how to address different types of conversations on Kaggle during competitions and community conversations.

Sample approaches to discussions

It is completely normal to feel lost in a competition at some point: you came in, tried a few ideas, got some traction on the leaderboard, and then hit the Kaggle version of a runner's wall. This is the moment when discussion forums are the place to go. In the following sections, we will look at a few examples of how you can leverage these forums.

Discussions on common challenges

First, let's look at the *Optiver Realized Volatility Prediction* competition (`https://www.kaggle.com/c/optiver-realized-volatility-prediction`), characterized by the organizers as follows:

"In the first three months of this competition, you'll build models that predict short-term volatility for hundreds of stocks across different sectors. You will have hundreds of millions of rows of highly granular financial data at your fingertips, with which you'll design your model forecasting volatility over 10-minute periods. Your models will be evaluated against real market data collected in the three-month evaluation period after training."

There is quite a lot to unpack here, so we will walk through the main components of this challenge and show how they can be approached via the discussion forums. First, participation in this competition requires some level of financial knowledge: not quite experienced trader level maybe, but understanding the different manners of calculating volatility is not trivial for a layman (which most Kagglers are in this specific context). Luckily for the participants, the organizers were very active during the competition and provided guidance to resources intended to help newcomers to the field: `https://www.kaggle.com/c/optiver-realized-volatility-prediction/discussion/273923`.

If the entry knowledge still proves insufficient to get started, don't be shy about figuring things out in public and asking for help, like here: `https://www.kaggle.com/c/optiver-realized-volatility-prediction/discussion/263039`.

As the competition went on, people started developing increasingly sophisticated models to handle the problem. There is a balance to strike here: on one hand, you might want to give something back if you have learned from veterans sharing their findings. On the other hand, you do not want to give away your (potential) advantage by publishing all your great code as a notebook.

A reasonable compromise is discussing, for example, your feature ideas in a post in the competition forum – along the lines of this one: https://www.kaggle.com/c/optiver-realized-volatility-prediction/discussion/273915.

In recent years, more competitions have started to move away from the fixed test dataset format and introduce some sort of variation. Sometimes they enforce the usage of the Kaggle API (such competitions require submission from a notebook) or introduce a special timetable split into a training phase and evaluation against live data. This was the case with Optiver:

"Starting after the final submission deadline there will be periodic updates to the leaderboard to reflect market data updates that will be run against selected notebooks. Updates will take place roughly every two weeks, with an adjustment to avoid the winter holidays."

While straightforward to formulate, this setup generated a few challenges for re-training and updating the models. Should you encounter such a situation, feel free to ask about it – as participants in this competition did: https://www.kaggle.com/c/optiver-realized-volatility-prediction/discussion/249752.

The validation scheme for your trained model is always an important topic in a Kaggle competition – usually coupled with the perennial *"cross-validation results vs leaderboard scores"* discussion. The *Optiver* competition was no exception: https://www.kaggle.com/c/optiver-realized-volatility-prediction/discussion/250650.

Unless a similar thread is already present – and it's always a good idea to check, so that redundancy can be minimized – you might want to consider a related type of thread: single-model performance. Sooner or later, everybody starts using ensembles of models, but they are not very efficient without good single-model components. The collaborative quest for knowledge does not stop there: if you think you have found a better way of approaching the problem, it is probably a good idea to share it – either you will have done something useful for others, or you will find out why you were wrong (saving you time and effort). Either way, it is a win, as shown in this discussion: https://www.kaggle.com/c/optiver-realized-volatility-prediction/discussion/260694.

Apart from the obvious personal benefit (you get a sort of peek into how competitors are doing), such threads allow for information exchange in the community, facilitating the collaborative element and being helpful for beginners. An example of such a discussion can be found at https://www.kaggle.com/c/optiver-realized-volatility-prediction/discussion/250695.

If you have gone through topics such as the ones listed earlier, there is a possibility you still find yourself wondering: am I missing anything important? Kaggle is the kind of place where it is perfectly fine to ask that. Refer to https://www.kaggle.com/c/optiver-realized-volatility-prediction/discussion/262203.

Information leakage and overfitting

Let's now turn to some other competitions and different kinds of examples. We mentioned **validation** earlier, which always links – at least for a Kaggler – to the topics of **information leakage** and **overfitting**. Leaks are discussed extensively in *Chapter 7*, dedicated to designing validation schemes. Here, we touch briefly on how they are approached via discussions. With Kaggle being a community of inquisitive people, if there is suspicion of leakage, somebody is likely to raise the topic. Here are some examples of competitions in which participants faced such issues:

- Names of files or IDs of records may contain timestamps, which means they can be reverse-engineered to effectively peek into the future and produce an unrealistically low error metric value. Such a situation took place in the *Two Sigma Connect* competition https://www.kaggle.com/c/two-sigma-connect-rental-listing-inquiries/ – you can read up on the details in *Kazanova's* post: https://www.kaggle.com/c/two-sigma-connect-rental-listing-inquiries/discussion/31870#176513.

- Another example is the *Airbus Ship Detection Challenge* (https://www.kaggle.com/c/airbus-ship-detection) in which the participants were tasked with locating ships in satellite images. It turned out that a significant proportion of the test images were (random) crops of the images in the training images and matching the two was relatively straightforward: https://www.kaggle.com/c/airbus-ship-detection/discussion/64355#377037.

- A rather infamous series of competitions is the ones sponsored by *Santander*: of the three instances when the company organized a Kaggle contest, two involved data leakage: https://www.kaggle.com/c/santander-value-prediction-challenge/discussion/61172.

What happens next varies per competition: there have been instances when Kaggle decided to reset the competition with new/cleaned-up data, but also when they allowed it to continue (because they perceived the impact as minimal). An example of handling such a situation was the Red Hat competition: https://www.kaggle.com/competitions/predicting-red-hat-business-value/discussion/22807.

Although leaks in data can disturb a competition severely, the good news is that over the last few years, leakage has become rarer on Kaggle – so with any luck, this section will be read once but not become a staple of your experience on the platform.

The topic of experience on the platform is an excellent segue into a Grandmaster interview with *Yifan Xie*, who is a Discussions and Competitions Master, as well as the co-founder of Arion.ai. Here's what he had to say about competing in competitions and working with other Kagglers.

Interview: Yifan Xie

`https://www.kaggle.com/yifanxie`

What's your favorite kind of competition and why? In terms of techniques and solving approaches, what is your specialty on Kaggle?

I don't really have a favorite type; I like tackling problems of all kinds. In terms of techniques, I have built up a solid pipeline of machine learning modules that allow me to quickly apply typical techniques and algorithms to most data problems. I would say this is a kind of competitive advantage for me: a focus on standardizing, both in terms of work routine and technical artifacts over time. This allows quicker iteration and in turn helps improve efficiency when conducting data experiments, which is a core component of Kaggle.

How do you approach a Kaggle competition? How different is this approach to what you do in your day-to-day work?

Over time, I have developed a specific way of managing and gathering information for most of my major data endeavors. This is applicable to work, Kaggle competitions, and other side projects. Typically, I capture useful information such as bookmarks, data dictionaries, to-do lists, useful commands, and experiment results in a standardized format dedicated to each competition, and when competing in a team, I will share this info with my teammates.

Tell us about a particularly challenging competition you entered and what insights you used to tackle the task.

For me, it has always been useful to understand the wider context of the competition; for instance, what are the social/engineering/financial processes that underpin and bring about the data we are working on? For competitions in which one can meaningfully observe individual data points, such as the Deepfake Detection Challenge, I would build a specific dashboard (usually using Streamlit) that allows me to check individual data points (in this case, it was pairs of true and fake videos), as well as building simple stat gathering into the dashboard to allow me a better feel of the data.

Has Kaggle helped you in your career? If so, how?

I would say Kaggle is the platform that contributed the most to my current career path as a co-owner of a data science consultancy firm. It allowed me to build, over several years, the skillset and methodology to tackle data problems in different domains. I have both customers and colleagues who I got to know from forming teams in Kaggle competitions, and it has always served me very well as a source of knowledge, even though I am less active on it these days.

In your experience, what do inexperienced Kagglers often overlook? What do you know now that you wish you'd known when you first started?

For newcomers on Kaggle, the one error I see is overlooking critical non-technical matters: rules on teaming, data usage, sharing of private information, usage of multiple accounts for innocuous reasons, etc. These are the types of errors that could completely invalidate one's often multi-month competition efforts.

The one thing I wish I knew at the beginning would be not to worry about the day-to-day position on the public leaderboard – it puts unnecessary pressure on oneself and causes overfitting.

Are there any particular tools or libraries that you would recommend using for data analysis or machine learning?

The usual: scikit-learn, XGB/LGB, PyTorch, etc. The one tool I would recommend that people learn to master beyond basic usage is NumPy, especially for more advanced ways to sort and subset information – stuff that a lazy approach via pandas makes easy, but for which a more elaborate equivalent version in NumPy would bring much better efficiency.

What's the most important thing someone should keep in mind or do when they're entering a competition?

There are four reasons to do any data science-related stuff, in my view: for profit, for knowledge, for fun, and for good. Kaggle for me is always a great source of knowledge and very often a great memory to draw upon, so my recommendation would always be to remind oneself that ranking is temporary, but knowledge/memory are permanent!

Do you use other competition platforms? How do they compare to Kaggle?

I am a very active participant on Numerai. For me, based on my four reasons to do data science, it is more for profit, as they provide a payout via their cryptocurrency. It is more of a solitary effort, as there is not really an advantage to teaming; they don't encourage or forbid it, but it is just that more human resources don't always equate to better profit on a trading competition platform like Numerai.

Numerai for me is a more sustainable activity than Kaggle during busy periods of my working calendar, because the training data is usually unchanged at each round, and I can productionize to a high degree to automate the prediction and submission once the initial models are built.

The continuity feature of Numerai also makes it better suited for people who want to build dedicated machine learning pipelines for tabular datasets.

Netiquette

Anybody who has been online for longer than 15 minutes knows this: during a discussion, no matter how innocent the topic, there is always a possibility that people will become emotional, and a conversation will leave the civilized parts of the spectrum. Kaggle is no exception to the rule, so the community has guidelines for appropriate conduct:

`https://www.kaggle.com/community-guidelines`

These apply not just to discussions, but also to notebooks and other forms of communication. The main points you should keep in mind when interacting on Kaggle are:

- Don't slip into what Scott Adams calls the "mind-reading illusion," overestimating your ability to understand others. Kaggle is an extremely diverse community of people from all over the world (for many of them, English is not their first language), so maintaining nuance is a massive challenge. Don't make assumptions and try to clarify whenever possible.
- Do not take things personal, discussions can get out of control or off-topic; Godwin's law exists for a reason. In particular, references to protected immutable characteristics are an absolute no-go area.
- Your mileage might vary, but the fact remains: this is not the Internet Wild West of the 1990s, when telling somebody online to read the manual (RTFM) was completely normal; putdowns tend to alienate people.
- Do not attempt to manipulate the progression system (which is a basis for awarding Kaggle medals). This aspect covers an entire spectrum of platform abuse, from explicitly asking for upvotes, to collusion, to outright cheating.

In short, do unto others as you would have them do to you – and things should work out fine.

Summary

In this chapter we talked about discussion forums, the primary manner of communication on the Kaggle platform. We demonstrated the forum mechanics, showed an example of how discussions can be leveraged in more advanced competitions, and briefly summarized the netiquette.

By practicing our suggestions, you will be able to better put to work your communication skills and community engagement and thus enhance your networking opportunities, which is essential for a successful career in data science, and foster collaboration, another critical quality in data science teams.

This concludes the first, introductory part of this book. The next chapter marks the start of a more in-depth exploration of how to maximize what you get out of Kaggle and looks at getting to grips with the huge variety of different tasks and metrics you must wrestle with in competitions.

Get This Book's PDF Version and Exclusive Extras

UNLOCK NOW

Scan the QR code (or go to packtpub.com/unlock). Search for this book by name, confirm the edition, and then follow the steps on the page.

Note: Keep your invoice handy. Purchases made directly from Packt don't require an invoice.

Join our book's Discord space

Join our community's Discord space for discussions with the authors and other readers:

https://packt.link/kaggle

Part 2

Elevating Your Game: Advanced Techniques for Competitive Success

This part dives deep into the art and science of competitive data science, equipping you with the advanced techniques necessary to excel. You'll learn how to dissect competition tasks and metrics, design robust validation strategies, and master modeling for diverse data types, including tabular, image, and text. We'll also cover crucial skills such as hyperparameter optimization, ensembling, and an introduction to newer frontiers, including generative AI and simulation competitions. Upon completing this part, you'll possess a comprehensive toolkit of advanced strategies and technical skills to significantly boost your performance and tackle a wide array of Kaggle challenges.

This part of the book includes the following chapters:

- *Chapter 6, Competition Tasks and Metrics*
- *Chapter 7, Designing Good Validation*
- *Chapter 8, Modeling for Tabular Competitions*
- *Chapter 9, Hyperparameter Optimization*
- *Chapter 10, Ensembling with Blending and Stacking Solutions*
- *Chapter 11, Modeling for Computer Vision*

- *Chapter 12, Modeling for NLP*
- *Chapter 13, Generative AI in Kaggle Competitions*
- *Chapter 14, Simulation and Optimization Competitions*

6
Competition Tasks and Metrics

In a competition, you start by examining the target metric. Understanding how your model's errors are evaluated is key to scoring highly in every competition. When your predictions are submitted to the Kaggle platform, they are compared to a ground truth based on the target metric.

For instance, in the *Titanic* competition (`https://www.kaggle.com/c/titanic/`), all your submissions are evaluated based on **accuracy**, or the percentage of surviving passengers you correctly predict. The organizers decided upon this metric because the aim of the competition is to find a model that estimates the probability of survival of a passenger under similar circumstances. In another knowledge competition, *House Prices – Advanced Regression Techniques* (`https://www.kaggle.com/c/house-prices-advanced-regression-techniques`), your work is evaluated based on an **average difference** between your prediction and the ground truth. This involves computing the logarithm, squaring, and taking the square root because the model is expected to be able to quantify the order of the price of a house on sale as correctly as possible.

In real-world data science, target metrics are also key for the success of your project, though there are certainly differences between the real world and a Kaggle competition. We could easily summarize by saying that there are more complexities in the real world. In real-world projects, you will often have not just one but multiple metrics against which your model will be evaluated. Frequently, some of the evaluation metrics won't even be related to how your predictions perform against the ground truth you are using for testing. For instance, the domain of knowledge you are working in, the scope of the project, the number of features considered by your model, the overall memory usage, any requirements for special hardware (such as a GPU, for instance), the latency of the prediction process, the complexity of the predicting model, and many other aspects may end up counting more than the mere predictive performance.

Real-world problems are indeed dominated by business, and tech infrastructure concerns much more than you may imagine before being involved in any of them. However, you cannot escape the fact that the basic principle at the core of both real-world projects and Kaggle competitions is the same. Your work will be evaluated according to some criteria, and understanding the details of such criteria, optimizing the fit of your model in a smart way, or selecting its parameters according to the criteria will bring you success. If you can learn more about how model evaluation occurs on Kaggle, your real-world data science job will also benefit.

In this chapter, we will detail how evaluation metrics for certain kinds of problems strongly influence how you can operate when building your model solution in a data science competition. We will also address the variety of metrics available in Kaggle competitions to give you an idea of what matters most, and, in the margins, we will discuss the different effects of metrics on predictive performance and how to translate them into your projects correctly. We will cover the following topics:

- Evaluation metrics and objective functions
- Basic types of tasks – regression, classification, and ordinal
- The Meta Kaggle dataset
- Handling never-before-seen metrics
- Metrics for regression (standard and ordinal)
- Metrics for binary classification (label prediction and probability)
- Metrics for multi-class classification
- Metrics for object detection problems
- Metrics for multi-label classification and recommendation problems
- Optimizing evaluation metrics

Evaluation metrics and objective functions

In a Kaggle competition, you can find out about the evaluation metric in the **Overview** section of the competition. Reading the **Evaluation** part of the **Overview** section will give you details about the evaluation metric. Sometimes, you will find the metric formula, the code to reproduce it, and some discussion about the metric. On the same page, you will also get an explanation about the submission file format, providing you with the header of the file and a few example rows.

The association between the evaluation metric and the submission file is important because you must consider that the metric essentially works after training your model and producing some predictions. Consequently, as a first step, you have to take the difference between an **evaluation metric** and an **objective function** into account.

Boiling everything down to the basics, an objective function serves your model during training because it is involved in the process of error minimization (or score maximization, depending on the problem). In contrast, an evaluation metric serves your model *after* it has been trained by providing a score. Therefore, it cannot influence how the model fits the data. Still, it does influence it indirectly by helping you select the most well-performing hyperparameter settings within a model and the best models among competing ones. Before proceeding with the rest of the chapter, which will show you how this can affect a Kaggle competition and why the analysis of the Kaggle evaluation metric should be your first act in a competition, let's first discuss some terminology you may encounter in the discussion forums.

You will often hear talk about objective functions, cost functions, and loss functions, sometimes interchangeably. They are not exactly the same thing, however, and we will explain the distinction here:

- A **loss function** is a function that is defined on a single data point and, by considering the difference between the prediction of the model and the ground truth for the data point, it quantifies this error as a numerical value.

- A **cost function** considers the whole dataset used for training (or a batch from it), computing a sum or average over the loss values of its data points. For instance, it can comprise further constraints, such as the L1 or L2 penalties. The cost function directly affects how the training happens.

- An **objective function** is the most general (and safe-to-use) term related to the scope of optimization during **machine learning** (**ML**) training: it comprises cost functions, but it is not limited to them. An objective function can, in fact, also take goals unrelated to the target into account, for instance, requiring sparse coefficients of the estimated model or minimization of the coefficients' values, such as in L1 and L2 regularizations. Moreover, whereas loss and cost functions imply an optimization based on minimization, an objective function is neutral and can imply either a maximization or a minimization activity performed by the learning algorithm.

Likewise, you'll hear about scoring and error functions when it comes to evaluation metrics. Distinguishing between them is easy:

- A **scoring function** assigns a numerical value to evaluate the performance of a model, often in the context of probabilistic predictions. In many cases, higher scores indicate better performance. However, there are exceptions, like the Brier score. For the Brier score, a score of zero represents perfect accuracy, and a score of one represents perfect inaccuracy.

- An **error function** instead measures the difference between the forecast probabilities and the actual outcomes.

Basic types of tasks

Not all objective functions are suitable for all problems. From a general point of view, you'll find two kinds of problems in Kaggle competitions: **regression** tasks and **classification** tasks. Recently, there have also been **reinforcement learning** (**RL**) tasks, but RL doesn't use metrics for evaluation. Instead, it relies on a ranking derived from direct match-ups against other competitors whose solutions are assumed to be as well-performing as yours (performing better in this match-up than your peers will raise your ranking while performing worse will lower it). Since RL doesn't use metrics, we will keep referring to the regression-classification dichotomy. However, **ordinal tasks**, wherein you predict ordered labels represented by integer numbers, may elude such categorization and can be dealt with successfully using a regression or classification approach.

Regression

Regression requires you to build a model that can predict a real number, often a positive number, although there have been examples of negative number prediction, too. A classic example of a regression problem is *House Prices – Advanced Regression Techniques* because you have to guess the value of a house. Evaluating a regression task involves computing a distance between your predictions and the ground truth values. This difference can be evaluated in different ways, for instance, by squaring it to punish larger errors or by applying a log to it to penalize predictions of the wrong scale.

Classification

When facing a **classification** task on Kaggle, there are more nuances to consider. The classification, in fact, could be **binary**, **multi-class**, or **multi-label**:

- In **binary** problems, you have to guess if an example should be classified into a specific class (usually called the *positive* class and compared to the *negative* one) or not. Here, the evaluation could involve the straightforward prediction of the class ownership itself or an estimation of the probability of such ownership. A typical example is the *Titanic* competition, wherein you have to guess a binary outcome: survival or no survival. In this case, the requirement of the competition is just the prediction. Still, in many cases, it is necessary to provide a probability because in certain fields, especially for medical applications, it is necessary to rank positive predictions across different options and situations in order to make the best decision.

  Though counting the exact number of correct matches in a binary classification may seem like a valid approach, this won't actually work well when there is an imbalance, that is, a different number of examples, between the positive and negative classes. Classification based on an imbalanced distribution of classes requires evaluation metrics that consider the imbalance if you want to track improvements in your model correctly.

- When you have more than two classes, you have a **multi-class** prediction problem. This also requires using suitable functions for evaluation since it is necessary to keep track of the model's overall performance and ensure that the performance across the classes is comparable (for instance, your model could underperform with respect to certain classes). Here, each case can be exclusively in one class and not in others. A good example is *Leaf Classification* (https://www.kaggle.com/c/leaf-classification), wherein each image of a leaf specimen has to be associated with the correct plant species.

- Finally, when your class predictions are not exclusive, and you can predict multiple-class ownership for each example, you have a **multi-label** problem requiring further evaluations to control whether your model predicts the correct classes and the correct number and mix of classes. For instance, you have to associate each article with all its topics in *Greek Media Monitoring Multilabel Classification (WISE 2014)* (https://www.kaggle.com/c/wise-2014).

Ordinal

In a problem involving a prediction on an ordinal scale, you have to guess integer numeric labels, which are naturally ordered. For example, an earthquake's magnitude is on an ordinal scale. In addition, data from marketing research questionnaires is often recorded on ordinal scales (for instance, consumers' preferences or opinion agreements). Since an ordinal scale is made of ordered values, ordinal tasks can be considered somewhat halfway between regression and classification, and you can solve them in both ways:

- The most common way is to treat your ordinal task as a **multi-class** problem. In this case, you will get a prediction of an integer value (the class label), but the prediction will not take the fact into account that the classes have a certain order. If you look at the prediction probability for the classes, you can tell that there is something wrong with approaching the problem as a multi-class problem. Often, probabilities will be distributed across the entire range of possible values, depicting a multi-modal and often asymmetric distribution (whereas you should expect a Gaussian distribution around the maximum probability class).

- The other way to solve the ordinal prediction problem is to treat it as a **regression** problem and then post-process your result. This way, the order among classes will be considered, though the prediction output won't immediately be useful for scoring on the evaluation metric. In fact, in a regression, you get a float number as an output, not an integer representing an ordinal class. Moreover, the result will include the full range of values between the integers of your ordinal distribution and possibly also values outside of it. Cropping the output values and casting them into integers by unit rounding may do the trick, but this might lead to inaccuracies requiring more sophisticated post-processing (we'll discuss this later in the chapter).

Now, you may be wondering what kind of evaluation you should master to succeed on Kaggle. Clearly, you always have to master the evaluation metric of the competition you have taken on. However, some metrics are more common than others, which is information you can use to your advantage. What are the most common metrics? How can we figure out where to look for insights in competitions that have used similar evaluation metrics? The answer is to consult the Meta Kaggle dataset.

The Meta Kaggle dataset

The *Meta Kaggle dataset* (`https://www.kaggle.com/kaggle/meta-kaggle`) is a collection of rich data about Kaggle's community and activity, published by Kaggle itself as a public dataset. It contains CSV tables filled with public activity from Competitions, Datasets, Notebooks, and Discussions. All you have to do is start a Kaggle Notebook (as you saw in *Chapters 2* and *3*), add to it the Meta Kaggle dataset, and start analyzing the data. The CSV tables are updated daily, so you'll have to refresh your analysis often, but that's worth it given the insights you can extract.

We will sometimes refer to the Meta Kaggle dataset in this book, both as inspiration for many interesting examples of the dynamics in a competition and as a way to pick up useful examples for your learning and competition strategies. Here, we are going to use it in order to figure out what evaluation metrics have been used most frequently for competitions in the last seven years. By looking at the most common ones in this chapter, you'll be able to start any competition from solid ground and then refine your knowledge of the metric, picking up competition-specific nuances using the discussion you find in the forums.

Here, we will introduce the necessary code to produce a data table of metrics and their counts per year. It is designed to run directly on the Kaggle platform:

```python
import numpy as np
import pandas as pd
comps = pd.read_csv("/kaggle/input/meta-kaggle/Competitions.csv")
evaluation = ['EvaluationAlgorithmAbbreviation',
              'EvaluationAlgorithmName',
              'EvaluationAlgorithmDescription',]
compt = ['Title', 'EnabledDate', 'HostSegmentTitle']
df = comps[compt + evaluation].copy()
df['year'] = pd.to_datetime(df.EnabledDate).dt.year.values
df['comps'] = 1
time_select = (df.year >= 2017) & (df.year <= 2023)
competition_type_select = df.HostSegmentTitle.isin(['Featured',
                                                    'Research'])
pd.pivot_table(df[time_select&competition_type_select],
               values='comps',
               index=['EvaluationAlgorithmAbbreviation'],
               columns=['year'],
               fill_value=0.0,
```

```
      aggfunc=np.sum,
      margins=True
   ).sort_values(
      by=('All'), ascending=False).iloc[1:,:].head(20)
```

In this code, we read the CSV table containing the data relating to the competitions. We focus on the columns representing the evaluation and the columns informing us of the competition's name, start date, and type. We limit the rows to the competitions held from 2017 until 2023 and to the Featured or Research type (the most common ones). To modify the considered time range, simply adjust the boundary years in `time_select = (df.year >= 2017) & (df.year <= 2023)`. We complete the analysis by creating a pandas pivot table, combining the evaluation algorithm with the year, and counting the number of competitions using it. We just display the top 20 algorithms.

Here is the resulting table (based on data from mid 2024):

Year **Evaluation Algorithm**	2017	2018	2019	2020	2021	2022	2023	Tot
LogLoss	5	2	3	2	0	2	0	14
Area under the curve (AUC)	1	3	3	2	3	1	0	13
MAP@{K}	0	4	1	0	2	3	1	11
FScoreBetaMicro	1	2	1	2	1	1	0	8
CategorizationAccuracy	4	0	1	2	0	1	0	8
MeanBestErrorAtK	2	2	1	1	0	1	0	7
GoogleGlobalAP	1	2	1	1	1	0	0	6
MCRMSLE	1	0	0	5	0	0	0	6
FScoreMacro	0	1	0	2	1	1	1	6
PostProcessorKernelDesc	0	0	0	2	0	1	3	6
RMSE	0	3	0	0	2	0	0	5
Dice	1	0	2	1	0	1	0	5
RMSLE	3	1	1	0	0	0	0	5
QuadraticWeightedKappa	0	1	2	1	0	0	0	4
MCAUC	1	0	0	3	0	0	0	4
OpenImagesObjectDetectionAP	0	1	1	1	1	0	0	4

MAE	0	0	1	0	1	1	1	4
MulticlassLoss	2	0	1	0	0	1	0	4
FScoreMicro	1	0	0	0	2	0	0	3
MWCRMSE	0	0	0	1	0	1	1	3

Table 6.1: Frequency of evaluation algorithms in Featured and Research competitions from 2017 to 2023

Using the same variables we just instantiated in order to generate the table, you can also check the data to find the competitions where the metric of your choice has been adopted:

```
metric = 'AUC'
metric_select = df['EvaluationAlgorithmAbbreviation']==metric
print(df[time_select&competition_type_select&metric_select]
        [['Title', 'year']])
```

In the preceding snippet, we decided to represent the competitions that have been using the AUC metric. You just have to change the string representing the chosen metric and the resulting list will be updated accordingly.

Now let's examine the most popular evaluation metrics used in competitions hosted on Kaggle at the time of writing:

- The two top metrics are closely related to each other and to binary probability classification problems:

    - **Log Loss** helps to measure how far your predicted probabilities are from the ground truth

    - The **AUC** metric helps to measure if your model's predicted probabilities tend to predict positive cases with high probabilities

    Consider that as you optimize for Log Loss, you generally also optimize for the AUC metric.

- In the third position, we find **MAP@{K}**, a common metric in recommender systems and search engines. In Kaggle competitions, this metric has been used mostly for information retrieval evaluations, such as in the *Humpback Whale Identification* competition (https://www.kaggle.com/c/humpback-whale-identification), where you have to identify a whale precisely, and you have five possible guesses. Another example of MAP@{K} usage is in the *Quick, Draw! Doodle Recognition Challenge* (https://www.kaggle.com/c/quickdraw-doodle-recognition/), where you aim to guess the content of a drawn sketch, and you are allowed three attempts. In essence, when MAP@{K} is the evaluation metric, you can score not just if you can guess correctly but also if your correct guess is among a certain number (the "K" in the name of the function) of other incorrect predictions.

- Only in the 10th position can we find a regression metric, the **root mean squared logarithmic error (RMSLE)**, next to the **root mean squared error (RMSE)**. In the 15th place is the **quadratic weighted kappa**, a metric particularly useful for estimating model performance on problems that involve guessing a progressive integer number (an ordinal scale problem).

As you skim through the list of top metrics, you will keep finding metrics commonly discussed in ML textbooks. In the next few sections, after first discussing what to do when you encounter a never-before-seen metric (something that happens more frequently in Kaggle competitions than you may expect), we will review some of the most common metrics found in regression and classification competitions.

Handling never-before-seen metrics

Before proceeding, we have to consider that the top 20 table doesn't cover all the metrics used in competitions. We should be aware that there are metrics that have only been used once in recent years.

Let's keep using the results from the previous code to find out what they are:

```
sel_df = df[time_select & competition_type_select]
counts = sel_df.groupby('EvaluationAlgorithmName')
total_comps_per_year = sel_df.groupby('year').count()[['comps']]
single_metrics_per_year = (
    counts.sum()[counts.sum().comps == 1]
            .groupby(['year', 'EvaluationAlgorithmName'])
            .count()[['comps']])
tot_single_metrics_per_year = (
    single_metrics_per_year.reset_index()
                            .groupby('year')
                            .count()['comps'])
table =  tot_single_metrics_per_year / total_comps_per_year['comps']
print(table)
```

As a result, we get the following table showing, for each year, how many competitions used a metric that has never been used afterward (n_comps), and the proportion of these competitions per year (pct_comps):

```
year    n_comps    comps
2017       34       0.147059
2018       35       0.200000
2019       36       0.305556
2020       44       0.295455
2021       31       0.354839
2022       38       0.421053
2023       36       0.666667
```

Observing the relative share of competitions with a never-to-be-seen-afterward metric, we immediately notice how it is growing year by year and that it reached over 66% in recent years, implying that typically, two competitions out of every three require you to study and understand a metric from scratch.

You can get the list of such metrics that have occurred in the past with a simple code snippet:

```
print(counts.sum()[counts.sum().comps==1].index.values)
```

By executing the code, you will get a long list similar to this one:

```
['34817366' '35896185' '36031993' '37085174' '38195349' '39078087'
 '39243534' '39243586' '39244032' '39244492' '40581166' '41308515'
 '42009344' '42595776' '42603795' '43391374' '45372968' '48030576'
 'AI4CodeKendallTau' 'AmexGiniAndPercentageCaptureX' 'BenetechMixedMatch'
 'CRPS' 'CSIROObjectDetectionFBeta' 'CVPRAutoDrivingAveragePrecision'
 'DFLEventDetectionAP' 'Dice3DHausdorff' 'DiceFBeta'
 'FScore_1 (deprecated)' 'GroupMeanLogMAE' 'ImageMatchingChallengeMaa'
 'ImageNetObjectLocalization' 'IndoorLocalization'
 'IntersectionOverUnionObjectSegmentationBeta'
 'IntersectionOverUnionObjectSegmentationWithClassification'
 'IntersectionOverUnionObjectSegmentationWithF1' 'JPXSharpe'
 'JaccardFbeta' 'JaneStreetPnl' 'JigsawAgreementWithAnnotators'
 'JigsawBiasAUC' 'LaplaceLogLikelihood' 'LevenshteinMean'
 'Lyft3DObjectDetectionAP' 'M5_WRMSSE' 'MCSpearmanR' 'MSE'
 'MeanAngularError' 'MeanColumnwiseLogLoss' 'MeanCosineSimilarity'
 'MeanPearson' 'MeanPearsonOld' 'MedicalBoardFBeta' 'NDCG@{K}'
 'NFLHelmetIdentification' 'NQMicroF1' 'NWRMSLE' 'NormalizedGini'
 'NvidiaDefconWeightedCategorizationAccuracy' 'PKUAutoDrivingAP'
 'PearsonCorrelationCoefficient' 'ProbFScoreBetaMicro' 'R2Score'
 'RootMeanSquarePercentageError' 'SIIMDice' 'SantaResident'
```

```
'SantaWorkshopSchedule2019' 'SantasPrintShop2022'
'SantasSuperpermutations2021' 'SpearmanR' 'TextOverlapFBeta' 'TrackML'
'TravelingSanta2' 'TwoSigmaNews' 'WMAE' 'WRRMSE' 'WeightedAUC'
'WeightedCorrelationCoefficient' 'WeightedPinballLoss'
'WeightedRecall@{K}' 'WeightedRowwisePinballLoss'
'YT8M_MeanAveragePrecisionAtK' 'ZillowMAE' 'football' 'halite'
'kore_fleets' 'lux_ai_2021' 'mab']
```

Upon close inspection, you can find many metrics relating to deep learning and RL competitions.

What do you do when you encounter a metric that has never been used before? Of course, you can rely on the discussions in the Kaggle discussion forums, where you can always find good inspiration and many Kagglers who will help you. However, suppose you want to build up your own knowledge about the metric aside from Googling it. In that case, we advise that you try to experiment with it by coding the evaluation function by yourself, even imperfectly, and try to simulate how the metric reacts to different types of error produced by the model. You could also directly test how it functions on a sample from the competition training data or synthetic data that you have prepared.

We can quote a few examples of this approach as used by Kagglers:

- *Carlo Lepelaars* with Spearman's Rho: `https://www.kaggle.com/carlolepelaars/understanding-the-metric-spearman-s-rho`
- *Carlo Lepelaars* with Quadratic Weighted Kappa: `https://www.kaggle.com/carlolepelaars/understanding-the-metric-quadratic-weighted-kappa`
- *Rohan Rao* with Laplace Log Likelihood: `https://www.kaggle.com/rohanrao/osic-understanding-laplace-log-likelihood`

This can give you increased insight into the evaluation and an advantage over other competitors relying only on answers from Googling and Kaggle forums.

Before we start exploring different metrics, let's catch up with *Rohan Rao* (AKA Vopani) himself, quadruple Grandmaster and head of AI (products) at Analytics Vidhya, about his successes on Kaggle and the wisdom he has to share with us.

Interview: Rohan Rao

```
https://www.kaggle.com/rohanrao
```

What's your favorite kind of competition and why? In terms of techniques and solving approaches, what is your specialty on Kaggle?

I like to dabble with different types of competitions, but my favorites would certainly be time series ones. I don't quite like the typical approaches to or concepts of time series in the industry, so I tend to innovate and think out of the box by building solutions in an unorthodox way, which has ended up being very successful for me.

How do you approach a Kaggle competition? How different is this approach to what you do in your day-to-day work?

For any Kaggle competition, my typical workflow would look like this:

1. Understand the problem statement and read all the information related to rules, format, timelines, datasets, metrics, and deliverables.

2. Dive deeply into the data. Slice and dice it in every way possible and explore/visualize it to be able to answer any question about it.

3. Build a simple pipeline with a baseline model and make a submission to confirm the process works.

4. Engineer features, tune hyperparameters, and experiment with multiple models to get a sense of what's generally working and what's not.

5. Constantly go back to analyzing the data, reading discussions on the forum, and tweaking the features and models to the fullest. Maybe team up at some point.

6. Ensemble multiple models and decide which submissions to make final.

In my day-to-day work in data science, most of this happens too. However, there are two crucial elements that are additionally required:

- Curating and preparing datasets for the problem statement
- Deploying the final model or solution into production

The majority of my time has been spent in these two activities for most of the projects I've worked on in the past.

Has Kaggle helped you in your career? If so, how?

The vast majority of everything I've learned in ML has come from Kaggle. The community, the platform, and the content are pure gold and there is an incredible amount of stuff you can learn.

What has benefitted me the most is the experience of competing in Kaggle competitions; it has given me immense confidence in understanding, structuring, and solving problems across domains, which I have been able to apply successfully in many of the companies and projects I worked on outside Kaggle.

Many recruiters have contacted me for opportunities after looking at my Kaggle achievements, primarily in competitions. It gives a fairly good indication of a candidate's ability to solve data science problems and hence, it is a great platform to showcase your skills and build a portfolio.

What mistakes have you made in competitions in the past?

I've made some mistakes in every competition! That's how you learn and improve. Sometimes it's a coding bug, sometimes a flawed validation setup, and sometimes an incorrect submission selection!

What's important is to learn from these and ensure you don't repeat them. Iterating over this process automatically helps to improve your overall performance on Kaggle.

Are there any particular tools or libraries that you would recommend using for data analysis/ML?

I strongly believe in never marrying a technology. Use whatever works best, and whatever is most comfortable and effective, but constantly be open to learning about new tools and libraries.

Metrics for regression (standard and ordinal)

When working with regression problems, that is, problems that involve estimating a continuous value (that could range from minus infinity to infinity), the most commonly used error measures are **RMSE** and **mean absolute error (MAE)**. Still, you can also find slightly different error measures, such as **RMSLE** or **mean column-wise RMSLE**, useful.

Mean squared error (MSE) and R squared

The RMSE is the square root of the **mean squared error** (**MSE**), which is nothing but the mean of the good old **sum of squared errors** (**SSE**) that you learned about when you studied how a regression works.

Here is the formula for the MSE:

$$MSE = \frac{1}{n}SSE = \frac{1}{n}\sum_{i=1}^{n}(\hat{y}_i - y_i)^2$$

Here, n indicates the number of cases, y_i is the ground truth, and $\hat{y}_i$ is the prediction.

Let's start by explaining how the formula works:

1. You first get the difference between your predictions and your real values.

2. You square the differences (so they become positive or simply zero), and then you sum them all, resulting in your SSE.

3. Then, you just have to divide this measure by the number of predictions to obtain the average value, the MSE.

Usually, all regression models minimize the SSE, so you won't have great problems trying to minimize the MSE or its direct derivatives such as **R squared** (also called the **coefficient of determination**), which is given by the following formula:

$$R^2 = 1 - \frac{SSE}{SST} = 1 - \frac{\sum_{i=1}^{n}(y_i - \hat{y}_i)^2}{\sum_{i=1}^{n}(y_i - \bar{y})^2}$$

Here, the complement to one of SSE is compared to the complement to one of the **sum of squares total** (**SST**), which is just the variance of the response. In statistics, in fact, SST is defined as the squared difference between your target values and their average:

$$SST = \sum_{i=1}^{n}(y_i - \bar{y})^2$$

To put it another way, R squared compares the squared errors of the model to the squared errors from the simplest model possible, the average of the response. Since both the SSE and the SST have the same scale, R squared can help you to determine whether transforming your target will help to obtain better predictions.

R-squared values typically fall within the range of 0 to 1, but can also be negative in some cases, especially when the model is a poor fit to the data. An R-squared value of 1 signifies that the model accurately accounts for all the variance in the dependent variable. Conversely, a value of 0 implies that the model fails to explain any variance in the data. It's worth noting that R-squared can also assume negative values, indicating that the chosen model performs worse than a simple horizontal line (representing the mean of the dependent variable) in predicting the target variable.

Traditionally, R-squared has been employed in statistics to assess the fit of a model, particularly in the context of linear models when predicting training data. In ML scenarios, however, R-squared can be applied to test data to evaluate how well a model performs compared to a naive predictor, such as the mean. Furthermore, it proves valuable for making comparisons between different models.

The MSE is a great instrument for comparing regression models applied to the same problem. The bad news is that the MSE assigns excessive weight to errors related to outliers, which usually and inherently pose challenges for accurate predictions. Hence, by squaring outliers' errors, the assessment or optimization of a model using the MSE has a tendency to diminish the impact of smaller prediction errors and assign a disproportionate weight to outliers.

In addition, the MSE is seldom used in Kaggle competitions since the RMSE is preferred because it makes it much easier to understand the magnitude of the errors. In fact, when taking the root of the MSE, the resulting value will resemble the original scale of your target, and it will be easier at a glance to figure out if your model is doing a good job or not.

To summarize, if you are considering the same regression model across different data problems (for instance, across various datasets or data competitions), R squared is a better evaluation measure because it is perfectly correlated with the MSE, and its values typically range between 0 and 1, making all comparisons easier.

RMSE

The RMSE is just the square root of the MSE, but this implies a subtle difference. Here is its formula:

$$RMSE = \sqrt{\sum_{i=1}^{n} \frac{(\hat{y}_i - y_i)^2}{n}}$$

In the preceding formula, n indicates the number of cases, y_i is the ground truth, and $\hat{y}_i$ is the prediction. When using MSE, large prediction errors are greatly penalized because of the squaring activity. In the case of the RMSE, this dominance is lessened because of the root effect (however, as mentioned with MSE, you should always pay attention to outliers; they can affect your model evaluation a lot, no matter whether you are using the MSE or the RMSE).

Consequently, depending on the problem, you can get a better fit with an algorithm using the MSE as an objective function by first applying the square root to your target (if possible, because it requires positive values) and then squaring the results. Functions such as `TransformedTargetRegressor` in scikit-learn help you appropriately transform your regression target to get better-fitting results with respect to your evaluation metric.

Notable competitions where the RMSE has been used include the following:

- *Avito Demand Prediction Challenge*: `https://www.kaggle.com/c/avito-demand-prediction`
- *PetFinder.my - Pawpularity Contest*: `https://www.kaggle.com/c/petfinder-pawpularity-score`
- *CommonLit Readability Prize*: `https://www.kaggle.com/competitions/commonlitreadabilityprize`

RMSLE

Another common transformation of MSE is the **root mean squared log error (RMSLE)**. MCRMSLE (where MC stands for mean column-wise) is just a variant made popular by the COVID-19 forecasting competitions (for instance, COVID19 Global Forecasting Week 1, which you can learn about at `https://www.kaggle.com/competitions/covid19-global-forecasting-week-1`), and it is the column-wise average of the RMSLE values of every single target when there are multiple ones. Here is the formula for RMSLE:

$$RMSLE = \sqrt{\frac{1}{n}\sum_{i=1}^{n}(\log(\hat{y}_i + 1) - \log(y_i + 1))^2}$$

In the formula, n indicates the number of cases, y_i is the ground truth, and $\hat{y}_i$ is the prediction. Since you are applying a **logarithmic transformation** to your predictions and your ground truth before all the other squaring, averaging, and rooting operations, you don't penalize huge differences between the predicted and the actual values, especially when both are large numbers. In other words, what you care the most about when using RMSLE is *the scale of your predictions with respect to the scale of the ground truth*.

Generally speaking, please remember that **linear transformations**, such as minmax (https://scikit-learn.org/stable/modules/generated/sklearn.preprocessing.MinMaxScaler.html) or standardization (https://scikit-learn.org/stable/modules/generated/sklearn.preprocessing.StandardScaler.html), do not change the performance of any regressor, since they are linear transformations of the target.

Non-linear transformations, such as the square root, the cubic root, the logarithm, the exponentiation, and their combinations, should instead definitely modify the performance of your regression model on the evaluation metric (hopefully for the better, if you decide on the right transformation).

As with RMSE, ML algorithms for regression can better optimize for RMSLE if you apply a logarithmic transformation to the target before fitting the model to the data (and then reverse the effect at prediction time by using the exponential function).

Notable competitions using RMSLE as an evaluation metric are as follows:

- *ASHRAE – Great Energy Predictor III*: https://www.kaggle.com/c/ashrae-energy-prediction
- *Santander Value Prediction Challenge*: https://www.kaggle.com/c/santander-value-prediction-challenge
- *Mercari Price Suggestion Challenge*: https://www.kaggle.com/c/mercari-price-suggestion-challenge
- *Sberbank Russian Housing Market*: https://www.kaggle.com/c/sberbank-russian-housing-market
- *Recruit Restaurant Visitor Forecasting*: https://www.kaggle.com/c/recruit-restaurant-visitor-forecasting

By far, at the moment, RMSLE is the most used evaluation metric for regression in Kaggle competitions.

MAE

The **MAE** evaluation metric is the absolute value of the difference between the predictions and the targets. Here is the formula for the MAE:

$$MAE = \frac{1}{n}\sum_{i=1}^{n}|\hat{y}_i - y_i|$$

In the preceding formula, n stands for the number of cases, y_i is the ground truth, and $\hat{y}_i$ is the prediction. The MAE is not particularly sensitive to outliers (unlike the MSE, where errors are squared). Hence, you may find that it is an evaluation metric in many competitions whose datasets present outliers. Moreover, you can easily work with it since many algorithms can directly use it as an objective function; otherwise, you can optimize for it indirectly by following these simple steps:

1. You just train on the square root transformation of your target.
2. Then you just square the resulting predictions.

In terms of the downside, using MAE as an objective function results in much slower convergence since you are actually optimizing for predicting the median of the target (also called the L1 norm) instead of the mean (also called the L2 norm), as occurs by MSE minimization. This results in more complex computations for the optimizer, so the training time can even grow exponentially based on your number of training cases (see, for instance, this Stack Overflow question: `https://stackoverflow.com/questions/57243267/why-is-training-a-random-forest-regressor-with-mae-criterion-so-slow-compared-to`).

Notable competitions that used MAE as an evaluation metric are as follows:

- *LANL Earthquake Prediction*: `https://www.kaggle.com/c/LANL-Earthquake-Prediction`
- *Google Brain – Ventilator Pressure Prediction*: `https://www.kaggle.com/competitions/ventilator-pressure-prediction`
- *Enefit – Predict Energy Behavior of Prosumers*: `https://www.kaggle.com/competitions/predict-energy-behavior-of-prosumers`

Having mentioned the ASHRAE competition earlier, we should also mention that regression evaluation measures are quite relevant to forecasting competitions. For instance, the M5 forecasting competition was held a few years ago (`https://mofc.unic.ac.cy/m5-competition/`) and data from all the other M competitions is available, too. If you are interested in forecasting competitions, of which there are a few on Kaggle, please see `https://robjhyndman.com/hyndsight/forecasting-competitions/` for an overview of M competitions and how valuable Kaggle is for obtaining better practical and theoretical results from such competitions.

Essentially, forecasting competitions do not require a very different evaluation from regression competitions. When dealing with forecasting tasks, it is true that you can get some unusual evaluation metrics such as the following:

- The **weighted root mean squared scaled error** (`https://www.kaggle.com/c/m5-forecasting-accuracy/overview/evaluation`)

- The **symmetric mean absolute percentage error (sMAPE)** (`https://www.kaggle.com/c/demand-forecasting-kernels-only/overview/evaluation`)

In particular, the sMAPE is not at all a variation of the RMSE or the MAE, but a percentage metric, expressing the accuracy of the forecast as a percentage of the actual values. To learn more about the distinction between scale-dependent, percentage-error, and scale-free metrics, you can read this article from the Australian statistician and forecasting expert *Robin J. Hyndman*:`https://robjhyndman.com/papers/foresight.pdf`

Metrics for classification (label prediction and probability)

Having discussed the metrics for regression problems, we are going now to illustrate the metrics for classification problems, starting from the binary classification problems (when you have to predict one of two distinct classes), moving to the multi-class (when you have more than two mutually exclusive classes), and then to the multi-label ones (when the classes overlap and are not mutually exclusive).

Accuracy

When analyzing the performance of a binary classifier, the most common and accessible metric that is used is **accuracy**. A misclassification error is when your model predicts the wrong class, for example. The accuracy is just the complement of the misclassification error. It can be calculated as the ratio between the number of correct numbers divided by the number of answers:

$$Accuracy = \frac{correct\ answers}{total\ answers}$$

This metric has been used, for instance, in *Cassava Leaf Disease Classification* (https://www.kaggle.com/c/cassava-leaf-disease-classification) and *Text Normalization Challenge – English Language* (https://www.kaggle.com/c/text-normalization-challenge-english-language), where you scored a correct prediction only if your predicted text matched the actual string.

As a metric, accuracy is focused strongly on the effective performance of the model in a real setting: it tells you if the model works as expected. However, suppose your purpose is to evaluate, compare, and have a clear picture of how effective your approach really is. In that case, you must be cautious when using the accuracy because it can lead to wrong conclusions when the classes are imbalanced (when they have different frequencies). For instance, if a certain class makes up just 10% of the data, a predictor that predicts nothing but the majority class will be 90% accurate, proving itself quite useless in spite of the high accuracy.

How can you spot such a problem? You can do this easily by using a **confusion matrix**. In a confusion matrix, you create a two-way table comparing the actual classes on the rows against the predicted classes on the columns. You can create a straightforward one using the scikit-learn confusion_matrix function:

```
sklearn.metrics.confusion_matrix(
    y_true, y_pred, *, labels=None,
    sample_weight=None, normalize=None
)
```

Providing the y_true and y_pred vectors will suffice to return a meaningful table. Still, you can also provide row/column labels and sample weights for the considered examples and normalize (set the marginals to sum to 1) over the true examples (the rows), the predicted examples (the columns), or all the examples. A perfect classifier will have all the cases on the principal diagonal of the matrix. Serious problems with the validity of the predictor are highlighted if there are few or no cases on one of the diagonal cells.

To give you a better idea of how it works, you can try the graphical example offered by scikit-learn at `https://scikit-learn.org/stable/auto_examples/model_selection/plot_confusion_matrix.html#sphx-glr-auto-examples-model-selection-plot-confusion-matrix-py`, shown in the following figure:

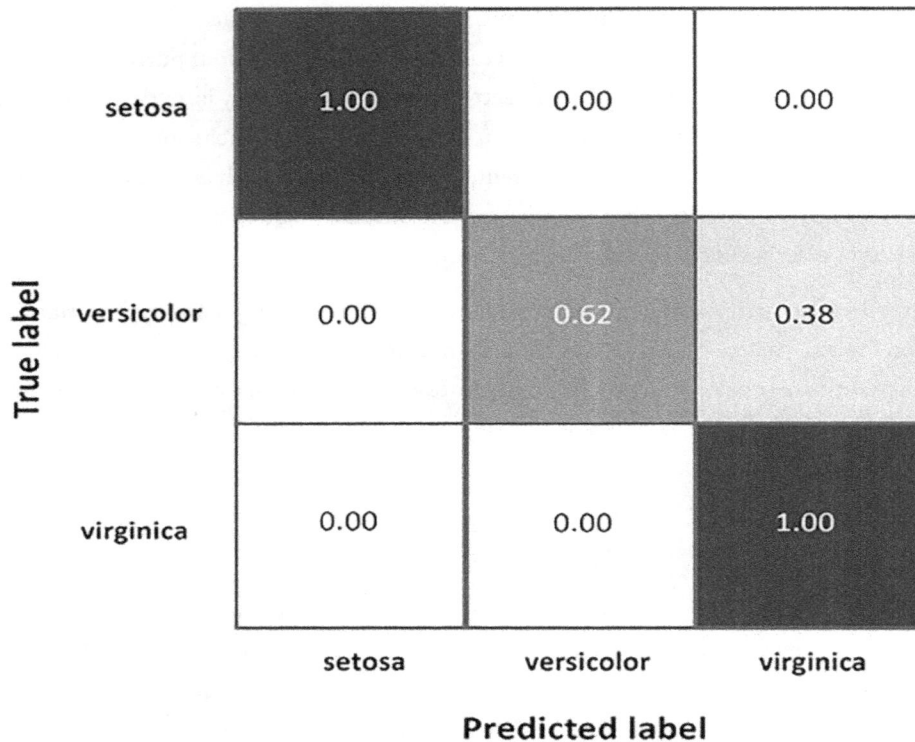

Figure 6.1: Confusion matrix, with each cell normalized to 1.00, to represent the share of matches

In the preceding confusion matrix, a 3x3 grid, you examine the intersections generated by the rows showing the actual flower type (true class) and the columns representing the predicted labels. Since the intersections are expressed as percentages of the dataset cases, look on the diagonal for high or low values, indicating how certain classes are predicted better or worse than others. The off-diagonal values indicate misclassified cases, providing insight into which species were confused with each other.

You can attempt to improve the usability of the accuracy score by considering the accuracy relative to each of the classes and averaging them, but you will find it more useful to rely on other metrics such as **precision**, **recall**, and the **F1 score**.

Precision and recall

To obtain the precision and recall metrics, we start from the confusion matrix again. First, we have to name each of the cells:

		Predicted	
		Negative	Positive
Actual	Negative	TNs (true negatives)	FPs (false positives)
	Positive	FNs (false negatives)	TPs (true positives)

Table 6.2: Confusion matrix with cell names

Here is how we define the cells:

- **TPs**: These are located in the lower-right cell, containing examples that have correctly been predicted as positive ones
- **FPs**: These are located in the upper-right cell, containing examples that have been predicted as positive but are actually negative
- **FNs**: These are located in the lower-left cell, containing examples that have been predicted as negative but are actually positive
- **TNs**: These are located in the upper-left cell, containing examples that have been correctly predicted as negative ones

Using these cells, you can get more precise information about how your classifier works and how you can tune your model better. First, we can easily revise the accuracy formula:

$$Accuracy = \frac{(TP + TN)}{(TP + TN + FP + FN)}$$

Then, the first informative metric is called **precision** (or **specificity**) and it is actually the accuracy of the positive cases:

$$Precision = \frac{TP}{TP + FP}$$

In the computation, only the numbers of TPs and FPs are involved. In essence, the metric tells you how often you are correct when you predict a positive.

Clearly, your model could get high scores by predicting positives for only the examples in which it has high confidence. That is actually the measure's purpose: to force models to predict a positive class only when they are sure and it is safe to do so.

However, if it is in your interest also to predict as many positives as possible, then you'll also need to watch over the **recall** (or **coverage**, **sensitivity**, or even **TP rate**) metric:

$$Recall = \frac{TP}{TP + FN}$$

Here, you will also need to know about FNs. The interesting thing about these two metrics is that since they are based on example classification and a classification is actually based on probability (which is usually set between the positive and negative class at the 0.5 threshold), you can change the threshold and have one of the two metrics be improved at the expense of the other.

For instance, if you increase the threshold, you will get more precision (the classifier is more confident of the prediction) but less recall. If you decrease the threshold, you get less precision but more recall. This is also called the **precision/recall trade-off**.

The scikit-learn website offers a simple and practical overview of this trade-off (`https://scikit-learn.org/stable/auto_examples/model_selection/plot_precision_recall.html`), helping you to trace a **precision/recall curve** and thus understand how these two measures can be exchanged to obtain a result that better fits your needs:

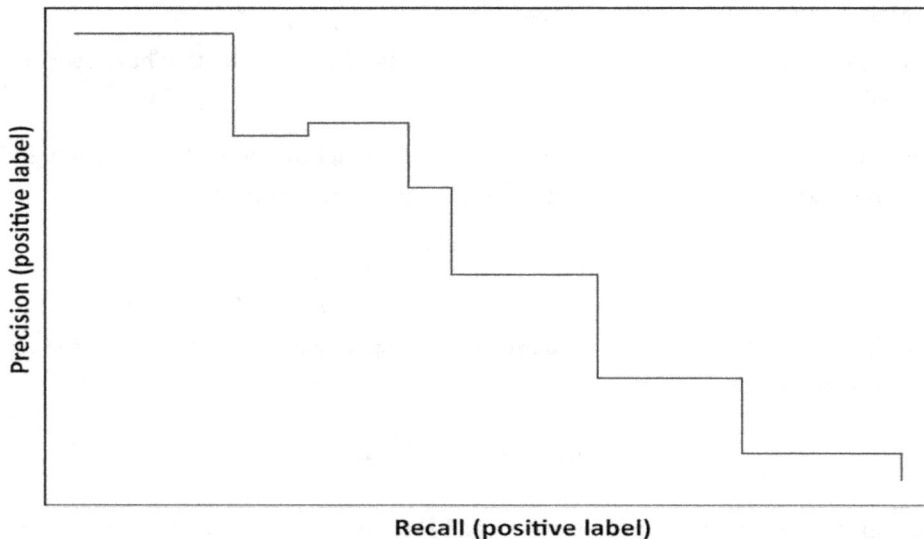

Figure 6.2: A two-class precision-recall curve with its characteristic steps

One metric associated with the precision/recall trade-off is **average precision**. Average precision computes the mean precision for recall values from 0 to 1 (basically, as you vary the threshold from 1 to 0). Average precision is very popular for tasks related to object detection, which we will discuss a bit later on, but it is also very useful for classification in tabular data. In practice, it proves valuable when you want to monitor model performance on a very rare class (when the data is extremely imbalanced) in a more exact way, which is often the case with fraud detection problems.

For more specific insights on this, read *Gael Varoquaux's* discussion at http://gael-varoquaux. info/interpreting_ml_tuto/content/01_how_well/01_metrics.html#average-precision.

The F1 score

At this point, you have probably already figured out that using precision or recall as an evaluation metric is not an ideal choice because you can only optimize one at the expense of the other. For this reason, there are no Kaggle competitions that use only one of the two metrics. You should combine them (as in the average precision). A single metric, the **F1 score**, which is the harmonic mean of precision and recall, is commonly considered to be the best solution:

$$F1 = 2 * \frac{precision * recall}{precision + recall}$$

If you get a high **F1** score, it is because your model has improved in precision, recall, or both. You can find a fine example of the usage of this metric in the *Quora Insincere Questions Classification* competition (https://www.kaggle.com/c/quora-insincere-questions-classification).

In some competitions, you also get the **F-beta** score. This is simply the weighted harmonic mean between precision and recall, and the beta value decides the weight of the precision and the recall in the combined score:

$$F_\beta = \frac{(1 + \beta^2) * (precision * recall)}{(\beta^2 * precision + recall)}$$

A beta value over 1 gives more weight to recall, while a beta value of less than 1 gives more importance to precision. The value of the beta value determines the optimization strategies you should take during a competition.

Since we have already introduced the concept of threshold and classification probability, we can now discuss the log loss and **receiver operating characteristic** (**ROC**)-AUC, both of which are quite common classification metrics.

Log loss

Let's proceed with **log loss**, also known as **cross-entropy** in deep learning models. The log loss is the difference between the predicted probability and the ground truth probability:

$$LogLoss = -\frac{1}{n}\sum_{i=1}^{n}[y_i \log(\hat{y}_i) + (1 - y_i)\log(1 - \hat{y}_i)]$$

In the preceding formula, n stands for the number of examples, y_i is the ground truth for the ith case, and $\hat{y}_i$ is the prediction.

If a competition uses log loss, it is implied that the objective is to estimate the probability of an example being of a positive class as correctly as possible. You can actually find log loss in quite a lot of competitions. In fact, apart from its intuitiveness, since a lower log loss value is better, it works well with unbalanced distributions without being skewed by the majority class dominance, always providing a meaningful evaluation of the model's performance. On the other hand, it is sensitive to outliers and does not directly relate to accuracy, as a decrease in log loss does not automatically translate into an increase in the model's accuracy.

We suggest you have a look, for instance, at the *Deepfake Detection Challenge* (https://www.kaggle.com/c/deepfake-detection-challenge) or at the older *Quora Question Pairs* (https://www.kaggle.com/c/quora-question-pairs).

ROC-AUC

The **ROC curve** is a graphical chart used to evaluate a binary classifier's performance and compare multiple classifiers. It is the building block of the ROC-AUC metric because it is simply the area delimited under the ROC curve. The ROC curve consists of the TP rate (the recall), which is plotted by changing the classification threshold against the FP rate (the ratio of negative instances that are incorrectly classified as positive ones). The FP rate is equivalent to one minus the TN rate (the ratio of negative examples that are correctly classified). Here are a few examples:

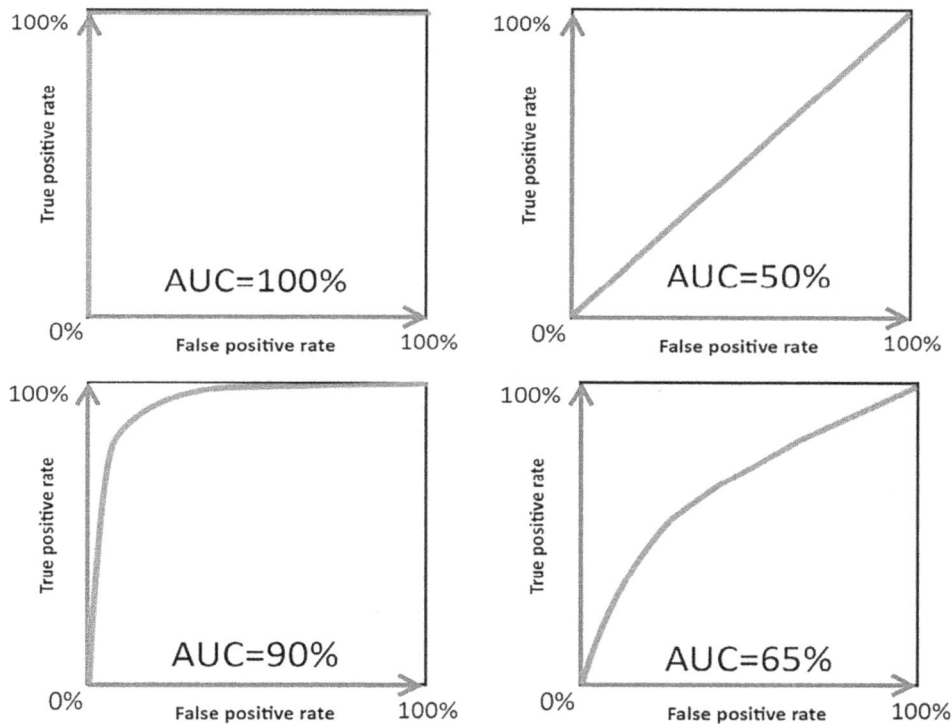

Figure 6.3: Different ROC curves and their AUCs

Ideally, the ROC curve of a well-performing classifier should quickly climb up the TP rate (recall) at low values of the FP rate. A ROC-AUC value between 0.9 and 1.0 is considered very good.

A bad classifier can be spotted by the ROC curve appearing very similar, if not identical, to the diagonal of the chart, which represents the performance of a purely random classifier, as in the top right of the preceding figure. ROC-AUC scores near 0.5 are considered to be almost random results. If you are comparing different classifiers, and you are using the **AUC**, the classifier with the higher AUC score is the more performant one.

If the classes are balanced, or not too imbalanced, increases in the AUC are proportional to the effectiveness of the trained model. They can be intuitively thought of as the ability of the model to output higher probabilities for TPs. We also think of it as the ability to order the examples more properly from positive to negative. However, when the positive class is rare, the AUC starts high and its increments may mean very little in terms of predicting the rare class better. As we mentioned before, in such a case, average precision is a more helpful metric.

AUC has been used for quite a lot of different competitions. We suggest that you have a look at these three:

- *IEEE-CIS Fraud Detection*: https://www.kaggle.com/c/ieee-fraud-detection
- *Riiid Answer Correctness Prediction*: https://www.kaggle.com/c/riiid-test-answer-prediction
- *Jigsaw Multilingual Toxic Comment Classification*: https://www.kaggle.com/c/jigsaw-multilingual-toxic-comment-classification/

You can read a detailed treatise in the paper *A relationship between the average precision and the area under the ROC curve* (https://dl.acm.org/doi/abs/10.1145/2808194.2809481).

Matthews correlation coefficient (MCC)

We complete our overview of binary classification metrics with the **Matthews correlation coefficient (MCC)**, which made its appearance in *VSB Power Line Fault Detection* (https://www.kaggle.com/c/vsb-power-line-fault-detection) and *Bosch Production Line Performance* (https://www.kaggle.com/c/bosch-production-line-performance). The MCC is useful for assessing the effectiveness of binary and multiclass classifications because it considers both correct and incorrect positive and negative predictions. Also, it's considered a well-balanced measure, suitable for scenarios where classes vary significantly in size. Its behavior is analogous to the correlation coefficient in regression problems.

The formula for the MCC is as follows:

$$MCC = \frac{(TP * TN) - (FP * FN)}{\sqrt{(TP + FP) * (TP + FN) * (TN + FP) * (TN + FN)}}$$

The preceding formula uses the same nomenclature as we saw when discussing precision and recall.

Since it behaves as a correlation coefficient, in other words, ranging from +1 (perfect prediction) to -1 (inverse prediction), this metric can be considered a measure of the quality of the classification even when the classes are quite imbalanced.

Despite its complexity, the formula can be rewritten and simplified, as demonstrated by Neuron Engineer (https://www.kaggle.com/ratthachat) in his Notebook:

www.kaggle.com/ratthachat/demythifying-matthew-correlation-coefficients-mcc

The work done by Neuron Engineer in understanding the ratio of the evaluation metric is indeed exemplary. In fact, his reformulated MCC becomes the following:

$$MCC = \left(Pos_{precision} + Neg_{precision} - 1\right) * PosNegRatio$$

Each element of the formula is as follows:

$$Pos_{precision} = \frac{TP}{TP + FP}$$

$$Neg_{precision} = \frac{TN}{TN + FN}$$

$$PosNegRatio = \sqrt{\frac{PosPredictionCount * NegPredictionCount}{PosLabelCount * NegLabelCount}}$$

$$PosPredictionCount = TP + FP$$

$$NegPredictionCount = TN + FN$$

The reformulation helps to clarify, in a more intelligible form than the original, that you can get higher performance from improving both positive and negative class precision, but that's not enough. You also have to have positive and negative predictions in proportion to the ground truth, or your submission will be greatly penalized.

Metrics for multi-class classification

When moving to multi-class classification, you simply use the binary classification metrics that we have just seen, applied to each class. Then, you summarize them using some of the averaging strategies that are commonly used for multi-class situations.

For instance, if you want to evaluate your solution based on the *F1* score, you have three possible averaging choices:

- **Macro averaging**: Simply calculate the *F1* score for each class and then average all the results. In this way, each class will count as much as the others, no matter how frequent its positive cases are or how important they are for your problem, resulting in equal penalizations when the model doesn't perform well with any class:

$$macro = \frac{F1_{class1} + F1_{class2} + \cdots + F1_{classN}}{N}$$

- **Micro averaging**: This approach will sum all the contributions from each class to compute an aggregated *F1* score. It results in no particular favor to or penalization of any class, since all the computations are made regardless of each class, so it can more accurately account for class imbalances:

$$micro = F1_{class1+class2+\cdots classN}$$

- **Weighting**: As with macro averaging, you first calculate the *F1* score for each class, but then you make a weighted average mean of all of them using a weight that depends on the number of true labels of each class. By using such a set of weights, you can take the frequency of positive cases from each class or the relevance of that class into account for your problem. This approach clearly favors the majority classes, which will be weighted more in the computations:

$$weighted = F1_{class1} * W_1 + F1_{class2} * W_2 + \cdots + F1_{classN} * W_n$$

$$W_1 + W_2 + \cdots + W_N = 1.0$$

Common multi-class metrics that you may encounter in Kaggle competitions are as follows:

- **Multiclass accuracy (weighted)**: See *Bengali.AI Handwritten Grapheme Classification* (`https://www.kaggle.com/c/bengaliai-cv19`). The final weighted accuracy is calculated by considering the predictions for each of its components separately and then combining them with specific weights.

- **Multiclass log loss (MeanColumnwiseLogLoss)**: See *Mechanisms of Action (MoA) Prediction* (`https://www.kaggle.com/c/lish-moa/`). The log loss is calculated for each class (or column) separately and then averaged.

- **Macro-F1** and **Micro-F1 (NQMicroF1)**: See the following:

  - *University of Liverpool – Ion Switching* (`https://www.kaggle.com/c/liverpool-ion-switching`)

  - *Human Protein Atlas Image Classification* (`https://www.kaggle.com/c/human-protein-atlas-image-classification/`)

  - *TensorFlow 2.0 Question Answering* (`https://www.kaggle.com/c/tensorflow2-question-answering`)

Macro-F1 is the arithmetic mean of the F1 scores of each class. It treats all classes equally, regardless of their frequency, and it helps to avoid biases introduced by imbalanced datasets. Micro-F1 (also referred to as NQMicroF1 in some contexts) aggregates the contributions of all classes to compute the average metric. It gives equal weight to each instance, which is helpful when the overall classification accuracy is the primary concern, but it is also influenced by the most frequent classes in imbalanced datasets.

- **Mean-F1**: See *Shopee – Price Match Guarantee* (`https://www.kaggle.com/c/shopee-product-matching/`). Here, the *F1* score is calculated for every predicted row and then averaged, whereas the Macro-F1 score is defined as the mean of class-wise/label-wise *F1* scores.

Then there is **Quadratic Weighted Kappa**, which is also a smart evaluation metric for ordinal prediction problems. Quadratic Weighted Kappa builds upon the idea of Cohen's Kappa, which is a simpler measure of agreement for categorical data. The **Cohen Kappa** score just measures the agreement between your predictions and the ground truth. The metric was created to measure **inter-annotation agreement**, but it is versatile and has found even better uses.

What is an inter-annotation agreement? Let's imagine that you have a labeling task: classifying some photos based on whether they contain an image of a cat, a dog, or neither. If you ask a set of people to do the task for you, you may incur erroneous labels because someone (called the *judge* in this kind of task) may misinterpret a dog as a cat or vice versa. The smart way to do this job correctly is to divide the work among multiple judges, have them label the same photos, and then measure their level of agreement based on the Cohen Kappa score.

Therefore, the Cohen Kappa is devised as a score expressing the level of agreement between two annotators on a labeling (classification) problem:

$$k = (p_0 - p_e)/(1 - p_e)$$

In the formula, p_0 is the relative observed agreement among raters, and p_e is the hypothetical probability of chance agreement. Using the confusion matrix nomenclature, this can be rewritten as follows:

$$k = \frac{2 * (TP * TN - FN * FP)}{(TP + FP) * (FP + TN) + (TP + FN) * (FN + TN)}$$

The interesting aspect of this formula is that the score takes the empirical probability that the agreement has happened just by chance into account. Hence, the measure has a correction for all the most probable classifications. The metric ranges from 1, meaning complete agreement, to -1, meaning the judges completely oppose each other (total disagreement).

Values around 0 signify that agreement and disagreement among the judges is happening by mere chance. This helps you determine if the model performs better than chance in most situations.

Now let's move on to our second interview of the chapter, which is with *Andrey Lukyanenko*, a Notebooks and Discussions Grandmaster and Competitions Master. In his day job, he is a Machine Learning Engineer at Meta. He had many interesting things to say about his Kaggle experiences!

Interview: Andrey Lukyanenko

`https://www.kaggle.com/artgor`

What's your favorite kind of competition and why? In terms of techniques and solving approaches, what is your specialty on Kaggle?

I prefer competitions where solutions can be general enough to be transferable to other datasets/ domains. I'm interested in trying various neural net architectures, state-of-the-art approaches, and post-processing tricks. I don't favor those competitions that require reverse engineering or creating some "golden features," as these approaches won't be applicable to other datasets.

While you were competing on Kaggle, you also became a Grandmaster in Notebooks (and ranked number one) and Discussions. Have you invested in these two objectives?

I was very motivated by this success and continued writing Notebooks. At first, I simply wanted to share my analysis and get feedback because I wanted to try to compare my analytics and visualization skills with other people to see what I could do and what people thought of it. People started liking my kernels and I wanted to improve my skills even further. Another motivation was a desire to improve my skills at making a quick **minimum viable product (MVP)**. When a new competition starts, many people begin writing Notebooks, and if you want to be one of the first, you have to be able to do it fast without sacrificing quality. This is challenging but fun and rewarding.

I was able to get the Notebook Grandmaster rank in February of 2019; after some time, I reached first place and held it for more than a year. Now I write Notebooks less frequently, but I still enjoy doing it.

As for discussions, I think it kind of happened on its own. I answered the comments about my Notebooks and shared and discussed ideas about competitions in which I took part, and my discussion ranking steadily increased.

Tell us about a particularly challenging competition you entered, and what insights you used to tackle the task.

It was the Predicting Molecular Properties competition. I have written a blog post about it in more detail (`https://medium.com/data-science/a-story-of-my-first-gold-medal-in-one-kaggle-competition-things-done-and-lessons-learned-c269d9c233d1`). It was a domain-specific competition aimed at predicting interactions between atoms in molecules. **Nuclear Magnetic Resonance (NMR)** is a technology that uses similar principles to MRI to understand the structure and dynamics of proteins and molecules. Researchers around the world conduct NMR experiments to further understand the structure and dynamics of molecules, across areas like environmental science, pharmaceutical science, and materials science. In this competition, we tried to predict the magnetic interaction between two atoms in a molecule (the scalar coupling constant). State-of-the-art methods from quantum mechanics can calculate these coupling constants given only a 3D molecular structure as input. However, these calculations are very resource-intensive, so they can't be always used. If ML approaches could predict these values, it would really help medicinal chemists to gain structural insights faster and more cheaply.

I usually write EDA kernels for new Kaggle competitions, and this one was no exception. A common approach for tabular data in Kaggle competitions is extensive feature engineering and using gradient-boosting models. I used LGBM too in my early attempts, but I knew that there should be better ways to work with graphs. I realized that domain expertise would provide a serious advantage, so I hunted for every piece of such information. Of course, I noticed that there were several active experts who wrote on the forum and created kernels, so I read everything they wrote.

One day, I received an e-mail from an expert in this domain who thought that our skills could complement each other. Usually, I prefer to work on competitions by myself for some time, but in this case, combining forces seemed to be a good idea to me. This decision turned out to be a great one! With time, we were able to gather an amazing team.

After some time, we noticed a potential for neural nets in the competition: a well-known Kaggler, Heng, posted an example of a **Message Passing Neural Network (MPNN)** model. After some time, I was even able to run it, but the results were worse compared to our models. Nevertheless, our team knew that we would need to work with these neural nets if we wanted to aim high. It was amazing to see how Christof was able to build new neural nets extremely fast. Soon, we focused only on developing those models.

After that, my role switched to a support one. I did a lot of experiments with our neural nets: trying various hyperparameters, different architectures, little tweaks to training schedules, and so on. Sometimes I did EDA on our predictions to find our interesting or wrong cases, and later we used this information to improve our models even further.

We got 8th place and I learned a lot during this competition.

Has Kaggle helped you in your career? If so, how?

Kaggle has definitely helped me a lot, especially with my skills and my personal brand. Writing and publishing Kaggle Notebooks not only taught me EDA and ML skills but also forced me to become adaptable, to be able to understand new topics and tasks quickly, and to iterate more efficiently between approaches. At the same time, it provided a measure of visibility for me, because people appreciated my work.

My first portfolio (`https://erlemar.github.io/`) had a lot of different Notebooks, and half of them were based on old Kaggle competitions. It was definitely helpful in getting my first few jobs. My Kaggle achievements also helped me attract recruiters from good companies, sometimes even skipping steps of the interview process, and even led me to several consulting gigs.

In your experience, what do inexperienced Kagglers often overlook? What do you know now that you wish you'd known when you first started?

I think we need to separate inexperienced Kagglers into two groups: those who are inexperienced in data science in general and those who are inexperienced on Kaggle.

Those who are inexperienced in general make a number of different mistakes (and it is okay, everyone started somewhere):

- One of the most serious problems is lack of critical thinking and not knowing how to do their own research
- Not knowing when and what tools/approaches to use
- Blindly taking public Notebooks and using them without understanding how they work
- Fixating on a certain idea and spending too much time pursuing it, even when it doesn't work
- Despairing and losing motivation when their experiments fail

As for those people who have experience in data science but don't have experience with Kaggle, I'd say that the most serious thing they overlook is that they underestimate Kaggle's difficulty. They don't expect Kaggle to be very competitive, that you need to try many different things to succeed, that there are a lot of tricks that work only in competitions, or that there are people who professionally participate in competitions.

Also, people often overestimate domain expertise. I admit that there have been a number of competitions when the teams with domain experts in them won gold medals and prizes, but in most cases, experienced Kagglers triumph.

Also, I have seen the following situation many times: some person proclaims that winning Kaggle is easy and that they (or their group of people) will get a gold medal or many gold medals in the near future. In most cases, they silently fail.

What mistakes have you made in competitions in the past?

Here are some of the main ones:

- Not spending enough time looking at the data. Sometimes, I wasn't able to generate better features or apply better postprocessing due to this. Reserve engineering and "golden features" is a whole additional topic.
- Spending too much time on a single idea because I hoped it would work is another major one. This is called the sunk-cost fallacy.
- Not enough experiments are also a problem. The effort pays off – if you don't spend enough time and resources on the competition, you won't get a high place on a leaderboard.
- Entering "wrong" competitions is also a big one. There were competitions with leaks, reverse engineering, and so on. There were competitions with an unreasonable split between public and private test data and a shake-up ensued. There were competitions that weren't interesting enough for me and I shouldn't have started participating in them.
- Lastly, there's teaming up with the wrong people. There were cases when my teammates weren't as active as I expected them to be and it led to a worse team score.

What's the most important thing someone should keep in mind or do when they're entering a competition?

I think it is important to remember your goal, know what are you ready to invest into this competition, and think about the possible outcomes. There are many possible goals that people have while entering a competition, including the following:

- Winning money or getting a medal
- Getting new skills or improving existing ones
- Working with a new task/domain
- Networking
- PR

Of course, it is possible to have multiple motivations.

As for what are you ready to invest, it is usually about the amount of time and effort you are ready to spend as well as the hardware that you have.

When I speak about the outcomes, I mean what will happen when the competition ends. It is possible that you will invest a lot into this competition and win, but you could also lose. Are you ready for this reality? Is winning a particular competition critical to you? Maybe you need to be prepared to invest more effort; on the other hand, maybe you have long-term goals and one failed competition won't hurt much.

Metrics for object detection and segmentation problems

In recent years, deep learning competitions have become more and more common on Kaggle. Most of these competitions, focused on image recognition or natural language processing tasks, have not required evaluation metrics that are much different from the ones we have explored. However, a couple of specific problems have required some special metric to be evaluated correctly: those relating to **object detection** and **segmentation**.

In **object detection**, you don't have to classify an image, but instead find relevant portions of a picture and label them accordingly. For instance, in *Figure 6.4*, an object detection classifier has been entrusted to locate the portions of the picture where either dogs or cats are present within a photo and classify each of them with a proper label.

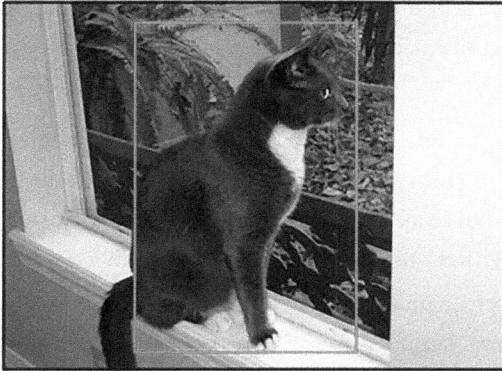

Classification + localization (cat)   Object detection (dog, cat)

Figure 6.4: Computer vision tasks (source: `https://cocodataset.` `org/#explore?id=38282,` `https://cocodataset.org/#explore?id=68717)`

The image on the left illustrates how a box can be used to isolate a cat. The image on the right shows multiple cats and dogs being detected, each enclosed in its own bounding box and correctly classified. (called a **bounding box**). The image on the right shows multiple cats and dogs being detected, each enclosed in its own bounding box and correctly classified.

In order to describe the spatial location of an object, in object detection we use **bounding boxes**, which define a rectangular area in which the object lies. A bounding box is usually specified using two (x, y) coordinates: the upper-left and lower-right corners. In terms of a ML algorithm, finding the coordinates of bounding boxes corresponds to applying a regression problem to multiple targets. However, you probably won't frame the problem from scratch and will instead rely on pre-built and often pre-trained models such as the following:

- Mask R-CNN: `https://arxiv.org/abs/1703.06870`
- RetinaNet: `https://arxiv.org/abs/2106.05624v1`
- FPN (Feature Pyramid Networks for Object Detection): `https://arxiv.org/abs/1612.03144`
- YOLO (You Only Look Once) architectures: `https://arxiv.org/abs/2304.00501`
- Faster R-CNN: `https://arxiv.org/abs/1506.01497v1`
- SSD (Single Shot Detection): `https://arxiv.org/abs/1512.02325`

In **segmentation**, you instead have a classification at the *pixel* level, so if you have a 320x200 image, you actually have to make 64,000 pixel classifications. Depending on the task, you can have either of the following:

- **Semantic segmentation:** For each pixel in a given image, the model identifies the object category to which it belongs based on a previously defined set of labels. For example, semantic segmentation assigns the same color to all instances of the same category, such as marking all the cats in a photo in purple without distinguishing between individual cats.

- **Instance segmentation:** For each pixel in a given image, the model distinguishes between different units of the same predefined class. Unlike semantic segmentation, instance segmentation distinguishes between individual objects within the same category. For example, it differentiates between different cats by assigning each cat a unique color.

The following figure shows the examples of detecting cats in a photo:

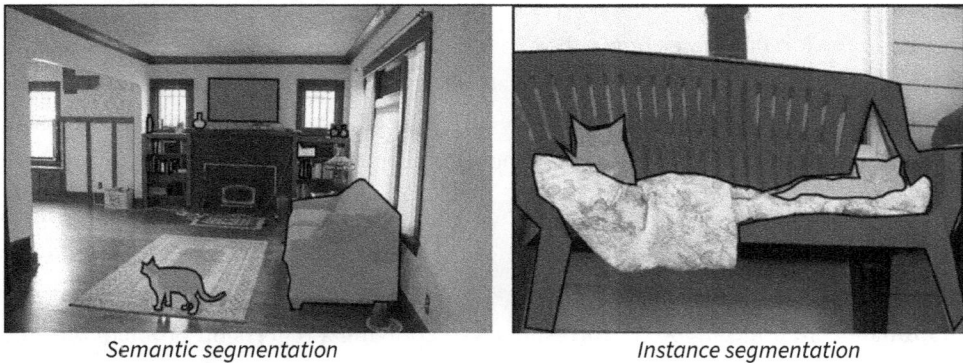

Semantic segmentation   Instance segmentation

Figure 6.5: Semantic segmentation focuses on distinguishing specific types of objects in the room and instance segmentation focuses on distinguishing each individual cat in the photo (sources: https://cocodataset.org/#explore?id=503294 *and* https://cocodataset.org/#explore?id=308739)

Let's start with an overview of the specific metrics for these tasks. These metrics can work well for both problems since, in both cases, you are predicting entire areas (rectangular ones in object detection, polygonal ones in segmentation) of a picture. You have to compare your predictions against a ground truth, which is expressed as areas.

On the side of segmentation, the easiest metric is **pixel accuracy**, which, as the name suggests, is the accuracy of the pixel classification.

It is not a great metric because, as happens with accuracy on binary and multi-class problems, your score may look great if the relevant pixels do not take up very much of the image (you just predict the majority claim; thus, you don't segment). Therefore, two metrics are used much more, especially in competitions: the **intersection over union** (IoU) and the **dice coefficient**. Let's discuss these next.

IoU

The **IoU** is also known as the **Jaccard index**. When used in segmentation problems, using the IoU implies that you have two images to compare: one is your prediction, and the other is the mask revealing the ground truth, which is usually a binary matrix where the value 1 stands for the ground truth and 0 is otherwise used. In the case of multiple objects, you have multiple masks, each labeled with the object's class.

When used in object detection problems, you have the boundaries of two rectangular areas (those of the prediction and the ground truth), expressed by the coordinates of their vertices. For each classified class, you compute the area of overlap between your prediction and the ground truth mask. Then, you divide this by the area of the union between your prediction and the ground truth, a sum that considers any overlap. In this way, you are proportionally penalized both if you predict a larger area than what it should be (the denominator will be larger) and if you predict a smaller one (the numerator will be smaller):

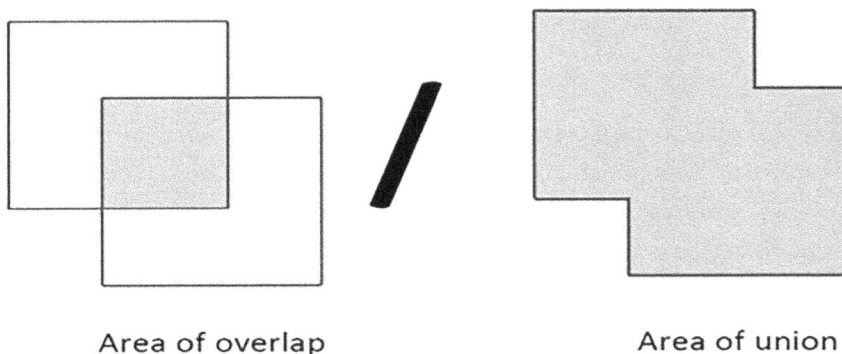

Area of overlap / Area of union

Figure 6.6: Visual representation of the IoU calculation

In *Figure 6.6*, you can see a visual representation of the areas involved in the computation. By imagining the squares overlapping more, you can figure out how the metric effectively penalizes your solution when your prediction, even if covering the ground truth, exceeds it (the area of union becomes larger).

Here are some examples of competitions where IoU has been used:

- *TGS Salt Identification Challenge* (`https://www.kaggle.com/c/tgs-salt-identification-challenge/`) with IoU object segmentation
- *iMaterialist (Fashion) 2019 at FGVC6* (`https://www.kaggle.com/c/imaterialist-fashion-2019-FGVC6`) with IoU object segmentation with classification
- *Airbus Ship Detection Challenge* (`https://www.kaggle.com/c/airbus-ship-detection`) with IoU object segmentation beta

Dice

The other useful metric is the **Dice coefficient**, commonly used in both object detection and segmentation problems, which is the area of overlap between the prediction and ground truth, doubled and then divided by the sum of the prediction and ground truth areas:

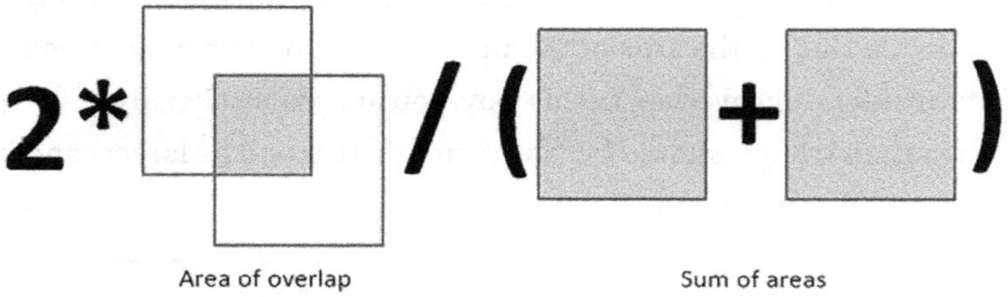

Figure 6.7: Visual representation of the Dice calculation

In this case, with respect to the Jaccard index, you do not take the overlap of the prediction with the ground truth in the denominator into account. Here, the expectation is that, as you maximize the area of overlap, you predict the correct area size. Again, you are penalized if you predict areas larger than you should be predicting. In fact, the two metrics are positively correlated, and they produce almost the same results for a single classification problem.

The differences actually arise when you are working with *multiple classes*. In fact, with both the IoU and the Dice coefficient, when you have multiple classes, you average the result of all of them. However, in doing so, the IoU metric penalizes the overall average more if a single class prediction is wrong. In contrast, the Dice coefficient is more lenient and tends to represent the average performance.

Examples of Kaggle competitions using the Dice coefficient (it is often encountered in competitions with medical purposes, but not necessarily only there because it can also be used for clouds and cars) include the following:

- *HuBMAP – Hacking the Kidney*: https://www.kaggle.com/c/hubmap-kidney-segmentation
- *Ultrasound Nerve Segmentation*: https://www.kaggle.com/c/ultrasound-nerve-segmentation
- *Understanding Clouds from Satellite Images*: https://www.kaggle.com/c/understanding_cloud_organization
- *Carvana Image Masking Challenge*: https://www.kaggle.com/c/carvana-image-masking-challenge

The IoU and Dice constitute the basis for all the more complex metrics in segmentation and object detection. By choosing an appropriate threshold level for the IoU or Dice (usually 0.5), you can decide whether or not to confirm a detection, and therefore a classification. At this point, you can use previously discussed metrics for classification, such as precision, recall, and *F1*, as is done in popular object detection and segmentation challenges such as Pascal VOC (http://host.robots.ox.ac.uk/pascal/VOC/voc2012) or COCO (https://cocodataset.org).

Metrics for multi-label classification and recommendation problems

Recommender systems are one of the most popular applications of data analysis and ML, and there are quite a few competitions on Kaggle that have used the recommendation approach. For instance, *Quick, Draw! Doodle Recognition Challenge* was a prediction evaluated as a recommender system. Some other competitions on Kaggle, however, truly strived to build effective recommender systems, such as *Expedia Hotel Recommendations* (https://www.kaggle.com/c/expedia-hotel-recommendations). RecSYS, the conference on recommender systems (https://recsys.acm.org/), even hosted one of its yearly contests on Kaggle: *RecSYS 2013* (https://www.kaggle.com/c/yelp-recsys-2013).

Mean average precision at k (MAP@K) is typically the metric of choice for evaluating the performance of recommender systems, and it is the most common metric you will encounter on Kaggle in all the competitions that try to build or approach a problem as a recommender system.

There are also some other metrics, such as **precision at k (P@K)** and **average precision at k (AP@K)**, which are loss functions. In other words, they're computed at the level of every single prediction. Understanding how they work can help you better understand MAP@K and how it can perform both in recommendations and in multi-label classification.

In fact, analogous to recommender systems, multi-label classifications imply that your model outputs a series of class predictions. Such results could be evaluated using some average of some binary classification metrics (such as in *Greek Media Monitoring Multilabel Classification* (*WISE 2014*), which used the mean *F1* score (`https://www.kaggle.com/c/wise-2014`)), as well as metrics that are more typical of recommender systems, such as MAP@K. In the end, you can deal with both recommendations and multi-label predictions as *ranking tasks*, which translates into a set of ranked suggestions in a recommender system and into a set of labels (without a precise order) in multi-label classification.

MAP@K is a complex metric and it derives from many computations. In order to understand the MAP@K metric fully, let's start with its simplest component, the **precision at k (P@K)**. In this case, since the prediction for an example is a ranked sequence of predictions (from the most probable to the least), the function only takes the top *k* predictions into account, then computes how many matches it got with respect to the ground truth and divides that number by *k*. In a few words, it is quite similar to an accuracy measure averaged over *k* predictions.

A bit more complex in terms of computation, but conceptually simple, the **AP@K** is the average of precision at *k* computed over all the values ranging from *1* to *k*. In this way, the metric evaluates how well the prediction works overall, using the top prediction, then the top two predictions, and so on until the top *k* predictions.

Finally, the **MAP@K** is calculated as the mean of the **AP@K** for all users in your dataset. MAP@K is a metric because it comprises all the predictions in its evaluation. Here is the MAP@5 formulation you can find in the *Expedia Hotel Recommendations* competition (`https://www.kaggle.com/c/expedia-hotel-recommendations`):

$$MAP@5 = \frac{1}{|U|} \sum_{u=1}^{|U|} \sum_{k=1}^{min(5,n)} P(k)$$

In the formula, $|U|$ is the number of user recommendations, $P(k)$ is the precision at cutoff *k*, and *n* is the number of predicted hotel clusters (you could predict up to five hotels for each recommendation).

It is clearly a bit more daunting than our explanation shows, but the formula just expresses that the MAP@K is the mean of all the AP@K evaluations over all the predictions.

Having completed this overview of specific metrics for different regression and classification metrics, let's discuss how to deal with evaluation metrics in a Kaggle competition.

Optimizing evaluation metrics

Summing up what we have discussed so far, an objective function is a function inside your learning algorithm that measures how well the model fits the provided data. The objective function also provides feedback to the algorithm for it to improve its fit across successive iterations. Clearly, since the entire algorithm's efforts are recruited to perform well based on the objective function, if the Kaggle evaluation metric perfectly matches the objective function of your algorithm, you will get the best results.

Unfortunately, this is not frequently the case. Often, the evaluation metric provided can only be approximated by existing objective functions. Getting a good approximation, or striving to get your predictions to perform better with respect to the evaluation criteria, is the secret to performing well in Kaggle competitions. When your objective function does not match your evaluation metric, you have a few alternatives:

- Modify your learning algorithm and have it incorporate an objective function that matches your evaluation metric, though this is not possible for all algorithms (for instance, algorithms such as LightGBM and XGBoost allow you to set custom objective functions, but most scikit-learn models don't allow this).

- Tune your model's hyperparameters, choosing the ones that make the result shine the most when using the evaluation metric.

- Post-process your results so they match the evaluation criteria more closely. For instance, you could code an optimizer that performs transformations on your predictions (probability calibration algorithms are an example; we will discuss them at the end of the chapter).

Incorporating the competition metric into your ML algorithm is the most effective method to achieve better predictions, though only a few algorithms can be hacked into using the competition metric as your objective function. The second approach is therefore the more common one, and many competitors end up in a struggle to get the best hyperparameters for their models to perform on the evaluation metric.

If you already have your evaluation function coded, then doing the right cross-validation or choosing the appropriate test set plays the lion's share. If you don't have the coded function at hand, you have to first code it in a suitable way, following the formulas provided by Kaggle.

Invariably, doing the following will make the difference:

- Looking for all the relevant information about the evaluation metric and its coded function on a search engine
- Browsing through the most common packages (such as scikit-learn: `https://scikit-learn.org/stable/modules/model_evaluation.html#model-evaluation` or TensorFlow: `https://www.tensorflow.org/api_docs/python/tf/keras/losses`)
- Browsing GitHub projects (for instance, *Ben Hammer's* Metrics project: `https://github.com/benhamner/Metrics`)
- Asking or looking around in the forums and available Kaggle Notebooks (both for the current competition and for similar competitions)
- In addition, as we mentioned before, querying the Meta Kaggle dataset (`https://www.kaggle.com/kaggle/meta-kaggle`) and looking in the **Competitions** table will help you find out which other Kaggle competitions used that same evaluation metric, and immediately provides you with useful code and ideas to try out

Let's discuss the alternatives you have when your evaluation metric doesn't match your algorithm's objective function in greater detail. We'll start by exploring custom metrics.

Custom metrics and custom objective functions

As a first option when your objective function does not match your evaluation metric, we have learned that you can solve this by creating your own custom objective function, but only a few ML algorithms can easily be modified to incorporate a specific objective function. Creating your own custom objective function, according to the specifications of the evaluation metric of the competition, can help you score better on the leaderboard.

The good news is that the few algorithms that allow this are among the most effective ones in Kaggle competitions and data science projects. Of course, creating your own custom objective function may sound a little bit tricky, but it is an incredibly rewarding approach to increasing your score in a competition. For instance, there are options to do this when using gradient boosting algorithms such as XGBoost, CatBoost, and LightGBM, as well as with all deep learning models based on TensorFlow or PyTorch.

You can find great tutorials for custom metrics and objective functions in TensorFlow and PyTorch here:

- Creating custom metrics in Keras: `https://medium.com/@vishvaselvam2000/custom-metrics-in-keras-9ce7584644d0`
- More examples and use cases to help you create your own loss functions tailored to specific needs: `https://petamind.com/advanced-keras-custom-loss-functions/`
- Extending PyTorch's built-in loss functions to incorporate custom metrics: `https://kevinmusgrave.github.io/pytorch-metric-learning/extend/losses/`

These will provide you with the basic function templates and some useful suggestions about how to code a custom objective or evaluation function.

If you want just to get straight to the custom objective function you need, you can try this Notebook by RNA (`https://www.kaggle.com/bigironsphere`): `https://www.kaggle.com/bigironsphere/loss-function-library-keras-pytorch/notebook`

It contains a large range of custom loss functions for both TensorFlow and PyTorch that have appeared in different competitions.

Suppose you need to create a custom loss function in LightGBM, XGBoost, or CatBoost, as indicated in their respective documentation. In that case, you have to code a function that takes as inputs the prediction and the ground truth and returns, as outputs, the gradient and the Hessian.

You can consult this post on Stack Overflow for a better understanding of what gradients and Hessians are: `https://stats.stackexchange.com/questions/231220/how-to-compute-the-gradient-and-hessian-of-logarithmic-loss-question-is-based`.

From a code implementation perspective, all you have to do is create a function using closures if you need to pass more parameters beyond just the vector of predicted and true labels. Here is a simple example of a **focal loss** (a loss that aims to heavily weight the minority class in the loss computations as described by Lin, T-Y. et al. in their paper *Focal loss for dense object detection,* available at `https://arxiv.org/abs/1708.02002`) function that you can use as a model for your own custom functions:

```
from scipy.differentiate import derivative
import xgboost as xgb
def focal_loss(alpha, gamma):
    def loss_func(y_pred, y_true):
        a, g = alpha, gamma
```

```
        def get_loss(y_pred, y_true):
            p = 1 / (1 + np.exp(-y_pred))
            loss = (-(a * y_true + (1 - a)*(1 - y_true)) *
                    ((1 - (y_true * p + (1 - y_true) *
                    (1 - p)))**g) * (y_true * np.log(p) +
                    (1 - y_true) * np.log(1 - p)))
            return loss
        partial_focal = lambda y_pred: get_loss(y_pred, y_true)
        grad = derivative(partial_focal, y_pred, n=1, dx=1e-6)
        hess = derivative(partial_focal, y_pred, n=2, dx=1e-6)
        return grad, hess
    return loss_func
xgb = xgb.XGBClassifier(objective=focal_loss(alpha=0.25, gamma=1))
```

In the preceding code snippet, we have defined a new cost function, focal_loss, which is then fed into an XGBoost instance's object parameters. The example is worth showing because the focal loss requires the specification of some parameters in order to work properly on your problem (alpha and gamma). The more simplistic solution of having their values directly coded into the function is not ideal, since you may have to change them systematically as you are tuning your model. Instead, in the proposed function, when you input the parameters into the focal_loss function, they reside in memory and are referenced by the loss_func function that is returned to XGBoost. The returned cost function will therefore work, referring to the alpha and gamma values that you initially instantiated.

Another interesting aspect of the example is that it really makes it easy to compute the gradient and the Hessian of the cost function by means of the derivative function from SciPy. If your cost function is differentiable, you don't have to worry about doing any calculations by hand. However, creating a custom objective function requires some mathematical knowledge and quite a lot of effort to make sure it works properly for your purposes. You can read about the difficulties that *Max Halford* experienced while implementing a focal loss for the LightGBM algorithm, and how he overcame them, at https://maxhalford.github.io/blog/lightgbm-focal-loss/. Despite the difficulty, being able to conjure up a custom loss function can really determine your success in a Kaggle competition where you have to extract the maximum possible result from your model.

If building your own objective function isn't working out, you can simply lower your ambitions, give up building your function as an objective function used by the optimizer, and instead code it as a custom *evaluation metric*. Though your model won't be directly optimized to perform against this function, you can still improve its predictive performance with hyperparameter optimization based on it. This is the second option we talked about in the previous section.

Just remember: if you are writing a metric from scratch, sometimes you may need to abide by certain code conventions for your function to work properly. For instance, if you use scikit-learn, you have to convert your functions using the make_scorer function. The make_scorer function is actually a wrapper that makes your evaluation function suitable for working with the scikit-learn API. It will wrap your function while considering some meta information, such as whether to use probability estimates or predictions, whether you need to specify a threshold for prediction, and, last but not least, the directionality of the optimization, that is, whether you want to maximize or minimize the score it returns:

```
from sklearn.metrics import make_scorer
from sklearn.metrics import average_precision_score
scorer = make_scorer(average_precision_score,
average='weighted', greater_is_better=True, needs_proba=False)
```

In the preceding example, you prepare a scorer based on the average precision metric, specifying that it should use a weighted computation when dealing with multi-class classification problems.

If you are optimizing for your evaluation metric, you can apply grid search, random search, or some more sophisticated optimization such as Bayesian optimization, and find the set of parameters that makes your algorithm perform optimally for your evaluation metric, even if it works with a different cost function. We will explore how to best arrange parameter optimization and obtain the best results in Kaggle competitions after having discussed model validation, specifically in the chapter dealing with tabular data problems.

Post-processing your predictions

Post-processing tuning implies that your predictions are transformed, by means of a function, into something else in order to present a better evaluation. After building your custom loss or optimizing for your evaluation metric, you can also improve your results by leveraging the characteristics of your evaluation metric using a specific function applied to your predictions.

Let's take the Quadratic Weighted Kappa, for instance. We previously mentioned that this metric is useful when you have to deal with predicting an ordinal value. To recap, the original Kappa coefficient is a chance-adjusted index of agreement between the algorithm and the ground truth. It is a kind of accuracy measurement corrected by the probability that the match between the prediction and the ground truth is due to a fortunate chance.

Here is the original version of the Kappa coefficient, as seen before:

$$k = (p_0 - p_e)/(1 - p_e)$$

In the formula, p_0 is the relative observed agreement among raters, and p_e is the hypothetical probability of chance agreement. Here, you need just two matrices: the one with the observed scores and the one with the expected scores based on chance agreement. When the Kappa coefficient is weighted, you also consider a weight matrix, and the formula turns into this:

$$k = (p_0 - p_e)/(1 - p_p)$$

The matrix p_p contains the penalizations to weight errors differently, which is very useful for ordinal predictions since this matrix can penalize much more when the predictions deviate further from the ground truths. Using the quadratic form, that is, squaring the resulting k, makes the penalization even more severe. However, optimizing for such a metric is really not easy, since it is very difficult to implement it as a cost function. Post-processing can help you.

An example can be found in the *PetFinder.my Adoption Prediction* competition (`https://www.kaggle.com/c/petfinder-adoption-prediction`). In this competition, given that the results could have five possible ratings (0, 1, 2, 3, or 4), you could deal with them either using a classification or a regression. If you used a regression, a post-processing transformation of the regression output could improve the model's performance against the Quadratic Weighted Kappa metric, outperforming the results you could get from a classification directly outputting discrete predictions.

In the case of the PetFinder competition, the post-processing consisted of an optimization process that started by transforming the regression results into integers, first using the boundaries [0.5, 1.5, 2.5, 3.5] as thresholds and, by an iterative fine-tuning process, finding a better set of boundaries that maximized the performance. The fine-tuning of the boundaries required the computations of an optimizer such as SciPy's `optimize.minimize`, which is based on the Nelder-Mead algorithm. The boundaries found by the optimizer were validated by a cross-validation scheme. You can read more details about this post-processing directly from the post made by *Abhishek Thakur* during the competition:

`https://www.kaggle.com/c/petfinder-adoption-prediction/discussion/76107`

Aside from the PetFinder competition, many other competitions have demonstrated that smart post-processing can lead to improved results and rankings. We'll point out a few examples here:

- Post-Processing Technique (c.f. 1st Place Jigsaw): `https://www.kaggle.com/khoongweihao/post-processing-technique-c-f-1st-place-jigsaw`
- Postprocessing based on leakage: `https://www.kaggle.com/tomooinubushi/postprocessing-based-on-leakage`
- indoor – Post-processing by Cost Minimization: `https://www.kaggle.com/saitodevel01/indoor-post-processing-by-cost-minimization`

Another interesting example of post-processing, based on the characteristics of the proposed task, has been shown in the community competition *Predict the LLM Kaggle Hackathon* (`https://www.kaggle.com/competitions/h2oai-predict-the-llm`) by H2O.ai. The code presented by *Kha Vo* (`https://www.kaggle.com/khahuras`) and Binga (`https://www.kaggle.com/phanisrikanth`) as part of their solution (`https://www.kaggle.com/competitions/h2oai-predict-the-llm/discussion/453809`) strives to optimize batches of predictions so that each one predicts unique classes. In fact, based on the training set, there are seven targets, and each row of a batch represents a single class. The procedure first predicts a single class for each example in a batch using the Hungarian method solver for the assignment problem (see `https://www.kaggle.com/competitions/h2oai-predict-the-llm/discussion/453809`). It then adjusts the original predictions, which are probabilities, by weighing them with the results from the optimization. By incorporating the constraints from the problem, the resulting probability predictions perform much better on the private and public leaderboards.

Unfortunately, post-processing is often very dependent on the metric you are using (understanding the metric is imperative for devising any good post-processing) and often also data-specific, such as in the case of time series data and leakages. Hence, it is challenging to generalize any procedure for figuring out the right post-processing for any competition. Nevertheless, always be aware of this possibility and be on the lookout for any hint that post-processing results are favorable in a competition. You can always get hints about post-processing from previous competitions that have been similar, and by engaging with forum discussions – eventually, someone will raise the topic.

Probabilistic adjustments of the predictions

To complete the preceding discussion on metrics optimization (post-processing of predictions), we will discuss situations where it is paramount to predict correct probabilities, but you are not sure if the algorithm you are using is doing a good job. As we detailed previously, classification probabilities concern both binary and multiclass classification problems, and they are commonly evaluated using the logarithmic loss (AKA log loss, logistic loss, or cross-entropy loss) in its binary or multi-class version (for more details, see the previous sections, *Metrics for classification (label prediction and probability)* and *Metrics for multi-class classification)*.

However, evaluating or optimizing for the log loss may not prove enough. The main problems to be on the lookout for when striving to achieve correct probabilistic predictions with your model are as follows:

- Models that do not return a truly probabilistic estimate
- Unbalanced distribution of classes in your problem
- Different class distribution between your training data and your test data (on both public and private leaderboards)

The first point alone provides a reason to check and verify the quality of classification predictions in terms of modeled uncertainty. In fact, even if many algorithms are provided in the scikit-learn package together with a `predict_proba` method, this is a very weak assurance that they will return a true probability.

Let's take, for instance, **decision trees**, which are the basis of many effective methods to model tabular data. The probability outputted by a classification decision tree (`https://scikit-learn. org/stable/modules/generated/sklearn.tree.DecisionTreeClassifier.html`) is based on terminal leaves; that is, it depends on the distribution of classes on the leaf that contains the case to be predicted. If the tree is fully grown, the case is likely in a small leaf with very few other cases, so the predicted probability will be very high. If you change parameters such as `max_depth`, `max_leaf_nodes`, or `min_samples_leaf`, the resulting probability will drastically change from higher values to lower ones depending on the growth of the tree.

Decision trees are the most common base model for ensembles such as bagging models and random forests, as well as boosted models such as gradient boosting (with its high-performing implementations XGBoost, LightGBM, and CatBoost). However, for the same reasons – probability estimates that are not truly based on solid probabilistic estimations – the problem affects many other commonly used models, such as support vector machines and *k*-nearest neighbors. Such aspects were mostly unknown to Kagglers until the *Otto Group Product Classification Challenge* (https://www.kaggle.com/c/otto-group-product-classification-challenge/overview/) when it was raised by *Christophe Bourguignat* and others during the competition (see https://www.kaggle.com/cbourguignat/why-calibration-works), and easily solved using the calibration functions that had recently been added to scikit-learn (read about CalibratedClassifierCV directly from the scikit-learn website at https://scikit-learn.org/stable/modules/calibration.html).

Aside from the model you will be using, the presence of an imbalance between classes in your problem may also result in models that are not at all reliable. Hence, a good approach in the case of unbalanced classification problems is to rebalance the classes using undersampling or oversampling strategies, or different custom weights for each class to be applied when the loss is computed by the algorithm. All these strategies may render your model more performant; however, they will surely distort the probability estimates and you may have to adjust them in order to obtain an even better model score on the leaderboard.

Finally, a third point of concern is related to how the test set is distributed. This kind of information is usually concealed, but there are often ways to estimate it and figure it out (for instance, by trial and error based on the public leaderboard results, as we mentioned in *Chapter 1, Introducing Kaggle and Other Data Science Competitions*).

For instance, this happened in the *iMaterialist Furniture Challenge* (https://www.kaggle.com/c/imaterialist-challenge-furniture-2018/) and the more popular *Quora Question Pairs* (https://www.kaggle.com/c/quora-question-pairs). Both competitions gave rise to various discussions on how to post-process in order to adjust probabilities to test expectations (see https://swarbrickjones.wordpress.com/2017/03/28/cross-entropy-and-training-test-class-imbalance/ and https://www.kaggle.com/dowakin/probability-calibration-0-005-to-lb for more details on the method used). From a general point of view, assuming that you do not have an idea of the test distribution of classes to be predicted, it is still very beneficial to correctly predict probability based on the priors you get from the training data (and until you get evidence to the contrary, that is the probability distribution that your model should mimic). In fact, it will be much easier to correct your predicted probabilities if your predicted probability distribution matches those in the training set.

The solution, when your predicted probabilities are misaligned with the training distribution of the target, is to use the **calibration function** provided by scikit-learn, CalibratedClassifierCV:

```
sklearn.calibration.CalibratedClassifierCV(base_estimator=None, *,
    method='sigmoid', cv=None, n_jobs=None, ensemble=True)
```

The purpose of the calibration function is to apply a post-processing function to your predicted probabilities in order to make them adhere more closely to the empirical probabilities seen in the ground truth. Provided that your model is a scikit-learn model or behaves similarly to one, the function will act as a wrapper for your model and directly pipe its predictions into a post-processing function. You have the choice between using two methods for post-processing:

- The **sigmoid** method (also called Plat's scaling) is nothing more than a logistic regression.
- **Isotonic regression** is a non-parametric regression method; beware that it tends to overfit if there are few examples

You also have to choose how to fit this calibrator. Remember that it is a model that is applied to the results of your model, so you have to avoid overfitting by systematically reworking predictions. You could use **cross-validation** (more on this in the following chapter, *Designing Good Validation*) and then produce a number of models that, once averaged, will provide your predictions (`ensemble=True`). Otherwise, and this is our usual choice, resort to an **out-of-fold prediction** (more on this in the following chapters) and calibrate on that using all the data available (`ensemble=False`).

Even if `CalibratedClassifierCV` can handle most situations, you can also figure out some empirical way to fix probability estimates for the best performance at test time. You can use any transformation function, from a handmade one to a sophisticated one derived from genetic algorithms, for instance. Your only limit is simply that you should cross-validate it and possibly have a good final result from the public leaderboard (but not necessarily because you should trust your local cross-validation score more, as we are going to discuss in the next chapter). A good example of such a strategy is provided by Silogram (`https://www.kaggle.com/psilogram`), who, in the *Microsoft Malware Classification Challenge*, found a way to tune the unreliable probabilistic outputs of random forests into probabilistic ones simply by raising the output to a power determined by grid search (see `https://www.kaggle.com/c/malware-classification/discussion/13509`).

In our final interview of the chapter, we speak to *Sudalai Rajkumar* (AKA SRK), a Grandmaster in Competitions, Datasets, and Notebooks, and a Discussion Master. He has been ranked #1 on the Analytics Vidhya data science platform and works for Tiger Analytics as an AI/ML advisor for start-ups.

Interview: Sudalai Rajkumar

`https://www.kaggle.com/sudalairajkumar`

What's your favorite kind of competition and why? In terms of techniques and solving approaches, what is your specialty on Kaggle?

My favorite kinds of competition are ones that involve a good amount of feature engineering. I think that is my strength as well. I am generally interested in data exploration to get a deep understanding of the data (which you can infer from my series of simple exploration Notebooks (`https://www.kaggle.com/sudalairajkumar/code`)) and then creating features based on it.

How do you approach a Kaggle competition? How different is this approach to what you do in your day-to-day work?

The framework for competitions involves exploring data, finding the right validation method, feature engineering, model building, and ensembling/stacking. All of these are involved in my day job as well. In addition to this, there is a good amount of stakeholder discussion, data collection, data tagging, model deployment, model monitoring, and data storytelling that is involved in my daily job.

Tell us about a particularly challenging competition you entered, and what insights you used to tackle the task.

Santander Product Recommendation is a memorable competition that we entered. Rohan and I did a lot of feature engineering and built multiple models. When we did the final ensembling, we used different weights for different products and some of them did not add up to 1. From the data exploration and understanding, we hand-picked these weights, which helped us. This made us realize the domain/data importance in solving problems and how data science is an art as much as a science.

Has Kaggle helped you in your career? If so, how?

Kaggle played a very important role in my career. I was able to secure my last two jobs mainly because of Kaggle. Also, the success from Kaggle helps to connect with other stalwarts in the data science field easily and learn from them. It also helps a lot in my current role as an AI/ML advisor for start-ups, as it gives credibility.

In your experience, what do inexperienced Kagglers often overlook? What do you know now that you wish you'd known when you first started?

Understanding the data in depth is crucial. Often, this is overlooked, and people get into model-building right away. Exploring the data plays a very important role in the success of any Kaggle competition. This helps to create proper cross-validation and better features, as well as to extract more value from the data.

What mistakes have you made in competitions in the past?

It is a very big list, and I would say that they are learning opportunities. In every competition, out of 20-30 ideas that I try, only one may work. These mistakes/failures give much more learning than the actual success or things that worked.

For example, I learned about overfitting the very hard way by falling from top deciles to bottom deciles in one of my very first competitions. However, that learning stayed with me forever thereafter.

Are there any particular tools or libraries that you would recommend using for data analysis/ML?

I primarily use XGBoost/LightGBM in the case of tabular data. I also use open source AutoML libraries and Driverless AI to get early benchmarks these days. I use Keras, Transformers, and PyTorch for deep learning models.

What's the most important thing someone should keep in mind or do when they're entering a competition?

Consistency is the key. Each competition will have its own ups and downs. There will be multiple days without any progress, but we should not give up and keep trying. I think this is applicable to anything and not just Kaggle competitions.

Do you use other competition platforms? How do they compare to Kaggle?

I have also taken part on other platforms like the Analytics Vidhya DataHack platform, Driven Data, CrowdAnalytix, and so on. They are good too, but Kaggle is more widely adopted and global in nature, so the amount of competition on Kaggle is much higher compared to other platforms.

Summary

In this chapter, we discussed evaluation metrics in Kaggle competitions. First, we explained how an evaluation metric can differ from an objective function. We also remarked on the differences between regression and classification problems. For each type of problem, we analyzed the most common metrics that you can find in a Kaggle competition.

After that, we discussed the metrics that have never previously been seen in a competition and that you won't likely see again. Finally, we explored and studied different common metrics, giving examples of where they have been used in previous Kaggle competitions. We then proposed a few strategies for optimizing an evaluation metric. In particular, we recommended trying to code your own custom cost functions and provided suggestions on possible useful post-processing steps.

You should now have grasped the role of an evaluation metric in a Kaggle competition. You should also have a strategy to deal with every common or uncommon metric by retracing past competitions and gaining a full understanding of how a metric works.

In the next chapter, we are going to discuss how to use evaluation metrics and properly estimate the performance of your Kaggle solution by means of a validation strategy.

Join our book's Discord space

Join our community's Discord space for discussions with the authors and other readers:

`https://packt.link/kaggle`

7

Designing Good Validation

In a Kaggle competition, it may seem like enough to take the results you get back from the leaderboard at face value in the heat of modeling and submitting results. In the end, you may think that what counts in a competition is your ranking. This is a common error that is made repeatedly in competitions. You won't know what the actual leaderboard (the private one) looks like until after the competition has closed, and trusting the public part of it is not advisable because it is often misleading.

This chapter will introduce you to the importance of **validation** in data competitions. You will learn about:

- What overfitting is and how a public leaderboard can be misleading
- The dreadful shake-ups
- The different kinds of validation strategies
- Adversarial validation
- How to spot and leverage leakages
- What your strategies should be when choosing your final submissions

Monitoring your performances when modeling and distinguishing when overfitting happens is a key competency in data science competitions and projects. Properly validating your models is one of the most important skills you can learn from a Kaggle competition and that you can resell professionally.

Snooping on the leaderboard

Kaggle divides the test set in each competition into:

- A **public part**: This part is used to calculate the public leaderboard score, which is visible to all participants.
- A **private part**: This part is used to compute the final score, which is only revealed after the competition ends.

These test parts are usually randomly determined (although in time series competitions, they are determined based on time), and the entire test set is released without any distinction made between public and private. To determine whether a specific test case is part of the public or private part, you can experiment by making predictions and observing how your public leaderboard score changes. If your score changes, the prediction is likely part of the public part; if not, it's probably part of the private part. This procedure is not foolproof and it requires wasting submissions to get the score feedback, however, in some cases, it has proved to be an advantage in the competition to the Kagglers that resorted to it.

In recent years, in order to avoid the participants scrutinizing test data in certain competitions, Kaggle has even held back the test data, providing only some examples and replacing them with the actual test set when the submission is made. These are called **code** competitions because you are not providing the predictions but a notebook containing the code to generate them.

Therefore, a submission derived from a model will cover the entire test set. Still, only the public part will immediately be scored, leaving the scoring of the private part until after the competition has closed.

Given this, three considerations arise:

- Training and test data should be from the same distribution for a competition to work correctly. Moreover, the private and public parts of the test data should resemble each other in terms of distribution.
- Even if the training and test data are apparently from the same distribution, the **lack of sufficient examples** in either set could make obtaining aligned results between the training data and the public and private test data challenging.

- The public test data should be regarded as a holdout test in a data science project and used only for final validation. Hence, it should not be queried much to avoid **adaptive overfitting**, which implies a model that works well on a specific test set but underperforms on others.

Considering these three considerations is paramount to understanding the dynamics of a competition. In most competitions, there are always quite a few questions in the discussion forums about how the training, public, and private test data relate to each other, and it is pretty common to see submissions of hundreds of solutions that have only been evaluated based on their efficacy on the public leaderboard.

It is also common to hear discussions about **shake-ups** that revolutionize the rankings between the public and private leaderboards. These can occur when the private leaderboard is revealed. Shake-ups often deeply disappoint many participants who previously held better positions on the public leaderboard. While shake-ups can potentially happen in any competition, their frequency and magnitude actually vary. Some competitions may have multiple re-rankings throughout the competition as more data or evaluation metrics are introduced. Others may have just one major shake-up when the final private leaderboard is revealed.

Anecdotally, shake-ups are often attributed to differences between the training and test set or between the private and public parts of the test data. They are measured *ex-ante* based on how competitors have seen their expected local scores correlate with the leaderboard feedback and *ex-post* by a series of analyses based on two figures:

- A general shake-up figure based on mean(abs(private_rank-public_rank)/number_of_ teams)
- A top leaderboard shake-up figure, taking into account only the top 10% of public ranks

These *ex-post* figures were first devised by *Steve Donoho* (https://www.kaggle.com/ breakfastpirate), who compiled a ranking of the worst Kaggle shake-ups (see https://www. kaggle.com/c/recruit-restaurant-visitor-forecasting/discussion/49106#278831). They are nowadays readily available, recreated by many notebooks based on the Meta Kaggle dataset we discussed in *Chapter 6, Competition Tasks and Metrics* (see https://www.kaggle.com/jtrotman/ meta-kaggle-competition-shake-up). For instance, by consulting these figures, you may discover how dreadful the *RSNA Intracranial Hemorrhage Detection* competition was for many because of its shake-ups, especially in the top positions.

However, aside from an *ex-post* evaluation, there are quite a few lessons that we can get from previous shake-ups that can help you in your Kaggle competitions. A few researchers from UC Berkeley think so, too. In their paper presented at the 2019 Conference on **Neural Information Processing Systems (NIPS)**, *A meta-analysis of overfitting in machine learning*, Roelofs et al. study in detail 120 Kaggle competitions to gain insight into the public-private leaderboard dynamics in Kaggle competitions. Although they focus on a limited subset of competitions (120, above a certain number of participants, focused on binary classification), they obtained some interesting findings:

- There is little adaptive overfitting; in other words, public standings usually hold in the unveiled private leaderboard.

- Most shake-ups are due to random fluctuations and overcrowded rankings where competitors are too near each other, and any slight change in the performance in the private test sets causes significant changes in the rankings.

- Shake-ups happen when the training set is tiny or the training data is not **independent and identically distributed (i.i.d.)**.

> The full paper can be found at https://papers.nips.cc/paper/2019/file/ ee39e503b6bedf0c98c388b7e8589aca-Paper.pdf

In our long experience of Kaggle competitions, however, we have seen many problems with adaptive overfitting since the beginning. For instance, you can read Greg Park's analysis of one of the first competitions we ever participated in: https://gregpark.io/blog/Kaggle-Psychopathy-Postmortem. Since this is quite a common and persistent problem for many Kagglers, we suggest a strategy that is a bit more sophisticated than simply following what happens on the public leaderboard:

- Always build reliable cross-validation systems for local scoring.

- Always try to control non-i.i.d distributions using the best validation scheme dictated by the situation. Unless clearly stated in the competition description, it is not easy to spot non-i.i.d. distributions. Still, you can get hints from discussion or by experimenting using stratified validation schemes (when stratifying according to a particular feature, the results improve decisively, for instance).

- Correlate local scoring with the public leaderboard to determine whether they go in the same direction.

- Test using adversarial validation, revealing whether or not the test distribution is similar to the training data.

- Make your solutions more robust using ensembling, especially when working with small datasets.

In the following sections, we will explore these ideas (except for ensembling, which will be discussed in detail in *Chapter 10: Ensembling with Blending and Stacking Solutions*) and provide you with all the best tools and strategies to obtain the best results, especially on the private dataset.

The importance of validation in competitions

If you think about a competition carefully, you can imagine it as a vast system of experiments. Whoever can create the most systematic and efficient way to run these experiments wins.

In fact, despite all your theoretical knowledge, you will compete with hundreds or thousands of data professionals who have more or less the same competencies as you.

In addition, they will use precisely the same data as you and roughly the same tools for learning from the data (TensorFlow, PyTorch, scikit-learn, and so on). Some will indeed have better access to computational resources, although the availability of Kaggle Notebooks and generally decreasing cloud computing prices mean the gap is no longer so vast. Consequently, if you look at differences in knowledge, data, models, and available computers, you won't find many discriminating factors between you and the other competitors that could explain huge performance differences in a competition. Yet, some participants consistently outperform others, implying there is some underlying success factor.

In interviews and meet-ups, some Kagglers describe this success factor as "grit," some others as "trying everything," and some others again as a "willingness to put everything you have into a competition." These may sound a bit obscure and magical. Instead, we call it **systematic experimentation**. In our opinion, the key to successful participation resides in the number of experiments you conduct and how you run them. The more experiments you undertake, the more chances you will have to crack the problem better than other participants. This number certainly depends on a few factors, such as the time you have available, your computing resources (the faster, the better, but as we previously mentioned, this is not such a strong differentiator *per se*), your team size, and their involvement in the task. This aligns with the commonly reported grit and engagement as keys to success.

However, these are not the only factors affecting the result. You have to consider that how you run your experiments also has an impact. *Failing fast and learning from it* is an essential factor in a competition. Of course, you need to reflect carefully both when you fail and when you succeed to learn something from your experiences, or your competition will just turn into a random sequence of attempts in the hope of picking the right solution.

Therefore, *ceteris paribus*, having a proper **validation strategy,** is the great discriminator between successful Kaggle competitors and those who just overfit the leaderboard and end up in lower-than-expected rankings after a competition.

Validation is the method you use to correctly evaluate the errors your model produces and measure how its performance improves or decreases based on your experiments.

Generally, the impact of choosing proper validation is too often overlooked in favor of more quantitative factors, such as having the latest, most powerful GPU or a larger team producing submissions.

Nevertheless, if you count only on the firepower of experiments and their results on the leaderboard, it will be like "throwing mud at the wall and hoping something will stick" (see https://gregpark. io/blog/Kaggle-Psychopathy-Postmortem). Sometimes, such a strategy will work, but most often, it won't because you will miss important opportunities to experiment in the right direction, and you won't even be able to see the shining gem you managed to produce in the middle of all that mud. For instance, if you concentrate too much on trying your luck on the public leaderboard using a random, unsystematic strategy, even if you produce great solutions, you may not choose your final submission correctly and miss the best scoring one on the private leaderboard.

Having a proper validation strategy can help you decide which of your models should be submitted for ranking on the private test set. Though the temptation to submit your top public leaderboard models may be high, *always consider your own validation scores*. For your final submissions, depending on the situation and whether or not you trust the leaderboard, choose one submission based on the best score on the leaderboard and one submission having your best score from your local validation results. If you don't trust the leaderboard (especially when the training sample is small or the examples are non-i.i.d.), submit models with two of the best validation scores, picking two very different models or ensembles. This way, you will reduce the risk of choosing solutions that won't perform on the private test set.

Having pointed out the importance of having a method of experimenting, what is left is all a matter of the practicalities of validation. In fact, when you model a solution, you make a series of interrelated decisions:

- How to process your data
- What model to apply
- How to change the model's architecture (this is especially true for deep learning models)
- How to set the model's hyperparameters
- How to post-process the predictions

Even if the public leaderboard is perfectly correlated with the private one, the limited number of daily submissions (a limitation present in all competitions) prevents you from scratching the surface of possible tests you could do in all the areas, as mentioned earlier. A proper validation system tells you beforehand if your actions could work on the leaderboard.

Before we go into further detail, we would like to share our interview with *Dmitry Larko*, a Kaggle Competitions Grandmaster and principal AI architect at Grid Dynamics. He has over a decade of experience in ML and data science. He discovered Kaggle in December 2012 and participated in his first competition a few months later. He is a strong advocate of validation in Kaggle competitions, as he told us in his interview.

Interview: Dmitry Larko

https://www.kaggle.com/dmitrylarko

What's your favorite kind of competition and why? In terms of techniques and solving approaches, what is your specialty on Kaggle?

I have mostly participated in competitions for tabular datasets but also enjoy competitions for computer vision.

How do you approach a Kaggle competition? How different is this approach to what you do in your day-to-day work?

I always try to start simple and build a submission pipeline for smaller/simpler models first. A major step here is to create a proper validation scheme so you can validate your ideas in a robust way. Also, it is always a good idea to spend as much time as you can looking at the data and analyzing it.

In my day-to-day work, I am building an AutoML platform, so a lot of things I try on Kaggle end up being implemented as a part of this platform.

Tell us about a particularly challenging competition you entered, and what insights you used to tackle the task.

Nothing comes to my mind, and it doesn't matter, because what is technically challenging for me could be a piece of cake for somebody else. Technical challenges are not that important; what's important is to remember that a competition is somewhat like a marathon, not a sprint. Or, you can see it as a marathon of sprints if you like. So, it is important not to get exhausted, sleep well, exercise, and take a walk in a park to regenerate your brain for new ideas. To win a Kaggle competition, you will need all your creativity and expertise and sometimes even a bit of luck.

Has Kaggle helped you in your career? If so, how?

I got my current job thanks to the fact I was a Kaggle Competition Grandmaster. For my current employer, this fact was evidence enough of my expertise in the field.

In your experience, what do inexperienced Kagglers often overlook? What do you know now that you wish you'd known when you first started?

Mostly, they overlook the right validation scheme and follow the feedback from the public leaderboard. That ends badly in most cases, leading to something known as a "shake-up" on Kaggle.

Also, they rush to skip exploratory data analysis and build models right away, which leads to simplistic solutions and mediocre leaderboard scores.

What mistakes have you made in competitions in the past?

My main mistake is really the same that an inexperienced person will make – following the leaderboard score and not my internal validation. Every time I decided to do so, it cost me several places on the leaderboard.

Are there any particular tools or libraries that you would recommend using for data analysis or machine learning?

That would be the usual suspects. For tabular data: LightGBM, XGBoost, CatBoost; for deep learning: PyTorch, PyTorch-Lightning, timm; and scikit-learn for everyone.

What's the most important thing someone should keep in mind or do when they're entering a competition?

Start simple, always validate; believe in your validation score and not the leaderboard score.

Bias and variance

A sound validation system based on the evaluation metrics is a more reliable piece of guidance in a competition than the error measures you get from your training set. In fact, metrics obtained on the training set are affected by the capacity and complexity of each model. You can think of the **capacity** or expressiveness of a model as its memory that it can use to learn complex patterns from data.

Each model has a set of internal parameters that help the model record the patterns taken from the data. Every model has its own skills for acquiring patterns, and some models will spot certain rules or associations, whereas others will spot others. As a model extracts patterns from data, it records them in its "memory."

You also hear about the capacity or expressiveness of a model as a matter of **bias and variance**. In this case, the bias and variance of a model refer to the predictions, but the underlying principle is strictly related to the expressiveness of a model. Models can be reduced to mathematical functions that map an input (the observed data) to a result (the predictions). Some mathematical functions are more complex than others in the number of internal parameters they have and in the ways they use them:

- Suppose the mathematical function of a model is not complex or is not expressive enough to capture the complexity of the problem you are trying to solve. In that case, we talk of **bias** because your predictions will be constrained ("biased") by the model's limits.

- Suppose the mathematical function at the core of a model is too complex for the problem at hand. In that case, we have a **variance** problem because the model will record more details and noise in the training data than needed, and its predictions will be deeply influenced and erratic.

Nowadays, given the advances in **machine learning** (**ML**) and the available computation resources, the problem is always due to variance since deep neural networks and gradient boosting, the most commonly used solutions, often have a mathematical expressiveness that exceeds what most of the problems you will face need in order to be solved.

When all the useful patterns a certain model can extract have been captured, if the model has not exhausted its learning capacity, it will start memorizing data characteristics and signals unrelated to the prediction. Such patterns are usually called **noise**. While the initially extracted patterns will help the model generalize to a test dataset and predict more correctly, not everything it learns specifically about the training set will help; instead, it may damage its performance. Learning elements of the training set that have no generalization value is commonly called **overfitting**.

The core purpose of validation is explicitly defining a score or loss value that separates the generalizable part of that value from the part due to overfitting the training set characteristics. This is the **validation loss**. The validation loss is calculated independently for each training step using a dedicated validation dataset. Using learning curves, you can visualize how the validation loss relates to the training loss when overfitting:

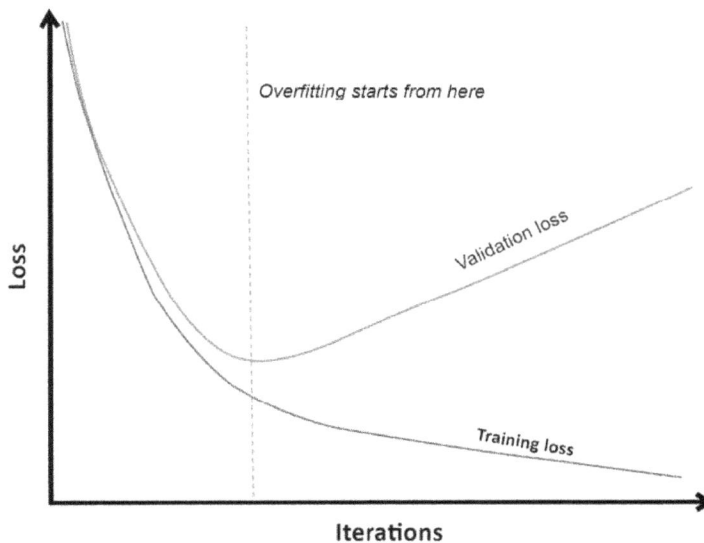

Figure 7.1: Learning more from the training data does not always mean learning to predict

If you graph the loss measure on the y-axis against some measure of the learning effort of the model (this could be epochs for neural networks or rounds for gradient boosting) on the x-axis, you will notice that learning always seems to happen on the training dataset, but this is not always true on other data.

The same happens even if you change the hyperparameters, process the data, or decide on a different model altogether. The curves will change shape, but you'll always have a sweet point where overfitting starts. That point can differ across models and between the various choices you make in your modeling efforts. Thanks to a correct validation strategy, suppose you have properly estimated the point when overfitting starts and stopped the model from learning further. In that case, your model's performance will indeed correlate with the leaderboard results (both public and private), and your validation metrics will provide you with a proxy to evaluate your work without making any submissions.

You can hear about overfitting at various levels:

- At the level of the training data, when you use a model that is too complex for the problem. For instance, when a complex model with many parameters is applied to a small dataset, and the model learns to memorize the training examples rather than learning the general patterns in the data.

- At the level of the validation set itself, when you tune your model too much with respect to a specific validation set. For example, by repeatedly adjusting the learning rate or regularization strength based on the validation loss, the model will overfit to the validation set.

- At the level of the public leaderboard, when your results are far from what you would expect from your training. Typically, in this case, you are overfitting at the training level and the model cannot generalize at all.

- At the level of the private leaderboard, when your private scores will be disappointing despite the good results on the public leaderboard. This might happen if the public and private test data have different distributions, and your model is not generalizing well to the unseen data.

Though slightly different in meaning, they all equally imply that your model is not generalizable, as we have described in this section.

Trying different splitting strategies

As previously discussed, the validation loss is based on a data sample that is not part of the training set. It is an empirical measure that tells you how good your model is at predicting, and a more correct one than the score you get from your training, which will tell you mostly how much your model has memorized the training data patterns. Correctly choosing the data sample you use for validation constitutes your validation strategy.

To summarize the strategies for validating your model and measuring its performance correctly, you have a couple of choices:

- The first choice is to **work with a holdout sample**, incurring the risk of not properly choosing a representative sample of the data or overfitting your validation holdout. A validation holdout is a portion of the training data that is set aside and not used for training the model. The percentage of training data reserved as holdout typically ranges from 10% to 20%, but it can vary depending on the specific dataset and problem.

- The second option is to **use a probabilistic approach** and rely on a series of samples to draw your conclusions on your models. You have cross-validation and bootstrap among the probabilistic approaches. Among the cross-validation strategies, there are different nuances depending on the sampling strategies you take based on the characteristics of your data (simple random sampling, stratified sampling, sampling by groups, time sampling).

What these strategies have in common is that they are **sampling strategies**. It means that they help you to infer a general measure (the performance of your model) based on a small part of your data, randomly selected. Sampling is at the root of statistics, and it is not an exact procedure because, based on your sampling method, your available data, and the randomness of picking up some instances as part of your sample, you will experience a certain degree of error.

For instance, if you rely on a biased sample, your evaluation metric may be estimated incorrectly (over- or underestimated). However, if properly designed and implemented, sampling strategies generally give you a good estimate of your general measure.

The other aspect that all these strategies have in common is that they are **partitions**, which divide cases exclusively as either part of the training or validation. In fact, as we discussed, since most models have a certain memorization capability, using the same cases in both training and validation leads to inflated estimates because it allows the model to demonstrate its memorization abilities; instead, we want it to be evaluated on its ability to derive patterns and functions that work on *unseen* examples.

The basic train-test split

The first strategy that we will analyze is the **train-test split**. In this strategy, you sample a portion of your training set (also known as the **holdout**) and use it as a test set for all the models you train using the remaining part of the data.

The great advantage of this strategy is that it is very simple: you pick up a part of your data and check your work on it. You usually split the data 80/20 in favor of the training partition.

In scikit-learn, splitting data between training and test sets is implemented in the train_test_ split function. We'll draw your attention to a few aspects of the method:

- Given a certain percentage of data for train or testing, you can always draw the same sets of data because the selection is based on the original ordering of the data. However, if your data is ordered somehow, you may introduce unknown **bias** in your validation system. However, based on a random seed (to be set with the random_state parameter), hence in a replicable way, the function can shuffle your data (setting the shuffle parameter to true), helping you pick up truly random samples for train and testing. By testing different random seeds, you can draw different test sets and find the one that fits the competition.

- A test set serves the purpose of evaluating your model in an independent way from the public leaderboard. It is not for checking on the best data transformations or model hyper-parameters. For that purpose, you need a **validation set**, which you extract from the training data by reapplying the train_test_split function on the train data only. Using the test set instead of a validation set for anything other than testing your model's overall performance will lead to overestimating.

- When you have large amounts of data, you can expect that the test data you extract is similar to (representative of) the original distribution on the entire dataset. However, since the extraction process is based on randomness, you always have the chance of extracting a non-representative sample. In particular, the chance increases if the training sample you start from is small. Comparing the extracted holdout partition using **adversarial validation** (more about this in a few sections) against the public test data from Kaggle can help you evaluate your efforts correctly.

- In addition, to ensure that your test sampling is representative, especially concerning how the training data relates to the target variable, you can use **stratification**, which ensures that the proportions of certain features are respected in the sampled data. You can use the stratify parameter in the train_test_split function and provide an array containing the class distribution to preserve.

- Finally, the definitive way to assess if you are using the right test set for assessing your work before submitting it to the leaderboard is to check how your evaluations on the test set compare to the ones in the leaderboard. This doesn't imply that they should be the same, though the more similar, the better. However, they can differ somehow, but what matters more is that they are **correlated**, pointing out that improvements on your model correspond to improvements on the leaderboard. This golden rule is valid for train-test splitting strategies and any other validation strategy presented in this chapter. Your local test result should be strongly correlated with the results on the leaderboard.

Even if you have a representative holdout available, a simple train-test split is sometimes insufficient to ensure correct tracking of your efforts in a competition. Even if there is a good correlation between your tests and the public leaderboard, the private leaderboard may be a different story. To be successful, your model should generalize beyond the test distributions you have access to.

In fact, as you keep checking on a specific test set, even if it is correlated to the public leaderboard, you may drive your choices to some kind of adaptation overfitting (in other words, erroneously picking up the noise of the training set as signals), as happens when you frequently evaluate on the public leaderboard. For this reason, a probabilistic evaluation, though more computationally expensive, is always more suited for a competition. You can achieve this using multiple test holdout sets or better decide on a more structured probabilistic approach.

Probabilistic evaluation methods

Probabilistic evaluation of the performance of an ML model is based on the statistical properties of a sample from a distribution. By sampling, you create a smaller set of your original data that is expected to have the same characteristics. In addition, what is left untouched from the sampling constitutes a sample in itself and is expected to have the same characteristics as the original data. By training and testing your model on this sampled data and repeating this procedure a large number of times, you are basically creating a statistical estimator measuring the performance of your model. Every sample may have some "error;" it may not fully represent the original data's actual distribution. However, as you sample more, the mean of your estimators on these multiple samples will converge to the true mean of the measure you are estimating (this is an observed outcome that, in probability, is explained by a theorem called the **law of large numbers**).

Probabilistic estimators naturally require more computations than a simple train-test split. Still, they offer more confidence that you are correctly estimating the right measure: the general performance of your model.

k-fold cross-validation

The most used probabilistic validation method is **k-fold cross-validation**, which is recognized as having the ability to correctly estimate your model's performance on unseen test data drawn from the same distribution.

> This is clearly explained in the paper by Bates et al., *Cross-validation: what does it estimate and how well does it do it?* (https://arxiv.org/pdf/2104.00673.pdf).

K-fold cross-validation can be successfully used to compare predictive models and when selecting the hyperparameters for your model that will perform the best on the test set.

Understanding how k-fold cross-validation works

There are quite a few different variations of k-fold cross-validation. In this section, we will start discussing the basic, vanilla, variation. The simplest k-fold cross-validation approach, implemented in the KFold function in scikit-learn, is based on splitting your available training data into k partitions. After that, for k iterations, one of the k partitions is taken as a test set while the others are used to train the model.

The k validation scores are then averaged, and that averaged score value is the k-fold validation score, which will tell you the estimated average model performance on any unseen data. The standard deviation of the scores will inform you about the uncertainty of the estimate. *Figure 7.2* demonstrates how five-fold cross-validation is structured:

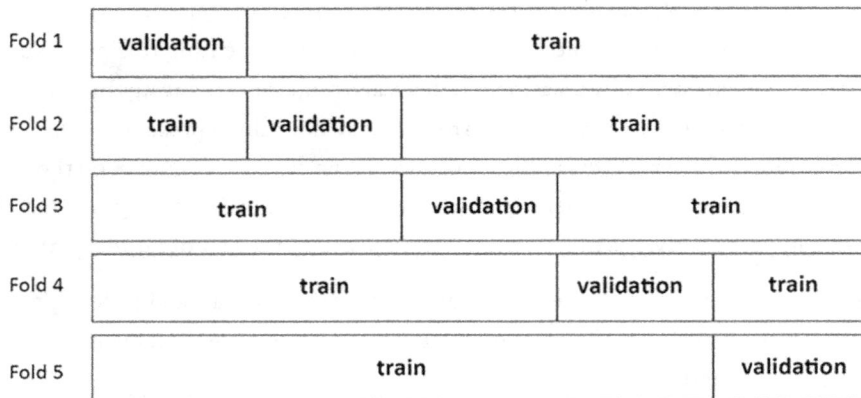

Figure 7.2: How a five-fold validation scheme is structured

One important aspect of the k-fold cross-validation score you must remember is that it estimates the average score of a model trained on the same quantity of data as k - 1 folds. If, afterward, you train your model on all your data, the previous validation estimate no longer holds. As k approaches the number n of examples, you have an increasingly correct estimate of the model derived on the entire training set. However, due to the growing correlation between the estimates you obtain from each fold, you will lose all the probabilistic estimates of the validation. In this case, you'll end up having a number showing you your model's performance on your training data (which is still a useful estimate for comparison reasons, but it won't help you correctly estimate your model's generalization power).

When you reach $k = n$, you have the **leave one out** (**LOO**) validation method, which is useful when you have a few cases available. The method is mostly an unbiased fitting measure since it uses almost all the available data for training and just one example for testing. Yet, it is not a reasonable estimate of the expected performance on unseen data. Its repeated tests over the whole dataset are highly correlated, and the resulting LOO metric represents the model's performance on the dataset rather than the model's performance on unknown data. LOO cannot replace CV when evaluating the generalization of a model. However, it is important to note that the LOO technique can be valuable in understanding how each individual data point influences the model's fit. This approach proves beneficial for detecting outliers in the dataset and assessing their impact on the chosen metric.

The correct k number of partitions to choose is decided based on a few aspects relative to the data you have available:

- The smaller the k (the minimum is 2), the smaller each fold will be, and consequently, the more bias in learning there will be for a model trained on k - *1* folds: your model validated on a smaller k will be less well-performing with respect to a model trained on a larger k.

- The higher the k, the more the data, yet the more correlated your validation estimates: you will lose the interesting properties of k-fold cross-validation in estimating the performance on unseen data.

Commonly, k is set to 5, 7, or 10, more seldom to 20 folds. We usually regard $k = 5$ or $k = 10$ as a good choice for a competition, with the latter using more data for each training (90% of the available data) and hence being more suitable for figuring out the performance of your model when you retrain on the full dataset.

When deciding what k to choose for a specific dataset in a competition, we find it helpful to reflect on two perspectives:

- The choice of the number of folds should reflect your goals:

    - If your purpose is performance estimation, you need models with low-bias estimates (which means no systematic distortion of estimates). You can achieve this using more folds, usually between 10 and 20.

    - If your aim is parameter tuning, you need a mix of bias and variance, so it is advisable to use a medium number of folds, usually between five and seven.

- If your purpose is just to apply variable selection and simplify your dataset, you need models with low variance estimates (or you will have disagreement). Hence, fewer folds will suffice, usually between three and five.

When the size of the available data is quite large, you can safely stay on the lower side of the suggested bands.

- If you are just aiming for performance estimation, consider that the more folds you use, the fewer cases you will have in your validation set, so the more the estimates of each fold will be correlated. Beyond a certain point, increasing k renders your cross-validation estimates less predictive of unseen test sets and more representative of an estimate of how well-performing your model is on your training set. This also means that, with more folds, you can get the perfect **out-of-fold** (**OOF**) prediction for stacking purposes, as we will explain in detail in *Chapter 9, Ensembling with Blending and Stacking Solutions*.

In Kaggle competitions, k-fold cross-validation is often applied not only for validating your solution approach and figuring out your model's performance but also to produce your prediction. When you cross-validate, you are subsampling, and averaging the results of multiple models built on subsamples of the data is an effective strategy for fighting against variance and often more effective than training on all the data available (we will discuss this more in *Chapter 9*). Hence, many Kaggle competitors use the models built during cross-validation to provide a series of predictions on the test set that, averaged, will provide them with the solution.

K-fold variations

Since it is based on random sampling, k-fold can provide unsuitable splits when:

- You have to preserve the proportion of small classes, both at a target level and at the level of features. This is typical when your target is highly imbalanced. Typical examples are spam datasets (because spam is a small fraction of the normal email volume) or any credit risk dataset where you have to predict the not-so-frequent event of a defaulted loan.
- You have to preserve the distribution of a numeric variable, both at a target level and at the level of features. This is typical of regression problems where the distribution is quite skewed, or you have heavy, long tails. A common example is house price prediction, where you have a consistently small portion of houses on sale that will cost much more than the average house.
- Your cases are non-i.i.d, particularly when dealing with time series forecasting.

In the first two scenarios, the solution is the **stratified k-fold**, where the sampling is done in a controlled way that preserves the distribution you want to preserve. Suppose you need to preserve the distribution of a single class. In that case, you can use StratifiedKFold from scikit-learn, using a stratification variable, usually your target variable, but also any other feature whose distribution you need to preserve. The function will produce a set of indexes that will help you to partition your data accordingly. After discretizing it, you can also obtain the same result with a numeric variable using pandas.cut or scikit-learn's KBinsDiscretizer.

Stratifying based on multiple variables or overlapping labels is a bit more complicated, such as in multi-label classification. You can find a solution for multilabel problems in the **scikit-multilearn** package (http://scikit.ml/), in particular, the IterativeStratification command that helps you to control the order (the number of combined proportions of multiple variables) that you want to preserve (http://scikit.ml/api/skmultilearn.model_selection.iterative_stratification.html). It implements the algorithm explained in the papers, *On the stratification of multi-label data* (http://lpis.csd.auth.gr/publications/sechidis-ecmlpkdd-2011.pdf), by Sechidis et al., and *A Network Perspective on Stratification of Multi-Label Data* (http://proceedings.mlr.press/v74/szyma%C5%84ski17a.html), by Szymański and Kajdanowicz.

You can actually make good use of stratification even when your problem is not a classification but a regression. Using stratification in regression problems helps your regressor to fit during cross-validation on a similar distribution of the target (or of the predictors) to the one found in the entire sample. In these cases, to have StratifiedKFold working correctly, you have to use a discrete proxy for your target instead of your continuous target. There are two ways of achieving this:

- The first, simplest way of achieving this is to use the pandas cut function and divide your target into a large enough number of bins, such as 10 or 20:

```
import pandas as pd
y_proxy = pd.cut(y_train, bins=10, labels=False)
```

To determine the number of bins to be used, *Abhishek Thakur* prefers to use **Sturges' rule** based on the number of examples available and provide that number to the pandas cut function (see https://www.kaggle.com/abhishek/step-1-create-folds):

```
import numpy as np
bins = int(np.floor(1 + np.log2(len(X_train))))
```

Keep in mind, however, that statistician Rob Hyndman (`https://robjhyndman.com/`) criticized Sturges' rule as inadequate for non-normal data and larger datasets (see `https://robjhyndman.com/papers/sturges.pdf`). While still commonly used, there are now alternatives like Scott's rule or the Freedman-Diaconis rule that offer even more accurate bin sizes for modern data analysis:

- **Scott's rule** calculates the bin width based on the standard deviation of the data.
- The **Freedman-Diaconis rule** uses the **interquartile range (IQR)** to determine the bin width, making it robust against outliers.

NumPy includes built-in support for both the Freedman-Diaconis and Scott's rules through the `numpy.histogram_bin_edges` function. You can specify the binning method using the bins parameter:

```
import numpy as np
data = np.random.normal(size=1000)
# Using Freedman-Diaconis rule
fd_bins = np.histogram_bin_edges(data, bins='fd')
# Using Scott's rule
scott_bins = np.histogram_bin_edges(data, bins='scott')
```

- An alternative approach is to focus on the distributions of the features in the training set and aim to reproduce them. This requires using **cluster analysis** (an unsupervised approach) on the features of the training set, thus excluding the target variable and any identifier, and then using the predicted clusters as strata. You can see an example in this notebook (`https://www.kaggle.com/lucamassaron/are-you-doing-cross-validation-the-best-way`), where first **principal component analysis (PCA)** is performed to remove correlations, and then a k-means cluster analysis is performed. By running empirical tests, you can decide on the number of clusters to use.

Proceeding with our discussion of the cases where k-fold can provide unsuitable splits, things get tricky in the third scenario discussed earlier in this section, when you have non-i.i.d. data, such as in the case of some grouping happening among examples. The problem with non-i.i.d. examples is that the features and target are correlated between the examples (hence, it is easier to predict all the examples if you know just one example among them).

In fact, if you happen to have the same group divided between training and testing, your model may learn to distinguish the groups and not the target itself, producing a good validation score but very bad results on the leaderboard. The solution here is to use GroupKFold: by providing a grouping variable, you will be assured that each group will be placed either in the training folds or the validation ones but never split between the two.

Discovering groupings in the data that render your data non-i.i.d. is not an easy task to accomplish. Unless stated by the competition problem, you will have to rely on your ability to investigate the data (using unsupervised learning techniques, such as cluster analysis) and the domain of the problem. For instance, if your data is about mobile telephone usage, you may realize that some examples are from the same user by noticing sequences of similar values in the features.

Sequential cross-validation in time series

Time series analysis presents the same problem, and since data is non-i.i.d., you cannot validate by random sampling because you will mix different time frames, and following time frames could bear traces of the previous ones, a characteristic called **auto-correlation** in statistics.

In the most basic approach to validation in time series, you can use a training and validation split based on time, as illustrated by *Figure 7.3*:

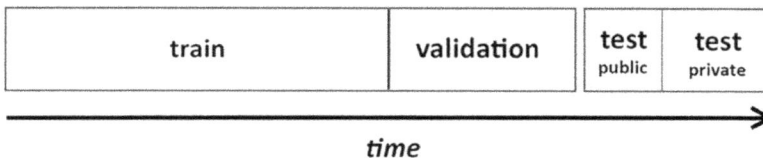

Figure 7.3: Training and validation splits are based on time

However, your validation capabilities will be limited since your validation will be anchored to a specific time. For a more complex approach, you can use time split validation, TimeSeriesSplit, as provided by the scikit-learn package (sklearn.model_selection.TimeSeriesSplit). TimeSeriesSplit can help you set the timeframe of your training and testing portions of the time series.

Regarding the training timeframe, the TimeSeriesSplit function can help you set your training data. Hence, it involves all the past data before the test timeframe or limiting it to a fixed period lookback (for instance, always using the data from three months before the test timeframe for training).

In *Figure 7.4*, you can see the structure of a time-based sequential validation strategy involving a growing training set and a moving validation set:

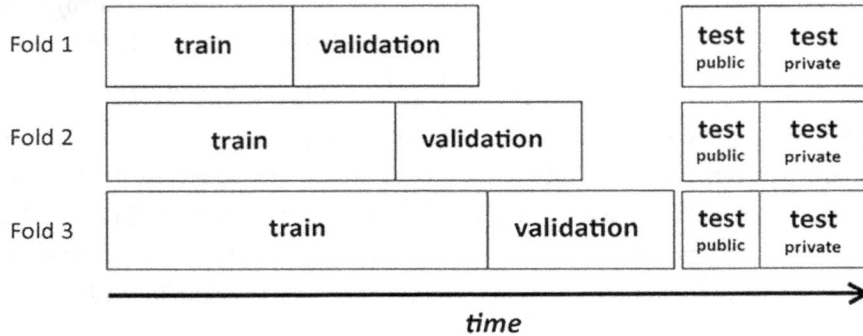

Figure 7.4: The training set is growing over time

In *Figure 7.5*, you can instead see how the strategy changes if you stipulate that the training set has a fixed lookback:

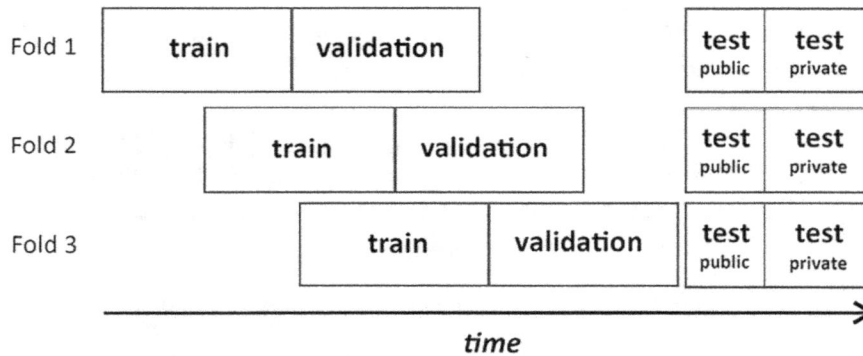

Figure 7.5: Training and validation splits are moving over time

In our experience, going by a fixed lookback helps to provide a fairer evaluation of time series models since you are always counting on the same training set size.

By instead using a growing training set size over time, you confuse the effects of your model performance across time slices with the decreasing bias in your model (since more examples mean less bias).

Finally, remember that `TimeSeriesSplit` can be set to keep a pre-defined gap between your training and test time. This is extremely useful when you are told that the test set is a certain amount of time in the future (for instance, a month after the training data), and you want to test whether your model is robust enough to predict that far into the future.

Validation in financial time series

In recent years, a number of competitions on financial data have taken place on Kaggle:

- *Jane Street Market Prediction* (https://www.kaggle.com/competitions/jane-street-market-prediction)

- *Optiver Realized Volatility Prediction* (https://www.kaggle.com/competitions/optiver-realized-volatility-prediction)

- *Ubiquant Market Prediction* (https://www.kaggle.com/competitions/ubiquant-market-prediction)

- *JPX Tokyo Stock Exchange Prediction* (https://www.kaggle.com/competitions/jpx-tokyo-stock-exchange-prediction)

- *Optiver – Trading at the Close*

- (https://www.kaggle.com/competitions/optiver-trading-at-the-close)

Sequential validation, where the folds are generated sequentially, ensuring that data in later folds occurs strictly after data in earlier folds, is a valid strategy for financial data problems. However, financial data also present characteristics that may make other options viable. For instance, as explained in the book *Advances in Financial Machine Learning* by M. L. de Prado, using a sequential approach (called the "walk-forward" method by the author), a specific temporal sequence may not be enough for an ML algorithm to learn the underlying financial dynamics and it can also be easily overfit. The solution is to use combinations of adjacent and non-adjacent folds to create new sequential paths, taking care to:

- Avoid time leakage by eliminating those situations where certain financial happenings span their effects and overlap with the test set. This is mitigated by "purging," which systematically cuts a certain amount of the temporarily terminal examples in folds.

- Avoid serial correlation effects by cutting the initial part of a training set adjacent to a previous validation fold. This censoring is called **embargo**.

The following figure should make clear where actual purging and embargoing are applied:

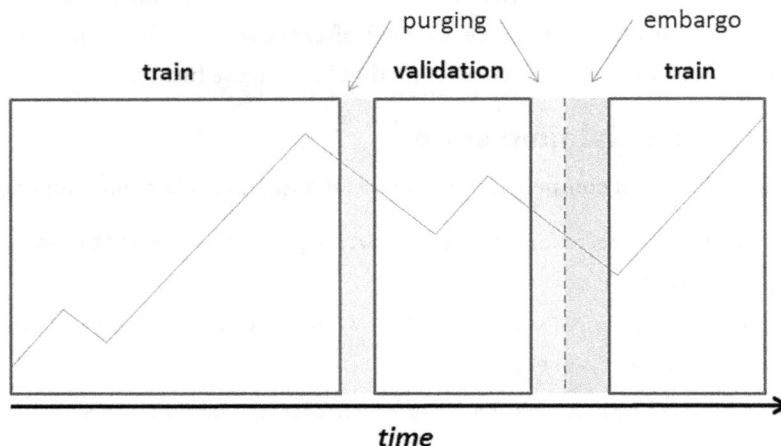

Figure 7.6: Purging and embargoing

Figure 7.6 illustrates the purging and embargo processes. The training data is divided into folds. Purging involves removing a portion of the data near the end of each fold to prevent time leakage. Embargoing involves excluding the initial part of a training fold that follows a validation fold to avoid serial correlation. By combining these techniques, we can obtain more reliable and unbiased sequential paths for time series analysis.

Yirun Zhang, the previous winner of another finance competition (the Jane Street Market Prediction), during the Ubiquant Market Prediction competition, prepared and successfully applied to different groups the combinatorial purged cross-validation method based on the discussion found in de Prado's book (see: https://www.kaggle.com/competitions/ubiquant-market-prediction/discussion/305118 and https://www.kaggle.com/code/gogo827jz/combinatorial-purged-group-k-fold/notebook). This was done using the following strategy:

```
cv = CombinatorialPurgedGroupKFold(n_splits=6, n_test_splits=2, purge=10, pctEmbargo=0.01):
```

In *Figure 7.7*, we visually represent the Combinatorial Purged Group K-Fold Cross-Validation strategy:

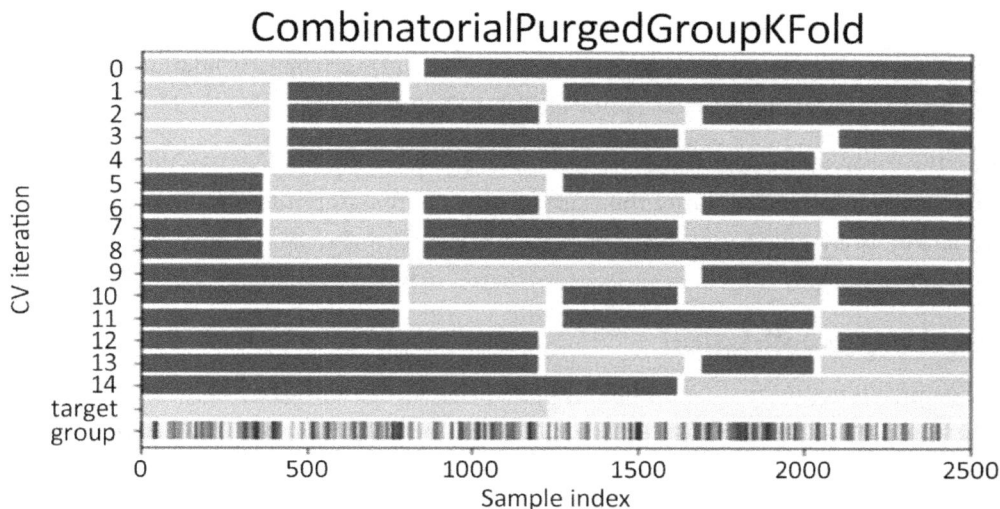

Figure 7.7: Combinatorial Purged Group K-Fold

Here, the X-axis represents the sample index, ranging from 0 to 2500 in this example, and the Y-axis represents the cross-validation iteration, numbered from 0 to 14. Each colored block represents a portion of the data used in a specific cross-validation iteration. In particular:

- The blue or darker blocks indicate data used for training.

- The light gray blocks indicate data used for validation.

- The white gaps represent data that is excluded due to purging or embargo.

This visualization helps understand how the data is divided into training and validation sets for each cross-validation iteration, considering the purging and embargo techniques to avoid time leakage and serial correlation.

It is also important to note that:

- **Time_ids** (trading temporal units) are the groups kept distinct in folds to avoid leakage of information from the same time unit. You can see the distinct cluster groups represented by the lowest charted bar named "group."

- In respect of every combination possible given the constraints, at every testing round, six time splits are created, with four of them used for training and two for validation

- Some Time_ids are purged from training data that have overlapping time with the test
- There is a further slack (embargo) between train and test when a validation fold precedes train folds to reduce any serial correlation leakage

Besides the inconvenient truth that no resampling or combinatorial technique will magically generate more information out of the existing one, Combinatorial Purged Group K-Fold may turn out to be an effective way to help your model at the price of an increased computational cost to learn in a better way from data and perform better in financial time series data. Give it a try!

Nested cross-validation

At this point, it is important to introduce **nested cross-validation**. So far, we have only discussed testing models with respect to their final performance, but often, you also need to test their intermediate performance when tuning their hyperparameters. In fact, you cannot test how specific model parameters work on your test set and then use the same data to evaluate the final performance. Since you have specifically found the best parameters that work on the test set, your evaluation measure on the same test set will be too optimistic; on a different test set, you will probably not obtain the same result. In this case, you must distinguish between a **validation set** used to evaluate the performance of various models and hyperparameters and a **test set** that will help you estimate the model's final performance.

Using a test-train split, this is achieved by splitting the test part into two new parts. The usual split is 70/20/10 for training, validation, and testing, respectively (but you can decide differently). If you are using cross-validation, you need nested cross-validation; that is, you do cross-validation based on the split of another cross-validation. You run your usual cross-validation, but when evaluating different models or parameters, you run cross-validation based on the fold split.

The example in *Figure 7.8* demonstrates this internal and external cross-validation structure. Within the external part, you determine the portion of the data used to test your evaluation metric. Within the inner part, which is fed by the training data from the external part, you arrange training/validation splits to evaluate and optimize specific model choices, such as deciding which model or hyperparameter values to pick:

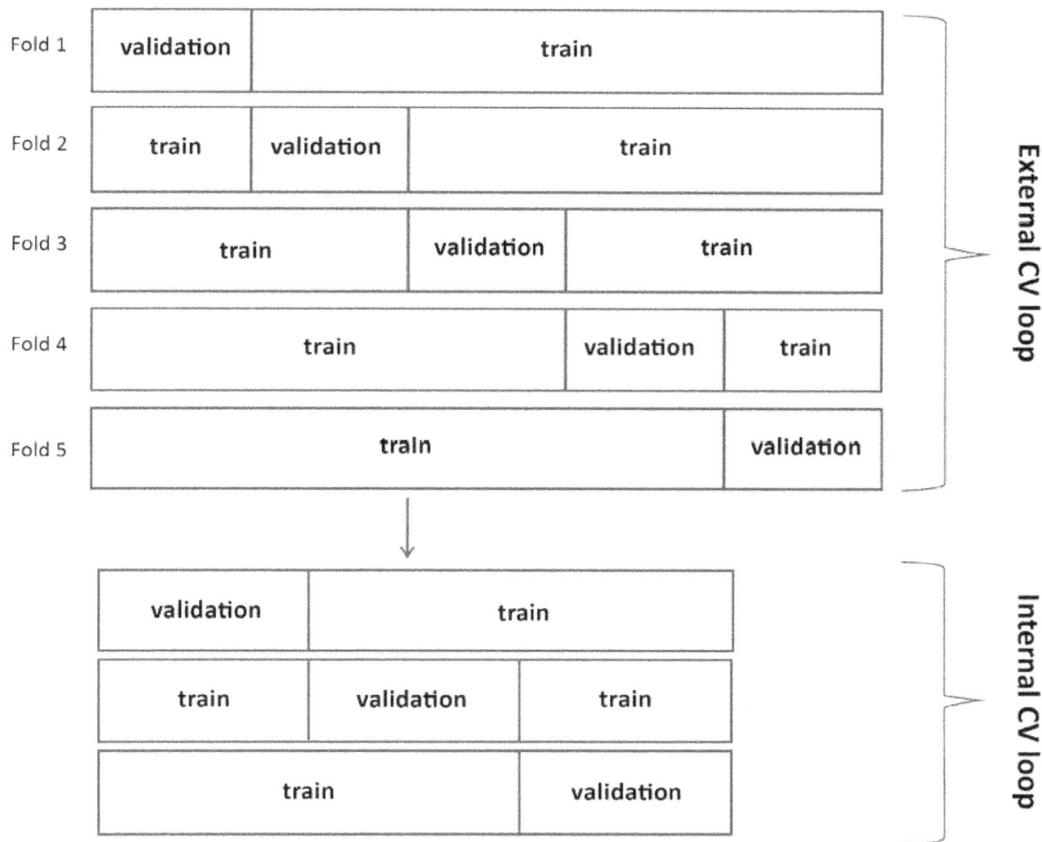

Figure 7.8: How nested cross-validation is structured in an external and an internal loop

This approach has the advantage of making your test and parameter search fully reliable, but in doing so, you incur a couple of problems:

- A reduced training set, since you first split by cross-validation, and then you split again
- More importantly, it requires a huge amount of model building: if you run two nested 10-fold cross-validations, you'll need to run 100 models

Especially for this last reason, some Kagglers tend to ignore nested cross-validation and risk some adaptive overfitting by using the same cross-validation for both model/parameter search and performance evaluation or using a fixed test sample for the final evaluation. In our experience, this approach can work as well. However, it may result in overestimating model performance and overfitting if you generate OOF predictions for successive modeling, which we will discuss in the next section. We always suggest you try the most suitable methodology for testing your models. If you aim to correctly estimate your model's performance and reuse its predictions in other models, remember that using nested cross-validation, whenever possible, can provide you with a less overfitting solution and could make a difference in certain competitions.

Producing OOF predictions

Besides estimating your evaluation metric performance, an interesting application of cross-validation is producing test and OOF predictions. In fact, as you train on portions of your training data and predict the remaining ones, you can:

- **Predict on the test set**: The average of all the predictions is often more effective than re-training the same model on all the data. This is an ensembling technique related to blending, which will be dealt with in *Chapter 9, Ensembling with Blending and Stacking Solutions*.

- **Predict on the validation set**: In the end, you will have predictions for the entire training set and can re-order them in the same order as the original training data. These predictions are commonly called OOF predictions and can be extremely useful.

Now, let's look at some scenarios in which we would use OOF predictions:

- The first use of OOF predictions is to estimate your performance since you can compute your evaluation metric directly on the OOF predictions. The performance obtained is different from the cross-validated estimates (based on sampling); it doesn't have the same probabilistic characteristics, so it is not a valid way to measure generalization performance, but it can inform you about the performance of your model on the specific set you are training on.

- A second use is to produce a plot and visualize the predictions against the ground truth values or other predictions from different models. This will help you understand how each model works and if their predictions are correlated.

- The last use is to create meta-features or meta-predictors. This will also be fully explored in *Chapter 9, Ensembling with Blending and Stacking Solutions*. Still, it is important to remark on now, as OOF predictions are a byproduct of cross-validation, and they work because, during cross-validation, your model is always predicting examples that it has not seen during training time.

Since every prediction in your OOF predictions has been generated by a model trained on different data, these predictions are unbiased. You can use them without fearing overfitting (though some caveats will be discussed in the next chapter).

Generating OOF predictions can be done in two ways:

- By coding a procedure that stores the validation predictions into a prediction vector, taking care to arrange them in the same index position as the examples in the training data
- By using the scikit-learn function `cross_val_predict`, which will automatically generate the OOF predictions for you

While both methods are viable, the `cross_val_predict` function is generally preferred due to its convenience and efficiency. It handles the complexities of cross-validation, including splitting the data, training the model, making predictions, and arranging the results, thus rendering the process more straightforward. We will see this technique in action when we look at adversarial validation later in this chapter.

Subsampling

There are other validation strategies aside from k-fold cross-validation, but they do not have the same generalization properties. We have already discussed LOO, which is the case when $k = n$ (where n is the number of examples). Another choice is **subsampling**. Subsampling is similar to k-fold, but you do not have fixed folds; you use as many as you think are necessary (in other words, take an educated guess). You repetitively subsample your data, using the data you sampled as training data and the data that has been left unsampled for your validation. By averaging the evaluation metrics of all the subsamples, you will get a validation estimate of the performances of your model.

Since you are systematically testing all your examples, as in k-fold, you actually need quite a lot of trials to have a good chance of testing all of them. For the same reason, some cases may be tested more than others if you do not apply enough subsamples. You can run this sort of validation using `ShuffleSplit` from scikit-learn.

The bootstrap

Another option is to try the **bootstrap**, devised in statistics to conclude the error distribution of an estimate; for the same reasons, it can be used for performance estimation. The bootstrap requires you to draw a sample, *with replacement*, that is the same size as the available data.

At this point, you can use the bootstrap in two different ways:

- As in statistics, you can bootstrap multiple times, train your model on the samples, and compute your evaluation metric on the training data itself. The average of the bootstraps will provide your final evaluation.

- Otherwise, as in subsampling, you can use the bootstrapped sample for your training and what is not sampled from the data as your test set.

In our experience, the first method of calculating the evaluation metric on the bootstrapped training data, often used in statistics for linear models to estimate the value of the model's coefficients and their error distributions, is much less useful in ML. This is because many ML algorithms tend to overfit the training data. Hence, you can never have a valid metric evaluation on your training data, even if you bootstrap it. For this reason, Efron and Tibshirani (see the paper *Improvements on cross-validation: the 632+ bootstrap method* at https://www.jstor.org/stable/2965703) proposed the **632+ estimator** as a final validation metric.

At first, they proposed a simple version called the 632 bootstrap:

$$Err_{.632} = 0.368 * err_{fit} + 0.632 * err_{bootstrap}$$

In this formula, given your evaluation metric *err*, err_{fit} is your metric computed on the training data, and $err_{bootstrap}$ is the metric computed on the bootstrapped data. However, in an overfitting training model, err_{fit} would tend to zero, rendering the estimator useless. Therefore, they developed a second version, the 632+ bootstrap:

$$Err_{.632} + (1 - w) * err_{fit} + w * err_{bootstrap}$$

Here, *w* is:

$$w = \frac{0.632}{1 - 0.632R}$$

$$R = \frac{err_{bootstrap} - err_{fit}}{\gamma - err_{fit}}$$

Here, you have a new parameter, γ, which is the **no-information error rate**, estimated by evaluating the prediction model on all possible combinations of targets and predictors. Calculating γ is indeed intractable, as discussed by the developers of scikit-learn (https://github.com/scikit-learn/scikit-learn/issues/9153).

Given the limits and intractability of using the bootstrap as in classical statistics for ML applications, you can use the second method instead, the 632+ bootstrap, when getting your evaluation from the examples left not sampled by the bootstrap.

In this form, the bootstrap is an alternative to cross-validation, but as with subsampling, it requires building many more models and testing them than for cross-validation. However, it makes sense to know about such alternatives in case your cross-validation is showing too high a variance in the evaluation metric and you need more intensive checking through testing and re-testing.

Previously, this method has been implemented in scikit-learn (`https://github.com/scikit-learn/scikit-learn/blob/0.16.X/sklearn/cross_validation.py#L613`) but was then removed. Since you cannot find the bootstrap anymore on scikit-learn, and it bootstrapped even the test data, you can use our own implementation. Here is our example:

```
import random
def Bootstrap(n, n_iter=3, random_state=None):
    """
    Random sampling with replacement cross-validation generator.
    For each iter a sample bootstrap of the indexes [0, n) is
    generated and the function returns the obtained sample
    and a list of all the excluded indexes.
    """
    if random_state:
        random.seed(random_state)
    for j in range(n_iter):
        bs = [random.randint(0, n-1) for i in range(n)]
        out_bs = list({i for i in range(n)} - set(bs))
        yield bs, out_bs
```

In conclusion, the bootstrap is indeed an alternative to cross-validation. It is certainly more widely used in statistics and finance. In ML, the golden rule is to use the k-fold cross-validation approach. However, we suggest not forgetting about the bootstrap in all those situations where you have a large standard error of the evaluation metric in cross-validation due to outliers or a few examples that are too heterogeneous. The bootstrap will prove much more useful in validating your models in these cases.

Our second interview of the chapter is with *Ryan Chesler*, a Discussions, Notebooks, and Competitions Grandmaster. He is a principal machine learning engineer at Deep Sentinel and one of the organizers of the San Diego Machine Learning group on Meetup (`https://www.meetup.com/San-Diego-Machine-Learning/`). The importance of validation came up in a few of his answers.

Interview: Ryan Chesler

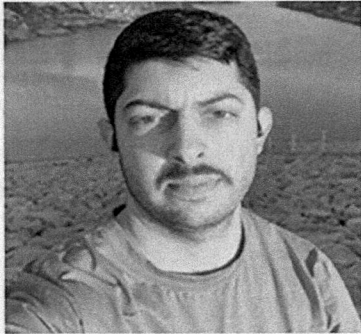

`https://www.kaggle.com/ryches`

What's your favorite kind of competition and why? In terms of techniques and solving approaches, what is your specialty on Kaggle?

I tend to dabble in all kinds of competitions. It is more interesting to sample varied problems than specialize in a specific niche like computer vision or natural language processing. The ones I find most interesting are the ones where there are deep insights that can be derived from the data and error of predictions. For me, error analysis is one of the most illuminating processes; understanding where the model is failing and trying to find some way to improve the model or input data representation to address the weakness.

How do you approach a Kaggle competition? How different is this approach to what you do in your day-to-day work?

My approach is similar in both cases. Many people seem to favor exploratory data analysis before any modeling efforts, but I find that the process of preparing the data for modeling is usually sufficient. My typical approach is to manually view the data and make some preliminary decisions about how I think I can best model the data and different options to explore. After this, I build the model and evaluate performance, and then focus on analyzing errors and reason about the next modeling steps based on where I see the model making errors.

Has Kaggle helped you in your career? If so, how?

Yes, it helped me get my current job. I work at H2O and they greatly value Kaggle achievements. My previous job also liked that I performed well in competitions.

You are also the organizer of a meetup in San Diego with over two thousand participants. Is this related to your experience with Kaggle?

Yes, it is absolutely related. I started out with very little knowledge and tried out a Kaggle competition without much success at first. I went to a local meetup and found people to team up with and learn from. At the time, I got to work with people of a much higher skill level than me and we did really well in a competition, 3rd/4500+ teams.

After this, the group stopped being as consistent and I wanted to keep the community going, so I made my own group and started organizing my own events. I've been doing that for a few years and I get to be on the opposite side of the table teaching people and helping them get started. We originally just focused on Kaggle competitions and trying to form teams, but have slowly started branching off to doing book clubs and lectures on various topics of interest. I attribute a lot of my success to having this dedicated weekly time to study and think about machine learning.

In your experience, what do inexperienced Kagglers often overlook? What do you know now that you wish you'd known when you first started?

In my experience, a lot of people overstate the importance of bias-variance trade-offs and overfitting. This is something I have seen people consistently worry about too much. The focus should not be on making training and validation performance close but on making validation performance as good as possible.

What mistakes have you made in competitions in the past?

My consistent mistake is not exploring enough. Sometimes, I have ideas that I discount too early and turn out to be important for improving performance. Very often, I can get close to competitive performance on the first try, but iterating and continuing to improve as I try new things takes a slightly different skill that I am still working on mastering.

Are there any particular tools or libraries that you would recommend using for data analysis or machine learning?

I use a lot of the standard tools: XGBoost, LightGBM, Pytorch, TensorFlow, scikit-learn. I don't have any strong affinity for a specific tool or library, just whatever is relevant to the problem.

What's the most important thing someone should keep in mind or do when they're entering a competition?

I think the most important thing people have to keep in mind is good validation. Very often I see people fooling themselves thinking their performance is improving but then submitting to the leaderboard and realizing it didn't actually go how they expected. It is an important skill to understand how to match assumptions with your new unseen data and build a model that is robust to new conditions.

Tuning your model validation system

At this point, you should have a complete overview of all possible validation strategies. When you approach a competition, you devise your validation strategy and you implement it. Then, you test whether the strategy you have chosen is correct.

As a golden rule, be guided in devising your validation strategy by the idea that you have to replicate the same approach used by the competition organizers to split the data into training, private, and public test sets. Ask yourself how the organizers have arranged those splits. Did they draw a random sample? Did they try to preserve some specific distribution in the data? Are the test sets actually drawn from the same distribution as the training data?

These are not the questions you would ask yourself in a real-world project. Contrary to a real-world project where you have to generalize at all costs, a competition has a much narrower focus on having a model that performs on the given test set (especially the private one). If you focus on this idea from the beginning, you will have more of a chance of finding the best validation strategy, which will help you rank more highly in the competition.

Since this is a trial-and-error process, as you try to find the best validation strategy for the competition, you can systematically apply the following two consistency checks to figure out if you are on the right path:

1. First, you have to check whether your local tests are consistent, that is, that the single cross-validation fold errors are not so different from each other or, when you opt for a simple train-test split, that the same results are reproducible using different train-test splits.

2. Then, you have to check if your local validation error is consistent with the results on the public leaderboard.

If you're failing the first check, you have a few options depending on the following possible origins of the problem:

*   You don't have much training data
*   The data is too diverse, and every training partition is very different from every other (for instance, if you have too many **high cardinality** features, that is, features with too many levels – like ZIP codes – or if you have multivariate outliers)

In both cases, the point is that you lack data for the model you want to implement. Even when the problem just appears to be that the data is too diverse, plotting learning curves will make it evident that your model needs more data.

In this case, unless you find out that moving to a simpler algorithm works on the evaluation metric (in which case trading variance for bias may worsen your model's performance, but not always), your best choice is to use an extensive validation approach. This can be implemented by:

- Using larger k values (thus approaching LOO where $k = n$). Your validation results will be less about the capability of your model to perform on unseen data. Still, using larger training portions will give you the advantage of more stable evaluations.

- Averaging the results of multiple k-fold validations (based on different data partitions picked by different random seed initializations).

- Using repetitive bootstrapping.

Remember that when you find unstable local validation results, you won't be the only one to suffer from the problem. Usually, this is a common problem due to the data's origin and characteristics. You may get hints at possible solutions by keeping tuned in to the discussion forums. For instance, a good solution for high cardinality features is target encoding; stratification can help with outliers, and so on.

The situation differs when you've passed the first check but failed the second; your local cross-validation is consistent but doesn't hold on the leaderboard. To anticipate this problem, you must carefully note all your experiments, validation test types, random seeds used, and leaderboard results if you submit the resulting predictions. In this way, you can draw a simple scatterplot and try fitting a linear regression or, even simpler, compute a correlation between your local results and the associated public leaderboard scores. It takes some time and patience to annotate and analyze all of these, but it is the most important meta-analysis of your competition performances that you can keep track of.

When the mismatch is because your validation score is systematically lower or higher than the leaderboard score, you actually have a strong signal that something is missing from your validation strategy. Still, this problem does not prevent you from improving your model. In fact, you can keep on working on your model and expect improvements to be reflected on the leaderboard, though not in a proportional way. However, systematic differences are always a red flag, implying something is different between what you are doing and what the organizers have arranged for testing the model.

An even worse scenario occurs when your local cross-validation scores correlate poorly with the leaderboard feedback. This is indeed a red flag. When you realize this is the case, you should immediately run a series of tests and investigations to figure out the reason because, regardless of whether it is a common problem or not, the situation poses a serious threat to your final rankings. There are a few possibilities in such a scenario:

- You figure out that the test set is drawn from a different distribution to the training set. The adversarial validation test (which we will discuss in the next section) is the method that can enlighten you in such a situation.

- The data is non-i.i.d. but this is not explicit. For instance, in *The Nature Conservancy Fisheries Monitoring* competition (`https://www.kaggle.com/c/the-nature-conservancy-fisheries-monitoring`), images in the training set were taken from similar situations (fishing boats). You had to figure out by yourself how to arrange them to avoid the model learning to identify the context of the images rather than the target (see, for instance, this work by Anokas: `https://www.kaggle.com/anokas/finding-boatids`).

- The multivariate distribution of the features is the same, but some groups are distributed differently in the test set. Determining the differences allows you to set your training and validation accordingly and gain an edge. You need to probe the public leaderboard to work this out.

- The test data is drifted or trended, which is usually the case in time series predictions. Again, you need to probe the public leaderboard to get some insight about possible post-processing that could help your score, for instance, applying a multiplier to your predictions, thus mimicking a decreasing or increasing trend in the test data.

As we've discussed before, probing the leaderboard is the act of making specifically devised submissions to get insights about the composition of the public test set. It works particularly well if the private test set is similar to the public one. There are no general methods for probing, so you have to devise a probing methodology according to the type of competition and problem.

For instance, in the paper *Climbing the Kaggle Leaderboard by Exploiting the Log-Loss Oracle* (`https://export.arxiv.org/pdf/1707.01825`), Jacob Whitehill explains how to get a fourth position in a competition without even downloading the training data.

About regression problems, in the recent *30 Days of ML* organized by Kaggle, Hung Khoi explained how probing the leaderboard helped him understand the differences in the mean and standard deviation of the target column between the training dataset and the public test data (see: `https://www.kaggle.com/c/30-days-of-ml/discussion/269541`).

He used the following equation:

$$RMSE^2 = MSE = variance + (mean - guessed_{value})^2$$

Essentially, you need just two submissions to solve for the mean and variance of the test target since there are two unknown terms – variance and mean. This equation is a quadratic equation. It can be solved using techniques like the quadratic formula, completing the square, or factoring.

You can also get some other ideas about leaderboard probing from *Chris Deotte* (`https://www.kaggle.com/cdeotte`) from this post, `https://www.kaggle.com/cdeotte/lb-probing-strategies-0-890-2nd-place`, relevant to the *Don't Overfit II competition* (`https://www.kaggle.com/c/dont-overfit-ii`).

If you want to get a feeling about how probing information from the leaderboard is a double-edged sword, you can read about how *Zahar Chikishev* managed to probe information from the *LANL Earthquake Prediction* competition, ending up in 87th place in the private leaderboard after leading in the public one: `https://medium.com/data-science/how-to-lb-probe-on-kaggle-c0aa21458bfe`.

Using adversarial validation

As we have discussed, cross-validation allows you to test your model's ability to generalize to unseen datasets from the same distribution as your training data. Hopefully, since in a Kaggle competition, you are asked to create a model that can predict the public and private datasets, you should expect that such test data is from the same distribution as the training data. In reality, this is not always the case.

Even if you do not overfit the test data because you have based your decision not only on the leaderboard results but also on your cross-validation, you may still be surprised by the results. This could happen in the event that the test set is slightly different from the training set on which you have based your model. In fact, the target probability and its distribution, as well as how the predictive variables relate to it, inform your model during training about certain expectations that cannot be satisfied if the test data is different from the training data.

Hence, it is not enough to avoid overfitting the leaderboard, as we have discussed up to now, but, in the first place, it is also advisable to find out whether your test data is comparable to the training data. Then, if they differ, you have to figure out if there is any chance that you can mitigate the different distributions between training and test data and build a model that performs on that test set.

Adversarial validation has been developed just for this purpose. This technique allows you to easily estimate the degree of difference between your training and test data. This technique was long rumored among Kaggle participants and transmitted from team to team until it emerged publicly thanks to a post by *Zygmunt Zając* (`https://www.kaggle.com/zygmunt`) on his FastML blog.

The idea is simple:

1. Take your training data and remove the target.
2. Assemble your training data together with your test data.
3. Create a new binary classification target where the positive label is assigned to the test data.
4. At this point, run an ML classifier and evaluate for the ROC-AUC evaluation metric (we discussed this metric in the previous chapter on *Detailing Competition Tasks and Metrics*).

If your ROC-AUC is around 0.5, the training and test data are not easily distinguishable and apparently from the same distribution. ROC-AUC values substantially higher than 0.5 and nearing 1.0 signal that it is easy for the algorithm to figure out what is from the training set and what is from the test set. In such a case, don't expect to be able to easily generalize to the test set because it clearly comes from a different distribution.

> You can find an example notebook written for the *Sberbank Russian Housing Market* competition (`https://www.kaggle.com/c/sberbank-russian-housing-market`) that demonstrates a practical example of adversarial validation and its usage in a competition here: `https://www.kaggle.com/konradb/adversarial-validation-and-other-scary-terms`.

Since your data may be of different types (numeric or string labels) and you may have missing cases, you'll need some data processing before being able to run the classifier successfully. Our suggestion is to use the random forest classifier because:

*   It doesn't output true probabilities, but their predictions can be interpreted as ordinal scores. These ordinal scores are sufficient for calculating ROC-AUC, which only considers the relative ranking of predictions, not the exact probabilities.
*   Random forests is a flexible algorithm based on decision trees that can do feature selection by itself and operate on different types of features without any pre-processing while rendering all the data numeric. It is also quite robust to overfitting, and you don't have to think too much about fixing its hyperparameters.

- You don't need much data processing because of its tree-based nature. For missing data, you can simply replace the values with an improbable negative value such as -999, and you can deal with string variables by converting their strings into numbers (for instance, using the scikit-learn ordinal encoder, `sklearn.preprocessing.OrdinalEncoder`). As a solution, it performs less well than one-hot encoding, but it is very speedy and will work properly for the problem.

Although building a classification model is the most direct way to validate your test set adversarially, you can also use other approaches. One approach is to map both training and test data into a lower-dimensional space, as in this post (`https://www.kaggle.com/nanomathias/distribution-of-test-vs-training-data`) by *NanoMathias* (`https://www.kaggle.com/nanomathias`). Although requiring more tuning work, such an approach based on t-SNE and PCA has the great advantage of being graphically representable, appealing, and understandable.

Don't forget that our brains are more adept at spotting patterns in visual representations than numeric ones (for an articulate discussion about our visual abilities, see `https://onlinelibrary.wiley.com/doi/full/10.1002/qua.24480`).

PCA and t-SNE are not the only tools that can help you reduce the dimensionality of your data and allow you to visualize it. **Uniform Manifold Approximation and Projection (UMAP)** (`https://github.com/lmcinnes/umap`) can often provide a faster low-dimensionality solution with clear and distinct data clusters. Variational autoencoders (discussed in *Chapter 8*, *Modeling for Tabular Competitions*) can also deal with non-linear dimensionality reduction and offer a more useful representation than PCA; however, they are more complicated to set up and tune.

Example implementation

While you can find examples of adversarial validation in the original article by Zygmunt and the notebook we linked, we have created a fresh example for you based on the Playground competition, *Tabular Playground Series – Jan 2021* (`https://www.kaggle.com/c/tabular-playground-series-jan-2021`):

1. You start by importing some Python packages and getting the training and test data from the competition:

```
import numpy as np
import pandas as pd
from sklearn.ensemble import RandomForestClassifier
from sklearn.model_selection import cross_val_predict
from sklearn.metrics import roc_auc_score
```

```
train = pd.read_csv(
    "../input/tabular-playground-series-jan-2021/train.csv")
test = pd.read_csv(
    "../input/tabular-playground-series-jan-2021/test.csv")
```

2. Data preparation is short and to the point. Since all features are numeric, you won't need any label encoding, but you do have to fill any missing values with a negative number (-1 usually works fine) and drop the target and also any identifiers; when the identifier is progressive, the adversarial validation may return a high ROC-AUC score:

```
train = train.fillna(-1).drop(["id", "target"], axis=1)
test = test.fillna(-1).drop(["id"], axis=1)
X = pd.concat([train, test], ignore_index=True)
y = [0] * len(train) + [1] * len(test)
```

3. At this point, you just need to generate `RandomForestClassifier` predictions for your data using the `cross_val_predict` function, which automatically creates a cross-validation scheme and stores the predictions on the validation fold:

```
model = RandomForestClassifier(random_state=0)
cv_preds = cross_val_predict(
    model, X, y, cv=5,
    n_jobs=-1, method='predict_proba'
)
```

Depending on the size of the data, this procedure may take a bit longer; for instance, in our example, it requires about 45 minutes to execute on a Kaggle notebook.

4. As a result, you obtain unbiased predictions (they are not overfitted as you did not predict the same examples you have trained on), which can be used for error estimation. Please note that `cross_val_predict` won't fit your instantiated model, so you won't get any information from it, such as what the important features used by the model are. If you need such information, you must first fit it by calling `model.fit(X, y)`.

5. Finally, you can query the ROC-AUC score for your predictions:

```
print(roc_auc_score(y_true=y, y_score=cv_preds[:,1]))
```

You should obtain a value of around 0.49-0.50 (`cross_val_predict` won't be deterministic unless you use cross-validation with a fixed `random_seed`). This means that you cannot easily distinguish training from test data. Hence, they come from the same distribution.

Handling different distributions of training and test data

ROC-AUC scores of 0.8 or more would alert you that the test set is peculiar and quite distinguishable from the training data. In these cases, what can you do? You actually have a few strategies at hand:

- Suppression
- Training on cases most similar to the test set
- Validating by mimicking the test set

With **suppression**, you remove the variables that most influence the result in the adversarial test set until the distribution result is the same. To do so, you require an iterative approach. This time, you fit your model to all your data, and then you check the importance measures (provided, for instance, by the feature_importances_ method in the scikit-learn RandomForest classifier) and the ROC-AUC fit score. At this point, you remove the most important variable for the model from your data and run everything again. You repeat this cycle where you train, measure the ROC-AUC fit, and drop the most important variable from your data until the fitted ROC-AUC score decreases to around 0.5.

The only problem with this method is that you may actually be forced to remove the majority of important variables from your data, and any model you then build on such variable-censored data won't be able to predict sufficiently correctly due to the lack of informative features.

When you **train on the examples most similar to the test set**, you take a different approach, focusing not on the variables but on the samples you use for training. In this case, you pick only the samples that fit the test distribution from the training set. Any trained model then suits the testing distribution (but it won't be generalizable to anything else), allowing you to test the best on the competition problem.

The limitation of this approach is that you are cutting down the size of your dataset and, depending on the number of samples that fit the test distribution, you may suffer from a very biased resulting model due to the lack of training examples. In our previous example, picking up just the adversarial predictions on the training data that exceed a score estimate of 0.5 and summing them results in picking only 1,495 cases (the number is so small because the test set is not very different from the training set):

```
print(np.sum(cv_preds[:len(X), 1] > 0.5))
```

Finally, with the strategy of **validating by mimicking the test set**, you keep training on all the data. Still, for validation purposes, you pick your examples only from the adversarial predictions on the training set that exceed a score of 0.5 (or an even higher threshold, such as 0.9). As a limitation of this approach, consider that the results of this strategy can be quite sensitive to the choice of threshold. If the threshold is too low, you may include many false positives in your validation set, leading to an overestimation of your model's performance. If the threshold is too high, you may miss valuable examples, leading to an underestimation of your model's performance.

Having a validation set tuned to the test set will allow you to pick all the possible hyperparameters and model choices that will favor a better result on the leaderboard.

In our example, by analyzing the feature importance scores obtained from the model, we can identify the features that exhibit the most significant differences between the training and test sets:

```
model.fit(X, y)
ranks = sorted(list(zip(X.columns, model.feature_importances_)),
               key=lambda x: x[1], reverse=True)
for feature, score in ranks:
    print(f"{feature:10} : {score:0.4f}")
```

To conclude, we have a few more remarks on adversarial validation:

- Using it will generally help you perform better in competitions, but not always. Kaggle's code competitions and other competitions where you cannot fully access the test set cannot be inspected by adversarial validation.

- Adversarial validation can inform you about the test data as a whole. Still, it cannot advise you on the split between the private and the public test data, which is the cause of the most common form of public leaderboard overfitting and consequent shake-up.

- Finally, adversarial validation, though a very specific method devised for competitions, has several practical use cases in the real world: how often have you picked the wrong test set to validate your models? The method we have presented here can enlighten you about whether you properly use the test data and any validation data in your projects. Moreover, data changes and models in production may be affected by such changes and produce bad predictions if you don't retrain them. This is due to both **concept drift**, which occurs when the underlying relationship between the input features (X) and the target variable (Y) changes over time, and **data drift**, which occurs when the distribution of the input features (X) changes over time.

- By using adversarial validation, you can immediately understand if you have to retrain new models to put into production or if you can leave the previous ones in operation.

Before moving on to explore how we can handle data leakage, let's see what *Giuliano Janson* has to say. Giuliano is a Competitions Grandmaster and principal applied scientist for ML and NLP at Zillow Group. He spoke to us about his competition wins, the importance of cross-validation, and data leakages, the subject of the upcoming section.

Interview: Giuliano Janson

`https://www.kaggle.com/adjgiulio`

What's your favorite kind of competition and why? In terms of techniques and solving approaches, what is your specialty on Kaggle?

My perfect competition is made up of a) an interesting problem to solve, b) a mid-size dataset that is small enough to fit in memory but not too small to become an overfitting headache, and c) an opportunity to be creative from a feature engineering perspective. The combination of those three dimensions is where I'm at my best in competitive ML because I feel I have the means to use rigor and creativity without having to worry about engineering constraints.

How do you approach a Kaggle competition? How different is this approach to what you do in your day-to-day work?

A Kaggle competition is a marathon. Going into a competition, I know I can get 90 to 95% of my best final score with a couple of days of work. The rest is a slow grind. The only success metric is your score; nothing else matters.

My daily work looks more like a series of sprints. Model performance is only a small portion of what I need to consider. A go-live date might be just as important, or other aspects such as interpretability, scalability, and maintainability could tip the scale in a totally different direction. After each sprint, priorities are reassessed and the end product might look totally different from what was originally envisioned. Also, modeling is a small part of my day. I spend far more time talking to people, managing priorities, building use cases, scrubbing data, and thinking about everything that it takes to make a prototype model a successful production solution.

Tell us about a particularly challenging competition you entered, and what insights you used to tackle the task.

One of the two competitions I won, the Genentech Cancer competition, was a Masters-only competition. The data provided was raw transactional data. There was no nice tabular dataset to start from. This is the type of work I love because feature engineering is actually one of my favorite parts of ML. Since I had worked in healthcare for a decade at the time of the competition, I had business and clinical insights on the data, but most of all, I had engineering insights on the complexity of correctly handling this type of data and about all the things that can go wrong when this type of transactional raw data is not handled carefully. That turned out to be key to winning, as one of the initial hypotheses regarding a possible source of leakage turned out to be true, and provided a "golden feature" that gave the final boost to our model. The insight from the competition is to always be extra careful when doing feature engineering or setting up validation approaches. Leakage can be very hard to detect and the usual train/validation/test approach to model validation will provide no help in identifying leakage in most cases, thus putting a model at risk of underperforming in production.

Has Kaggle helped you in your career? If so, how?

Kaggle has helped me in two ways. First, it provided a low-barrier entry point to modern ML, a ton of exposure to cutting-edge modeling techniques, and forced me to truly understand the art and science of professional-grade model validation techniques. Second, Kaggle provided access to some of the brightest minds in applied ML. What I learned from teaming up with some of the top Kaggle participants are lessons I cherish and try to share with my teammates every day.

How have you built up your portfolio thanks to Kaggle?

My professional career hasn't been directly impacted much by my Kaggle résumé. By that, I mean I haven't got job offers or interviews as a result of my Kaggle standings. I started Kaggle when I was already in a senior data science role, albeit with not much of an ML focus. Thanks to what I learned on Kaggle, I was able to better advocate a change in my career to move into an ML-focused job.

To this date, many folks I work with enjoy chatting about competitive ML and are curious about tips and tricks from my Kaggle experience, but it is also true that a large portion of the ML community might not even know what Kaggle is.

In your experience, what do inexperienced Kagglers often overlook? What do you know now that you wish you'd known when you first started?

The importance of proper cross-validation is easily overlooked by participants new to competitive ML. A solid cross-validation framework allows you to measure improvement reliably and objectively. In a competition that might be as long as six months, the best models do not usually come from those who have the best initial ideas, but from those who are willing to iterate and adjust based on empirical feedback from the data. A great validation framework is at the foundation of it all.

What mistakes have you made in competitions in the past?

One of the lessons learned that I always share with people new to ML is to "never get over-enamored with overly complex ideas." When facing a new complex problem, it is easy to be tempted to build complex solutions. Complex solutions usually require time to develop. But the main issue is that complex solutions are often of marginal value, conditional on robust baselines. For example, imagine you want to model the outcome of an election and start thinking about a series of features to capture complex conditional relationships among observable and latent geographic, socio-economic, and temporal features. You could spend weeks developing these features, under the assumption that because they are so well thought out, they will be impactful.

The mistake is that while often those complex features could be very powerful on their own, conditional on a series of simple features and on a model that can already build highly optimized, data-driven deep interaction, all of a sudden, the complex features we built with time and effort may lead to little to no marginal improvement. My advice is to stick to Occam's razor and try easy things before being tempted by more complex approaches.

Are there any particular tools or libraries that you would recommend using for data analysis or machine learning?

I'm a pandas and scikit-learn person. I love how pandas enables easy data manipulation and exploration and how I can quickly prototype models using scikit-learn in a matter of minutes. Most of my prototype work is done using these two libraries. That said, my final models are often based on XGBoost. For deep learning, I love using Keras.

Handling data leakage

Data leakage is a common issue in Kaggle competitions that can affect the outcome of the challenge. **Data leakage**, often mentioned simply as **leakage** or with other fancy names (such as the so-called *golden features* or *magic features*), involves information in the training phase that won't be available at prediction time. The presence of such information (leakage) will make your model over-perform in training and testing, allowing you to rank highly in the competition. Still, it will render unusable or, at best, suboptimal any solution based on it from the sponsor's point of view.

We can define leakage as "when information concerning the ground truth is artificially and unintentionally introduced within the training feature data, or training metadata," as stated by Michael Kim (`https://www.kaggle.com/mikeskim`) in his presentation at *Kaggle Days San Francisco* in 2019.

Leakage is often found in Kaggle competitions despite careful checking from both the sponsor and the Kaggle team. Such situations are due to the subtle and sneaky nature of leakage, which can unexpectedly appear due to the intense searching undertaken by Kagglers, who always look for any way to score better in a competition.

Don't confuse data leakage with a **leaky validation strategy**. In a leaky validation strategy, the problem is that you have arranged your validation strategy to favor better validation scores because some information leaks from the training data. It has nothing to do with the competition itself but relates to how you handle your validation. It occurs if you run any pre-processing modifying your data (normalization, dimensionality reduction, missing value imputation) before separating training and validation or test data.

To prevent leaky validation, if you are using scikit-learn to manipulate and process your data, you absolutely have to exclude your validation data from any fitting operation. Fitting operations tend to create leakage if applied to any data you use for validation. The best way to avoid this is to use scikit-learn pipelines (`https://scikit-learn.org/stable/modules/generated/sklearn.pipeline.Pipeline.html`), which will enclose both your data processing and model together, thereby avoiding any risk of inadvertently applying any leaking transformation to your data.

Data leakage is, therefore, not strictly related to validation operations, though it affects them deeply. Even though this chapter is principally devoted to validation strategies, at this point, we consider it necessary to discuss data leakage since this issue can profoundly affect how you evaluate your models and their ability to generalize beyond the competition test sets.

Generally speaking, leakage can originate at a feature or example level.

Feature leakage

Feature leakage is by far the most common. It can be caused by the existence of a proxy for the target or by a feature that is posterior to the target itself. A target proxy could be anything derived from processing the label itself or from the test split process; for instance, when defining identifiers, specific identifiers (a numeration arc, for instance) may be associated with certain target responses, making it easier for a model to guess if properly fed with the information processed in the right way. A more subtle way data processing can cause leakage is when the competition organizers have processed the training and test set together before splitting it. Historically, leakages in Kaggle competitions have been found in:

- Mishandled data preparation from organizers, especially when they operate on a combination of training and test data (for example, in *Loan Default Prediction* (https://www.kaggle.com/c/loan-default-prediction), organizers initially used features with aggregated historical data that leaked future information).

- Row order when it is connected to a time index or to specific data groups (for instance, in *Telstra Network Disruptions* (https://www.kaggle.com/c/telstra-recruiting-network), the order of records in a feature hinted at proxy information, the location, which was not present in the data and which was very predictive).

- Column order when it is connected to a time index (you get hints by using the columns as rows).

- Feature duplication in consecutive rows because it can hint at examples with correlated responses, such as in *Bosch Production Line Performance* (see the first-place solution by *Beluga* at https://www.kaggle.com/c/bosch-production-line-performance/discussion/25434).

- Image metadata (as in *Two Sigma Connect: Rental Listing Inquiries* (https://www.kaggle.com/c/two-sigma-connect-rental-listing-inquiries)).

- Hashes or other easily crackable anonymization practices of encodings and identifiers.

The trouble with posterior information originates from how we deal with information when we do not consider the effects of time and the sequence of cause and effect across time. Since we are looking back at the past, we often forget that certain variables that make sense at the present moment do not have value in the past. For instance, if you have to calculate a credit score for a loan to a new company, knowing that payments of the borrowed money are often late is a great indicator of the lower reliability and higher risk represented by the debtor. Still, you cannot know this before you lend the money. This is also a problem you will commonly find when analyzing company databases in your projects: your query data will represent present situations, not past ones. Reconstructing past information can also be difficult if you cannot specify that you wish to retrieve only the information that was present at a certain time. For this reason, great effort has to be spent on finding these leaking features and excluding or adjusting them before building any model.

Similar problems are also common in Kaggle competitions based on the same kind of data (banking or insurance, for instance). However, since much care is put into preparing the data for the competition, they appear in more subtle ways and forms. In general, it is easy to spot these leaking features since they strongly correlate with the target, and a domain expert can figure out why (for instance, knowing at what stage the data is recorded in the databases). Therefore, in competitions, you never find such obvious features but derivatives of them, often transformed or processed features that have slipped away from the control of the sponsor. Since the features are often anonymized to preserve the sponsor's business, they lurk among the others. This has given rise to a series of hunts for the golden/magic features, a search to combine existing features in the dataset to have the leakage emerge.

You can read an enlightening post by Corey Levison here: https://www.linkedin.com/pulse/ winning-13th-place-kaggles-magic-competition-corey-levinson/. It tells the story of how the *Santander Customer Transaction Prediction* competition turned into a hunt for magic features for his team.

Another good example is provided by *dune_dweller* here: https://www.kaggle.com/c/telstra- recruiting-network/discussion/19239#109766. By looking at how the data was ordered, dune_ dweller found that the data was likely in time order. Putting this information in a new feature increased the score.

Leakage at the example level

The other way leakage can occur is by **leakage in the training examples**. This happens especially with non-i.i.d. data. This means that some cases correlate between themselves because they are from the same period (or from contiguous ones) or the same group. Suppose such cases are not all together in the training or test data but separated. In that case, there is a high chance that the ML algorithm will learn how to spot the cases (and derive the predictions) rather than using general rules. An often-cited example of such a situation involves the team of Prof. Andrew Ng (see https://twitter.com/nizkroberts/status/931121395748270080). In 2017, they wrote a paper using a dataset of 100,000 X-rays from 30,000 patients. They used a random split to separate training and test data, not realizing that the X-rays of the same patient could end up partly in the training set and partly in the test set. Practitioners such as Nick Roberts spotted this fact, pointing out a possible leakage that could have inflated the performances of the model and that led to a substantial revision of the paper itself.

Handling data leakage in Kaggle

What happens when there is a data leakage in a Kaggle competition? Kaggle has clear policies about it and will either:

- Let the competition continue as is (especially if the leakage only has a small impact)
- Remove the leakage from the set and relaunch the competition
- Generate a new test set that does not have the leakage present

In particular, Kaggle recommends making any leakage found public, though this is not compulsory or sanctioned if it doesn't happen. However, in our experience, if there is any leakage in a competition, it will soon become very apparent. The discussion forums will start lighting up with a discussion about magic stuff and the like. You will soon know if you are attentive to what is being said in the forums and can put together all the hints provided by different Kagglers.

However, please beware that some players may even discuss **magic features** to distract other competitors from serious modeling. For instance, in *Santander Customer Transaction Prediction*, there was a famous situation involving some Kagglers who fueled in other participants an interest in magic features that weren't actually so magic, directing their efforts in the wrong direction (see the discussion here: `https://www.kaggle.com/c/santander-customer-transaction-prediction/discussion/87057#502362`).

Our suggestion is to carefully read the discussions around leakage and magic features that arise in the competition's forum, and decide whether to pursue the research and use any leakage found based on your own interest and motivations for participating in the competition.

Not exploiting any leakage may really damage your final rankings. However, it will surely spoil your learning experience (because leakage is a distortion, and you cannot claim anything about the models using it). If you are not participating in a competition in order to gain a reputation or later to approach the sponsor for an opportunity to be hired, it is perfectly fine to use any leakage you come across. Otherwise, ignore it and keep working hard on your models because, who knows; maybe Kaggle will reset or fix the competition by the end, rendering the leakage ineffective, to the great disappointment of the many who used it.

Leakages are very different from competition to competition. If you want to get an idea of a few real leakages that have happened in Kaggle competitions, you can have a look at these three memorable ones:

- `https://www.kaggle.com/c/predicting-red-hat-business-value/discussion/22807` from *Predicting Red Hat Business Value* (`https://www.kaggle.com/c/predicting-red-hat-business-value`), where the problem arose because of an imperfect train/test split methodology of the competition.
- `https://www.kaggle.com/c/talkingdata-mobile-user-demographics/discussion/23403` from *TalkingData Mobile User Demographics* (`https://www.kaggle.com/c/talkingdata-mobile-user-demographics`), where a series of problems and non-i.i.d cases affected the correct train/test split of the competition.
- `https://www.kaggle.com/c/two-sigma-connect-rental-listing-inquiries/discussion/31870` from *Two Sigma Connect: Rental Listing Inquiries* (`https://www.kaggle.com/c/two-sigma-connect-rental-listing-inquiries`), where metadata (the creation time of each folder) did the trick.

Summary

Having arrived at the end of the chapter, we will summarize the advice we have discussed along the way so you can organize your validation strategy and reach the end of a competition with a few suitable models to submit.

In this chapter, we first analyzed the dynamics of the public leaderboard, exploring problems such as adaptive overfitting and shake-ups. We then discussed the importance of validation in a data science competition, building a reliable system, tuning it to the leaderboard, and keeping track of your efforts.

Having discussed the various validation strategies, we also saw the best way of tuning your hyperparameters and checking your test data or validation partitions using adversarial validation. We concluded by discussing some of the various leakages that have been experienced in Kaggle competitions, and we provided advice about how to deal with them.

Here are our closing suggested best practices:

- Always spend the first part of the competition building a reliable validation scheme, favoring a *k*-fold over a train-test split, given its probabilistic nature and ability to generalize to unseen data.

- If your validation scheme is unstable, use more folds or run it multiple times with different data partitions. Always check your test set using adversarial validation.

- Keep track of results based on both your validation scheme and the leaderboard. Trust your validation score more to explore possible optimizations and breakthroughs (such as magic features or leakages).

- As we explained at the beginning of the chapter, use your validation scores when deciding your final submissions to the competition. For your final submissions, depending on the situation and whether or not you trust the leaderboard, choose among your best local cross-validated models and good-scoring submissions on the leaderboard, favoring the first over the second.

At this point of our journey, we are ready to discuss how to tackle competitions using tabular data, which is numeric or categorical data arranged in matrices (with rows representing the examples and columns the features). In the next chapter, we will discuss the Tabular Playground Series, a monthly contest by Kaggle using tabular data (organized by *Inversion*: `https://www.kaggle.com/inversion`).

In addition, we will introduce you to some specific techniques to help you shine in these competitions, such as feature engineering, target encoding, denoising autoencoders, and some neural networks for tabular data, as an alternative to the recognized state-of-the-art learning algorithms in tabular data problems (the gradient boosting algorithms, such as XGBoost, LightGBM, or CatBoost).

Get This Book's PDF Version and Exclusive Extras

UNLOCK NOW

Scan the QR code (or go to `packtpub.com/unlock`). Search for this book by name, confirm the edition, and then follow the steps on the page.

Note: Keep your invoice handy. Purchases made directly from Packt don't require an invoice.

Join our book's Discord space

Join our community's Discord space for discussions with the authors and other readers:

`https://packt.link/kaggle`

8

Modeling for Tabular Competitions

Until 2017, there was no need to distinguish too much between competition types, and since the vast majority of competitions were based on tabular data, you could not even find mention of "tabular competitions" on Kaggle forums. Then, suddenly, something changed. After a relative shortage of competitions (see https://www.kaggle.com/general/49904), deep learning competitions took the upper hand, and tabular competitions became rarer, disappointing many. They became so rare that Kaggle launched a series of tabular competitions based on synthetic data in 2021. What happened?

By 2017-2018, data science had fully matured, and many companies had initiated their data journeys. Data science was still a hot topic, but no longer uncommon. Solutions to problems similar to those that had populated Kaggle for years at the time had become standard practice in many companies. Under these circumstances, sponsors were less motivated to launch external tabular competitions since they were already internally dealing with the same problems. By contrast, deep learning, particularly in the areas of vision and text, including **large language models (LLMs)**, is a rapidly evolving field. While significant advancements have been made, there is still much room for innovation. Kaggle competitions can serve as a valuable platform to push the boundaries of these technologies and encourage the development of novel approaches.

In this chapter, we will discuss tabular competitions. We will touch on some famous historical ones and also focus on the more recent reality of the Tabular Playground Series because tabular problems are standard practice for most data scientists worldwide, and there really is a lot to learn from Kaggle. We will discuss **exploratory data analysis (EDA)** and **feature engineering**, two common activities in these competitions.

After presenting key strategies for feature engineering, we will expand to many related topics, such as categorical encoding, feature selection, target transformations, and pseudo-labeling. We will end by touching on deep learning methodologies for tabular data, presenting a few specialized deep neural networks such as TabNet, and illustrating a **denoising autoencoder (DAE)**. We will explain why autoencoders have become so relevant for recent Kaggle competitions despite still being marginal in real-world applications.

We will cover the following topics:

- The Tabular Playground Series
- Setting a random state for reproducibility
- The importance of EDA
- Reducing the size of your data
- Applying feature engineering
- Pseudo-labeling
- Denoising with autoencoders
- Neural networks for tabular competitions

> It's important to note that data science is a vast field. This chapter will only present a range of specialized techniques and approaches that are characteristic of tabular competitions on Kaggle and that are not easily found outside of Kaggle forums. While this chapter covers some advanced topics specific to tabular competitions, it won't address every topic related to tabular data.

The Tabular Playground Series

Due to the large demand for tabular problems, Kaggle staff started an experiment in 2021, launching a monthly contest called the Tabular Playground Series. The contests were based on synthetic datasets replicating public data or previous competitions' data. The synthetic data was created thanks to a deep learning generative network called the **conditional tabular generative adversarial network (CTGAN)**. *Synthetic Data Vault* (https://sdv.dev/), an MIT initiative, created the technology behind CTGAN and several tools around it. You can find the CTGAN code at https://github.com/sdv-dev/CTGAN. The corresponding paper explains how it works by modeling the probability distribution of rows in tabular data and then generating realistic synthetic data (see https://arxiv.org/pdf/1907.00503v2.pdf).

The result is a set of open source software systems built to help enterprises generate synthetic data that mimics real data; it can help data scientists create anonymous datasets based on real ones and augment existing ones for modeling purposes.

Kaggle launched 13 fairly successful competitions in 2021, which have attracted many Kagglers despite not offering points, medals, or prizes (only some merchandise). The 2021 Tabular Playground competitions offer an interesting overview of many fundamental problems in tabular data. You can also use these tabular competitions to identify specific problems by type or metric and find related resources such as focused discussions or notebooks:

Month	Problem	Variables	Metric	Missing data
January 2021	Regression on an unspecified problem	Numeric	Root mean squared error (RMSE)	No
February 2021	Regression predicting the value of an insurance claim	Numeric and categorical	RMSE	No
March 2021	Binary classification predicting an insurance claim	Numeric and categorical	AUC	No
April 2021	Binary classification on a replica very similar to the original Titanic dataset	Numeric and categorical	Accuracy	Yes
May 2021	Multiclass classification predicting the category of an e-commerce product given various attributes about the listing	Categorical	Multiclass LogLoss	No
June 2021	Multiclass classification predicting the category of an e-commerce product given various attributes about the listing	Numeric and categorical	Multiclass LogLoss	No

July 2021	Multiple regression predicting air pollution in a city via various input sensor values (for example, a time series)	Numeric, time	RMSLE	Yes
August 2021	Regression calculating the loss associated with a loan default	Numeric	RMSE	No
30 Days of ML	Regression on the value of an insurance claim	Numeric and categorical	RMSE	No
September 2021	Binary classification predicting whether a claim will be made on an insurance policy	Numeric	AUC	Yes
October 2021	Binary classification predicting the biological response of molecules given various chemical properties	Numeric and categorical	AUC	No
November 2021	Binary classification identifying spam emails via various features extracted from the email	Numeric	AUC	No
December 2021	Multiclass classification based on the original Forest Cover Type Prediction competition	Numeric and categorical	Multiclass classification accuracy	No

Table 8.1: Tabular Playground Series competitions in 2021

The Tabular Playground competitions continued in 2022, with even more sophisticated and challenging problems:

Month	Problem	Variables	Metric	Missing data
January 2022	Forecasting the sales of Kaggle merchandise from two fictitious independent store chains	Dates and categorical	Symmetric mean absolute percentage error (SMAPE)	No
February 2022	Classifying 10 different bacteria species using data from a genomic analysis technique that contains some data compression and data loss	Numeric	Categorization accuracy	No
March 2022	Forecasting twelve hours of traffic flow in a U.S. metropolis (time series)	Numeric and categorical	Mean absolute error	No
April 2022	Sequences of biological sensor data recorded from several hundred participants who could have been in either of two possible activity states (time series)	Numeric	The area under the ROC curve	No
May 2022	Binary classification problem that includes a number of different feature interactions	Normalized continuous data and categorical data	The area under the ROC curve	No
June 2022	Data imputation challenge	Normalized continuous data and categorical data	RMSE	Yes

July 2022	Unsupervised clustering challenge on manufacturing control data where you have to predict the cluster each row belongs to without having a ground truth to train on	Continuous and categorical data	Adjusted rand index	No
August 2022	Build a model that predicts product failures based on the results of a large product testing study	Continuous and categorical data	The area under the ROC curve	Yes
September 2022	Forecast four book sales in six countries during the tumultuous year 2021	Dates, plus numerical and categorical data	SMAPE	No
October 2022	Predict the probability of each team scoring within the next 10 seconds of the game given a snapshot from a Rocket League match	Numeric	Log Loss	No
November 2022	Given a folder of submissions that contain predictions for a binary classification task, blend (ensemble) the various submission files to produce better model predictions	Binary model predictions	Log loss	No

Table 8.2: Tabular Playground Series competitions in 2022

In 2022, the tabular competitions shifted to more general and creative problems than regression and classification, such as ensembling predictions, figuring out natural clusters in data, or imputing missing data. After the closing of 2022, the series rebooted, dropping the "Tabular" designation in the title in order to accommodate the expansion beyond tabular competitions. Challenges are now identified by season and edition numbers. The Season-Edition structure offers greater flexibility, allowing Kaggle to organize challenges without strict adherence to monthly schedules. Competition durations also tend to vary, and multiple challenges may run concurrently, enhancing participants' overall learning and engagement opportunities. Season 3, which spanned the entirety of 2023, consisted of 26 episodes with varying durations. Season 4 (2024), with its 12 episodes, reverted to being a monthly competition.

Season 5 (2025) is ongoing at the time of writing this book, and it continues in the spirit of previous playgrounds, providing interesting and approachable datasets every month for the community to practice their **machine learning (ML)** skills in the context of tabular data problems.

Much of this chapter has been written by observing the code and discussion that emerged in all these competitions instead of analyzing more glorious competitions from the past. As we mentioned, we believe that tabular competitions in general are indeed gone for good, given the changed professional landscape. However, skills in tabular data analysis are as relevant as ever, and you will find it more beneficial to focus on suggestions and hints relating to the current Tabular Playground Series rather than those from past competitions.

As in other fully-fledged competitions with Kaggle points and medals, in tabular competitions, we recommend you follow a simple yet very effective pipeline:

1. Start with EDA.
2. Data preparation comes next.
3. Move on to modeling (using a cross-validation strategy for model validation).
4. Begin post-processing.
5. Submission is the last step.

As a rule, you also have to take care to maintain reproducibility and save all the models (from every fold), the list of the parameters used, all the fold predictions, all the out-of-fold predictions, and all predictions from models trained on all the data.

You should save all this information in a way that makes it easy to recover and reconstruct – for instance, using appropriate labeling, taking note of MD5 hashing values (you can refer to this Stack Overflow answer for details: `https://stackoverflow.com/questions/16874598/how-do-i-calculate-the-md5-checksum-of-a-file-in-python`), and tracking the CV scores and leaderboard results from each experiment. Most Kagglers do this with simple tools such as `.txt` files or Excel spreadsheets. Still, there exist more sophisticated ways, such as using the following:

- **Weights and Biases** (`https://wandb.ai/site`)
- **MLflow** (`https://mlflow.org/`)
- **Neptune** (`https://neptune.ai/experiment-tracking`)
- **DVC** (`https://dvc.org/`)

In the next chapter, we will have a concluding section dealing with one of these tools, *Weight and Biases*, applied to artifact (data and models) storage, results logging, and automatic hyperparameter optimization. However, let us emphasize that, in the end, what matters are the results, not the tool you use, so try your best to keep order in your experiments and models, even in the heat of a competition and even with just the help of a simple notebook or spreadsheet.

Before we proceed, consider also thinking about the technology that Kaggle used to generate the data for these competitions; if you can properly understand how the data has been generated, you get an important advantage. In addition, understanding how synthetic data works can impact how you do data science in the real world because it allows you to easily obtain more varied data for training.

For instance, let's take the *Google Brain – Ventilator Pressure Prediction* competition (`https://www.kaggle.com/c/ventilator-pressure-prediction`). In this competition, you had to develop ML for mechanical ventilation control. Although you could obtain good results by modeling the data provided with deep learning, given the synthetic origin of the data, you could also reverse engineer its generative process and obtain a top leaderboard result, as *Jun Koda* (`https://www.kaggle.com/junkoda`) did and explains in his post:

`https://www.kaggle.com/c/ventilator-pressure-prediction/discussion/285278`

Generating artificial data by yourself and understanding synthetic data has never been so easy, as you can verify from this notebook (`https://www.kaggle.com/lucamassaron/how-to-use-ctgan-to-generate-more-data`), derived from a notebook originally coded and tested by Dariush Bahrami (`https://www.kaggle.com/dariushbahrami`).

Setting a random state for reproducibility

Before we start discussing the steps and models you may use in a tabular competition, returning to the reproducibility theme we discussed earlier will be useful.

In most of the commands in the code you see in Kaggle Notebooks, you will find a parameter declaring a number, a **seed**, as the random state. This setting is important for the reproducibility of your results. Since many algorithms are not deterministic but based on randomness, by setting a seed, you can influence the behavior of the random generator, making it *predictable* in its randomness. This is because the same random seed corresponds to the same sequence of random numbers. In other words, it allows you to obtain the same results after every run of the same code.

That is why you find a random seed-setting parameter in all ML algorithms in scikit-learn, as well as in all scikit-learn-compatible models (for instance, XGBoost, LightGBM, and CatBoost, to name the most popular ones).

The reproducibility of results is important in real-world projects, as well as in Kaggle competitions. In the real world, having a reproducible model allows for better tracking of model development and consistency. In Kaggle competitions, reproducibility helps test hypotheses better because you control any source of variation in your models. For instance, if you created a new feature, putting it into a reproducible pipeline would help you understand whether it was advantageous. You would be sure that any improvement or deterioration in the model could be attributed only to the feature and not to the effects of some random process that had changed since the last time you ran the model.

Again, reproducibility can be used to your advantage when dealing with public notebooks. These notebooks will often have a fixed seed that could be 0, 1, or 42. The value 42 is quite popular because it is a reference to Douglas Adams' *The Hitchhiker's Guide to the Galaxy*, in which it is the "answer to the ultimate question of life, the universe, and everything," calculated by an enormous supercomputer named Deep Thought over a period of 7.5 million years. The use of 42 as a random seed is so widespread that LLMs, having learned from code repositories and other sources, tend to frequently return 42 when asked to generate a random number (see, for example, this Reddit discussion: `https://www.reddit.com/r/ChatGPT/comments/1796evg/so_apparently_asking_chatgpt_to_give_you_a_random/`). Now, if everyone in a competition is using the same random seed, it could have a double effect:

- The random seed might be working too well with your local validation scheme or the public leaderboard, which means overfitting and a lack of generalizing properties to the private leaderboard

- A lot of Kagglers will produce similar results that will influence their standings in the private leaderboard in the same way – for instance, experiencing the same directionality in a shake-up when the private leaderboard is revealed

By changing the random seed, you are avoiding overfitting and also breaking rank; in other words, you are getting different results from everyone else, which could put you at an advantage in the end. In particular, more than when using classical ML models, when you leverage deep learning, setting one random seed instead of another can make a big difference in your results.

In deep learning, various random factors influence the training process. These factors include the random initialization of model weights, random selection of batches for training, and random application of data augmentation techniques. This inherent randomness becomes more pronounced when fine-tuning models on smaller datasets and dealing with regression tasks, which typically exhibit more volatile validation metrics. It is therefore normal to have widely different performances of a deep learning model just because of different random seeds, and it may also happen that a specific random seed may just make your model perform better on your local test.

The suggestion is never to treat the random seed as a hyperparameter to optimize because it is not. A random seed is simply a source of randomness. Instead, the right approach is to try as many random seeds as possible to determine whether your solution is robust (it brings comparable results in local tests and the public leaderboard) and, hence, can successfully generalize to the private leaderboard. Another strategy is to ensemble the same model trained under different random seeds to stabilize it. You can explore this topic by reading an interesting post by Philipp Singer (https://www.kaggle.com/philippsinger) on X about his experience using multiple random seeds to obtain models that can get top scores in Kaggle competitions: https://twitter.com/ph_singer/status/1696877654497013835, or you can read *The Kaggle Grandmaster Playbook* from NVIDIA technical blog (https://developer.nvidia.com/blog/the-kaggle-grandmasters-playbook-7-battle-tested-modeling-techniques-for-tabular-data), where Kazuki Onodera, Théo Viel, Gilberto Titericz, and Chris Deotte distil years of experience in competitions and quote different random seeds as a way to build stronger blending ensembles of the same model. Fascinating is Figure 5 in the post (labelled *Benefit of Ensembling 100 XGBoost with itself different seed!* and also to be found in Chris Deotte's solution for the playground competition *Predicting Optimal Fertilizers*: https://www.kaggle.com/competitions/playground-series-s5e6/writeups/chris-deotte-1st-place-fast-gpu-experimentation-wi), which shows how ensembling XGBoost models with different random seeds steadily improves the evaluation metric (MAP@5 in the Playground competition *Predicting Optimal Fertilizers* at https://www.kaggle.com/competitions/playground-series-s5e6) compared to single-seed averages, based on XGBoost's internal random sampling.

As a concluding remark about the necessity of rendering your models on Kaggle reproducible, keep in mind that if you end up winning a Kaggle competition, you will probably need to demonstrate how your models produced the winning submission. In this case, it is paramount that everything is completely reproducible if you want to obtain your prize quickly and smoothly. In real-life data science projects, reproducibility is always paramount because of model accountability and sometimes because of specific regulations in fields such as banking/finance, insurance, and medicine.

TensorFlow and PyTorch models don't explicitly use a random seed parameter, so it is more challenging to ensure their complete reproducibility. The following code snippet, when run, sets the same random seed for the TensorFlow and PyTorch models:

```
def seed_everything(seed,
                    tensorflow_init=True,
                    pytorch_init=True):
    """
    Seeds basic parameters for reproducibility of results
    """
    random.seed(seed)
    os.environ["PYTHONHASHSEED"] = str(seed)
    np.random.seed(seed)
    if tensorflow_init is True:
        tf.random.set_seed(seed)
    if pytorch_init is True:
        torch.manual_seed(seed)
        torch.cuda.manual_seed(seed)
        torch.backends.cudnn.deterministic = True
        torch.backends.cudnn.benchmark = False
```

As for scikit-learn, it is instead advisable to set the random seed directly – when it is allowed by the class or the function – using the `random_state` parameter.

The importance of EDA

The term **EDA** comes from the work of *John W. Tukey*, one of the most prominent exponents of modern statistical methodology. In his 1977 book *Exploratory Data Analysis* (hence the acronym EDA), Tukey thinks of EDA as a way to explore data, uncover evidence, and develop hypotheses that statistical tests can later confirm.

His idea was that how we define statistical hypotheses could be based more on observation and reasoning than just sequential tests based on mathematical computations. This idea translates well to the world of ML because, as we will discuss in the next section, data can be improved and pre-digested so that learning algorithms can work better and more efficiently.

Performing EDA in Kaggle

In an EDA for a Kaggle competition, you will be looking for the following:

- Missing values and, most importantly, missing value patterns correlated with the target
- Skewed numeric variables and their possible transformations
- Rare categories in categorical variables that can be grouped together
- Potential outliers, both univariate and multivariate
- Highly correlated (or even duplicated) features; for categorical variables, focus on categories that overlap
- The most predictive features for the problem

You achieve this by several descriptive analyses, graphs, and charts, first examining each distinct feature (**univariate analysis**, in statistical terms), then matching a couple of variables (**bivariate** analysis, such as in a scatterplot), and finally considering more features together at once (a **multivariate** approach).

As part of the EDA, since your models have to be robust to distribution changes when shifting from your local validation to the public and private leaderboard, you also need to:

- Ensure that the test data is from a similar distribution to the train data by resorting to adversarial validation, as we described in the previous chapter, in the section on *Using Adversarial Validation*.
- If time is among the predictors, ensure that the target is not affected by the flow of time. For such verification, you need to plot the target against time and look for trends and seasonal patterns. If anything comes up, you know that you must validate based on time (using time-based splits) and develop models that account for the role of time (for instance, creating new features from the time variable itself, such as lag features based on the previous time step or moving averages).

If you are feeling lazy or unsure about how and where to start, relying on automated strategies can help you initially. For instance, you may find that **AutoViz** (`https://github.com/AutoViML/AutoViz`), a popular rapid EDA freeware tool, can save you a lot of time. You can install it on your notebook by running the following command:

```
!pip install git+git://github.com/AutoViML/AutoViz.git
```

You can obtain a clearer understanding of what AutoViz can do for you by reading the Medium article by Dan Roth at https://towardsdatascience.com/autoviz-a-new-tool-for-automated-visualization-ec9c1744a6ad or browsing a few interesting public Notebooks such as https://www.kaggle.com/gvyshnya/automating-eda-and-feature-importance-detection by Georgii Vyshnia (https://www.kaggle.com/gvyshnya).

At the latter link, you will also find references to another tool, **Sweetviz** (https://github.com/fbdesignpro/sweetviz). Sweetviz has an overview article and tutorial based on the Titanic dataset at https://towardsdatascience.com/powerful-eda-exploratory-data-analysis-in-just-two-lines-of-code-using-sweetviz-6c943d32f34.

Another popular tool that you may find useful is **pandas Profiling** (https://github.com/pandas-profiling/pandas-profiling), which is more reliant on classical statistical descriptive statistics and visualization, as explained by this article:

https://medium.com/analytics-vidhya/pandas-profiling-5ecd0b977ecd

Waiting for other Kagglers to publish interesting EDA notebooks could also be a solution, so always keep an eye on the Notebooks sections; sometimes, precious hints may appear. This should kick-start your modeling phase and help you understand the basic dos and don'ts of the competition. However, remember that EDA stops being a commodity and becomes an asset for the competition when it is highly specific to the problem at hand; this is something that you will never find in automated solutions and seldom in public notebooks. You have to do your EDA by yourself and gather key winning insights.

All things considered, our suggestion is to look into the automated tools a bit because they are really easy to learn and run. You will save a lot of time that you can instead spend looking at charts and reasoning about possible insights, and that will certainly help your competition performance. However, after doing that, you need to pick up Matplotlib and Seaborn and try something by yourself on not-so-standard plots that depend on the type of data provided and the problem.

For example, if you are given a series of measurements performed over time, plotting the continuous function based on time is as useful as plotting the single recorded points in time, such as showing different lags between one observation and another, a fact that may point to revealing insights for better predictions.

Dimensionality reduction with t-SNE and UMAP

There are many possible plots you can create when doing EDA, and it is not our intention to list them all here. However, there are a couple of dimensionality reduction plots that are worth spending a few words on because they can provide as much information as very specific and data-tailored charts. These are t-distributed stochastic neighbor embedding (**t-SNE**; `https://lvdmaaten.github.io/tsne/`) and uniform manifold approximation and projection (**UMAP**; `https://github.com/lmcinnes/umap`).

t-SNE and UMAP are two techniques often used by data scientists that enable the projection of multivariate data into lower dimensions. They are often used to represent complex sets of data in two dimensions. 2D UMAP and t-SNE plots can reveal the presence of outliers and relevant clusters for your data problem. In fact, if you can plot the scatter graph of the resulting 2D projection and color it by target value, the plot may give hints about possible strategies for dealing with subgroups.

Although it is related to an image competition, an excellent example of how UMAP and t-SNE can help you understand your data better is Chris Deotte's analysis for the *SIIM-ISIC Melanoma Classification* competition (see `https://www.kaggle.com/c/siim-isic-melanoma-classification/discussion/168028`). Chris has related training and test data on the same low-dimensionality projections in this example, highlighting portions where only test examples were present.

Though UMAP and t-SNE offer invaluable help in discovering hard-to-find data patterns, you can still use them as features in your modeling efforts. An interesting example of this usage was demonstrated in the *Otto Group Product Classification Challenge*, where Mike Kim used t-SNE projections as training features for the competition (`https://www.kaggle.com/c/otto-group-product-classification-challenge/discussion/14295`).

As stated by the article *How to t-SNE Effectively* (`https://distill.pub/2016/misread-tsne/`), you have to use these techniques properly because it is easy to spot clusters and patterns where there are none. The same warning is valid for UMAP because it can also produce plots that can be misread. Guides such as `https://pair-code.github.io/understanding-umap/` offer sound advice on the performance of both UMAP and t-SNE on real-world data, providing suggestions and caveats.

Despite these dangers, in our experience, these approaches are certainly more revealing than the classical methods based on variance restructuring by linear combination, such as PCA or SVD. Compared to these approaches, UMAP and t-SNE manage to reduce the dimensionality in a more meaningful way, allowing for the visual charting of the results while maintaining the topography of the data. As a side effect, they are much slower to fit. However, NVIDIA has released its **RAPIDS** suite (`https://developer.nvidia.com/rapids`) based on CUDA, which, using a GPU-powered notebook or script, returns the results of both UMAP and t-SNE in a very reasonable timeframe, allowing their effective use as an EDA tool.

You can find a useful example of applying both UMAP and t-SNE with a RAPIDS implementation and a GPU for data exploration purposes for the *30 Days of ML* competition at `https://www.kaggle.com/lucamassaron/interesting-eda-tsne-umap/`. In the following figure, which is the output of this example notebook, you can see how multiple clusters populate the dataset, but none of them could be deemed to reveal a particular relationship with the target:

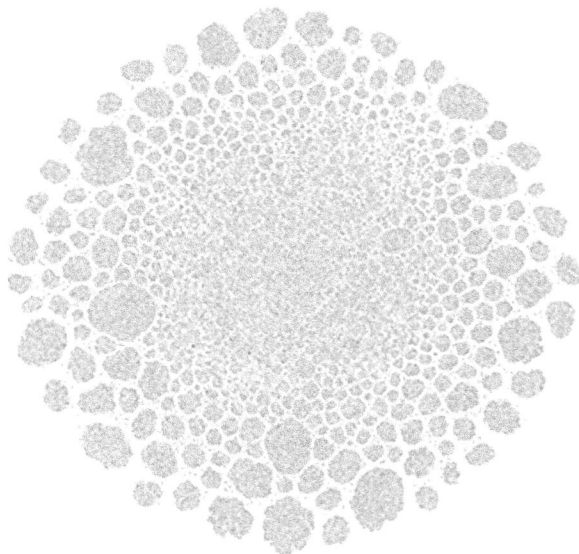

Figure 8.1: Multiple clusters appearing in a t-SNE plot

In another notebook (`https://www.kaggle.com/lucamassaron/really-not-missing-at-random`), the same techniques are applied to the binary indicators for missing samples instead, revealing evocative figures that hint at specific and separate areas dominated by a certain type of response. Indeed, in that example, missing samples did not occur at random, and they were quite predictive:

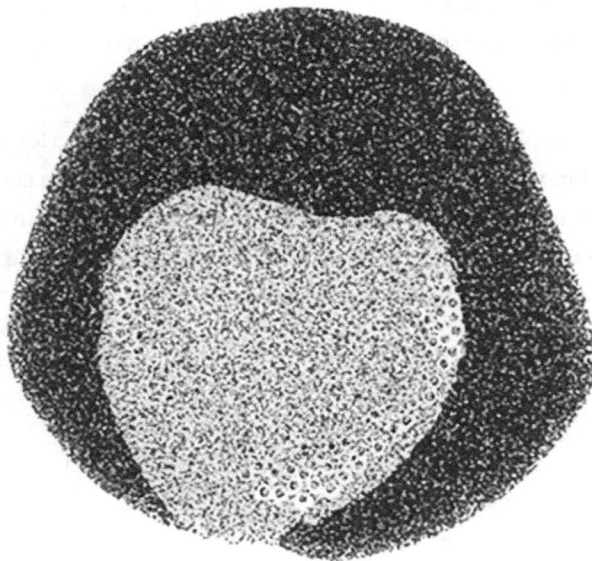

Figure 8.2: This t-SNE plot easily reveals areas where the positive target is predominant

Reducing the size of your data

If you are working directly on Kaggle Notebooks, you will likely find their limitations to be quite annoying and dealing with them to be a timesink. One of these limitations is the **out-of-memory errors** that will stop the execution and force you to restart the script from the beginning. This is quite common in many competitions. However, unlike deep learning competitions based on text or images, where you can retrieve the data from disk in small batches and have them processed, most of the algorithms that work with tabular data require handling all the data in memory.

The most common situation is when you have uploaded the data from a CSV file using pandas' `read_csv`. Still, the DataFrame is too large to be handled for feature engineering and ML in a Kaggle notebook. The solution is to compress the size of the pandas DataFrame you are using without losing any information (**lossless compression**). This can easily be achieved using the following script derived from the work by Guillaume Martin (you can find the original notebook at `https://www.kaggle.com/gemartin/load-data-reduce-memory-usage`):

```python
def reduce_mem_usage(df, verbose=True):
    numerics = ['int16', 'int32', 'int64',
                'float16', 'float32', 'float64']
    start_mem = df.memory_usage().sum() / 1024**2
    for col in df.columns:
        col_type = df[col].dtypes
        if col_type in numerics:
            c_min = df[col].min()
            c_max = df[col].max()
            if str(col_type)[:3] == 'int':
                if (c_min > np.iinfo(np.int8).min and
                        c_max < np.iinfo(np.int8).max):
                    df[col] = df[col].astype(np.int8)
                elif (c_min > np.iinfo(np.int16).min and
                          c_max < np.iinfo(np.int16).max):
                    df[col] = df[col].astype(np.int16)
                elif (c_min > np.iinfo(np.int32).min and
                          c_max < np.iinfo(np.int32).max):
                    df[col] = df[col].astype(np.int32)
                elif (c_min > np.iinfo(np.int64).min and
                          c_max < np.iinfo(np.int64).max):
                    df[col] = df[col].astype(np.int64)
            else:
                if (c_min > np.finfo(np.float16).min and
                        c_max < np.finfo(np.float16).max):
                    df[col] = df[col].astype(np.float16)
                elif (c_min > np.finfo(np.float32).min and
                          c_max < np.finfo(np.float32).max):
                    df[col] = df[col].astype(np.float32)
                else:
                    df[col] = df[col].astype(np.float64)
    end_mem = df.memory_usage().sum() / 1024**2
    if verbose:
        print('Mem. usage decreased to {:5.2f} Mb ({:.1f}% reduction)'
              .format(end_mem, 100 * (start_mem - end_mem) / start_mem))
    return df
```

Guillaume Martin was not the first to propose an idea like this on Kaggle. The very first Kaggler with the idea of compressing a pandas DataFrame was Arjan Groen, who wrote a reducing function during the Zillow competition (https://www.kaggle.com/arjanso/reducing-dataframe-memory-size-by-65).

This script leverages the fact that all the numeric features in a dataset reside in a specific range of values. Since we have different types of integer and floating-point numeric variables in Python, based on the number of bytes they occupy in memory, the script compares the range of values found in each feature to the maximum and minimum values that each numeric type can accept. This is done to set the feature to the numeric type that works with its range of values and requires the lowest memory.

The approach works like a breeze on Kaggle Notebooks, but with some caveats. Once you have set the best-fitting numeric type for each feature by compression, you cannot apply any feature engineering that may result in values exceeding the capacity of the selected numeric types, because such an operation will produce erroneous results. We suggest applying it after feature engineering or major transformations that do not rescale your existing data. Combining it with the garbage collection library gc and the gc.collect() method will improve the memory situation of your Kaggle notebook.

Another way to reduce the size of your data (among other things) is to use feature engineering (in particular, feature selection and data compression).

Speeding up data processing

With the incredible recent growth of deep learning in data science, GPUs have become an indispensable tool, now easily accessible for both local and cloud-based computing and naturally available in Kaggle notebooks. Previously associated with tasks like 3D gaming and graphic rendering, GPUs have found a natural fit in ML due to their efficiency in performing large-scale matrix operations. Their cost-effectiveness and speed have made them essential for training neural networks, prompting widespread adoption in academia and industry alike. NVIDIA's RAPIDS, a suite of GPU-accelerated libraries, extends beyond deep learning to support the entire ML workflow. The NVIDIA KGMoN team (the Grandmasters' team) often uses it in competitions, because using GPUs can increase the number of high-quality experiments they can run during a competition, which means discovering more patterns and catching faster when a model is failing, drifting, or overfitting.

There are several usage examples of how Chris Deotte (`https://www.kaggle.com/cdeotte`), Giba (`https://www.kaggle.com/titericz`), and other KGMoN team members (`https://www.nvidia.com/en-us/ai-data-science/kaggle-grandmasters`) have leveraged RAPIDS in tabular data problems for fast feature engineering and **gradient-boosted decision trees (GBDT)** inference. Basically, there are two possible approaches using GPUs for accelerating your work on Kaggle:

- To accelerate operations on dataframes using GPU drop-in replacements for pandas or Polars, and scale data preparation and feature engineering
- To train models more rapidly using GPU backends of XGBoost, LightGBM, and Catboost, or the NVIDIA cuML library for other machine learning models.

For instance, you can take a look at examples in Kaggle notebooks, showing how to harness GPU power, such as the following:

- Chris Deotte has used RAPIDS cuDF for fast feature engineering and RAPIDS FIL for fast GBDT inference. Both helped achieve a 14th place gold: `https://www.kaggle.com/code/cdeotte/xgboost-starter-0-793`.

  You can also find the discussion about its usage at `https://www.kaggle.com/competitions/amex-default-prediction/discussion/347641`.

- Giba in Optiver Trading used RAPIDS cuDF for fast feature engineering, achieving a top-50 solution in the time series forecast: `https://www.kaggle.com/code/titericz/rapids-gpu-accelerated-from-scratch-solution`

This toolkit optimizes not only neural network computations but also traditional ML tasks, such as those involving large datasets with noisy or complex feature sets. **Support vector machines (SVMs)**, for instance, benefit significantly from GPU acceleration, enabling faster processing of high-dimensional and sparse data. The RAPIDS libraries are designed to integrate seamlessly with familiar APIs, making it easy for practitioners to adopt the technology without needing to learn entirely new frameworks—an advantage for Kaggle competitors seeking to optimize their models without a steep learning curve. Here is a quick overview:

RAPIDS package	Task	API mimicked
cuPy	Array operations	NumPy
cuDF	Data processing	pandas
cuML	ML	scikit-learn

Table 7.3: An overview of available RAPIDS packages

In addition, using a GPU and integrating RAPIDS algorithms can revitalize classic ML models for tabular data, making them competitive with more modern approaches. For instance, SVMs, which are not really nicely scalable to larger datasets, have been used in a few solutions from the KGMoN team, demonstrating how you can gain an edge in tabular data problems (and even in image competitions, too, when using embeddings) by leveraging a good old classic ML algorithm:

- Chris Deotte in *Feedback Prize* NLP, using a RAPIDS SVR on NLP regression without fine-tuning any LLM. Here are two examples, the second of which led to third place: `https://www.kaggle.com/code/cdeotte/rapids-svr-starter-cv-0-830-lb-0-804` and `https://www.kaggle.com/code/cdeotte/rapids-svr-cv-0-450-lb-0-44x`. You can find more discussion about the methods at `https://www.kaggle.com/competitions/feedback-prize-english-language-learning/discussion/369609`.

- Giba leveraged RAPIDS SVR in the *PetFinder* competition on computer vision without fine-tuning any image models. That won Giba first place: `https://www.kaggle.com/competitions/petfinder-pawpularity-score/discussion/301686`

  You can examine the code at `https://www.kaggle.com/code/titericz/imagenet-embeddings-plus-rapids-svr`.

- Kazuki Onodera (`https://www.kaggle.com/onodera`), in the *Trends* competition, successfully used RAPIDS SVR and a few other tabular ML algorithms and won second place: `https://www.kaggle.com/competitions/trends-assessment-prediction/discussion/162765`

By leveraging GPU acceleration, these traditional algorithms can handle larger datasets and complex computations much faster, narrowing the performance gap with deep learning methods.

Applying feature engineering

In real-world projects, what can make the difference between a successful ML model and a mediocre one is often the data, not the model. When we talk about data, the differentiator between bad, good, and excellent data is not just the lack of missing values and the reliability of the values (its "quality"), or the number of available examples (its "quantity"). In our experience, the real differentiator is the informational value of the content itself, which is represented by the type of features.

The features are the real clay to mold in a data science project because they contain the information that models use to separate the classes or estimate the values. Every model has expressiveness and the ability to transform features into predictions, but if you lack features, no model can bootstrap you and offer better predictions. *Models only make the value in data apparent. They are not magical in and of themselves.*

On Kaggle, apart from the rare competitions where you can look for further data to add, all participants have the same data available from the beginning. At that point, how you handle the data makes most of the difference. Overlooking the fact that you can improve the data you have is a common mistake made by many Kagglers. **Feature engineering**, a set of techniques for transforming data into more useful information for your models, is invariably the key to performing better in competitions. Even the more powerful models you can apply need you to process the data and render it into a more understandable form.

Feature engineering is also the way you embed any **prior knowledge** (usually specialist expertise on the problem) into the data: by summing, subtracting, or dividing the existing features, you obtain indicators or estimates that you know can better explain the problem you are dealing with. There are also other purposes of feature engineering, which are less valuable in a Kaggle competition but could prove important in a real-world project:

- The first is to reduce the training data size (this could also be useful in a Kaggle competition when working with notebooks, which have limits in memory)
- The second is to make the interpretation of the resulting model easier by using features that are understandable to humans

Each domain may have encoded specific variable transformations that are not necessarily self-evident but well-known to field experts. Just think of finance, where you must separate signals from noise for different sets of features representing market and company data by applying specific transformations like Kalman filters or wavelet transformations. Given the large number of possible fields and the complexity of many feature engineering procedures, in this section, we won't enter into specific domains of expertise and their particular ways of dealing with features. Instead, we will present you with the most common and most general techniques that you can apply in any tabular competition.

Easily derived features

Deriving features with transformations is the simplest approach, but often the most effective. For instance, computing feature ratios (dividing one feature by another) can prove quite effective because many algorithms cannot mimic divisions (for example, gradient boosting) or can have a hard time trying to (for example, deep neural networks). Here are the most common transformations to try out:

- **Time feature processing**: Several tasks can be performed using time feature processing techniques. The most common include splitting a date into its elements (year, month, and day), transforming it into the week of the year and the weekday, calculating differences between dates, and computing differences to key events (such as holidays).

  For dates, another common transformation is extracting time elements from a date or a time. Cyclic continuous transformations (based on sine and cosine transformations) are also useful for representing the continuity of time and creating periodic features:

  ```
  cycle = 7
  df['weekday_sin'] = np.sin(
      2 * np.pi * df['col1'].dt.dayofweek / cycle)
  df['weekday_cos'] = np.cos(
      2 * np.pi * df['col1'].dt.dayofweek / cycle)
  ```

- **Numeric feature transformations**: These include scaling; normalization; logarithmic or exponential transformations; separating the integer and decimal parts; and summing, subtracting, multiplying, or dividing two numeric features. Scaling obtained by standardization (the z-score method used in statistics) or by normalization (also called min-max scaling) of numeric features can make sense if you are using algorithms sensitive to the scale of features, such as any neural network.

- **Binning of numeric features**: This is used to transform continuous variables into discrete ones by distributing their values into a number of bins. Binning helps remove noise and errors in data, and it allows easy modeling of non-linear relationships between the binned features and the target variable when paired with **one-hot encoding**, a categorical data processing technique that merges two or three categorical features together (see the scikit-learn implementation, for instance: https://scikit-learn.org/stable/modules/generated/sklearn.preprocessing.KBinsDiscretizer.html).

- **Categorical feature encoding**: This involves one-hot encoding or the more sophisticated target encoding (more on this in the following sections).

- **Splitting and aggregating categorical features based on the levels**: For instance, in the *Titanic* competition (https://www.kaggle.com/c/titanic), you can split names and surnames, as well as their initials, to create new features.

- **Creating polynomial features**: Polynomial features are created by raising features to an exponent. See, for instance, this scikit-learn function: https://scikit-learn.org/stable/modules/generated/sklearn.preprocessing.PolynomialFeatures.html

Apart from these feature engineering techniques, there are more data cleaning techniques, as well as missing data and outlier treatments, that involve making changes to the data that nevertheless transform your features and can help make signals from the data emerge:

- **Missing values treatment:** Make binary features that point out missing values because sometimes missingness is not random, and a missing value could have some important reason behind it. Usually, missingness points out something about the way data is recorded, acting like a proxy variable for something else. It is just like in census surveys: if someone doesn't tell you their income, it likely means they are extremely poor or rich. If required by your learning algorithm, replace the missing values with the mean, median, or mode (it is seldom necessary to use more sophisticated methods).

  You can refer to this complete guide written by Parul Pandey (`https://www.kaggle.com/parulpandey`) as a reference:

  `https://www.kaggle.com/parulpandey/a-guide-to-handling-missing-values-in-python`

  Just remember that some models can handle missing values by themselves and do so fairly better than many standard approaches because the handling of missing values is part of their optimization procedure. The models that can handle missing values by themselves are all gradient-boosting models:

  - **XGBoost:** `https://xgboost.readthedocs.io/en/latest/faq.html`
  - **LightGBM:** `https://lightgbm.readthedocs.io/en/latest/Advanced-Topics.html`
  - **CatBoost:** `https://catboost.ai/docs/concepts/algorithm-missing-values-processing.html`

- **Outlier capping or removal:** Exclude, cap to a maximum or minimum value, or modify outlier values in your data. To do so, you can use sophisticated multivariate models, such as those present in scikit-learn (`https://scikit-learn.org/stable/modules/outlier_detection.html`).

  Otherwise, you can simply locate the outlying samples in a univariate fashion, basing your judgment on how many standard deviations they are from the mean or their distance from the boundaries of the **interquartile range (IQR)**. In this case, you might simply exclude any points above the value of `1.5 * IQR + Q3` (upper outliers) and any points below `Q1 - 1.5 * IQR` (lower outliers). Once you have found the outliers, you can also proceed by pointing them out with a binary variable.

All these data transformations can add predictive performance to your models, but they are seldom decisive in a competition. Though it is necessary, you cannot simply rely on basic feature engineering. In the following sections, we'll suggest more complex procedures for extracting value from your data.

Meta-features based on rows and columns

To perform competitively, you need trickier feature engineering. A good place to start is looking at features based on each **row**, considered separately:

- Compute the mean, median, sum, standard deviation, minimum, or maximum of the numeric values (or of a subset of them)
- Count the missing values
- Compute the frequencies of common values found in the rows (for instance, considering the binary features and counting the positive values)
- Assign each row to a cluster derived from a cluster analysis, such as k-means

These **meta-features** (called thus because they are features that are representative of a set of single features) help to distinguish the different kinds of samples found in your data by pointing out specific groups of samples to your algorithm.

Meta-features can also be built based on **columns**. Aggregation and summarization operations on single features instead aim to provide further information about the value of numeric and categorical features; *is this characteristic common or rare?* This is information that the model cannot grasp because it cannot count categorical instances in a feature.

As meta-features, you can use any kind of column statistic (such as mode, mean, median, sum, standard deviation, min, max, skewness, and kurtosis for numerical features). For column-wise meta-features, you can proceed in a few different ways:

- **Frequency encoding**: Simply count the frequency of the values in a categorical feature and then create a new feature to replace those values with their frequency. You can also apply frequency encoding to numeric features with frequently recurring values.
- **Frequencies and column statistics computed with respect to a relevant group**: In this case, you can create new features from the values of both numeric and categorical features because you are considering distinct groups in the data. A group could be a cluster you compute by cluster analysis or a group you can define using a feature (for instance, age may produce age groups, locality may provide areas, and so on).

The meta-features describing each group are then applied to each sample based on its group. For instance, using a pandas groupby function, you can create your meta-features, which are then merged with the original data based on the grouping variable. The trickiest part of this feature engineering technique is finding meaningful groups in the data to compute the features on.

- Further column frequencies and statistics can be derived by combining more groups together.

The list is certainly not exhaustive, but it should show you how to look for new features at the feature level and row level using frequencies and statistics.

Let's see a simple example based on the *Amazon Employee Access Challenge* data:

1. First, we will apply a frequency encoding on the ROLE_TITLE feature:

```
import pandas as pd
train = pd.read_csv(
    "../input/amazon-employee-access-challenge/train.csv")
# Frequency count of a feature
feature_counts = train.groupby('ROLE_TITLE').size()
print(train['ROLE_TITLE'].apply(lambda x: feature_counts[x]))
```

The result will show that the feature classes have been replaced by their observed frequency.

2. We now proceed to encode the ROLE_TITLE feature based on the groupings of ROLE_DEPTNAME because we expect that different titles may be more common in certain departments and rarer in others.

The result is a new feature composed of both, which we use to count the frequency of its values:

```
feature_counts = train.groupby([
    'ROLE_DEPTNAME', 'ROLE_TITLE'
]).size()
print(train[['ROLE_DEPTNAME', 'ROLE_TITLE']]
    .apply(lambda x: feature_counts[x[0]][x[1]], axis=1))
```

You can find all the working code and the results in this Kaggle notebook:

```
https://www.kaggle.com/lucamassaron/meta-features-and-target-encoding/
```

Target encoding

Categorical features are usually not a challenge to deal with, thanks to simple functions offered by scikit-learn, such as the following:

- LabelEncoder: This function is used to convert categorical labels into numeric values. It assigns a unique integer to each category. However, it is best suited for target variables (e.g., labels in classification tasks), but it can also be used to transform non-numerical labels.

- OneHotEncoder: This technique is useful for transforming categorical variables into a binary matrix. It creates a separate binary column for each category.

- OrdinalEncoder: Similar to LabelEncoder, this encoder assigns an integer to each category but is typically applied to feature variables. It assumes an inherent ordering among the categories, though it can also be used when assigning arbitrary orderings.

Such functions can transform categories into numeric features and then into binary features that are easily dealt with by ML algorithms. However, when the number of categories to deal with is too large, the dataset resulting from a one-hot encoding strategy becomes **sparse** (most values in it will be zero values) and cumbersome to handle for the memory and processor of your computer or notebook. In these situations, we talk about a **high-cardinality feature**, which requires special handling.

Since early Kaggle competitions, high-cardinality variables have been processed using an encoding function computed according to Daniele Micci-Barreca's paper, *A preprocessing scheme for high-cardinality categorical attributes in classification and prediction problems* (https://dl.acm.org/doi/10.1145/507533.507538). The idea behind this approach is to transform the many categories of a categorical feature into their corresponding expected target value, the definition of which would differ based on the method used:

- In the case of a regression, this is the average expected value for that category

- For a binary classification, it is the conditional probability given that category

- For a multi-class classification, you have the conditional probability for each possible outcome

For instance, in the *Titanic* GettingStarted competition (https://www.kaggle.com/competitions/titanic), where you have to figure out the survival probability of each passenger, target encoding a categorical feature, such as the gender feature, would mean replacing the gender value with its average probability of survival.

In this way, the categorical feature is transformed into a numeric one without having to convert the data into a larger and sparser dataset. In short, this is **target encoding**, and it is indeed very effective in many situations because it resembles a stacked prediction based on the high-cardinality feature. Like stacked predictions, however, where you are essentially using a prediction from another model as a feature, target encoding brings about the risk of overfitting. In fact, when some categories are too rare, using target encoding is almost equivalent to providing the target label. There are ways to mitigate this risk. Blending the encoded value with the average expected target value is a common approach, as we will demonstrate in the following implementation of this technique.

Before seeing the implementation, you can directly import it into your code. Let's see an actual code example of target encoding. This code was used for one of the top-scoring submissions of the *PetFinder.my Adoption Prediction* competition:

```python
import numpy as np
import pandas as pd
from sklearn.base import BaseEstimator, TransformerMixin
class TargetEncode(BaseEstimator, TransformerMixin):

    def __init__(self, categories='auto', k=1, f=1,
                 noise_level=0, random_state=None):
        if type(categories)==str and categories!='auto':
            self.categories = [categories]
        else:
            self.categories = categories
        self.k = k
        self.f = f
        self.noise_level = noise_level
        self.encodings = dict()
        self.prior = None
        self.random_state = random_state

    def add_noise(self, series, noise_level):
        return series * (1 + noise_level *
                          np.random.randn(len(series)))

    def fit(self, X, y=None):
        if type(self.categories)=='auto':
```

```python
        self.categories = np.where(X.dtypes == type(object()))[0]
    temp = X.loc[:, self.categories].copy()
    temp['target'] = y
    self.prior = np.mean(y)
    for variable in self.categories:
        avg = (temp.groupby(by=variable)['target']
                    .agg(['mean', 'count']))
        # Compute smoothing
        smoothing = (1 / (1 + np.exp(-(avg['count'] - self.k) /
                    self.f)))
        # The bigger the count the less full_avg is accounted
        self.encodings[variable] = dict(self.prior * (
            1 - smoothing) + avg['mean'] * smoothing)

    return self

def transform(self, X):
    Xt = X.copy()
    for variable in self.categories:
        Xt[variable].replace(self.encodings[variable],
                            inplace=True)
        unknown_value = {value:self.prior for value in
                        X[variable].unique()
                        if value not in
                        self.encodings[variable].keys()}
        if len(unknown_value) > 0:
            Xt[variable].replace(unknown_value, inplace=True)
        Xt[variable] = Xt[variable].astype(float)
        if self.noise_level > 0:
```

```
            if self.random_state is not None:
                np.random.seed(self.random_state)
            Xt[variable] = self.add_noise(Xt[variable],
                                            self.noise_level)
        return Xt

    def fit_transform(self, X, y=None):
        self.fit(X, y)
        return self.transform(X)
```

The input parameters of the function are as follows:

- categories: The column names of the features you want to target-encode. You can leave 'auto' on, and the class will pick the object strings.

- k: This integer number is the minimum number of samples to take a category average into account.

- f: This integer number has a smoothing effect to balance the category average versus the prior probability or the mean value relative to all the training examples.

- noise_level: This is the amount of noise you want to add to the target encoding to avoid overfitting. Start with very small numbers.

- random_state: This is the reproducibility seed to replicate the same target encoding when noise_level > 0.

Notice the presence of the k and the f parameters. In fact, for a level i of a categorical feature, we are looking for an approximate value that can help us better predict the target using a single encoded variable. Replacing the level with the observed conditional probability could be the solution, but it doesn't work well for levels with few observations. The solution is to blend the observed posterior probability on that level (the probability of the target given a certain value of the encoded feature) with the a priori probability (the probability of the target observed on the entire sample) using a lambda factor. This is called the **empirical Bayesian approach.**

In practical terms, we are using a function to determine whether, for a given level of a categorical variable, we are going to use the conditional target value, the average target value, or a blend of the two. This is dictated by the lambda factor, which, for a fixed k parameter (usually, it has a unit value, implying a minimum cell frequency of two samples), has different output values depending on the f value we choose. As shown by the following chart, where the x-axis represents the number of cases for a given categorical level and the y-axis the weight of the conditional target value, smaller f values tend to switch abruptly from using the average target to using the conditional value:

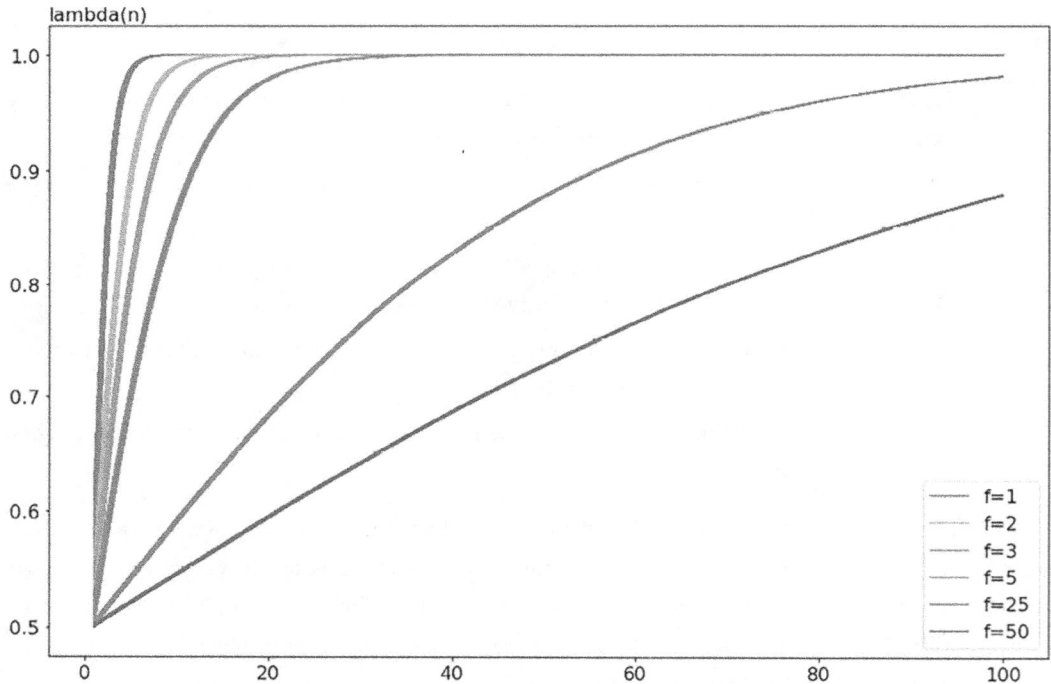

Figure 8.3: Plot of lambda values (on the y-axis) depending on f values and sample size of the categorical value (on the x-axis)

As you can see, higher values of f tend to blend the conditional value with the average unless we are dealing with a categorical level with a large sample size.

Therefore, for a fixed k, higher values of f dictate less trust in the observed empirical frequency and more reliance on the empirical probability for all cells. The right value for f is usually a matter of testing (supported by cross-validation) since you can consider the f parameter a hyperparameter in itself.

After all these explanations, the class is actually quite straightforward to use. Instantiate it with the names of the features you want to target-encode and the parameters you want to try on some training data. Then, you can transform any other piece of data, only target-encoding the fitted features:

```
te = TargetEncode(categories='ROLE_TITLE')
te.fit(train, train['ACTION'])
te.transform(train[['ROLE_TITLE']])
```

The example works on the same *Amazon Employee Access Challenge* data we used before, and it only target-encodes the ROLE_TITLE feature.

Instead of writing your own code, you can also use the package from https://github.com/scikit-learn-contrib/category_encoders and its target encoder (http://contrib.scikit-learn.org/category_encoders/targetencoder.html). It is an out-of-the-box solution that works exactly like the code in this section.

Using feature importance to evaluate your work

Applying too much feature engineering can have side effects. If you create too many features that are correlated or not important for the problem, models could take too long to complete their training, and you may get worse results. This may seem like a paradox, but it is explained by the fact that every variable carries some noise (a random component due to measurement or recording errors) that may be picked by mistake by the model. The more variables you use, the higher the chance your model may pick up noise instead of signals. Therefore, you should try to keep only the relevant features in the dataset you use for training; consider feature selection as a part of your feature engineering process (the pruning phase).

Figuring out the features you need to keep is a hard problem because, as the number of available features grows, the number of possible combinations grows, too. There are various ways to select features, but first, it is important to consider the stage in your data preparation pipeline where the selection must happen.

Based on our experiences, we suggest you consider placing feature selection at the *end* of your data preparation pipeline. Since features share a part of their variance with other features, you cannot evaluate their effectiveness by testing them one at a time; you have to consider them all at once to figure out which you should use correctly.

In addition, you should then test the effectiveness of your selected features using cross-validation. Therefore, after you have all the features prepared, a consistent pipeline, and a working model (it doesn't need to be a fully optimized model, but it should work properly and return acceptable results for the competition), you are ready to test what features should be retained and what could be discarded. At this point, there are various ways to perform feature selection:

- Classical approaches used in statistics resort to forward addition or backward elimination by testing each feature entering or leaving the set of predictors. Such an approach can be quite time-consuming, though, because it relies on some measure of the internal importance of variables or their effect on the model's performance with respect to a specific metric, which you have to recalculate for every feature at every step of the process.

- For regression models, using lasso selection can provide a hint about all the important yet correlated features (the procedure may, in fact, retain even highly correlated features) by using the **stability selection** procedure. In stability selection, you test what features should be retained multiple times (using a bagging procedure) – considering only the features whose coefficients are not zero at each test – and then you apply a voting system to keep the ones most frequently assigned non-zero coefficients. You can get more details about the procedure at this repository: https://github.com/scikit-learn-contrib/stability-selection.

- For tree-based models, such as random forests or gradient boosting, a decrease in impurity or a gain in the target metric based on splits are common ways to rank features. A threshold can cut away the least important ones.

- Test-based randomization of features (or simple comparisons with random features) helps to distinguish features that help the model predict correctly from features that are just noise or redundant. This is always true for tree-based models, but is also easily generalizable to other models.

Relative to this last point, an example of how randomizing features helps in selecting important features is proposed in this example by Chris Deotte in the *Ventilator Pressure Prediction* competition: https://www.kaggle.com/cdeotte/lstm-feature-importance

This notebook tests the role of features in an LSTM-based neural network. First, the model is built and the baseline performance is recorded. Then, one by one, features are shuffled, and the model is required to predict again. If the resulting prediction worsens, it suggests you shuffled an important feature that you shouldn't have touched. Instead, if the prediction performance stays the same or improves, the shuffled feature is not influential or detrimental to the model.

There is also no free lunch in importance evaluation. Shuffling doesn't require re-training, which is a great advantage when training a fresh model costs time. However, it can fail in certain situations:

- Shuffling can sometimes create unrealistic input combinations that make no sense to evaluate.

- In other cases, it can be fooled by the presence of highly correlated features (incorrectly determining that one is important and the other is not). In this case, the best solution is to remove the feature (instead of shuffling it), retrain the model, and then evaluate its performance against the baseline.

In another approach based on shuffled features, **Boruta** (`https://github.com/scikit-learn-contrib/boruta_py`) uses random features to test the model's validity in an iterative fashion. An alternative version of the Boruta selection procedure, **BorutaShap** (`https://github.com/Ekeany/Boruta-Shap`), leverages SHAP values to combine feature selection and for explainability reasons. The resulting selection is usually more reliable than simple rounds of removal or randomization of features because features are tested multiple times against random features until they can statistically prove their importance. Boruta or BorutaShap may take up to 100 iterations and can only be performed using tree-based ML algorithms.

If you are selecting features for a linear model, Boruta may actually overshoot. This is because it will consider the features important for their main effects and interactions with other features (but in a linear model, you only care about the main effects and a selected subset of interactions). You can still effectively use Boruta when selecting a linear model by using gradient boosting with max depth set to one tree, so you are only considering the main effects of the features and not their interactions.

You can have a look at how simple and quick it is to set up BorutaShap feature selection by following this tutorial notebook presented during the *30 Days of ML* competition:

`https://www.kaggle.com/lucamassaron/tutorial-feature-selection-with-boruta-shap`

Pseudo-labeling

In competitions where the number of examples used for training can make a difference, **pseudo-labeling** can boost your scores by providing further examples taken from the test set. The idea is to add examples from the test set whose predictions you are confident about to your training set.

First introduced in the *Santander Customer Transaction Prediction* competition by team Wizardry (read more at `https://www.kaggle.com/c/santander-customer-transaction-prediction/discussion/89003`), pseudo-labeling simply helps models to refine their coefficients thanks to having more data available, but it won't always work. First of all, it is not necessary in some competitions. Adding pseudo-labels won't change the result; it may even worsen it if there is some added noise in the pseudo-labeled data.

Unfortunately, you cannot know for sure beforehand whether or not pseudo-labeling will work in a competition (you have to test it empirically), though plotting learning curves may provide you with a hint as to whether having more data could be useful (see this example provided by scikit-learn: `https://scikit-learn.org/stable/auto_examples/model_selection/plot_learning_curve.html`).

Second, it is not easy to decide which parts of the test set predictions to add or how to tune the entire procedure for the best results. Generally, this is the procedure:

1. Train your model.
2. Predict on the test set.
3. Establish a confidence measure.
4. Select the test set elements to add.
5. Build a new model with the combined data.
6. Predict using this model and submit.

An excellent example of the complete procedure for obtaining pseudo-labeling has been offered by Chris Deotte in the *Instant Gratification* competition: `https://www.kaggle.com/cdeotte/pseudo-labeling-qda-0-969`

You don't need to know more than a few tricks to apply it.

You should consider a few caveats when applying pseudo-labeling:

*   You should have a very good model that produces good predictions for them to be usable in training. Otherwise, you will just add more noise.
*   Since it is impossible to have perfect predictions in the test set, you must distinguish the good ones from those you shouldn't use. If you are predicting using CV folds, check your predictions' standard deviation (this works with regression and classification problems) and pick only the test examples where the standard deviation is the lowest. If you are predicting probabilities, use only high-end or low-end predicted probabilities (the cases where the model is actually more confident).

- When you concatenate the training examples with the test ones in order to build the second model, do not put in more than 50% of the test examples. Ideally, a share of 70% original training examples and 30% pseudo-labeled examples is the best. Suppose you put in too many pseudo-labeled examples. In that case, your new model will risk learning little from the original data and more from the easier test examples, resulting in a distilled model that does not perform better than the original. In fact, as you are training, your model is also learning how to deal with noise in labels, but pseudo-labeled examples do not have this noise.

  Don't forget that you cannot completely trust your pseudo-labels, so keep in mind that you are also partially spoiling your data by using test predictions as training examples. The trick works when you get more benefits than adverse effects from doing so.

- If you depend on validation for early stopping, fixing hyperparameters, or simply evaluating your model, do not use pseudo-labels in the validation. They could be highly misleading. Always use the original training cases for the same reasons we quoted earlier.

- If possible, use a different kind of model when training to estimate the pseudo-labels and when training your final model using both the original labels and the pseudo-labels. This will ensure you are not simply enforcing the same information your previous model used, but also extracting new information from the pseudo-labels.

Clearly, pseudo-labeling is more of an art than a science. It can make a difference in certain competitions, but it needs to be executed very well to generate results. Consider it a resource, and always try one submission based on pseudo-labels.

Denoising with autoencoders

Autoencoders are a type of artificial neural network primarily used for unsupervised learning to efficiently encode and reconstruct data. They consist of three main components: an encoder, a bottleneck (or latent space), and a decoder. Initially better known for being fit for tasks such as non-linear data compression (a kind of non-linear PCA) and image denoising, autoencoders started being recognized as an interesting tool for tabular competitions after Michael Jahrer (https://www.kaggle.com/mjahrer) successfully used them to win the *Porto Seguro's Safe Driver Prediction* competition (https://www.kaggle.com/c/porto-seguro-safe-driver-prediction). *Porto Seguro* was a popular insurance-based risk analysis competition (with more than 5,000 participants) characterized by particularly noisy features.

Michael Jahrer describes how he found a better representation of the numeric data for subsequent neural net supervised learning using **DAEs**. A DAE can produce a new dataset with a considerable number of features based on the activations of the hidden layers at the center of the network, as well as the activations of the middle layers encoding the information.

In his famous post (`https://www.kaggle.com/c/porto-seguro-safe-driver-prediction/ discussion/44629`), Michael Jahrer describes how a DAE can not only remove noise but also automatically create new features, so the representation of the features is learned in a similar way to what happens in image competitions. In the post, he mentions the secret sauce for the DAE recipe, which is not simply the layers but the **noise** you put into the data in order to augment it. He also made it clear that the technique requires stacking together training and test data, implying that the method would not have applications beyond winning a Kaggle competition. In fact, after this winning exploit, the technique disappeared from the forums and most competitions until its recent re-emergence during the Tabular Playground Series.

DAEs are technically composed of the following parts:

- The **encoding** part takes the training data as input, followed by a few dense layers. Ideally, you have a hidden middle layer whose activations just encode all the training information. If the number of nodes in this middle layer is smaller than the original input shape, you have a **compression**, and hopefully, in statistical terms, you are representing some latent dimensionality that is behind the generative process of the input data. Otherwise, you are simply eliminating redundancies and separating noise from signals (which is not a bad result).

- In the second part of the layer, the **decoder** part, you are enlarging the layers again until they regain the shape of the original input. The output is compared with the input to compute an error loss to backpropagate to the network.

From these solutions, you can deduce that there are two types of DAEs:

- In **bottleneck DAEs**, mimicking the approach used in image processing, you take the activations from the middle layer as new features, the one separating the encoding part from the decoding part. These architectures have an hourglass shape, first reducing the number of neurons layer by layer until the middle bottleneck layer, then expanding it back in the second half. The number of hidden layers, which are part of the encoder and decoder, is always odd.

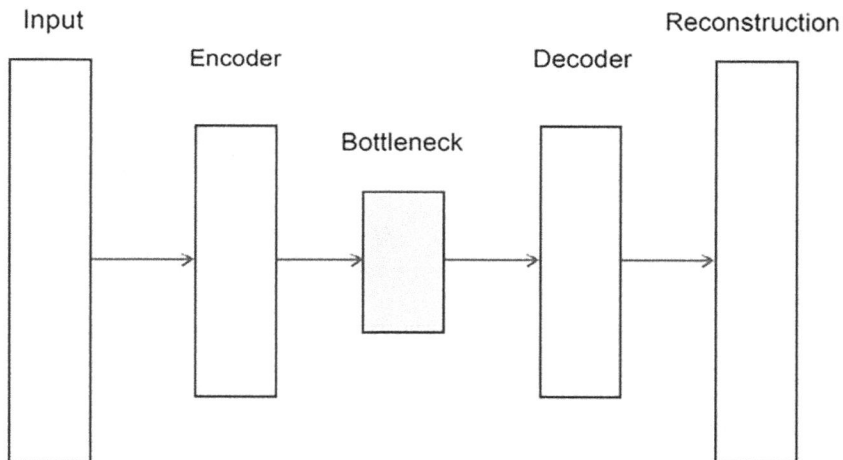

Figure 8.4: A bottleneck DAE, in which you take only the bottleneck layer weights as features

- In **deep stack DAEs,** you take all the activations from the hidden layers without distinguishing between the encoding, decoding, or middle layer. In these architectures, layers are the same size. The number of hidden layers can be even or odd.

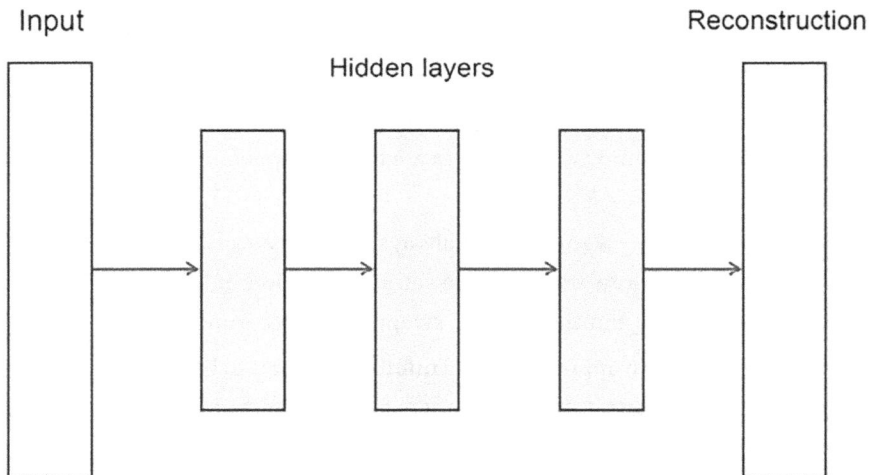

Figure 8.5: A deep stack DAE, in which you take all the stacked hidden layer weights as features

As we mentioned, an important aspect often discussed is adding some **random noise** to your DAE. To help train any kind of DAE, you need to inject noise that helps augment the training data and avoid the over-parameterized neural network just memorizing inputs (in other words, overfitting). In the *Porto Seguro* competition, Michael Jahrer added noise by using a technique called **noise swapping**, which he described as follows:

Here I sample from the feature itself with a certain probability "inputSwapNoise" in the table above. 0.15 means 15% of features replaced by values from another row.

What Jahrer describes is essentially a data augmentation technique similar to **mixup**, which is also commonly used in image augmentation (see https://arxiv.org/abs/1710.09412).

However, the difference between noise swapping and mixup lies in the mechanics:

- Noise swapping replaces a percentage of feature values from one instance with values from another instance in the same dataset, introducing noise but maintaining the individual labels.
- Mixup, on the other hand, generates new examples by creating a weighted interpolation of both the feature values and their corresponding labels from two examples. This effectively blends both the inputs and the outputs.

In his walkthrough (https://www.kaggle.com/springmanndaniel/1st-place-turn-your-data-into-daeta), Danzel describes three approaches to implement noise swapping:

- In **column-wise** noise swapping, you swap values in a certain number of columns. The proportion of columns whose values are to be swapped is decided based on your mixup probability.
- In **row-wise** noise swapping, you always swap a certain number of values in each row. Essentially, every row contains the same proportion of swapped values, based on the mixup probability, but the features swapped change from row to row.
- In **random** noise swapping, you fix a number of values to be swapped based on the mixup probability, and you randomly pick them from the entire dataset (this is somewhat similar to row-wise swapping in effect).

Like pseudo-labeling, DAE is also more of an art than a science, which is another way to say that it is all trial and error. It won't always work, and the details that make it work on one problem probably won't help for another. In order to obtain a good DAE for your competition, you need to keep an eye on a series of aspects that need to be tested and tuned:

- The architecture of the DAE (deep stack tends to work better, but you need to determine the number of units per layer and the number of layers)
- Learning rate and batch size
- Loss (also distinguishing between the loss of numeric and categorical features helps)
- Stopping point (the lowest loss is not always the best; use validation and early stopping if possible)

Depending on the problem, you should expect to face some difficulties in setting up the right architecture and adjusting it to work properly. Your efforts, however, could be rewarded by a top result on the final private leaderboard. In fact, in quite a few tabular competitions, DAE techniques appeared as part of the recipe of many winning submissions:

- Danzel (`https://www.kaggle.com/springmanndaniel`) reported in `https://www.kaggle.com/c/tabular-playground-series-jan-2021/discussion/216037` having used the hidden weights of three 1,500-neuron layers, expanding the original data from 14 columns to 4,500. This new, processed dataset was used as input in other neural networks and gradient-boosting models.

- Ren Zhang (`https://www.kaggle.com/ryanzhang`) discussed his solution (`https://www.kaggle.com/c/tabular-playground-series-feb-2021/discussion/222745`) and shared his code (`https://github.com/ryancheunggit/Denoise-Transformer-AutoEncoder`), revealing that he used stacked transformer encoders rather than your typical linear and ReLU-activated hidden layers (and that such an approach can mean it takes up to 20 hours to train a proper DAE). In his approach, he also suggested adding some random noise to the data (by using a noise mask) to be reconstructed and to compute the loss based not only on the error from reconstructing the original data but also on the noise mask. Using this combined loss helps the network to converge better. Studying the code provided at the GitHub link and the graph in the Kaggle discussion post will help you better understand and easily replicate this innovative approach.

- JianTT (`https://www.kaggle.com/jiangtt`) noticed how some techniques that are key to DAEs, in particular creating new observations by adding noise, can help train better algorithms without the need to create a complete DAE: `https://www.kaggle.com/c/tabular-playground-series-apr-2021/discussion/235739`

If you don't want to spend too much time building your own DAE, but you would like to explore whether something like it could work for the competition you are taking on, you can test out a couple of pre-prepared solutions. First, you can refer to a notebook for a PyTorch network from Hung Khoi (`https://www.kaggle.com/hungkhoi/train-denoise-transformer-autoencoder`) and re-adapt it to your needs, or you can use the Kaggler library from Jeong-Yoon Lee (`https://www.kaggle.com/jeongyoonlee`). In his notebook, Jeong-Yoon Lee presents how it works on one of the Tabular Playground competitions: `https://www.kaggle.com/jeongyoonlee/dae-with-2-lines-of-code-with-kaggler`.

AutoML for tabular competitions

Automated Machine Learning (AutoML) promises to automate the tedious and time-consuming aspects of machine learning model development, which enables rapid prototyping by testing multiple models and creating baseline models. AutoML solutions can also provide guidance on how to handle features and how to combine solutions effectively.

Designed to provide access to modeling for domain experts and non-coders, AutoML solutions draw strong inspiration from Kaggle competitions. Tools such as DataRobot (`https://www.datarobot.com`), H2O Driverless AI (`https://h2o.ai`), and AutoGluon (`https://auto.gluon.ai`) drew on the experience of Kaggle Grandmasters and took inspiration from their solutions to develop automatic pipelines that achieve optimal results, starting from a raw training dataset.

From the *Kaggle Days San Francisco Edition*, where participants struggled to beat the Google AutoML solution (for more insights, read this report from Google Research: `https://research.google/blog/an-end-to-end-automl-solution-for-tabular-data-at-kaggledays`), to the recent 2024 AutoML Grand Prix on Kaggle, Kagglers have consistently had mixed feelings about AutoML.

On the one hand, there is always the fear that AutoML solutions may replace personal abilities in modeling and data preparation for computing power (the one who can afford more extensive hyperparameter optimization often wins). However, even some recent claims about how LLMs can completely replace humans in competitions using grandmaster-level agents should be put in the proper perspective (see, for instance, the paper *Large language models orchestrating structured reasoning achieve Kaggle grandmaster level*, now renamed *Kolb-Based Experiential Learning for Generalist Agents with Human-Level Kaggle Data Science Performance*, `https://arxiv.org/abs/2411.03562`). In the end, all these automation tools have proven to be somewhat effective. However, they are still far from being capable of fully replacing human judgment and performing at a grandmaster level in competitions.

On the other hand, AutoML solutions can improve Kagglers' experience in competitions through:

- **Rapid baseline model generation**: Providing a strong baseline model to compare and learn from when building your own hand-crafted solutions.

- **Accelerated experimentation**: AutoML can rapidly test whether your feature engineering can make a difference with different models and ensembles.

- **Democratization of competitions**: It enables individuals with strong domain knowledge but limited machine learning expertise to participate and compete effectively.

- **Enhancing expert workflows**: Even experienced data scientists can benefit from AutoML, which handles the task of searching for optimal models and hyperparameters, thereby freeing up time for other value-added activities in tabular modeling, such as feature engineering.

Based on the results of the 2024 AutoML Grand Prix, it is a good idea to keep an eye on and try integrating your modeling routine with AutoML tools such as LightAutoML (`https://lightautoml.readthedocs.io`), AutoGluon (`https://auto.gluon.ai`), and H2O (`https://h2o.ai/platform/ai-cloud/make/h2o-driverless-ai`).

Neural networks for tabular competitions

Having discussed neural networks with DAEs, we will complete this chapter by discussing how neural networks can help you more generally in a tabular competition. Gradient boosting (particularly GBDT) solutions still clearly dominate tabular competitions, as well as real-world projects. As for the success of GBDT solutions in real-world applications, we think that the comment in the *TabArena* paper (TabArena is a recent benchmark for machine learning on tabular data, which you can read about at `https://arxiv.org/abs/2506.16791`) precisely represents our experience and what happens in large and smaller organizations: "we perceive the most common reasons for practitioners to rely on GBDTs instead of deep learning to be the insufficient code quality and maintenance of deep learning methods compared to GBDT frameworks".

You can get an idea of how dominant gradient boosting tools are in Kaggle tabular competitions by looking at the winning solutions that use one of the top GBDT tools, XGBoost: `https://github.com/dmlc/xgboost/tree/master/demo#machine-learning-challenge-winning-solutions`. In their paper, *Why do tree-based models still outperform deep learning on typical tabular data?* (`https://arxiv.org/abs/2207.08815`), Léo Grinsztajn, Edouard Oyallon, and Gaël Varoquaux discuss in an academic rather than practical way why GBDT excels over DNNs. They answer that gradient boosting is resistant to irrelevant features, thereby preserving the data's orientation and easily learning irregular functions of the form $y=f(x)$. The following chart showcases the performance of different models on tabular data while tuning hyperparameters:

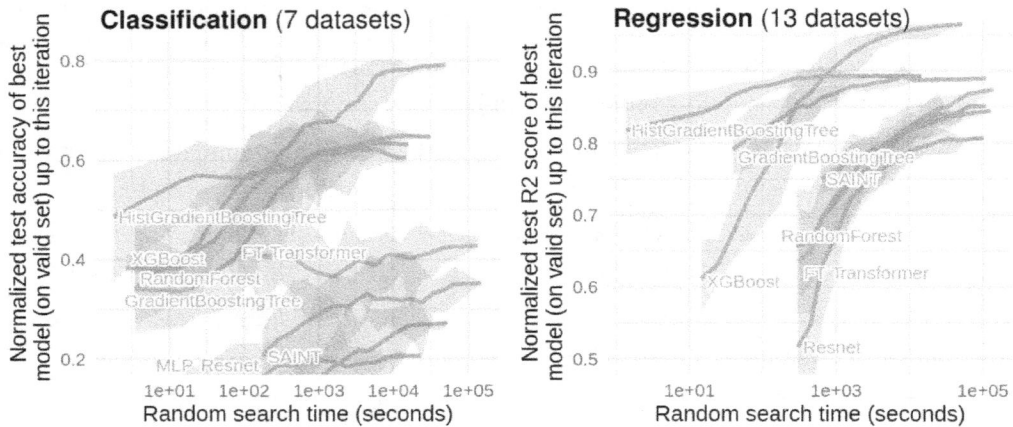

Figure 8.6: Performance of different models on tabular data (source: Grinsztajn et al., 2022)

Here, each value represents the test score of the best model (on the validation set) after a specific amount of time spent conducting a random search. The shaded ribbons represent the minimum and maximum scores across 15 shuffled runs. The HistGradientBoostingTree, GradientBoosting-Tree, and RandomForest models are from scikit-learn, while FT-transformer, Saint, ResNet, and MLP are all deep-learning architectures. The FT-transformer and Saint models are specifically designed for tabular data. Notably, with enough time to search for better hyperparameters, XG-Boost is shown in the chart to be able to gain an advantage.

There is much more involved in tabular competitions than simply stating how GBDT outperforms DNNs. As many past and present Grandmasters often state, combining diverse models (such as a neural network and a gradient boosting model) in a tabular data problem consistently yields better results than using a single model. Owen Zhang, who was previously ranked number one on Kaggle, discusses how neural networks and GBDTs can be effectively blended for improved results in a competition in the following interview: https://www.youtube.com/watch?v=LgLcfZjNF44. The empirical experience of Kaggle Grandmasters is now also formalized in papers such as *Tabular data: Deep learning is not all you need* (https://arxiv.org/abs/2106.03253) by Ravid Shwartz-Ziv and Amitai Armon, where the authors observed that on many benchmarks, a simple blend of a DNN and a GBDT systematically beats any single model.

Hence, although it is often true that "XGBoost is all you need" (or another GBDT solution, such as LightGBM or CatBoost), it is also our opinion that the debate over GBDT versus deep learning is a "false dichotomy," as both solutions can contribute to ensembles that outperform single models. Especially when there are sufficient examples, neural networks can capture signals that gradient boosting models cannot detect, and deep learning solutions can be effective as stand-alone models or as part of an ensemble.

Building a neural network quickly for a tabular competition is no longer a daunting challenge. Libraries such as TensorFlow, Keras, and PyTorch make it easier to build models from scratch. However, packages like *PyTabKit* (https://github.com/dholzmueller/pytabkit) or *PyTorch Tabular* (https://github.com/manujosephv/pytorch_tabular) make things even easier by providing immediate access to a large variety of pre-made architectures.

Recently, there has been a resurgence of various neural architectures for tabular data, and we have witnessed the emergence of pre-trained models, such as Prior Labs' TabPFN and INRIA's TabICL, which can perform well on small and medium-sized datasets without requiring prior training. While none of these models has won a Kaggle competition on its own, they have consistently appeared in top-performing ensembles:

- **TabNet**, a network developed by Google researchers S. O. Arık and T. Pfister, is described in their paper, *Tabnet: Attentive interpretable tabular learning* (https://cdn.aaai.org/ojs/16826/16826-13-20320-1-2-20210518.pdf). It is designed to select and process relevant features and intelligently handle both categorical and numerical data. It doesn't have many hyperparameters to tune, though the results may differ significantly between an untuned network and a tuned one (hence the necessity of spending some time to optimize it for best performance). There are a few implementations, such as the excellent pytorch-tabnet package (https://github.com/dreamquark-ai/tabnet) or the implementations coded by Yirun Zhang (https://www.kaggle.com/gogo827jz), found at https://www.kaggle.com/ludovick/introduction-to-tabnet-kfold-10-training and https://www.kaggle.com/ludovick/introduction-to-tabnet-kfold-10-inference. Both were devised for the *Mechanism of Action (MoA) Prediction* competition.

- **Neural Oblivious Decision Ensembles (NODE)** is an architecture that attempts to mimic in a neural network how a decision tree works (*Neural oblivious decision ensembles for deep learning on tabular data* by Popov et al., available at https://arxiv.org/abs/1909.06312). You can use the implementation offered by Yirun Zhang for TensorFlow at https://www.kaggle.com/gogo827jz/moa-neural-oblivious-decision-ensembles-tf-keras or for PyTorch at https://www.kaggle.com/gogo827jz/moa-public-pytorch-node.

- **RealMLP** is an enhanced MLP that incorporates a "bag of tricks" to improve performance on tabular data (see `https://openreview.net/forum?id=fwajDrDy89` for the paper by David Holzmüller, Léo Grinsztajn, and Ingo Steinwart, and the more technically detailed NeurIPS paper at `https://proceedings.neurips.cc/paper_files/paper/2024/file/2e e1c87245956e3eaa71aaba5f5753eb-Paper-Conference.pdf`). These improvements span architecture, preprocessing, training, and regularization. Key components include robust scaling and smooth clipping for preprocessing, a novel numerical embedding variant, a diagonal weight layer, new learning rate schedules, and different initialization methods. While RealMLP cannot boast any specific Kaggle win, it is often mentioned in ensemble solutions and ranks as the top model as part of an ensemble in the TabArena benchmark (`https://huggingface.co/spaces/TabArena/leaderboard`).

- **TabM** (short for **Tabular Deep Learning Model** that makes Multiple predictions: `https://arxiv.org/abs/2410.24210`) emulates an ensemble of MLPs within a single, efficient model. It achieves this through parameter-efficient ensembling, where the underlying MLPs are trained in parallel and share most of their weights. By training the ensemble members simultaneously within a single model, TabM can be optimized for the performance of the entire ensemble, not just individual models. This weight sharing also acts as an effective regularization technique. TabM has demonstrated significant success in Kaggle competitions. It was part of the winning solution in the CIBMTR competition (`https://www.kaggle.com/competitions/equity-post-HCT-survival-predictions`) and shone in many other ensembles in the top 25.

- **TabR** (from the creators of CatBoost: `https://arxiv.org/pdf/2307.14338`) is a retrieval-augmented deep learning model. It enhances a feed-forward network by adding a k-Nearest-Neighbors-like component. For a given data point, TabR retrieves similar data points (neighbors) from the training set and uses their features and labels to enrich the representation before making a prediction. The model consists of an encoder, a retrieval module, and a predictor.

- **ModernNCA** (`https://arxiv.org/abs/2407.03257v1`) is a modernized version of **Neighborhood Component Analysis (NCA)**, a differentiable k-NN method. It uses deep nonlinear embeddings through stacked MLP blocks to learn a representation of the data. It also employs stochastic neighborhood sampling during training for scalability and regularization.

- **TabPFNv2** (https://www.nature.com/articles/s41586-024-08328-6) and **TabICL** (https://arxiv.org/abs/2502.05564) are both tabular foundation models trained primarily on synthetic datasets, with the inclusion of some real-world data in the evaluation. TabPFNv2 has demonstrated state-of-the-art performance on small tabular datasets, outperforming strong baselines that have been tuned for hours in just seconds. While its applicability is limited to smaller datasets, TabPFNv2 offers a very powerful and fast solution for competitions that fit within its constraints. On large datasets (over 10,000 samples), TabICL has been shown to outperform both TabPFNv2 and CatBoost, demonstrating its potential for Kaggle competitions, especially those with larger datasets where previous ICL models would be computationally prohibitive.

You can use a wide range of other DNN models, such as Wide & Deep, DeepFM, xDeepFM, AutoInt, and many others based on factorization machines, which are primarily designed for click-through rate estimation. You don't have to build all these neural architectures by yourself; you can rely on packages such as DeepCTR (https://github.com/shenweichen/DeepCTR) or DeepTables (https://github.com/DataCanvasIO/deeptables) as suggested by Changhao Lee (https://www.kaggle.com/leechh) and Jian Yang (https://www.kaggle.com/jackguagua), who won second and first place, respectively, in the *Categorical Feature Encoding Challenge II* competition.

In conclusion, as a piece of advice based on our experience, don't expect a neural network to be the best model in a tabular competition, as this is seldom the case. Instead, produce a variety of solutions based on the DNN architectures we mentioned, and blend them with gradient boosting models. This is because DNNs tend to capture different signals from the data, which can be effectively integrated into a powerful ensemble.

To round off this chapter, we spoke to Jean-François Puget, AKA CPMP, about the importance of reproducibility, how to work with data, his best competition, and more. As a Kaggle Grandmaster in competitions and a distinguished engineer at RAPIDS, NVIDIA, he had many good insights to share with us. The author particularly likes what he has to say about the scientific method.

Interview: Jean-François Puget

```
https://www.kaggle.com/cpmpml
```

What's your favorite kind of competition and why? In terms of techniques and solving approaches, what is your specialty on Kaggle?

I like competitions with a scientific background or a background I can relate to. I dislike anonymous data and synthetic data unless the data is generated via a very precise physics simulation. More generally, I like Kaggle competitions on domains I don't know much about, as this is where I will learn the most. It is not the most effective way to get ranking points, but it is the one I entertain most.

How do you approach a Kaggle competition? How different is this approach from what you do in your day-to-day work?

I start by looking at data and understanding it as well as possible. I try to find patterns in it, especially predictive patterns. What I often do is plot samples using two features or derived features on the x- and y-axes, and a third feature for color coding samples. One of the three features can be the target. I use lots of visualization, as I believe that human vision is the best data analysis tool there is.

The second thing I spend time on is how to assess model or pipeline performance. Indeed, it is extremely important to be able to evaluate the performance of a model as accurately as possible. There is no surprise here; evaluation is often a variant of k-fold cross-validation. However, the fold definition can be tailored to the competition type (time-based folds for forecasting competitions, group k-fold when samples are linked together for some reason, e.g., actions with the same user ID, etc.).

I then create an end-to-end baseline that goes from data to submission and try it. If this is a code competition, then testing that you have gotten your pipeline right is key.

Then I try more complex models (if I'm using deep learning models) or more features (if I'm using XGBoost or other models from RAPIDS or sklearn). I submit these to see if there is a correlation between my local evaluation score and the public test score. If the correlation is good, then I submit less and less.

After a few weeks, I spend time doing hyperparameter tuning. However, I only do it once, or maybe twice, with a last tuning near the end of the competition. Indeed, hyperparameter tuning is one of the best ways to overfit, and I fear overfitting a lot.

Tell us about a particularly challenging competition you entered, and what insights you used to tackle the task.

One of the competitions I am most proud of is the *TalkingData AdTracking Fraud Detection Challenge* competition, where we had a very large volume of click history, and we had to predict which clicks led to some app downloads. There were very few features and a large number of rows (like half a billion). At the time, I only had a 64 GB machine, and I had to implement a very efficient way to create new features and evaluate them. I had a few insights into this competition. First, the click that led to an app download was the last click on the app download page for a user. Therefore, the "time to next click from the same user on the same app" was the most important feature. A derived insight was this: there were quite a number of clicks from the same user and app with the same timestamp. I hypothesized that the one with a download, if any, was the last one. A third insight was to use a matrix factorization approach to approximate feature value co-occurrences. I implemented the libFM model in Keras at the time, and adding the latent vectors as features helped. The only other team doing this was the top team. With this, I got a solo 6th place among teams of Grandmasters. I was not a Kaggle Grandmaster yet.

Has Kaggle helped you in your career? If so, how?

Kaggle helped me twice. At IBM, Kaggle was a great source of knowledge on SOTA ML practices. I used that knowledge to inform and guide the development of IBM ML tooling (IBM Watson Studio and IBM Watson ML).

For instance, I managed to have IBM support Python packages in 2016 at a time when IBM was a Java/Scala powerhouse. Without me, IBM would have bet on Spark and Scala for ML and would have missed the Python wave entirely. I also pushed for XGBoost very early, when IBM wanted to only support Spark ML or TensorFlow.

The second time Kaggle helped me was to get my current job. NVIDIA was looking for Kaggle competition Grandmasters with good social presence to help promote the NVIDIA stack, including the RAPIDS GPU-accelerated ML package.

In your experience, what do inexperienced Kagglers often overlook? What do you know now that you wish you'd known when you first started?

The one thing that differentiates Kagglers from other data scientists is the evaluation of model performance. Kagglers need to master this because if they don't, then they choose submissions that look great on the public leaderboard but perform poorly on the private leaderboard. Once a Kaggler knows how to build models that perform well on the private leaderboard, then they know how to build models that perform well on new data, that is, models that do not overfit.

The other thing that inexperienced Kagglers do is to ask if method/model X can work in a given competition. My answer to this is always, "Try it and see if it works or not." People often miss that ML is an experimental science. In order to build good models, one must follow the scientific method:

Make a hypothesis (e.g., adding this feature, or adding this NN layer, will improve pipeline performance).

Run an experiment to test the hypothesis (train the modified pipeline).

Analyze experiment results (is the CV score better than before? Where is it better? Where is it worse?).

Each experiment should be done so that it can confirm or reject a hypothesis. For this, an experiment should change only one thing at a time. Often, inexperienced people change many things, and then cannot conclude what worked or not.

Are there any particular tools or libraries that you would recommend using for data analysis and ML?

I use Matplotlib plots mostly for data exploration. I do data wrangling in pandas if the dataset is small, or in cuDF (from RAPIDS) if the dataset is large. For ML, I use cuML from RAPIDS, XGBoost with GPU acceleration, and PyTorch. If possible, I will use pre-trained models, for instance, NLP models from Hugging Face, or image classification models from the timm package.

What's the most important thing someone should keep in mind or do when they're entering a competition?

Make sure you can spend enough time on it.

Summary

In this chapter, we discussed tabular competitions on Kaggle. Since most of the knowledge applicable in a tabular competition overlaps with standard data science knowledge and practices, we focused our attention on techniques that are more specific to Kaggle in this chapter.

Starting from the Tabular Playground Series, we touched on topics relating to reproducibility, EDA, feature engineering, feature selection, target encoding, pseudo-labeling, and neural networks applied to tabular datasets.

EDA is a crucial phase if you want to get insights into how to win a competition. It is also quite unstructured and heavily dependent on the kind of data you have. Aside from giving you general advice on EDA, we brought your attention to techniques such as t-SNE and UMAP that can summarize your entire dataset at a glance. The next phase, feature engineering, is also strongly dependent on the kind of data you are working on. We therefore provided a series of possible feature engineering ideas that you can try applying to your specific case. As for feature selection, after a brief overview, we drew your attention to techniques based on feature importance and randomization, which can be applied to almost any ML algorithm.

After explaining target encoding, which we wanted to point out cannot be dealt with in an automated way, we moved on to special techniques that you probably won't apply in your real-world projects but that can work very well in Kaggle competitions: pseudo-labeling and DAEs for tabular competitions. Finally, after discussing how categorical features can also be dealt with using embedding layers in neural networks, we gave you a quick overview of the pre-made neural architectures that could work for tabular data.

In the next chapter, we will complete our overview of all the techniques that you need to take on tabular competitions by discussing how best to perform hyperparameter optimization.

Join our book's Discord space

Join our community's Discord space for discussions with the authors and other readers:

```
https://packt.link/kaggle
```

9

Hyperparameter Optimization

How a Kaggle solution performs is not simply determined by the type of learning algorithm you choose. Aside from the data and features you use, it is also strongly determined by the algorithm's **hyperparameters**, which must be fixed before training and cannot be learned during training. Choosing suitable features is most effective in tabular data competitions; however, **hyperparameter optimization** is effective in all competitions of any kind. In fact, given fixed data and an algorithm, hyperparameter optimization is the only sure way to enhance the predictive performance of the algorithm and climb the leaderboard. It also helps in ensembling because an ensemble of tuned models always outperforms an ensemble of untuned ones.

> This chapter is certainly one of the most challenging in the book, but mastering the techniques discussed here will yield significant rewards. It is addressed to readers with an intermediate level of competence in **machine learning** (**ML**), who are ready to deepen their understanding of how to optimize models effectively and gain a competitive edge both in Kaggle competitions and in real-world applications.

You may hear that tuning hyperparameters manually is possible if you know and understand the effects of your choices on the algorithm. Many Kaggle Grandmasters and Masters have declared that they often rely on directly tuning their models in competitions. They operate selectively on the most important hyperparameters in a bisection operation style, exploring smaller and smaller intervals of a parameter's values until they find the value that produces the best result. Then, they move on to another parameter. This works perfectly well if there is a single minimum for each parameter and if the parameters are independent of each other. In this case, the search is mostly driven by experience and knowledge of learning algorithms.

In our experience, however, that is not the case with most tasks you will encounter on Kaggle. The sophistication of the problems and the complexity of algorithms require a systematic approach to finding the correct hyperparameters that only a search algorithm can provide.

This chapter will explore extending your cross-validation approach to find the best hyperparameters to generalize to your test set. The idea is to deal with the pressure and scarcity of time and resources that you experience in competitions. For this reason, we will concentrate on **Bayesian optimization methods**, a proven way to optimize for complex models and data problems based on your available resources. We won't limit ourselves to searching for the best values for predefined hyperparameters; we will also delve into the problem of neural network architecture.

We will cover the following topics:

- Basic optimization techniques
- Key parameters and how to use them
- Bayesian optimization
- Experiment tracking

Let's start!

Basic optimization techniques

The core algorithms for hyperparameter optimization, found in the scikit-learn package, are **grid search** and **random search**. Recently, the scikit-learn contributors have also added the **halving algorithm** to improve the performances of both grid search and random search strategies.

> HalvingRandomSearchCV is still experimental as of scikit-learn version 1.5.2, hence the API might suddenly change without any previous deprecation warnings. To use it, you need to explicitly import enable_halving_search_cv: from sklearn. experimental import enable_halving_search_cv.

In this section, we will discuss all these basic techniques. By mastering them, you will obtain practical optimization tools for specific problems (for instance, support vector machines or SVMs are usually optimized by grid search) and become familiar with the basics of hyperparameter optimization.

To start with, it is crucial to figure out what the necessary ingredients are:

- A model whose hyperparameters have to be optimized

- A search space containing the boundaries of the values to search between for each hyperparameter
- A cross-validation scheme
- An evaluation metric and its score function

All these elements come together in the search method to determine the solution you are looking for. Let's see how it works.

Grid search

Grid search is a method that searches through the hyperparameters exhaustively and is not feasible in high-dimensional space. For every parameter, you pick a set of values you want to test. You then test all the possible combinations between these sets. That is why it is exhaustive: you try everything in terms of hyperparameters. It is a straightforward strategy, but clearly, it heavily suffers from computational issues. On the positive side, it's an *embarrassingly parallel* algorithm. An embarrassingly parallel algorithm is a type of parallel algorithm where tasks or sub-problems can be solved independently without requiring communication or coordination between processes. Each process works on its own part of the problem, and there is minimal need for data sharing during computation. Examples you may already know about in ML include methods like random forests or bagging. In grid search, the fact that it is embarrassingly parallel means you can obtain optimal tuning quickly if you have enough processors to run the search.

For example, SVMs for both classification and regression problems are probably the ML algorithms that you will use grid search for the most. Here, let's take a classification problem and **support vector classifier** (**SVC**). Using the `make_classification` function from scikit-learn, we can generate a classification dataset quickly:

```
from sklearn.datasets import make_classification
from sklearn.model_selection import train_test_split
X, y = make_classification(n_samples=300, n_features=50,
                           n_informative=10,
                           n_redundant=25, n_repeated=15,
                           n_clusters_per_class=5,
                           flip_y=0.05, class_sep=0.5,
                           random_state=0)
```

For our next step, we define a basic SVC algorithm and set the search space. Since the **kernel function** of the SVC (the internal function that transforms the input data in an SVM) determines the different hyperparameters to set, we provide a list containing two dictionaries of distinct search spaces for parameters to be used depending on the type of kernel chosen. We also set the evaluation metric (we use accuracy in this case since the target is perfectly balanced):

```
from sklearn import svm
svc = svm.SVC()
svc = svm.SVC(probability=True, random_state=1)
from sklearn import model_selection
search_grid = [
    {'C': [1, 10, 100, 1000], 'kernel': ['linear']},
    {'C': [1, 10, 100, 1000], 'gamma': [0.001, 0.0001],
    'kernel': ['rbf']}
]

scorer = 'accuracy'
```

In our example, a linear kernel doesn't require tuning the gamma parameter, which is, instead, very important for a radial basis function kernel. Gamma controls the influence of a single training example; a small gamma means far-reaching influence, while a large gamma makes the influence more localized. Proper tuning of gamma is crucial for controlling the complexity of the model and achieving good performance with a **radial basis function** (**RBF**) kernel.

Therefore, to address the role of gamma in an RBF kernel, we provide two dictionaries:

- The first contains the parameters for the linear kernel
- The second contains parameters for a radial basis function kernel

Each dictionary only contains a reference to the kernel it is relevant to and only the range of relevant parameters for that kernel.

It is important to note that the evaluation metric can differ from the cost function optimized by the algorithm. In fact, as discussed in *Chapter 6, Competition Tasks and Metrics*, you may encounter scenarios in which the evaluation metric for the competition is different, but you cannot modify the cost function of your algorithm. Under these circumstances, tuning the hyperparameters according to your evaluation metric can still help obtain a well-performing model. Though built around the algorithm's cost function, the optimal set of hyperparameters found will return the best evaluation metric under such constraints.

It probably won't be the theoretically best result that you could obtain for the problem, but it may often not be far from it.

All the ingredients (model, search space, evaluation metric, and cross-validation scheme) are combined into the GridSearchCV instance, and then the model is fit to the data:

```
search_func = model_selection.GridSearchCV(estimator=svc,
                                           param_grid=search_grid,
                                           scoring=scorer,
                                           n_jobs=-1,
                                           cv=5)
search_func.fit(X, y)
print (search_func.best_params_)
print (search_func.best_score_)
```

After a while, depending on the machine you are running the optimization on, you will obtain the best combination based on cross-validated results.

In conclusion, grid search is a very simple optimization algorithm that can leverage the availability of multi-core computers. It can work fine with ML algorithms that do not require many tunings (such as SVM and the ridge and lasso regressions), but its applicability is quite narrow in all other cases. First, it is limited to optimizing hyperparameters through discrete choices, meaning you need a limited set of values to cycle through. In addition, you cannot expect it to work effectively on algorithms requiring *multiple* hyperparameters to be tuned because it is not scalable and does not take into account the results of previous trials. With grid search, you are essentially applying brute force, blindly trying parameter values, many of which may not work well for the problem. This approach reaches its limits due to the exploding complexity of the search space and its inherent computational inefficiency.

Random search

Random search, which simply samples the search space randomly, is feasible in high-dimensional spaces (that is, in situations where you have lots of hyperparameters to tune) and is widely used in practice. The downside of random search, however, is that it doesn't use information from prior experiments to select the next setting, a problem shared by grid search, as we previously should have noted. In addition, to find the best solution as fast as possible, you cannot do anything except hope to be lucky and catch the right hyperparameters.

Random search works incredibly well and it is simple to understand. Although it relies on randomness, it isn't just based on blind luck, though it may initially appear to be. In fact, it works like *random sampling* in statistics: the main point of the technique is that if you do enough random tests, you have a good possibility of finding the right parameters without wasting energy on testing slightly different combinations of similarly performing combinations.

Many AutoML systems rely on random search when there are too many parameters to set (see Golovin et al. *Google Vizier: A Service for Black-Box Optimization*, 2017). As a rule of thumb, consider looking at the random search strategy when the dimensionality of your hyperparameter optimization problem is sufficiently high (for example, over 16).

Here, we run the previous example using random search:

```python
import scipy.stats as stats
search_dict = {'kernel': ['linear', 'rbf'],
               'C': stats.loguniform(1, 1000),
               'gamma': stats.loguniform(0.0001, 0.1)}
scorer = 'accuracy'
search_func = model_selection.RandomizedSearchCV(
    estimator=svc,
    param_distributions=search_dict,
    n_iter=6,
    scoring=scorer,
    n_jobs=-1,
    cv=5
)
search_func.fit(X, y)
print (search_func.best_params_)
print (search_func.best_score_)
```

Notice that we don't care about running the search on separate spaces for the different kernels. Contrary to grid search, where each parameter, even the ineffective ones, is systematically tested and requires computational time, the efficiency of random search is not affected by the set of hyperparameters tested here. The search doesn't depend on irrelevant parameters but is guided by chance; in such a sense, any trial is useful for the purpose, even if you are testing only one valid parameter among many for the chosen kernel.

Halving search

As we mentioned, both grid search and random search work in an uninformed way: if some tests find out that certain hyperparameters do not impact the result or that certain value intervals are ineffective, the information is not propagated to the following searches.

For this reason, scikit-learn has recently introduced the `HalvingGridSearchCV` and `HalvingRandomSearchCV` estimators, which can be used to search a parameter space using **successive halving** applied to the grid search and random search tuning strategies.

In halving, a large number of hyperparameter combinations are evaluated in an initial round of tests but using a small amount of computational resources. This is achieved by running the tests on a subsample of a few cases from your training data. A smaller training set needs fewer computations to be tested, so fewer resources (namely, time) are used at the cost of more imprecise performance estimations. This initial round allows the selection of a subset of candidate hyperparameter values, which have performed better on the problem, to be used for the second round when the training set size is increased.

The following rounds proceed in a similar way, allocating larger and larger subsets of the training set to be searched as the range of tested values is restricted (testing now requires more time to execute but returns a more precise performance estimation) while the number of candidates continues to be halved.

Here is an example applied to the previous problem:

```
from sklearn.experimental import enable_halving_search_cv
from sklearn.model_selection import HalvingRandomSearchCV
search_func = HalvingRandomSearchCV(estimator=svc,
                                    param_distributions=search_dict,
                                    resource='n_samples',
                                    max_resources=100,
                                    aggressive_elimination=True,
                                    scoring=scorer,
                                    n_jobs=-1,
                                    cv=5,
                                    random_state=0)
search_func.fit(X, y)
print (search_func.best_params_)
print (search_func.best_score_)
```

In this way, halving provides information for the successive optimization steps via selecting the candidates. In the next sections, we will discuss even smarter ways to achieve a more precise and efficient search through the space of hyperparameters.

Let's pause for an interview with another Kaggler. *Kazuki Onodera* is a Competitions Grandmaster and Discussions Master who has around 7 years of competition experience. He's also a senior deep learning data scientist at NVIDIA and a member of the NVIDIA KGMoN (Kaggle Grandmasters of NVIDIA) team.

Interview: Kazuki Onodera

`https://www.kaggle.com/onodera`

What's your favorite kind of competition and why? In terms of techniques and solving approaches, what is your specialty on Kaggle?

Instacart Market Basket Analysis. This competition proved quite challenging for the Kaggle community because of its use of anonymized data related to customer orders over time in order to predict which previously purchased products will be in a user's next order. The reason why I like it is that I love feature engineering and I could come up with a bunch of good and interesting features everyone else couldn't, which allowed me to get second place in the competition.

How do you approach a Kaggle competition? How different is this approach from what you do in your day-to-day work?

I try to imagine how a model works and delve into false negatives and false positives. The same as in my daily work.

Tell us about a particularly challenging competition you entered, and what insights you used to tackle the task.

Human Protein Atlas - Single Cell Classification. This competition was a kind of instance segmentation competition, but no masks were provided. So, it turned into a weakly supervised multi-label classification problem. I created a two-stage pipeline for removing label noise.

Has Kaggle helped you in your career? If so, how?

Yes. I'm now working in the NVIDIA KGMON (Kaggle Grandmasters of NVIDIA) team. Kaggle launches many different ML competitions, different with regard to data type, tabular, image, natural language, and signal, as well as with regard to sector and domain: industry, finance, astronomy, pathology, sports, retail, and so on. I bet nobody can access and have experience with all these kinds of data except Kagglers.

In your experience, what do inexperienced Kagglers often overlook? What do you know now that you wish you'd known when you first started?

Target analysis. Also, seed averaging is quite overlooked: always simple but powerful.

What mistakes have you made in competitions in the past?

Target analysis. Top teams always analyze the target better than others, so if I couldn't get a better place in a competition, I read about the top solutions, because they always describe to me the knowledge about the data that I missed during the competition.

Are there any particular tools or libraries that you would recommend using for data analysis or ML?

Just Python and Jupyter Notebook.

What's the most important thing someone should keep in mind or do when they're entering a competition?

If you can learn from a defeat, you haven't really lost.

Do you use other competition platforms? How do they compare to Kaggle?

KDD Cup and RecSys. Both meet the minimum requirements for being interesting and challenging.

Key parameters and how to use them

The next problem is using the right set of hyperparameters for each kind of model you use. In particular, to be efficient in your optimization, you need to know each hyperparameter's values so that it makes sense to test for each distinct algorithm.

In this section, we will examine the most common models used in Kaggle competitions, especially the tabular ones, and discuss the hyperparameters you need to tune in order to obtain the best results. We will distinguish between classical ML models and gradient boosting models (which are much more demanding regarding their space of parameters) for generic tabular data problems.

As for neural networks, we can give you an idea about specific parameters to tune when we present the standard models (for instance, the TabNet neural model has some specific parameters to set to work properly). However, most of the optimization on deep neural networks in Kaggle competitions is not performed on standard models but on *custom* ones. Consequently, apart from basic learning parameters such as the learning rate and the batch size, optimization in neural networks is based on the specific characteristics of the neural architecture of your model. You have to deal with the problem in an ad hoc way. Later in this chapter, we will discuss an example of **neural architecture search (NAS)** using KerasTuner (`https://keras.io/keras_tuner/`).

Linear models

The linear models that need to be tuned are usually linear regressions or logistic regressions with regularization:

- `C`: The range you should search is `np.logspace(-4, 4, 10)`; smaller values specify stronger regularization.
- `alpha`: You should search the range `np.logspace(-2, 2, 10)`; it is a constant that multiplies the regularization term, hence, larger values specify stronger regularization. Also, take note that higher values take more time to process when using lasso.
- `l1_ratio`: Though you can draw in the range from zero to one, you can save time just by picking from the list `[.1, .5, .7, .9, .95, .99, 1]`; it applies only to the elastic net algorithm.

In scikit-learn, depending on the algorithm, you find either the hyperparameter `C` (logistic regression) or `alpha` (lasso, ridge, elastic net).

Support vector machines

SVMs are a family of powerful and advanced supervised learning techniques for classification and regression that can automatically fit linear and non-linear models. Scikit-learn offers an implementation based on LIBSVM, a complete library of SVM classification and regression implementations, and LIBLINEAR, a scalable library for linear classification ideal for large datasets, especially sparse text-based ones. In their optimization, SVMs strive to separate target classes in classification problems using a decision boundary characterized by the largest possible margin between classes.

Though SVMs work fine with default parameters, they are often not optimal, and you need to test various value combinations using cross-validation to find the best ones. Listed in decreasing order of importance, you have to set the following parameters:

- C: The penalty value. Decreasing it makes the margin between classes larger, thus ignoring more noise but also making the model more generalizable.

  An optimal value is normally found in the range np.logspace(-3, 3, 7).

- kernel: This parameter will determine how non-linearity will be implemented in an SVM, and it can be set to 'linear', 'poly', 'rbf', 'sigmoid', or a custom kernel.

  The most commonly used value is certainly rbf.

- degree: Works with kernel='poly', signaling the dimensionality of the polynomial expansion. Other kernels ignore it.

  Usually, setting its values to between 2 and 5 works the best.

- gamma: A coefficient for 'rbf', 'poly', and 'sigmoid'. High values tend to fit data better but can lead to some overfitting. Intuitively, we can imagine gamma as the parameter ruling the influence that a single example exercises over the model. Low values make the influence of each example reach further. Since many points have to be considered, the SVM curve will tend to take a shape less influenced by local points, and the result will be a smoother decision contour curve. High values of gamma, instead, mean that the curve takes into account how points are arranged locally more, and, as a result, you get a more irregular and wiggly decision curve.

  This hyperparameter's suggested grid search range is np.logspace(-3, 3, 7).

- nu: For regression and classification with nuSVR and nuSVC, this parameter sets a tolerance for the training points near the margin that are not classified correctly. It helps in ignoring misclassified points just near or on the margin. Hence, it can render the classification decision curve smoother. Ultimately, it acts like C, with high proportions enlarging the margin.

  It should be in the range [0,1] since it is a proportion relative to your training set.

- epsilon: This parameter specifies the tolerance margin where no penalty is incurred for erroneous predictions within this range.

  The suggested search range is np.logspace(-4, 2, 7).

- penalty, loss, and dual: For LinearSVC, these parameters accept the ('l1', 'squared_hinge', False), ('l2', 'hinge', True), ('l2', 'squared_hinge', True), and ('l2', 'squared_hinge', False) combinations.

  The ('l2', 'hinge', True) combination is analogous to the SVC(kernel='linear') learner.

It may appear that an SVM has many hyperparameters to set, but many settings are specific only to implementations or to kernels, so you only have to select the relevant parameters.

Random forests and extremely randomized trees

Leo Breiman and *Adele Cutler* originally devised the idea at the core of the random forest algorithm, and the name of the algorithm remains a trademark of theirs today (though the algorithm is open source). Random forests are implemented in scikit-learn as RandomForestClassifier or RandomForestRegressor.

A random forest works in a similar way to bagging, also devised by Breiman, but operates only using binary split decision trees, which are left to grow to their extremes. Moreover, it samples the cases to be used in each model of the ensemble using **bootstrapping**. As the tree grows, at each branch split, the set of variables considered for the split is drawn randomly, too.

This is the secret at the heart of the algorithm: it ensembles trees that, due to different samples and variables considered at the splits, are very different from each other. As they are different, they are also uncorrelated. This is beneficial because when the results are ensembled, much variance is ruled out, as the extreme values on both sides of a distribution tend to balance out. In other words, bagging algorithms guarantee a certain level of *diversity* in the predictions, allowing them to develop rules that a single learner (such as a decision tree) might not come across. All this diversity is useful because it helps build a distribution whose average is a better predictor than any of the individual trees in the ensemble.

Extra trees (also known as **extremely randomized trees**), represented in scikit-learn by the ExtraTreesClassifier/ExtraTreesRegressor classes, are a more randomized kind of random forest that produces a lower variance in the estimates at the cost of greater bias of the estimators. However, when it comes to CPU efficiency, extra trees can deliver a considerable speed-up compared to random forests, so they can be ideal when working with large datasets in terms of examples and features. The reason for the resulting higher bias but better speed is the way splits are built in an extra tree. After drawing a random set of features to be considered for splitting a branch of a tree, random forests carefully search among them for the best values to assign to each branch. By contrast, in extra trees, both the set of candidate features for the split and the actual split value are decided completely randomly. So, there's no need for much computation, though the randomly chosen split may not be the most effective one (hence the bias).

For both algorithms, the key hyperparameters that should be set are as follows:

- max_features: This is the number of sampled features present at every split, which can determine the algorithm's performance. The lower the number, the speedier, but with higher bias.

  As a rule of thumb, try values such as sqrt, log2, and integer numbers representing 1/10th and 1/20th of the number of the features.

- max_depth: This controls the maximum depth of the trees. If set to None, the tree will continue to grow until another stopping criterion is met, such as when a branch has only a few samples left. Reducing the depth decreases variance but increases bias.

- min_samples_leaf: This parameter allows for more flexible control of the depth compared to max_depth, as it affects branches based on their sample size. Branches with more samples are allowed to grow deeper, while max_depth imposes a cut-off after a set number of levels. Larger values of min_samples_leaf reduce variance but increase bias. Usually set to 1, try values up to 30.

- bootstrap: This is a Boolean that allows bootstrapping. In certain cases, setting bootstrap=False might be beneficial when you want each tree to be built from the full dataset, leading to lower bias but potentially higher variance.

- n_estimators: This is the number of trees. Remember that the more trees, the better, though there is a threshold beyond which we get diminishing returns depending on the data problem. Also, this comes at a computational cost that you have to take into account based on the resources you have available. As a rule of thumb that works for most problems, start from 100 and grow up to 1,000.

Extra trees are a good alternative to random forests, especially when the data you have is particularly noisy. Since they trade some variance reduction for more bias, given their random choice of splits, they tend to overfit less on important yet noisy features that would otherwise dominate the splits in a random forest.

Gradient tree boosting

Gradient tree boosting or **gradient boosting decision trees (GBDT)** is an improved version of boosting (boosting works by fitting a sequence of weak learners on reweighted versions of the data). Like AdaBoost, GBDT is based on a gradient descent function. The algorithm has proven to be one of the most proficient ones from the family of models that are based on ensembles, though it is characterized by an increased variance of estimates and more sensitivity to noise in data (both problems can be mitigated by using subsampling) and high computational costs due to non-parallel operations.

Besides deep learning, gradient boosting is the most developed ML algorithm. Since AdaBoost and the initial gradient boosting implementation, as developed by *Jerome Friedman*, various other implementations of the algorithms appeared, the most recent ones being XGBoost, LightGBM, and CatBoost.

LightGBM

The high-performance LightGBM algorithm (`https://github.com/Microsoft/LightGBM`) can be distributed on multiple computers and quickly handle large amounts of data. It was developed by a team at Microsoft as an open-source project on GitHub.

> If you're interested in the finer details, you can refer to the following academic paper by Ke et al. available at `https://papers.nips.cc/paper/2017/hash/6449f44a1` `02fde848669bdd9eb6b76fa-Abstract.html`).

LightGBM is based on decision trees, like XGBoost, but it follows a different strategy. While XGBoost uses decision trees to split on a variable and explore different tree splits at that variable (the **level-wise** tree growth strategy), LightGBM concentrates on one split and goes on splitting from there to achieve a better fit (the **leaf-wise** tree growth strategy). This allows LightGBM to quickly reach a good fit of the data and generate alternative solutions compared to XGBoost (which is good if you expect to blend the two solutions together to reduce the variance of the estimates). Algorithmically speaking, if we consider the structure of splits operated by a decision tree as a graph, XGBoost pursues a *breadth-first* search (BFS) and LightGBM a *depth-first* search (DFS).

Tuning LightGBM may appear daunting; it has more than a hundred parameters to tune, which you can explore at https://github.com/Microsoft/LightGBM/blob/master/docs/Parameters. rst (also here: https://lightgbm.readthedocs.io/en/latest/Parameters.html).

As a rule of thumb, you should focus on the following hyperparameters, which usually have the most impact on the results:

- n_estimators: An integer between 100 and 10,000 that sets the number of iterations.
- learning_rate: A real number between 0.001 and 1.0, usually sampled from a log-uniform distribution. It represents the step size of the gradient descent procedure that computes the weights for the summed ensemble of all the iterations of the algorithm up to this point.
- max_depth: An integer usually between 1 and 12, representing the maximum number of splits on features. Setting it to a number below 0 allows the maximum possible number of splits, usually risking overfitting to data.
- num_leaves: An integer between 2 and 2^max_depth, representing the number of final leaves each tree will have at most.
- min_data_in_leaf: An integer between 0 and 300 that determines the minimum number of data points in one leaf and helps to deal with overfitting. Zero means no constraint.
- min_gain_to_split: A float between 0 and 15; it sets the minimum gain of the algorithm for tree partitioning. By setting this parameter, you can avoid unnecessary tree splits and thus reduce overfitting (it corresponds to the gamma parameter in XGBoost).
- max_bin: An integer between 32 and 512 that sets the maximum number of bins into which feature values will be bucketed. Having this parameter larger than the default value of 255 implies more risk of producing overfitting results.
- subsample: A real number, usually between 0.1 and 1.0, that represents the sample portion to be used in training.
- subsample_freq: An integer between 0 and 10 specifying the frequency, in terms of iterations, at which the algorithm will subsample the examples. It is set to 0 by default, but note that when subsample_freq is set to 0, the algorithm will ignore any value given to the subsample parameter.
- feature_fraction: A real number, usually between 0.1 and 1.0, that allows you to specify the subset of features to be subsampled. Subsampling the features is another way to allow more randomization to play a role in the training, fighting noise, and multicollinearity present in the features.

- `reg_lambda`: A real number between 0 and 100.0 that sets the L2 regularization. Since it is more sensitive to the scale than to the exact number of the parameter, it is usually sampled from a log-uniform distribution.

- `reg_alpha`: A real number between 0 and 100.0, usually sampled from a log-uniform distribution, which sets the L1 regularization.

- `scale_pos_weight`: A real number between 1e-6 and 500, better sampled from the log-uniform distribution. The parameter weights the positive cases (thus effectively upsampling or downsampling) against the negative cases, which are kept to the value of 1.

Although the number of hyperparameters to tune when using LightGBM may appear daunting, in reality, only a few of them matter a lot. Given a fixed number of iterations and learning rate, just a few are the most impactful (`feature_fraction`, `num_leaves`, `subsample`, `reg_lambda`, `reg_alpha`, and `min_data_in_leaf`), as explained in this blog article by *Kohei Ozaki*, a Kaggle Grandmaster: `https://medium.com/optuna/lightgbm-tuner-new-optuna-integration-for-hyperparameter-optimization-8b7095e99258`. Ozaki leverages this fact to create a fast-tuning procedure for Optuna (you'll find more on the Optuna optimizer at the end of this chapter).

XGBoost

XGBoost (`https://github.com/dmlc/XGBoost`) stands for **eXtreme gradient boosting**. It is an open-source project that is not part of scikit-learn, though it has been expanded with a scikit-learn wrapper interface making it easier to incorporate XGBoost into a scikit-learn-style data pipeline.

The XGBoost algorithm gained momentum and popularity in 2015 data science competitions, such as those on Kaggle and the KDD Cup 2015. As the creators *Tianqui Chen, Tong He*, and *Carlos Guestrin* report in papers they wrote on the algorithm, out of 29 challenges held on Kaggle during 2015, 17 winning solutions used XGBoost as a standalone solution or as part of an ensemble of multiple different models. Since then, the algorithm has always retained a strong appeal among the community of data scientists, though it struggled to keep pace with the innovation brought about by other **gradient boosting machine (GBM)** implementations such as LightGBM and CatBoost.

Aside from good performance in terms of accuracy and computational efficiency, XGBoost is also a scalable solution, using, at best, multi-core processors and distributed machines.

XGBoost represents a new generation of GBM algorithms thanks to important tweaks to the initial tree boost GBM algorithm:

- Sparsity awareness because the algorithm can directly utilize sparse matrices, saving both memory (no need for dense matrices) and computation time (zero values are handled in a special way).

- Approximate tree learning (weighted quantile sketch) produces similar results but in much less time than the classical complete explorations of possible branch cuts.

- Parallel computing on a single machine (using multi-threading during the search for the best split) and, similarly, distributed computations on multiple machines.

- Out-of-core computations on a single machine, leveraging a data storage solution called **column block**. This arranges data on a disk by columns, thus saving time by pulling data from the disk in the way the optimization algorithm (which works on column vectors) expects it.

XGBoost (as well as LightGBM and Catboost) can also deal with missing data in an effective way. In contrast, scikit-learn tree ensembles, which are based on standard decision trees, require missing data to be imputed first, often using an off-scale value such as a negative number, to create an appropriate branching of the tree for handling missing values.

As for XGBoost's parameters (`https://xgboost.readthedocs.io/en/latest/parameter.html`), we have decided to highlight a few key ones you will find across competitions and projects:

- `n_estimators`: Specifies the number of trees to be built in the model. It is an integer typically ranging from 100 to 5,000.

- `learning_rate`: A real number ranging from 0.001 to 1.0, better sampled from the log-uniform distribution.

- `min_child_weight`: The minimum sum of instance weight (hessian) needed in a child. The default is 1. The larger `min_child_weight` is, the more conservative the algorithm will be. Usually an integer between 1 and 10.

- `max_depth`: The maximum depth of a tree. Usually an integer between 1 and 12.

- `max_delta_step`: Usually an integer sampled between 0 and 10, representing the maximum delta step we allow for each leaf output. Set at 0 at default, meaning no constraints, if set to a positive number, it acts as a regularizer because it sets a limit to the updates. It is beneficial when there is an imbalance among the classes in a classification because it prevents the updates relative to a class from dominating the others.

- `subsample`: A real number from 0.1 to 1.0 indicating the proportion of subsampled examples.

- `colsample_bytree`: A real number from 0.1 to 1.0 indicating the subsample ratio of columns by tree.

- `colsample_bylevel`: A real number from 0.1 to 1.0 indicating the subsample ratio by level in trees.

- `max_bin`: The maximum number of bins used by histograms. We suggest an integer between 32 and 512.

- `reg_lambda`: A real number between 1e-9 and 100.0, preferably sampled from the log-uniform distribution. This parameter controls the L2 regularization.

- `reg_alpha`: A real number between 1e-9 and 100.0, preferably sampled from the log-uniform distribution. This parameter controls the L1 regularization.

- `gamma`: Specifying the minimum loss reduction for tree partitioning, this parameter requires a real number between 1e-9 and 0.5, preferably sampled from the log-uniform distribution.

- `scale_pos_weight`: A real number between 1e-6 and 500.0, preferably sampled from the log-uniform distribution, which represents a weight for the positive class.

Like LightGBM, XGBoost also has many similar hyperparameters to tune. Hence, all the considerations previously made for LightGBM are also valid for XGBoost.

CatBoost

In July 2017, Yandex, the company behind the popular Russian search engine, made another interesting GBM algorithm public, CatBoost (https://catboost.ai/), whose name comes from putting together the two words "category" and "boosting." In fact, its strong point is its ability to handle categorical variables, which comprise most of the information in most relational databases, by adopting a mixed strategy of the following techniques:

- **One-hot encoding**: An encoding technique where you transform categorical variables into binary vectors by representing each category by a separate binary feature.

- **Target encoding**: A way to express categorical levels by assigning them an appropriate numeric value for the problem at hand. Also known as likelihood encoding, impact coding, or mean encoding, this is simply a way to transform your labels into a number based on their association with the target variable. More on this can be found in *Chapter 7, Modeling for Tabular Competitions*.

The idea used by CatBoost to encode categorical variables is not new; it is a kind of feature engineering that has been used before, mostly in data science competitions. CatBoost uses one-hot encoding for categorical features with a small number of unique categories (e.g., less than a preset threshold like 10). When the number of categories is larger, the algorithm resorts to using target encoding. In regression tasks, labels can be transformed based on the mean target value for each category level. For classification tasks, target encoding involves using the probability of the target given that label, i.e., the probability of your target conditional on each category value.

It may appear to be a simple and smart feature engineering trick, but it has side effects, mostly in terms of overfitting, because you are taking information from the target into your predictors.

CatBoost has quite a few parameters (see `https://catboost.ai/en/docs/references/training-parameters/`). We have limited our discussion to the eight most important ones:

- `iterations`: This parameter specifies the number of boosting iterations, or trees, to be built during training. It is usually an integer between 10 and 1,000, but it can be increased based on the problem.

- `depth`: This parameter controls the depth of the individual trees, with deeper trees being able to capture more complex patterns in the data. You can set it as an integer between 1 and 8; usually, higher values require longer fitting times and do not produce better results.

- `learning_rate`: This parameter determines the step size at each iteration while moving toward a minimum of the loss function, helping to control how quickly the model learns. It is a real value drawn between 0.01 and 1.0, better sampled from the log-uniform distribution.

- `random_strength`: This parameter introduces randomness into the model's training process, helping to improve generalization by preventing overfitting. It can be controlled by a real number log-linearly sampled from 1e-9 to 10.0, specifying the randomness level for scoring splits.

- `bagging_temperature`: This parameter controls the degree of randomness in the selection of training data for each tree, with higher values leading to more diversity among the trees. It is a real value between 0.0 and 1.0 that sets the Bayesian bootstrap.

- `border_count`: This parameter determines how many discrete values (borders) will be used to convert continuous features into categorical features, affecting how the model interprets numeric data. It is an integer between 1 and 255, indicating the number of splits used to evaluate the numerical features.

- `l2_leaf_reg`: This parameter helps to prevent overfitting by adding a penalty for larger leaf weights in the model, promoting simpler models. It is an integer between 2 and 30: the value for L2 regularization.

- `scale_pos_weight`: This parameter is useful in cases of class imbalance, allowing the model to give more importance to the positive class during training. It is a real number between 0.01 and 10.0 representing the weight for the positive class.

Even if CatBoost may appear to be just another GBM implementation, it has quite a few differences (also highlighted by the different parameters being used) that may greatly help in a competition, both as a single-model solution and as a model integrated into a larger ensemble.

HistGradientBoosting

Recently, scikit-learn has introduced a new version of gradient boosting inspired by LightGBM's binned data and histograms (see this presentation at EuroPython by *Olivier Grisel*: `https://www.youtube.com/watch?v=urVUlKbQfQ4`). Either as a classifier (`HistGradientBoostingClassifier`) or a regressor (`HistGradientBoostingRegressor`), it can be used for enriching ensembles with different models because, despite many similarities with LightGBM, it builds trees differently from other GBDT algorithms. It also presents a much shorter and essential range of hyperparameters to be tuned:

- `learning_rate`: A real number between 0.01 and 1.0, usually sampled from a log-uniform distribution. This parameter controls the contribution of each tree to the final prediction. A smaller learning rate typically requires more iterations to converge but can lead to better performance by preventing overfitting.

- `max_iter`: An integer that can range from 10 to 10,000. This specifies the number of boosting iterations (trees) to be built. More iterations can increase the model's performance, but it also raises the risk of overfitting.

- `max_leaf_nodes`: An integer from 2 to 500. This parameter sets the maximum number of leaf nodes in each tree. It interacts with `max_depth`; hence, setting only one of the two and leaving the other set to None is advisable.

- `max_depth`: An integer between 2 and 12. This controls the maximum depth of each tree, limiting how complex the trees can become. Deeper trees can capture more intricate patterns in the data but may also lead to overfitting.

- `min_samples_leaf`: An integer between 2 and 300. This parameter specifies the minimum number of samples that must be present in a leaf node. Increasing this value can help prevent overfitting by ensuring that leaves have a sufficient number of samples to make reliable predictions.

- `l2_regularization`: A float between 0.0 and 100.0. This parameter applies L2 regularization to the model, helping to prevent overfitting by penalizing large weights in the decision trees.

- `max_bins`: An integer between 32 and 512. This parameter indicates the number of discrete bins used to represent continuous features. More bins can improve the model's ability to capture nuances in the data but can also increase computational complexity.

Even if scikit-learn's `HistGradientBoosting` is nothing too different from LightGBM or XGBoost, it does provide a different way to implement GBMs in a competition, and models built by `HistGradientBoosting` may provide a contribution when ensembling multiple predictions, such as in blending and stacking.

Having reached the end of this section, you should be more familiar with the most common ML algorithms (only deep learning solutions have not been discussed) and their most important hyperparameters to tune, which will help you build an outstanding solution in a Kaggle competition. Knowing the basic optimization strategies, usable algorithms, and their key hyperparameters is just a starting point. In the next section, we will begin an in-depth discussion about how to tune them more optimally using Bayesian optimization.

But before moving on, let's turn our attention to our second interview of the chapter, with *Alberto Danese*, Head of Data Science at Nexi, an Italian credit card and digital payments company. A Competitions Grandmaster who joined the platform in 2015, he obtained most of his gold medals as a solo competitor.

Interview: Alberto Danese

`https://www.kaggle.com/albedan`

What's your favorite kind of competition and why? In terms of techniques and solving approaches, what is your specialty on Kaggle?

I've always worked in the financial services industry, dealing mostly with structured data, and I prefer competitions that belong to this category. I enjoy being able to have a practical grasp of what the data is all about and doing some smart feature engineering in order to squeeze every bit of information out of the data.

Technically speaking, I've got good experience with classical ML libraries and especially with GBDT: the most common libraries (XGBoost, LightGBM, CatBoost) are always my first choice.

How do you approach a Kaggle competition? How different is this approach from what you do in your day-to-day work?

I always spend a lot of time just exploring the data and trying to figure out what the problem that the sponsor is actually trying to solve with ML is. Different from what newbies usually think about Kaggle, I don't spend so much time on all the "tweaking" of the specific ML algorithm – and apparently this approach has paid off!

In my daily job, understanding the data is also extremely important, but there are some additional phases that are completely missing in a Kaggle competition. I've got to:

- Define a business problem to be solved with ML (together with colleagues in the business departments)
- Find the data, sometimes also from external data providers
- When the ML part is done, understand how to put it in production and manage the evolutions

Tell us about a particularly challenging competition you entered, and what insights you used to tackle the task.

I enjoyed the TalkingData AdTracking Fraud Detection Challenge, with which I became a Grandmaster. Besides being on an extremely interesting topic (fighting fraud from click-farms), it really pushed me to do efficient feature engineering, as the volumes were huge (more than 100M labeled rows), and cutting on computation times was key in order to test different approaches. It also forced me to understand how to exploit lag/lead features (and other window functions) in the best way, in order to create a sort of time series in an otherwise classical ML problem.

Has Kaggle helped you in your career? If so, how?

Definitely! Being able to achieve great objective and verifiable results is no doubt something that makes a resume stand out. When I was hired by Cerved (a marketing intelligence service company) in 2016, the hiring manager was perfectly aware of what Kaggle was – and having some real-world projects to talk about during an interview is something extremely valuable. For sure, Kaggle had an important role in the evolution of my career.

In your experience, what do inexperienced Kagglers often overlook? What do you know now that you wish you'd known when you first started?

I think that everyone just starts coding, maybe forking a public kernel and just changing a few lines or parameters. This is perfectly fine at the beginning! But you do have to spend a decent amount of time not coding, but studying the data and understanding the problem.

What mistakes have you made in competitions in the past?

Not sure if it counts as a mistake, but I have often preferred to compete solo: on one hand, it's great as it forces you to handle every single aspect of a competition, and you're able to manage your time as you wish. But I've really enjoyed collaborating with teammates on a couple of competitions as well: I probably should consider teaming up more often, as you can learn a lot from collaborating.

Are there any particular tools or libraries that you would recommend using for data analysis or ML?

Besides the usual ones, I've always been a great fan of data.table (starting from the R version): I think it's not getting the credit it deserves! It's really a great package when you want to deal with huge data on a local machine.

What's the most important thing someone should keep in mind or do when they're entering a competition?

Understand the problem and the data first: don't start coding right away!

Bayesian optimization

Leaving behind grid search (feasible only when the space of experiments is limited), the usual choice for the practitioner is to apply random search optimization or try a **Bayesian optimization** (**BO**) technique, which requires a more complex setup.

Originally introduced in the paper *Practical Bayesian optimization of machine learning algorithms* by Snoek, J., Larochelle, H., and Adams, R. P. (http://export.arxiv.org/pdf/1206.2944), the key idea behind Bayesian optimization is that we optimize a **proxy function**, also called a **surrogate function,** rather than directly relying on the true objective function, something that grid search and random search both do.

In other words, a proxy function is a simpler or computationally cheaper model that approximates the true objective function. It allows us to make informed decisions about where to sample next in the parameter space without having to evaluate the actual objective function, which can be costly or time-consuming. We rely on this approach when gradients are unavailable, when testing the true objective function requires significant costs (if it is not, then we simply go for a random search), or when the search space is noisy and complex.

Bayesian search balances *exploration* with *exploitation*. Exploration refers to the process of trying out a variety of options or parameter values to gather information about their potential performance. In the context of hyperparameter tuning, exploration involves testing different combinations of hyperparameters that may not have been considered previously. Exploitation, on the other hand, focuses on leveraging the best-known options based on prior evaluations. In hyperparameter tuning, this means refining and selecting hyperparameter values that have already shown good results in earlier trials. In more detail, at the start, Bayesian search leverages pure exploration, hence it explores randomly, thus training the surrogate function as it goes. Based on that surrogate function, the search algorithm then exploits its initial approximate knowledge of how the predictor works in order to sample more useful examples and minimize the cost function. As the *Bayesian* part of the name suggests, we are using priors to make smarter decisions about sampling during optimization. Priors in Bayesian statistics represent our beliefs or assumptions about the parameters or the model before we observe any data. In Bayesian optimization, such priors initially consist of the ranges we provide for exploration, and as we proceed, they are updated based on the results of the various exploration and exploitation experiments we conduct. Consequently, through a series of updates regarding where to search for better results in our optimization, we reach a minimization more quickly by limiting the number of evaluations we need to make.

Bayesian optimization uses an **acquisition function**, which provides an evaluation measure of how useful it would be to try any given point; in other words, it tells us how promising an observation will be. In fact, the algorithm defines an acquisition function to manage the trade-off between exploration and exploitation.

Often, Bayesian optimization is powered by Gaussian processes. A **Gaussian process (GP)** is a statistical model used to model and predict unknown functions based on observed data. As requirements for doing so, GPs assume that any finite subset of observed values drawn from the function follows a normal (Gaussian) distribution. GPs perform better when the search space has a smooth and predictable response. An alternative when the search space is more complex is using tree algorithms (for instance, random forests) or a completely different approach called **Tree Parzen Estimators** or **Tree-structured Parzen Estimators (TPEs)**.

Instead of directly building a model that estimates the success of a set of parameters, thus acting like an oracle, TPEs estimate the parameters of a multivariate distribution that define the best-performing values of the parameters based on successive approximations provided by the experimentations. In this way, TPEs derive the best set of parameters by sampling them from a probabilistic distribution and not directly from a ML model like GPs do.

We will discuss each of these approaches, first by examining scikit-optimize and KerasTuner, both based on GPs (scikit-optimize can also use random forests and KerasTuner can use multi-armed bandits), and then Optuna, which is principally based on TPE (though it also offers different strategies: `https://optuna.readthedocs.io/en/stable/reference/samplers/`).

> Though Bayesian optimization is considered the state of the art for hyperparameter tuning, always remember that for more complex parameter spaces, using Bayesian optimization provides no advantage in terms of time and computation spent over a solution simply found by random search. For instance, in Google Cloud Machine Learning Engine services, the usage of Bayesian optimization is limited to problems involving, at most, sixteen parameters. For larger numbers of parameters, it resorts to random sampling.

Using scikit-optimize

Scikit-optimize (`skopt`) has been developed using the same API as scikit-learn, as well as making extensive use of NumPy and SciPy functions. In addition, it was created by some of the contributors to the scikit-learn project, such as *Gilles Louppe*. Recently, due to the increased success of packages such as Optuna (see `https://optuna.org/`, a package we will be discussing later) or Hyperopt (https://github.com/hyperopt/hyperopt), the project has been closed and the owner archived it on February 28, 2024. It is now read-only.

While newer libraries like Optuna and Hyperopt may have gained popularity, scikit-optimize remains a valuable tool for Bayesian optimization due to its strong community, maturity, and educational resources. Its clear documentation, integration with scikit-learn, and wide range of applications make it an excellent choice for inclusion in this book on Kaggle, providing learners with a solid foundation in Bayesian optimization and its practical applications.

In addition, the package has an intuitive API, and it is quite easy to hack and use its functions in custom optimization strategies. Scikit-optimize is also renowned for its useful graphical representations. In fact, by visualizing the results of an optimization process (using scikit-optimize's `plot_objective` function), you can determine whether you can redefine the search space for the problem and formulate an explanation of how optimization works for a problem.

The examples in the first part of this section on Bayesian optimization will use scikit-optimize to illustrate basic principles of optimization, customization, and extensions to neural architecture search. In the second part, we will instead turn our attention to other optimizers such as KerasTuner and Optuna. In our example, we will refer to the work that can be found in the following Kaggle Notebooks:

- https://www.kaggle.com/lucamassaron/tutorial-bayesian-optimization-with-lightgbm
- https://www.kaggle.com/lucamassaron/scikit-optimize-for-lightgbm

Our purpose here is to show you how to quickly handle an optimization problem for a competition such as *30 Days of ML*, a recent competition that involved many Kagglers learning new skills and applying them in a competition lasting 30 days. The goal of this competition is to predict the value of an insurance claim, so it is a regression problem. You can learn more about this initiative and download the data necessary for the example we will present (materials are always available to the public) by visiting https://www.kaggle.com/thirty-days-of-ml.

If you cannot access the data because you have not previously participated in the competition, you can use this Kaggle dataset: https://www.kaggle.com/lucamassaron/30-days-of-ml.

The following code will present how to load the data for this problem and then set up a Bayesian optimization process to improve the performance of a LightGBM model:

1. We start by loading the packages:

```
# Importing core libraries
import numpy as np
import pandas as pd
from time import time
import pprint
import joblib
# Suppressing warnings because of skopt verbosity
import warnings
warnings.filterwarnings("ignore")
# Classifiers
import lightgbm as lgb
# Model selection
from sklearn.model_selection import KFold
# Metrics
from sklearn.metrics import root_mean_squared_error
```

```
from sklearn.metrics import make_scorer
# Skopt functions
from skopt import BayesSearchCV
from skopt.callbacks import DeadlineStopper, DeltaYStopper
from skopt.space import Real, Categorical, Integer
```

2. As a next step, we load the data. The data doesn't need much processing, aside from turning some categorical features with alphabetical letters as levels into ordered numeric ones:

```
# Loading data
X = pd.read_csv("../input/30-days-of-ml/train.csv")
X_test = pd.read_csv("../input/30-days-of-ml/test.csv")
# Preparing data as a tabular matrix
y = X.target
X = X.set_index('id').drop('target', axis='columns')
X_test = X_test.set_index('id')
# Dealing with categorical data
categoricals = [item for item in X.columns if 'cat' in item]
cat_values = np.unique(X[categoricals].values)
cat_dict = dict(zip(cat_values, range(len(cat_values))))
X[categoricals] = X[categoricals]\
    .replace(cat_dict).astype('category')
X_test[categoricals] = X_test[categoricals]\
    .replace(cat_dict).astype('category')
```

3. After making the data available, we define a reporting function that can be used by scikit-optimize for various optimization tasks. The function takes the data and the optimizer as inputs. It can also handle **callback functions**, which are functions that perform actions such as reporting, early stopping based on having reached a certain threshold of time spent searching or performance not improving (for instance, not seeing improvements for a certain number of iterations), or saving the state of the processing after each optimization iteration:

```
# Reporting util for different optimizers
def report_perf(optimizer, X, y, title="model", callbacks=None):
    """
    A wrapper for measuring time and performance of optimizers
    optimizer = a sklearn or a skopt optimizer
    X = the training set
    y = our target
```

```
    title = a string label for the experiment
    """

    start = time()

    if callbacks is not None:
        optimizer.fit(X, y, callback=callbacks)
    else:
        optimizer.fit(X, y)

    d=pd.DataFrame(optimizer.cv_results_)
    best_score = optimizer.best_score_
    best_score_std = d.iloc[optimizer.best_index_].std_test_score
    best_params = optimizer.best_params_

    print((
        title + " took %.2f seconds, candidates checked: %d,
        best CV score: %.3f" + u" \u00B1"+" %.3f"
    ) % (
        time() - start,
        len(optimizer.cv_results_['params']),
        best_score,
        best_score_std
    ))
    print('Best parameters:')
    pprint.pprint(best_params)
    print()
    return best_params
```

4. We now have to prepare the scoring function (upon which the evaluation is based), the validation strategy (based on cross-validation), the model, and the search space. For the scoring function, which should be a root mean squared error metric, we refer to the practices in scikit-learn where you always minimize a function (if you have to maximize a measure, such as with the ROC-AUC score, the solution is to minimize the negative value of the measure).

The make_scorer wrapper can easily replicate such practices:

```
# Setting the scoring function
scoring = make_scorer(root_mean_squared_error,
                      squared=False),
```

```
                              greater_is_better=False)
# Setting the validation strategy
kf = KFold(n_splits=5, shuffle=True, random_state=0)
# Setting the basic regressor
reg = lgb.LGBMRegressor(boosting_type='gbdt',
                        metric='rmse',
                        objective='regression',
                        n_jobs=1,
                        verbose=-1,
                        random_state=0)
```

5. Setting the search space requires the use of different functions from scikit-optimize, such as `Real`, `Integer`, or `Choice`, each one sampling from a different kind of distribution that you define as a parameter (usually the uniform distribution, but the log-uniform distribution is also used when you are more interested in the scale effect of a parameter than its exact value):

```
# Setting the search space
search_spaces = {
    # Boosting learning rate
    'learning_rate': Real(0.01, 1.0, 'log-uniform'),
    # Number of boosted trees to fit
    'n_estimators': Integer(30, 5000),
    # Maximum tree leaves for base learners
    'num_leaves': Integer(2, 512),
    # Maximum tree depth for base learners
    'max_depth': Integer(-1, 256),
    # Minimal number of data in one leaf
    'min_child_samples': Integer(1, 256),
    # Max number of bins buckets
    'max_bin': Integer(100, 1000),
    # Subsample ratio of the training instance
    'subsample': Real(0.01, 1.0, 'uniform'),
    # Frequency of subsample
    'subsample_freq': Integer(0, 10),
    # Subsample ratio of columns
    'colsample_bytree': Real(0.01, 1.0, 'uniform'),
    # Minimum sum of instance weight
    'min_child_weight': Real(0.01, 10.0, 'uniform'),
```

```
        # L2 regularization
    'reg_lambda': Real(1e-9, 100.0, 'log-uniform'),
        # L1 regularization
    'reg_alpha': Real(1e-9, 100.0, 'log-uniform'),
}
```

So far, you have defined:

- Your cross-validation strategy
- Your evaluation metric
- Your base model
- Your hyperparameter search space

6. Now, all that is left is to feed them into your optimization function, `BayesSearchCV`. Based on the CV scheme provided, this function will look for the minimum of your scoring function based on values within the search space. You can set a maximum number of iterations performed, the kind of surrogate function (GP works on most occasions), and the random seed for reproducibility:

```
# Wrapping everything up into the Bayesian optimizer
opt = BayesSearchCV(estimator=reg,
                    search_spaces=search_spaces,
                    scoring=scoring,
                    cv=kf,
                    n_iter=60,              # max number of trials
                    n_jobs=-1,              # number of jobs
                    iid=False,
                    # if not iid it optimizes on the cv score
                    return_train_score=False,
                    refit=False,
                    # Gaussian Processes (GP)
                    optimizer_kwargs={'base_estimator': 'GP'},
                    # random state for replicability
                    random_state=0)
```

7.   At this point, you can start the search using the reporting function we defined previously. After a while, the function will return the best parameters for the problem:

```
# Running the optimizer
overdone_control = DeltaYStopper(delta=0.0001)
# We stop if the gain of the optimization becomes too small
time_limit_control = DeadlineStopper(total_time=60 * 60 * 6)
# We impose a time limit (6 hours)
best_params = report_perf(opt, X, y,'LightGBM_regression',
                          callbacks=[overdone_control,
                          time_limit_control])
```

In the preceding example, we set a limit on operations by specifying the maximum time allowed (6 hours) before stopping and reporting the best results. Since the Bayesian optimization approach blends together the exploration and exploitation of different combinations of hyperparameters, stopping at any time will always return the best solution found so far (but not necessarily the best one possible). This is because the acquisition function will always prioritize exploring the most promising parts of the search space based on the estimated performances the surrogate function returns and their uncertainty intervals. In the next section, we will learn how to customize a Bayesian optimization to include

Customizing a Bayesian optimization search

The BayesSearchCV function offered by scikit-optimize is certainly convenient because it wraps and arranges all the elements of a hyperparameter search by itself, but it also has limitations. For instance, you may find it useful in a competition to:

- Have more control over each search iteration, for instance, mixing random search and Bayesian search
- Be able to apply early stopping on algorithms
- Customize your validation strategy more
- Stop experiments that do not work early (for instance, immediately evaluating the performance of the single cross-validation folds when it is available instead of waiting to have all folds averaged at the end)
- Create clusters of hyperparameter sets that perform in a similar way (for instance, to create multiple models differing only in the hyperparameters used, to be used for a blending ensemble)

Each of these tasks would not be too complex if you could modify the `BayesSearchCV` internal procedure. Luckily, scikit-optimize lets you do just this. In fact, behind `BayesSearchCV`, as well as behind other wrappers from the package, there are specific minimizing functions that you can use as standalone parts of your own search function:

- `gp_minimize`: Bayesian optimization using GPs
- `forest_minimize`: Bayesian optimization using random forests or extremely randomized trees
- `gbrt_minimize`: Bayesian optimization using gradient boosting
- `dummy_minimize`: Just a random search strategy

In the following example, we will modify the previous search using our custom search function. You can find the example working in a Kaggle Notebook at `https://www.kaggle.com/lucamassaron/hacking-bayesian-optimization`.

The new custom function will accept early stopping during training and will prune experiments if one of the fold validation results is not top-performing. As in the previous example, we start by importing the necessary packages:

```python
# Importing core libraries
import numpy as np
import pandas as pd
from time import time
import pprint
import joblib
# Suppressing warnings because of skopt verbosity
import warnings
warnings.filterwarnings("ignore")
# Classifier/Regressor
from xgboost import XGBRegressor
# Model selection
from sklearn.model_selection import KFold, StratifiedKFold
from sklearn.model_selection import cross_val_score
from sklearn.model_selection import train_test_split
# Metrics
from sklearn.metrics import root_mean_squared_error
from sklearn.metrics import make_scorer
# Skopt functions
```

```
from skopt import BayesSearchCV
from skopt.callbacks import DeadlineStopper, DeltaYStopper
from skopt.space import Real, Categorical, Integer
from skopt import gp_minimize, forest_minimize
from skopt import gbrt_minimize, dummy_minimize
# Decorator to convert a list of parameters to named arguments
from skopt.utils import use_named_args
# Data processing
from sklearn.preprocessing import OrdinalEncoder
```

In the same way as before, we upload the data from the *30 Days of ML* competition:

```
# Loading data
X_train = pd.read_csv("../input/30-days-of-ml/train.csv")
X_test = pd.read_csv("../input/30-days-of-ml/test.csv")
# Preparing data as a tabular matrix
y_train = X_train.target
X_train = X_train.set_index('id').drop('target', axis='columns')
X_test = X_test.set_index('id')
# Pointing out categorical features
categoricals = [item for item in X_train.columns if 'cat' in item]
# Dealing with categorical data using OrdinalEncoder
ordinal_encoder = OrdinalEncoder()
X_train[categoricals] = ordinal_encoder.fit_transform(
    X_train[categoricals])
X_test[categoricals] = ordinal_encoder.transform(X_test[categoricals])
```

Now, we set all the necessary elements for a hyperparameter search: the scoring function, the validation strategy, the search space, and the ML model to be optimized. The scoring function and the validation strategy will later become the core elements constituting the objective function, the function that the Bayesian optimization will strive to minimize:

```
# Setting the scoring function
scoring = make_scorer(root_mean_squared_error,
                      greater_is_better=False)
# Setting the cv strategy
kf = KFold(n_splits=5, shuffle=True, random_state=0)
# Setting the search space
space = [Real(0.01, 1.0, 'uniform', name='learning_rate'),
```

```
            Integer(1, 8, name='max_depth'),
            Real(0.1, 1.0, 'uniform', name='subsample'),
            # Subsample ratio of columns by tree
            Real(0.1, 1.0, 'uniform', name='colsample_bytree'),
            # L2 regularization
            Real(0, 100., 'uniform', name='reg_lambda'),
            # L1 regularization
            Real(0, 100., 'uniform', name='reg_alpha'),
            # minimum sum of instance weight (hessian)
            Real(1, 30, 'uniform', name='min_child_weight')
    ]
    model = XGBRegressor(n_estimators=10_000,
                         booster='gbtree', random_state=0)
```

Notice this time that we have not included the number of estimators (the n_estimators parameter) in the search space. Instead, we set it when instantiating the model, and we enter a high value since we expect to stop the model early based on a validation set.

As a next step, you now need to create the objective function. The objective function should just accept the parameters to be optimized as input and return the resulting score. However, the objective function must also accept the elements necessary for the search you have just prepared. Naturally, you could refer to them from inside the function. However, it is a good practice to take them into the function itself, using its internal memory space. This has its advantages. For instance, you will make the elements immutable and carry them along with the objective function, either by pickling or by distributing the search task on a multi-processor level. You can obtain this second result by creating a make function that takes in the elements, with the modified objective function being returned by the make function. With this simple structure, your objective function will incorporate all the elements, such as the data and the model, and you will only need to pass in the parameters to be tested.

Let's start coding the function. We will stop along the way to discuss some relevant aspects. In this first part of the function, you simply create an objective function, doing cross-validation and fitting the data using early stopping:

```
# The objective function to be minimized
def make_objective(model, X, y, space, cv, scoring, validation=0.2):
    # This decorator converts your objective function
```

```python
# with named arguments into one that accepts a list as argument,
# while doing the conversion automatically.
@use_named_args(space)
def objective(**params):
    model.set_params(**params)
    print("\nTesting: ", params)
    validation_scores = list()
    for k, (train_index, test_index) in enumerate(kf.split(X, y)):
        val_index = list()
        train_examples = int(train_examples * (1 - validation))
        train_index, val_index = (train_index[:train_examples],
                                  train_index[train_examples:])

        start_time = time()
        model.fit(X.iloc[train_index,:], y[train_index],
                  early_stopping_rounds=50,
                  eval_set=[(X.iloc[val_index,:], y[val_index])],
                  verbose=0)
        end_time = time()

        rounds = model.best_iteration

        test_preds = model.predict(X.iloc[test_index,:])
        test_score = scoring(y[test_index], test_preds)
        print(f"CV Fold {k+1} rmse:{test_score:0.5f}-{rounds}
              rounds - it took {end_time-start_time:0.0f} secs")
        validation_scores.append(test_score)
```

Here, we have used an aggressive early-stopping strategy to save time, but you could raise the number of patient rounds if you believe that it might work better for your problem. Notice that the validation examples are sequentially taken out from the examples in the training folds (see how `train_index` and `val_index` are defined in the code), leaving the out-of-fold examples (`test_index` derived from the `kf` cross-validation splitting) untouched for the final validation. This is important if you do not want to incur adaptive overfitting on the data you use for early stopping.

In the next part, before moving on to the cross-validation loop and proceeding to the remaining cross-validation folds to be trained and tested, you analyze the result obtained by the fold on the out-of-fold set:

```
if len(history[k]) >= 10:
    threshold = np.percentile(history[k], q=25)
    if test_score > threshold:
        print(f"Early stopping for under-performing fold:
            threshold is {threshold:0.5f}")
        return np.mean(validation_scores)

history[k].append(test_score)
return np.mean(validation_scores)
return objective
```

Notice that we are keeping a global dictionary, `history`, containing the results obtained from each fold up to now. We can compare the results across multiple experiments and cross-validations; the cross-validation is reproducible due to the random seed, so the results of the same fold are perfectly comparable. If the result of the present fold is sub-par compared to the previously obtained folds in other iterations (using the bottom quartile as a reference), the idea is to stop and return the average of the folds tested so far. The rationale for this is that if one fold doesn't present acceptable results, then the whole cross-validation probably won't either. You can, therefore, just quit and move on to another set of more promising parameters. It is a kind of early stopping on cross-validation that should speed up your search and allow you to cover more experiments in less time.

Next, using our `make_objective` function, we put together all the elements (model, data, search space, validation strategy, and scoring function) into a single function, the objective function. As a result, we now have a function that only takes in the parameters to be optimized and returns a score, based on which the minimization engine of the optimization will decide the next experiments:

```
objective = make_objective(model,
                           X_train, y_train,
                           space=space,
                           cv=kf,
                           scoring=scoring)
```

Since we want to control each optimization step and save it for later use, we also prepare a callback function to save a list of the experiments executed and their results at every iteration of the minimization process. Simply by using these two pieces of information, the minimization engine can be halted at any time, and it can thereafter resume the optimization from the checkpoint:

```
def onstep(res):
    global counter
    x0 = res.x_iters    # List of input points
    y0 = res.func_vals # Evaluation of input points
    print('Last eval: ', x0[-1],
            ' - Score ', y0[-1])
    print('Current iter: ', counter,
            ' - Best Score ', res.fun,
            ' - Best Args: ', res.x)
    # Saving a checkpoint to disk
    joblib.dump((x0, y0), 'checkpoint.pkl')
    counter += 1
```

At this point, we are ready to start. Bayesian optimization needs some starting points to work properly. We create several experiments with random search (using the `dummy_minimize` function) and save their results:

```
counter = 0
history = {i:list() for i in range(5)}
used_time = 0
gp_round = dummy_minimize(func=objective,
                          dimensions=space,
                          n_calls=30,
                          callback=[onstep],
                          random_state=0)
```

We can then retrieve the saved experiments and print the sequence of sets of hyperparameters that the Bayesian optimization has tested, along with their results. In fact, we can find the set of parameters and their results contained in the x0 and y0 lists:

```
x0, y0 = joblib.load('checkpoint.pkl')
print(len(x0))
```

At this point, we can even resume the Bayesian optimization with some changes in the search space, the acquisition function, the number of calls, or the callbacks:

```
x0, y0 = joblib.load('checkpoint.pkl')
gp_round = gp_minimize(func=objective,
                       x0=x0,      # already examined values for x
                       y0=y0,      # observed values for x0
                       dimensions=space,
                       acq_func='gp_hedge',
                       n_calls=30,
                       n_initial_points=0,
                       callback=[onstep],
                       random_state=0)
```

Once we are satisfied that we don't need to continue calling the optimization function, we can print both the best score obtained (based on our inputs and validation scheme) and the set of best hyperparameters:

```
x0, y0 = joblib.load('checkpoint.pkl')
print(f"Best score: {gp_round.fun:0.5f}")
print("Best hyperparameters:")
for sp, x in zip(gp_round.space, gp_round.x):
    print(f"{sp.name:25} : {x}")
```

Based on the best result, we can retrain our model for use in the competition.

Now we have the set of parameters and their results (the x0 and y0 lists), we could also explore the different results and cluster together the ones that are similar in output but different in the set of parameters used. This will help us train more diverse models with similar performances but different optimization strategies. This is the ideal situation for **blending**, which is the averaging of multiple models to lower the variance of the estimates and obtain a better public and private leaderboard score.

> Refer to *Chapter 10, Ensembling with Blending and Stacking Solutions,* for a discussion on blending.

Extending Bayesian optimization to neural architecture search

Moving on to deep learning, neural networks also seem to have quite a few hyperparameters to fix:

- Batch size
- Learning rate
- The kind of optimizer and its internal parameters

All these parameters influence how the network learns and can make a big impact; just a slight difference in batch size or learning rate can determine whether a network can reduce its error beyond a certain threshold.

That being said, these learning parameters are not the only ones that you can optimize when working with **deep neural networks (DNNs)**. How the network is organized in layers, and the details of its architecture, can make even more of a difference.

In fact, technically speaking, an **architecture** implies the representational capacity of the DNN, which means that, depending on the layers you use, the network will either be able to read and process all the information available in the data or it will not. While you had a large but limited set of choices with other ML algorithms, with DNNs, your choices seem unlimited because the only apparent limit is your knowledge and experience in handling parts of neural networks and putting them together.

Common best practices for great deep learning practitioners when assembling well-performing DNNs mainly are:

- Relying on pretrained models (so you have to be very knowledgeable about the solutions available, such as those found on Hugging Face at `https://huggingface.co/models` or on GitHub)
- Reading cutting-edge papers
- Copying top Kaggle Notebooks from the same competition or previous ones
- Trial and error
- Ingenuity

In a famous lesson given by *Professor Geoffrey Hinton*, he states that you can achieve similar and often better results in deciding about a neural architecture and its hyperparameters by using automated methods such as Bayesian optimization.

Bayesian optimization will also prevent you from getting stuck because you cannot figure out the best combinations of hyperparameters among the many possible ones.

> For a recording of Professor Geoffrey Hinton's lesson, see `https://www.youtube.com/watch?v=i0cKa0di_lo`. For the slides, see `https://www.cs.toronto.edu/~hinton/coursera/lecture16/lec16.pdf`.

As we mentioned before, even in most sophisticated AutoML systems, when you have too many hyperparameters, relying on random optimization may produce better results or the same results in the same amount of time as Bayesian optimization. There's no definitive rule of thumb; a general guideline is to consider switching to random optimization when the number of hyperparameters exceeds 10–15. In addition, in this case, you also have to fight against an optimization landscape with sharp turns and surfaces. This means that the function you're trying to optimize has areas where the values change rapidly or unpredictably, which makes it harder for optimization algorithms to find the best solution. Instead of having a smooth, gradual surface where small steps lead to steady improvements, the landscape may have steep cliffs or sudden dips, causing the algorithm to struggle or get stuck in local optima (suboptimal solutions) instead of reaching the global best solution.

In addition, when working on DNN optimization, many of your parameters won't work with continuous values but use Boolean flags instead, and just one change to a flag could unexpectedly transform the performance of your network for the better or for the worse.

Our experience tells us that random optimization may *not* be suitable for a Kaggle competition because you have limited time and resources, and you can leverage your previous optimization results to find better solutions.

Bayesian optimization in this scenario is ideal due to the following reasons:

- You can set it to work based on the time and computational resources that you have and do it in stages, refining your settings through multiple sessions.

- Moreover, it is unlikely that you will easily be able to leverage parallelism for tuning DNNs since they use GPUs unless you have multiple very powerful machines at hand. By working sequentially, Bayesian optimization just needs one good machine to perform the task.

- Finally, even if it is hard to find optimal architectures by a search due to the optimization landscape, you leverage information from previous experiments, especially at the beginning, totally avoiding combinations of parameters that won't work. With random optimization, all combinations are always liable to be tested unless you change the search space along the way.

There are also drawbacks, however:

- Bayesian optimization models the hyperparameter space using a surrogate function built from previous trials, which is not an error-free process.

- It is not a remote possibility that the process concentrates only on a part of the search space while ignoring other parts (which may instead contain the minimum you are looking for).

The solution to this is to run many experiments to be safe or to alternate between random search and Bayesian optimization, challenging the Bayesian model with random trials that can force it to reshape its search model in a more optimal way.

For our example, we again use the data from the *30 Days of ML* initiative by Kaggle, a regression task. Our example is based on TensorFlow, but with small modifications, it can run on other deep learning frameworks such as PyTorch or MXNet.

As before, you can find the example on Kaggle here: `https://www.kaggle.com/lucamassaron/hacking-bayesian-optimization-for-dnns`.

Let's begin:

```
import tensorflow as tf
```

After importing the TensorFlow package, we leverage its `Dataset` function to create an iterable capable of feeding our neural network with batches of data:

```
def df_to_dataset(dataframe, shuffle=True, batch_size=32):
    dataframe = dataframe.copy()
    labels = dataframe.pop('target')
    ds = tf.data.Dataset.from_tensor_slices((dict(dataframe),
                                             labels))
    if shuffle:
        ds = ds.shuffle(buffer_size=len(dataframe))
    ds = ds.batch(batch_size)
    return ds
```

We proceed to code a function that creates our deep neural network model based on a set of hyperparameters:

```
def create_model(cat0_dim, cat1_dim, cat2_dim,
                 cat3_dim, cat4_dim, cat5_dim,
                 cat6_dim, cat7_dim, cat8_dim, cat9_dim,
```

```
            layers, layer_1, layer_2, layer_3, layer_4, layer_5,
            activation, dropout, batch_normalization, learning_rate,
            **others):
    categorical_labels = ['cat0', 'cat1', 'cat2', 'cat3', 'cat4', 'cat5',
                          'cat6', 'cat7', 'cat8', 'cat9']
numeric_labels = ['cont1', 'cont2', 'cont3', 'cont4',
                  'cont5', 'cont6', 'cont7', 'cont8',
                  'cont9', 'cont10', 'cont11', 'cont12', 'cont13']
dims = {'cat0': cat0_dim, 'cat1': cat1_dim, 'cat2': cat2_dim,
        'cat3': cat3_dim, 'cat4': cat4_dim, 'cat5': cat5_dim,
        'cat6': cat6_dim, 'cat7': cat7_dim, 'cat8': cat8_dim,
        'cat9': cat9_dim}
vocab = {h:X_train['cat4'].unique().astype(int)
         for h in categorical_labels}
layers = [layer_1, layer_2, layer_3, layer_4, layer_5][:layers]
feature_columns = list()
categorical_inputs  = list()
flattened_categorical_inputs = list()
for header in numeric_labels:
    input_layer = tf.keras.Input(shape=(1,), name=header)
    continuous_inputs.append(input_layer)
  feature_columns.append(tf.feature_column.numeric_column(header))
for header in categorical_labels:
    input_layer = tf.keras.Input(shape=(1,), name=header)
    embedded_input = tf.keras.layers.Embedding(
        input_dim=vocab[header] + 1,
        output_dim=dims[header],
        name=f'{header}_embedding'
    )(input_layer)
    flattened_input = tf.keras.layers.Flatten()(embedded_input)
    categorical_inputs.append(input_layer)
    flattened_categorical_inputs.append(flattened_input)
concatenated_inputs = tf.keras.layers.Concatenate()(
    continuous_inputs + flattened_categorical_inputs)
x = concatenated_inputs
for nodes in layers:
    x = tf.keras.layers.Dense(nodes, activation=activation)(x)
```

```
       if batch_normalization:
           x = tf.keras.layers.BatchNormalization()(x)
       if dropout > 0:
           x = tf.keras.layers.Dropout(dropout)(x)
   output = tf.keras.layers.Dense(1)(x)
   model = tf.keras.Model(inputs=continuous_inputs + categorical_inputs,
                          outputs=output)
   model.compile(
       optimizer=tf.keras.optimizers.Adam( `
           learning_rate=learning_rate),
       loss= tf.keras.losses.MeanSquaredError(),
       metrics=['mean_squared_error'])
   return model
```

Internally, the code in the create_model function customizes the neural network architecture based on the inputs provided. For instance, as parameters for the function, you can provide the dimensions of the embeddings for each categorical variable or define the structure and number of dense layers present in the network. All these parameters are related to the parameter space you want to be explored by Bayesian optimization. Hence, every input parameter of the function creating the model should be related to a **sampling function** defined in the search space. All you have to do is place the sampling functions in a list in the same order as expected by the create_model function:

```
# Setting the search space

space = [Integer(1, 2, name='cat0_dim'),
         Integer(1, 2, name='cat1_dim'),
         Integer(1, 2, name='cat2_dim'),
         Integer(1, 3, name='cat3_dim'),
         Integer(1, 3, name='cat4_dim'),
         Integer(1, 3, name='cat5_dim'),
         Integer(1, 4, name='cat6_dim'),
         Integer(1, 4, name='cat7_dim'),
         Integer(1, 6, name='cat8_dim'),
         Integer(1, 8, name='cat9_dim'),
         Integer(1, 5, name='layers'),
         Integer(2, 256, name='layer_1'),
         Integer(2, 256, name='layer_2'),
```

```
        Integer(2, 256, name='layer_3'),
        Integer(2, 256, name='layer_4'),
        Integer(2, 256, name='layer_5'),
        Categorical([leaky_relu, relu], name='activation'),
        Real(0.0, 0.5, 'uniform', name='dropout'),
        Categorical([True, False], name='batch_normalization'),
        Categorical([0.01, 0.005, 0.002, 0.001], name='learning_rate'),
        Integer(256, 1024, name='batch_size')
]
```

As previously illustrated, you now combine all the elements related to the search into an objective function to be created by a function incorporating your basic search elements, such as the data and the cross-validation strategy:

```
def make_objective(model_fn, X, space, cv, scoring, validation=0.2):
    # This decorator converts your objective function with named arguments
    # into one that accepts a list as argument, while doing the conversion
    # automatically.
    @use_named_args(space)
    def objective(**params):
        print("\nTesting: ", params)
        validation_scores = list()
        for k, (train_index, test_index) in enumerate(kf.split(X)):
            val_index = list()
            train_examples = len(train_index)
            train_examples = int(train_examples * (1 - validation))
            train_index, val_index = (
                train_index[:train_examples],
                train_index[train_examples:]
            )
            start_time = time()
            model = model_fn(**params)
            measure_to_monitor = 'val_mean_squared_error'
            modality='min'
            early_stopping = tf.keras.callbacks.EarlyStopping(
                monitor=measure_to_monitor,
                mode=modality,
                patience=5,
```

```
        verbose=0
    )
    model_checkpoint = tf.keras.callbacks.ModelCheckpoint(
        'best_model.keras',
        monitor=measure_to_monitor,
        mode=modality,
        save_best_only=True,
        verbose=0)
    run = model.fit(
        df_to_dataset(X_train.iloc[train_index, :],
                      batch_size=params['batch_size']),
        validation_data=df_to_dataset(X_train.iloc[val_index, :],
                                      batch_size=1024),
        epochs=1_000,
        callbacks=[model_checkpoint, early_stopping],
        verbose=0
    )
    end_time = time()

    rounds = np.argmin(
        run.history['val_mean_squared_error']) + 1
    model = tf.keras.models.load_model('best_model.keras')
    os.remove('best_model.keras')
    test_preds = model.predict(
        df_to_dataset(X.iloc[test_index, :],
                      shuffle=False, batch_size=1024)
    ).flatten()
    test_score = scoring(X.iloc[test_index, :]['target'],
                         test_preds)

    print(
        f"CV Fold {k+1} rmse:{test_score:0.5f} - {rounds} "
        f"rounds - it took {end_time-start_time:0.0f} secs"
    )
    validation_scores.append(test_score)

    if len(history[k]) >= 10:
```

```
            threshold = np.percentile(history[k], q=25)
            if test_score > threshold:
                print(
                    f"Early stopping for under-performing fold: "
                    f"threshold is {threshold:0.5f}")
                return np.mean(validation_scores)
        history[k].append(test_score)
    return np.mean(validation_scores)
return objective
```

The next step is to provide a sequence of random search runs (as a way to start building some feedback from the search space) and gather the results as a starting point. Then, we can feed them into a Bayesian optimization and proceed by using `forest_minimize` as a surrogate function:

```
counter = 0
history = {i:list() for i in range(5)}
used_time = 0
gp_round = dummy_minimize(func=objective,
                          dimensions=space,
                          n_calls=10,
                          callback=[onstep],
                          random_state=0)
gc.collect()
x0, y0 = joblib.load('checkpoint.pkl')
gp_round = gp_minimize(func=objective,
                       x0=x0,   # already examined values for x
                       y0=y0,   # observed values for x0
                       dimensions=space,
                       n_calls=30,
                       n_initial_points=0,
                       callback=[onstep],
                       random_state=0)
gc.collect()
```

Notice that after the first ten rounds of random search, we proceed with our search using a random forest algorithm as a surrogate function. That That will ensure better and faster results than using a **Gaussian Process (GP)**.

As before, in this process, we have to strive to make the optimization feasible within our time and resources (for instance, by setting a low number of n_calls). Hence, we can proceed with batches of search iterations by saving the state of the optimization, checking the results obtained, and deciding thereafter to proceed or conclude the optimization process and not invest more time and energy into looking for a better solution.

Creating lighter and faster models with KerasTuner

If the previous section has puzzled you because of its complexity, KerasTuner can offer you a fast solution for setting up an optimization without much hassle. Though it uses Bayesian optimization and GPs by default, the new idea behind KerasTuner is **hyperband optimization**. Hyperband optimization uses the bandit approach to determine the best parameters (see http://web.eecs. umich.edu/~mosharaf/Readings/HyperBand.pdf). The bandit approach balances exploration (trying different options to gather information) and exploitation (focusing on the best option based on what's been learned so far). It helps optimize efficiently by allocating more resources to promising solutions while still exploring new possibilities. This works quite well with neural networks, whose optimization landscape is quite irregular and discontinuous and thus not always suitable for GPs.

> Remember that you cannot avoid building the function that builds a custom network using input hyperparameters; KerasTuner just makes it much easier to handle.

Let's start from the beginning. KerasTuner (https://keras.io/keras_tuner/) was announced as a "flexible and efficient hyperparameter tuning for Keras models" by *François Chollet*, the creator of Keras.

The recipe proposed by Chollet for running KerasTuner is made up of simple steps, starting from your existing Keras model:

1. Wrap your model in a function with hp as the first parameter.
2. Define hyperparameters at the beginning of the function.
3. Replace DNN static values with hyperparameters.
4. Write the code that models a complex neural network from the given hyperparameters.
5. If necessary, dynamically define hyperparameters as you build the network.

We'll now explore how all these steps can work for you in a Kaggle competition by using an example. Currently, KerasTuner is part of the stack offered by any Kaggle Notebook. Hence, you don't need to install it. In addition, the TensorFlow add-ons are part of the Notebook's pre-installed packages.

If you are not using a Kaggle Notebook and you need to try KerasTuner, you can easily install both using the following commands:

```
!pip install -U keras-tuner
!pip install -U tensorflow-addons
```

This example is already set up on a Kaggle Notebook here: `https://www.kaggle.com/lucamassaron/kerastuner-for-imdb/`.

Our first step is to import the necessary packages (creating shortcuts for some commands, such as for `pad_sequences`) and to upload the data we will be using directly from Keras:

```
import numpy as np
import pandas as pd
import tensorflow as tf
from tensorflow import keras
from sklearn.model_selection import train_test_split
from tensorflow.keras.models import Sequential
from tensorflow.keras.layers import LeakyReLU
from tensorflow.keras.layers import Activation
from tensorflow.keras.optimizers import SGD, Adam
from tensorflow.keras.wrappers.scikit_learn import KerasClassifier
from tensorflow.keras.callbacks import EarlyStopping, ModelCheckpoint

pad_sequences = keras.preprocessing.sequence.pad_sequences
imdb = keras.datasets.imdb
(train_data, train_labels), (test_data, test_labels) = \
    imdb.load_data(num_words=10000)
train_data, val_data, train_labels, val_labels = train_test_split(
    train_data, train_labels, test_size=0.30,
    shuffle=True, random_state=0
)
```

This time, we are using the IMDb dataset, which is available in the Keras package (`https://keras.io/api/datasets/imdb/`). The dataset has some interesting characteristics:

- It is a dataset of 25,000 movie reviews from IMDb

- The reviews are labeled by sentiment (positive/negative)

- The target classes are balanced (hence, accuracy works as a scoring measure)

- Each review is encoded as a list of word indexes (integers)

- For convenience, words are indexed by overall frequency

In addition, it has been successfully used in a popular Kaggle competition on word embeddings (`https://www.kaggle.com/c/word2vec-nlp-tutorial/overview`).

This example involves **natural language processing (NLP)**. This problem is often solved using **recurrent neural networks (RNNs)** based on **long short-term memory (LSTM)** or **gated recurrent unit (GRU)** layers. BERT, RoBERTa, and the other transformer-based models often achieve better results – being pretrained models relying on large language corpora – but this is not necessarily true in all problems, and RNNs can prove a strong baseline to beat or a good addition to an ensemble of neural models. In our example, all words are already numerically indexed. We just add to the existing indices the numeric codes that denote padding (so we can easily normalize all the text to the phrase length), the start of the sentence, an unknown word, and an unused word:

```python
# A dictionary mapping words to an integer index
word_index = imdb.get_word_index()
# The first indices are reserved
word_index = {k:(v+3) for k,v in word_index.items()}
word_index["<PAD>"] = 0
word_index["<START>"] = 1
word_index["<UNK>"] = 2  # unknown
word_index["<UNUSED>"] = 3
reverse_word_index = dict([(value, key) for (key, value) in word_index.
items()])
def decode_review(text):
    return ' '.join([reverse_word_index.get(i, '?') for i in text])
```

The next step involves creating a custom layer for **attention**. Attention is the foundation of transformer models, and it is one of the most innovative ideas in neural NLP of recent times.

For all the details of how these kinds of layers work, see the seminal paper on attention by Vaswani et al. published in 2017, titled *Attention Is All You Need* (`https://proceedings.neurips.cc/paper/2017/file/3f5ee243547dee91fbd053c1c4a845aa-Paper.pdf`).

The idea of attention can be easily conveyed. LSTM and GRU layers output processed sequences, but not all the elements in these output sequences are necessarily important for your predictions.

Instead of averaging all the output sequences using a pool layer across the stratified sequences, you can actually take a *weighted average* of them (and learn the correct weights to be used during the training phase). This weighting process (**attention**) improves the results you will pass on further. Of course, you can make this approach even more sophisticated using multiple attention layers (we call this **multi-head attention**), but in our example, a single layer will suffice because we want to demonstrate that using attention is more effective in this problem than simply averaging or just concatenating all the results together:

```python
from tensorflow.keras.layers import Dense, Dropout
from tensorflow.keras.layers import Flatten, RepeatVector, dot, multiply,
Permute, Lambda
def attention(layer):
    # --- Attention is all you need --- #
    units = layer.shape[-1]
    attention = Dense(1, activation='tanh')(layer)
    attention = Flatten()(attention)
    attention = tf.keras.activations.softmax(attention)
    attention = RepeatVector(units)(attention)
    attention = Permute([2, 1])(attention)
    representation = multiply([layer, attention])
    representation = Lambda(lambda x: tf.reduce_sum (x, axis=-2),
                           output_shape=(units,))(representation)
    # --------------------------------- #
    return representation
```

As a further variation in our experiments on the architecture of the DNNs for this problem, we also want to test the effectiveness of using different kinds of optimizers. Here, we use:

- **Rectified Adam (RAdam):** This is an adaptive learning rate optimizer that builds on the popular Adam optimizer, addressing one of its key weaknesses: high variance in the early stages of training. This variance can cause instability, especially when training models with limited data or in the initial stages of learning. You can read about this optimizer in this post and learn more: https://lessw.medium.com/new-state-of-the-art-ai-optimizer-rectified-adam-radam-5d854730807b).

- **Stochastic Weighted Averaging (SWA):** SWA is a way to average the weights traversed during the optimization based on a modified learning rate schedule: if your model tends to overfit or overshoot, SWA helps in getting near to an optimal solution, and it is proven to work especially well in NLP problems.

Let's look at the code:

```
def get_optimizer(option=0, learning_rate=0.001):
    if option==0:
        return tf.keras.optimizers.Adam(learning_rate)
    elif option==1:
        return tf.keras.optimizers.SGD(learning_rate,
                                       momentum=0.9, nesterov=True)
    elif option==2:
        return tf.keras.optimizers.RMSprop(learning_rate)
    else:
        return tf.keras.optimizers.Adam(learning_rate)
```

Having defined two key functions, we now face the most important function to code: the one that will provide different neural architectures given the parameters. We don't encode all the various parameters we want to connect to the different architectural choices; we only provide the hp parameter, which should contain all the possible parameters we want to use, and KerasTuner will run that. Aside from hp in the function input, we fix the size of the vocabulary and the length to be padded (adding dummy values if the effective length is shorter or cutting the phrase if the length is longer):

```
layers = keras.layers
models = keras.models
def create_tunable_model(hp, vocab_size=10000, pad_length=256):
    # Instantiate model params
    embedding_size = hp.Int('embedding_size', min_value=8,
                            max_value=512, step=8)
    spatial_dropout = hp.Float('spatial_dropout', min_value=0,
                               max_value=0.5, step=0.05)
    conv_layers = hp.Int('conv_layers', min_value=1,
                         max_value=5, step=1)
    rnn_layers = hp.Int('rnn_layers', min_value=1,
                        max_value=5, step=1)
    dense_layers = hp.Int('dense_layers', min_value=1,
                          max_value=3, step=1)
    conv_filters = hp.Int('conv_filters', min_value=32,
                          max_value=512, step=32)
    conv_kernel = hp.Int('conv_kernel', min_value=1,
```

```
                                    max_value=8, step=1)
       concat_dropout = hp.Float('concat_dropout', min_value=0,
                                   max_value=0.5, step=0.05)
       dense_dropout = hp.Float('dense_dropout', min_value=0,
                                   max_value=0.5, step=0.05)
```

In the first part of the function, we simply recover all the settings from the hp parameter. We also make the range of the search space for each of them explicit. Contrary to the solutions we've seen so far, this part of the work is done *inside* the model function, not outside.

The function defines the different layers using the parameters extracted from hp. Sometimes, a parameter will switch on or off a part of the network performing certain data processing. For instance, in the code, we inserted a branch of the graph (conv_filters and conv_kernel) that processes the sequence of words using convolutional layers, which, in their 1D form, can also prove useful for NLP problems since they can catch local sequences of words and meanings that LSTMs may find harder to grasp.

Now, we can define the actual model:

```
inputs = layers.Input(name='inputs',shape=[pad_length])
layer  = layers.Embedding(vocab_size, embedding_size,
                          input_length=pad_length)(inputs)
layer  = layers.SpatialDropout1D(spatial_dropout)(layer)
for l in range(conv_layers):
    if l==0:
        conv = layers.Conv1D(
            filters=conv_filters,
            kernel_size=conv_kernel, padding='valid',
            kernel_initializer='he_uniform')(layer)
    else:
        conv = layers.Conv1D(
            filters=conv_filters,
            kernel_size=conv_kernel, padding='valid',
            kernel_initializer='he_uniform')(conv)
avg_pool_conv = layers.GlobalAveragePooling1D()(conv)
max_pool_conv = layers.GlobalMaxPooling1D()(conv)
representations = list()
for l in range(rnn_layers):
```

```
                use_bidirectional = hp.Choice(f'use_bidirectional_{l}',
                                               values=[0, 1])
            use_lstm = hp.Choice(f'use_lstm_{l}', values=[0, 1])
            units = hp.Int(f'units_{l}', min_value=8, max_value=512, step=8)
            if use_lstm == 1:
                rnl = layers.LSTM
            else:
                rnl = layers.GRU
            if use_bidirectional==1:
                layer = layers.Bidirectional(rnl(
                    units, return_sequences=True))(layer)
            else:
                layer = rnl(units, return_sequences=True)(layer)
            representations.append(attention(layer))
        layer = layers.concatenate(representations + [avg_pool_conv,
                                                      max_pool_conv])
        layer = layers.Dropout(concat_dropout)(layer)
        for l in range(dense_layers):
            dense_units = hp.Int(f'dense_units_{l}', min_value=8,
                                 max_value=512, step=8)
            layer = layers.Dense(dense_units)(layer)
            layer = layers.LeakyReLU()(layer)
            layer = layers.Dropout(dense_dropout)(layer)
        layer = layers.Dense(1, name='out_layer')(layer)
        outputs = layers.Activation('sigmoid')(layer)
        model = models.Model(inputs=inputs, outputs=outputs)
```

Here, we start by defining the input layer and transform it with a subsequent embedding layer that will encode the sequence values into dense layers. Some dropout regularization is applied to the process using SpatialDropout1D, which randomly drops entire columns of the output matrix (standard dropout drops random single elements in the matrix instead). After these initial phases, we split the network into one pipeline based on convolutions (Conv1D) and another based on recurrent layers (GRU or LSTM). It is after the recurrent layers that we apply the attention layer. Finally, the outputs of these two pipelines are concatenated, and after a few more dense layers, they arrive at the final output node, a sigmoid, since we have to represent a probability bounded in the range 0 to 1.

After the model definition, we set the learning parameters and compile the model before returning it:

```python
hp_learning_rate = hp.Choice('learning_rate',
                             values=[0.002, 0.001, 0.0005])
optimizer_type = hp.Choice('optimizer', values=list(range(6)))
optimizer = get_optimizer(option=optimizer_type,
                          learning_rate=hp_learning_rate)


model.compile(optimizer=optimizer,
              loss='binary_crossentropy',
              metrics=['acc'])


return model
```

Note that we have built the model using the functional API of Keras, not the sequential one. We would advise you to avoid the sequential one; it is easier to set up but severely restricts your potential architectures.

At this point, most of the work is already done. As a suggestion, having worked out many optimizations using KerasTuner, we prefer to first build a **non-parametric model**, using all the possible architectural features we want to test, with the mutually exclusive parts of the network set to the most complex solutions. After we have set up the generative function and our model seems to be working properly, we can, for instance, represent its graph and have it successfully fit some examples as a test. After that, we insert the parametric variables into the architecture and set up the hp parameter definitions.

In our experience, starting with a parametric function immediately will take more time and debugging effort. The idea behind KerasTuner is to let you think of your DNNs as a set of modular circuits and to help you optimize how the data flows inside them.

Now, we import KerasTuner. First, we set the tuner itself, and then we start the search:

```python
import keras_tuner as kt
tuner = kt.BayesianOptimization(hypermodel=create_tunable_model,
                                objective='val_acc',
                                max_trials=100,
                                num_initial_points=3,
                                directory='storage',
                                project_name='imdb',
```

```
                              seed=42)
tuner.search(train_data, train_labels,
             epochs=30,
             batch_size=64,
             validation_data=(val_data, val_labels),
             shuffle=True,
             verbose=2,
             callbacks = [EarlyStopping('val_acc',
                                        patience=3,
                                        restore_best_weights=True)])
```

As a tuner, we opt for the Bayesian optimization one, but you can also try the Hyperband tuner (https://keras.io/api/keras_tuner/tuners/hyperband/) and check if it works better for your problem. We provide our model function to the hypermodel parameter. Then, we set the objective using a string or a function, the maximum number of trials (KerasTuner will stop earlier if there is nothing more to be done), and the initial number of random trials – the more, the better – to inform the Bayesian process. Early stopping is a standard and well-performing practice in modeling DNNs that you absolutely cannot ignore. Finally, but importantly, we set the directory where we want to save our search and a seed number for the reproducibility of the optimization steps.

The search phase is run like a standard fit of a Keras model, and – this is quite important – it accepts callbacks. Therefore, you can easily add early stopping to your model. In this case, the given epoch number should be considered the maximum number of epochs. You may also want to optimize the batch size, which we haven't done in our example. This still requires some extra work, but you can get an idea of how to achieve it by reading this GitHub closed issue: https://github.com/keras-team/keras-tuner/issues/122.

After the optimization is complete, you can extract the best parameters and save the best model without any need to retrain it:

```
best_hps = tuner.get_best_hyperparameters()[0]
model = tuner.hypermodel.build(best_hps)
print(best_hps.values)
model.summary()
model.save("best_model.h5")
```

In this example, KerasTuner finds a solution that uses:

- A larger embedding layer
- Just plain GRU and LSTM layers (no bidirectional layers)
- Stacking of multiple one-dimensional convolution layers (Conv1D)
- More and larger dense layers

Interestingly, the solution is more effective, lighter, and faster than our previous attempts based on intuition and experience with the problem.

Chollet suggests using KerasTuner not just to make your DNNs perform better but also to shrink them to a more manageable size, which may make a difference in code competitions. This allows you to put together more models that work together within the limited inference time provided by the competition's sponsors.

If you would like to examine more examples of using KerasTuner, François Chollet also created a series of Notebooks for Kaggle competitions to showcase the workings and functionalities of his optimizer:

- `https://www.kaggle.com/fchollet/keras-kerastuner-best-practices` for the *Digit Recognizer* datasets
- `https://www.kaggle.com/fchollet/titanic-keras-kerastuner-best-practices` for the *Titanic* dataset
- `https://www.kaggle.com/fchollet/moa-keras-kerastuner-best-practices` for the *Mechanisms of Action (MoA) Prediction* competition

The TPE approach in Optuna

We complete our overview of Bayesian optimization with another interesting tool and approach to it. As we have discussed, scikit-optimize uses GPs (as well as tree algorithms) and directly models the surrogate and acquisition functions.

As a reminder of these topics, the **surrogate function** helps the optimization process to model the potential performance result when you try a set of hyperparameters. The surrogate function is built using the previous experiments and their results; it is just a predictive model applied to forecast the behavior of a specific ML algorithm on a specific problem. You get an expected performance output for each parameter input to the surrogate function. That's intuitive and also quite hackable, as we have seen.

The **acquisition function** instead points out what set of hyperparameters could be tested to improve the ability of the surrogate function to predict the performances of the ML algorithm. It is also useful for really testing if we can reach a top-performing result based on the surrogate function's forecasts. These two objectives represent the *explore* part (where you run experiments) and the *exploit* part (where you test the performance) of a Bayesian optimization process.

Instead, optimizers based on **TPE** tackle the problem by estimating the likelihood of success of the values of parameters. In other words, they model the success distribution of the parameters themselves using successive refinements, assigning a higher probability to more successful value combinations.

In this approach, the set of hyperparameters is divided into good and bad ones by these distributions. These take the role of the surrogate and acquisition functions in Bayesian optimization since the distributions tell you where to sample to get better performances or explore uncertainty.

> To explore the technical details of TPE, we suggest reading Bergstra et al. *Algorithms for hyper-parameter optimization* (`https://proceedings.neurips.cc/paper/2011/file/86e8f7ab32cfd12577bc2619bc635690-Paper.pdf`).

Therefore, TPE can model the search space and simultaneously suggest what the algorithm can try next by sampling from the adjusted probability distribution of parameters.

For a long time, **Hyperopt** was the option for those preferring to use TPE instead of Bayesian optimization based on GPs. In October 2018, however, Optuna appeared in the open source, and it has become the preferred choice for Kagglers due to its versatility (it also works out of the box for neural networks and even for ensembling), speed, and efficiency in finding better solutions compared to previous optimizers. Its latest version, Optuna 4.0, released in 2024, introduced several key features that have further enhanced its capabilities in hyperparameter optimization. These updates include large-scale optimization, which improves the tool's ability to handle large datasets and more complex optimization problems. Additionally, Optuna 4.0 offers enhanced algorithms like TPE with constraints, allowing for a more efficient and precise hyperparameter search. The release also focused on usability, introducing a simplified interface that makes integration and customization easier for users, solidifying Optuna's role as a versatile tool for both beginner and expert data scientists.

In this section, we will demonstrate just how easy is to set up a search, which is called a *study* in Optuna terminology. All you need to do is write an objective function that takes as input the parameters to be tested by Optuna and then returns an evaluation.

Validation and other algorithmic aspects can be handled in a straightforward manner inside the objective function, also using references to variables external to the function itself (both global variables and local ones). Optuna also allows **pruning**, i.e., early dropping of unuseful experiments. If a particular experiment is not going well, Optuna can stop and forget about it. Optuna provides a list of functions that activate this callback (see `https://optuna.readthedocs.io/en/stable/reference/integration.html`); the algorithm will run everything efficiently for you after that, which will significantly reduce the time needed for optimization.

All of this is in our next example. We return to optimizing for the *30 Days of ML* competition. This time, we are trying to figure out what parameters make XGBoost work for this competition.

You can find the Notebook for this example at `https://www.kaggle.com/lucamassaron/optuna-bayesian-optimization`.

As a first step, we upload the libraries and the data, as before:

```python
import pandas as pd
import numpy as np
from sklearn import preprocessing
from sklearn.metrics import mean_squared_error
from sklearn.model_selection import train_test_split
from sklearn.preprocessing import OrdinalEncoder
from xgboost import XGBRegressor
import optuna
import cupy as cp
from optuna_integration import XGBoostPruningCallback
# Loading data
X_train = pd.read_csv(
    "../input/30-days-of-ml/train.csv"
).iloc[:100_000, :]
X_test = pd.read_csv("../input/30-days-of-ml/test.csv")
# Preparing data as a tabular matrix
y_train = X_train.target
X_train = X_train.set_index('id').drop('target', axis='columns')
X_test = X_test.set_index('id')
# Pointing out categorical features
categoricals = [item for item in X_train.columns if 'cat' in item]
# Dealing with categorical data using OrdinalEncoder
ordinal_encoder = OrdinalEncoder()
```

```
X_train[categoricals] = ordinal_encoder.fit_transform(
    X_train[categoricals])
X_test[categoricals] = ordinal_encoder.transform(X_test[categoricals])
```

When using Optuna, you just have to define an objective function containing the model, the cross-validation logic, the evaluation measure, and the search space.

Naturally, for data, you can refer to objects outside the function itself, rendering the construction of the function much easier. As in KerasTuner, here, you need a special input parameter based on a class from Optuna:

```
def objective(trial):
    params = {
        'learning_rate': trial.suggest_float("learning_rate",
                                             0.01, 1.0, log=True),
        'reg_lambda': trial.suggest_float("reg_lambda",
                                          1e-9, 100.0, log=True),
        'reg_alpha': trial.suggest_float("reg_alpha",
                                         1e-9, 100.0, log=True),
        'subsample': trial.suggest_float("subsample", 0.1, 1.0),
        'colsample_bytree': trial.suggest_float(
                            "colsample_bytree", 0.1, 1.0),
        'max_depth': trial.suggest_int("max_depth", 1, 7),
        'min_child_weight': trial.suggest_int("min_child_weight",
                                              1, 7),
        'gamma': trial.suggest_float("gamma", 0.1, 1.0, step=0.1)
    }
    model = XGBRegressor(
        random_state=0,
        tree_method = "hist",
        device = "gpu",
        n_estimators=10_000,
        early_stopping_rounds=300,
        callbacks=[XGBoostPruningCallback(trial, 'validation_0-rmse')],
        **params
    )
```

```
        model.fit(x, y,
                  eval_set=[(x_val, y_val)], verbose=1000)
    preds = model.predict(cp.asarray(x_test.values))
    rmse = mean_squared_error(y_test, preds, squared=False)
    return rmse
```

In this example, we won't cross-validate for performance reasons but use one fixed dataset for training, one for validation (early stopping), and one for testing purposes. We are using a GPU in this example and subsetting the available data to fit the execution of 60 trials into a reasonable length of time. If you don't want to use a GPU, just remove the tree_method and device parameters from the XGBRegressor instantiation. Also, notice how we set a callback in the fit method to provide Optuna feedback on how the model is performing so the optimizer can stop an underperforming experiment early to give space to other attempts. Another notable aspect is that you can decide to optimize either for minimization or maximization, depending on your problem (Scikit-optimize works only on minimization problems).

```
    x, x_val, y, y_val = train_test_split(X_train, y_train, random_state=0,
                                          test_size=0.2)
    x, x_test, y, y_test = train_test_split(x, y, random_state=0,
                                            test_size=0.25)
    study = optuna.create_study(direction="minimize")
    study.optimize(objective, n_trials=100)
```

To complete the run, you just have to print or export the best test performance and the best parameters found by the optimization:

```
    print(study.best_value)
    print(study.best_params)
```

Before we conclude this dense chapter, let's look at one last interview. This time, we're speaking to *Ruchi Bhatia*, a Grandmaster in Datasets and Notebooks. Ruchi is currently a Product Marketing Manager at Amazon AWS.

Interview: Ruchi Bhatia

https://www.kaggle.com/ruchi798

What initially attracted you to Kaggle competitions, and how has your involvement evolved over time?

I joined Kaggle out of pure curiosity. It was the one place where learning, experimentation, and collaboration existed without boundaries. My first dataset came from a question I couldn't find an answer to during the pandemic: how do streaming platforms vary in popularity by age group? So I made the dataset myself. That small act of curiosity became my entry into the world of data science.

Over time, Kaggle evolved from being a competitive playground to a learning accelerator. It taught me the discipline of experimentation, reproducibility, and sharing knowledge publicly. Now, my involvement is less about leaderboard scores and more about community building creating resources, mentoring new Kagglers, and connecting real-world Gen AI use cases to the data science community.

Describe your process for selecting competitions to participate in. What criteria do you consider, and how do you prioritize your time and resources? How does this approach differ from your day-to-day work in data science, if applicable?

In 2025, Kaggle is more than competitions. It's a launchpad for multimodal and generative AI innovation. I look for challenges that push the boundaries of what's possible with new architectures, such as LLM fine-tuning, model evaluation, or structured reasoning tasks.

When choosing projects, I ask myself:

- Will this expand my understanding of how models reason or generalize?
- Can it help me explore scaling strategies, efficiency trade-offs, or evaluation gaps that show up in real-world systems?

- Is there an opportunity to contribute insights that others can build on?

I'm especially drawn to new and underexplored domains, areas where large models still struggle or where datasets are scarce. That includes geospatial modeling, scientific and materials data, synthetic biology, edge AI, and real-time multimodal streams like robotics or sensor fusion. These are spaces where model architectures aren't standardized yet, so there's room for real experimentation and innovation. Kaggle complements my professional work by keeping me close to how developers and data scientists actually build, iterate, and experiment. It's also a way to stay hands-on, even at scale. I make time to prototype ideas that connect research trends to applied use cases. I prioritize depth over breadth, choosing fewer projects, but going all-in when the technical learning curve is steep and the insights have long-term value.

Can you share a memorable competition experience where you encountered unexpected challenges? How did you adapt your approach to overcome these challenges?

One competition that stands out involved multimodal data, text combined with tabular and image features. My early models performed well locally but didn't hold up on the leaderboard. After digging deeper, I realized the problem wasn't the model itself; it was the validation design. The data splits weren't representative of the test distribution, which meant my cross-validation scores gave a false sense of progress. I ended up rebuilding the entire validation pipeline from scratch. That included checking for target drift, running adversarial validation, and creating grouped stratified folds that respected the structure across modalities. Once the validation reflected the real data behavior, the leaderboard performance stabilized, and my experiments started to make sense again. That experience completely changed how I approach competitions. It reinforced that validation is the backbone of trustworthy machine learning. A solid validation setup not only improves model reliability but also exposes deeper issues from data collection biases to feature leakage before training even begins. Once validation is sound, reaching medal level is largely a matter of disciplined experimentation and understanding the problem deeply. Climbing into the top one percent takes more than clean code or strong hardware. It comes from deep intuition built through hands-on experience with varied model architectures, an understanding of Kaggle's nuances around validation, blending, and post-processing, and the ability to recognize promising directions early enough to capitalize on them. Teaming up also makes a big difference. Collaborating with others exposes you to new ways of thinking, diversifies your modeling approaches, and often surfaces insights that you might overlook working alone.

In your opinion, what distinguishes a successful Kaggle competitor from an average one? Are there any specific skills or qualities that contribute to success in data science competitions?

Top Kagglers approach each competition like an applied research problem. They understand every layer of the stack from data engineering to model design, validation, and deployment. Their process starts with building a stable, versioned pipeline for preprocessing, feature extraction, and experiment tracking. Tools like MLflow and Weights & Biases keep runs reproducible and environments consistent. They have strong architectural intuition. They know how attention mechanisms scale, how to manage memory and compute trade-offs, and how to fine-tune efficiently with methods such as LoRA or QLoRA. Mixed-precision training and tensor parallelism are used routinely to maximize throughput. Their validation design is methodical and data-aware. They create folds that reflect the true test distribution, check for leakage with adversarial validation, and use permutation testing to measure robustness. They analyze calibration, residuals, and error clusters before trusting a model. In generative and multimodal tasks, advanced competitors build pipelines that connect text, vision, and structured data. Retrieval-augmented generation and synthetic data generation are common techniques to extend coverage, improve context, and balance underrepresented cases while monitoring bias drift. They also communicate clearly. Their notebooks read like concise technical reports with metric summaries and reasoning behind every architectural choice. Top Kagglers approach their work like engineers. They know how to turn a good idea into a repeatable process that runs cleanly, delivers consistent results, and could plug into a production setup without a rewrite.

In your experience, what are some common pitfalls or oversights that inexperienced Kagglers tend to make? If you could offer one piece of advice to aspiring Kagglers, what would it be?

A lot of new Kagglers start by stacking models before they truly understand their data. It's easy to replicate a public notebook, but much harder to question the assumptions behind it. The real progress begins when you stop executing code blindly and start treating each submission as an experiment. Define a clear hypothesis before running anything are you testing feature sensitivity, an augmentation strategy, or metric robustness? Build reproducible pipelines with version control, track every run, and log metrics consistently. That discipline pays off far more than adding another layer to a model. Validation design deserves the same attention as modeling. Most solutions fail not because the model is weak, but because the validation doesn't reflect the real data distribution. A strong cross-validation strategy is the single best safeguard against overfitting and false confidence. And finally, document your work as if someone else needs to reproduce it from scratch. Clean, transparent, and well-explained work always stands out. Kaggle consistently rewards clarity and reproducibility more than complexity. I've learned that consistency beats intensity. When I committed to contributing every day for a year, the compounding effect was massive my learning accelerated, my portfolio grew, and my visibility skyrocketed. Most people underestimate how much steady, small progress changes your trajectory.

Have you ever collaborated with other Kagglers or participated in team competitions? What are the benefits and challenges of collaboration in this context?

Yes, and collaboration is where the real learning happens. You're combining different strengths: one person may specialize in model architecture, another in data leakage detection, another in reproducibility automation, or feature engineering via embeddings. Together, you build something more robust than any individual could. It's also a reminder that data science isn't a solo sport. It's inherently collaborative and increasingly multidisciplinary.

How do you stay motivated and productive throughout the duration of a competition, especially during periods of uncertainty or setbacks?

I approach competitions the same way I approach long AI projects: through iteration, reflection, and balance.

I break them into small, measurable goals: improving validation stability, optimizing inference latency, testing a new augmentation strategy, or refining a prompt for an LLM. Those micro-milestones keep me engaged and help maintain momentum when progress slows. When a project hits a plateau, I change focus instead of forcing progress. Sometimes that means exploring error analysis, cleaning feature pipelines, or documenting experiments. Other times, it means stepping back completely to reset perspective.

What role do you think domain expertise plays in Kaggle competitions? How do you approach competitions in domains that are unfamiliar to you?

Domain expertise is still vital, but the definition has evolved. In the LLM era, domain knowledge is about understanding how models internalize and represent that domain. Knowing what signal the data carries, and what's missing, often matters more than knowing every domain detail. When tackling unfamiliar problems say, climate modeling or medical imaging I start by building intuition. I'll read a few research papers, explore community discussions, and study public notebooks to understand the context behind the data. I look for structure: what the inputs represent, what constraints exist, and what relationships should logically hold. That context shapes everything feature design, evaluation strategy, and error interpretation. Once you understand why the data behaves the way it does, the how of modeling becomes much easier.

Kaggle has been one of the most defining influences in my professional journey. It taught me to think like an applied researcher grounding every experiment in evidence and to communicate like a strategist who can bridge technical depth with real-world impact. It strengthened my ability to turn complex data insights into clear, actionable stories, a skill that now defines my work in AI product marketing at Amazon Web Services. Every dataset, notebook, and discussion I shared on Kaggle shaped how I approach innovation today. The discipline of reproducibility, validation design, and open documentation became second nature. Those same principles guide how I think about building trustworthy, efficient, and scalable AI systems in my current work. Kaggle also opened doors that shaped the next stages of my career. My early contributions helped me earn recognition as a Global Data Science Ambassador for Z by HP, and later as Leader of Data Science at OpenMined, where I led a team working on privacy-preserving AI. During that time, we partnered with X (previously Twitter) and the United Nations Privacy-Enhancing Technologies Lab to develop synthetic data pipelines and privacy-aware machine learning frameworks aimed at making responsible AI practical and scalable. Beyond technical roles, Kaggle helped me grow into a mentor and advocate. As a KaggleX BIPOC Mentor, I've guided emerging data scientists, emphasizing not only modeling and validation, but also storytelling, ethics, and community building lessons Kaggle itself instilled in me. Looking back, Kaggle showed me that meaningful innovation comes from the intersection of rigor, transparency, and collaboration. Everything I do today from mentoring to product marketing traces back to that foundation.

Looking ahead, what trends or developments do you anticipate shaping the future of Kaggle competitions and the broader field of data science? How do you envision the role of generative AI evolving in Kaggle competitions and the broader field of data science?

I think the next wave of Kaggle will revolve around agentic workflows, synthetic data, and evaluation for generative systems. We'll likely see competitions that focus on reinforcement learning from human feedback, model alignment, multimodal reasoning, and automated evaluation pipelines. The field is moving toward autonomous experimentation where AI agents assist participants in generating hypotheses, running experiments, and interpreting results. Human competitors will spend less time on manual tuning and more time designing systems, frameworks, and constraints that guide the AI's behavior. Compute efficiency will keep driving innovation. Methods like quantization, sparse attention, and model distillation are making frontier AI more accessible. Kaggle will remain an important space for democratizing this kind of research, giving people the chance to test cutting-edge ideas without massive infrastructure.

Generative AI will reshape how we compete. It's becoming a true collaborator a way to explore new design spaces, generate data, and analyze model behavior at scale. I see the most effective Kagglers using AI as a co-researcher: combining human judgment, intuition, and domain understanding with the speed and scale of machine experimentation.

Using Weights & Biases

Although optimization functions and tools such as Optuna offer the possibility to visualize, also interactively, the tested parameters and their effect on the evaluation metric, a Kaggle competition often requires tuning multiple algorithms, processing data differently, creating different batches of features, and ensembling everything together. All these operations can be considered as having multiple hyperparameters to fix, and as things get more complex, it is difficult to track everything by yourself with paper and pencil or a spreadsheet.

Experiment trackers are software or services that help you track everything that happens with your models and solutions, no matter what you do. There are a few notable ones that we encountered during our Kaggle competitions:

- **Neptune** (`https://neptune.ai`) is a tracking platform with a centralized hub for storing and managing your models, providing version control, organizing, and analyzing your ML experiments.
- **Weights & Biases** (`https://wandb.ai`) tracks all aspects of your experiments, stores your models, and visually represents your results.
- **Data Version Control (DVC;** `https://dvc.org`) is a data management tool that also includes experiment-tracking capabilities. It is a popular choice for ML practitioners who want to manage their data and experiments in a single tool.
- **MLflow** (`https://mlflow.org`) is a completely open-source platform that is part of the Apache Software Foundation and is used to manage the ML lifecycle, including experimentation.

Since it is predominant among Kagglers, we will cover a few examples of using Weights & Biases (W&B for short from now on) as an experiment tracker. We will follow the examples provided by `https://docs.wandb.ai/quickstart` and extend them with examples commonly found in Kaggle Notebooks.

Tracking experiments

As a first step, you need to access the website (`https://wandb.ai`) and sign up to create an account or use your existing account on GitHub, Google, or Microsoft. Once your account has been created, just go to the **User Settings** menu item under your profile and, on the **Settings** page, look for the **Danger Zone** section, where you can find your API keys, and copy them to the clipboard.

Once you get your API keys, you can start working on your experiments using any notebook, whether on your local machine, on the cloud, or in a Kaggle Notebook. If you are working with a Kaggle Notebook, don't forget to update to the latest version of the W&B package to have full compatibility with the service:

```
!pip install --upgrade -q wandb
```

After that, it is crucial that you log in to the service. Just import the library and run its initialization:

```
import wandb
wandb.login()
```

You will be asked for the API key, which you will have to input. You can also use Kaggle Secrets to store your API key and retrieve it for input when initializing with wandb.login. See `https://www.kaggle.com/discussions/product-feedback/114053` for a simple walk-through. You can find the **Secrets** menu in your Kaggle Notebook under the **Add-ons** menu:

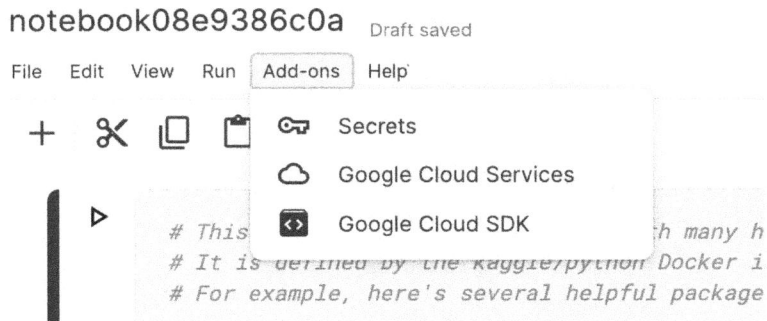

Figure 9.1: Add-ons menu

Once you have accessed it, just complete the *Label* and *Value* forms and click the *Save* button to add the secret for it to be immediately usable

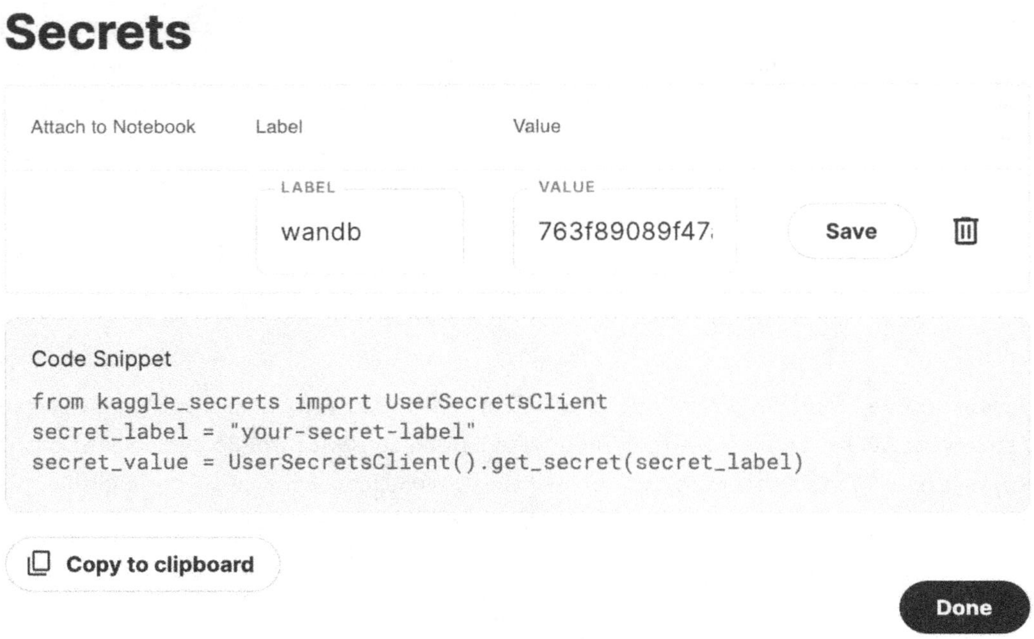

Secrets

Attach to Notebook	Label	Value		
	LABEL	VALUE		
	wandb	763f89089f47	**Save**	🗑

Code Snippet

```
from kaggle_secrets import UserSecretsClient
secret_label = "your-secret-label"
secret_value = UserSecretsClient().get_secret(secret_label)
```

📋 **Copy to clipboard**

Done

Figure 9.2: Setting secrets

Once the operation is completed, you can just use the kaggle_secrets import and the get_secret method to obtain the label-associated secret value:

```
from kaggle_secrets import UserSecretsClient
user_secrets = UserSecretsClient()
wandb.login(key=user_secrets.get_secret("wandb"))
```

At this point, you can use W&B to track your experiments. As a first step, you must establish a fresh W&B run using the wandb.init() function:

```
run = wandb.init(project='kaggle-competition',
                 config={'num_labels': 2,
                         'train_val_split': 0.25},
                 group='first-experiments',
                 job_type='train')
```

This approach allows you to track distinct stages of your ML training pipeline, such as the training and evaluation stages, as separate runs within W&B. The key arguments for wandb.init() are:

- **project**: Explain the name of the W&B project where you want to store the new run. The run will be placed in the default "Uncategorized" project if omitted.

- **entity**: Optionally, you can also designate the username or team name where you intend to send your runs. This helps organize runs within a specific context, such as a team or project.

- **config**: Set the `wandb.config` dictionary-like object to store inputs relevant to your job, encompassing hyperparameters for your model or settings for data preprocessing tasks. This facilitates reproducibility and facilitates understanding of the impact of specific choices.

- **group**: Utilize the group argument to organize individual runs into a broader experiment. This feature is particularly useful for managing multiple model architectures or training configurations.

- **job_type**: Specify the type of run, such as "train" or "evaluate," to categorize further and analyze your experiments. This distinction becomes crucial when grouping runs together using a group.

When you start W&B monitoring by variable assignation, you actually have to close the interaction at the end using the finish method:

```
run.finish()
```

However, you can deal more smoothly with the interactions using the with command and just operating under it:

```
with wandb.init(project='kaggle-competition',
                config={'num_labels': 2,
                        'train_val_split': 0.25},
                group='first-experiments',
                job_type='train') as run:
```

At this point, you can run your experiment, typically with a batch/online neural network, where you can manually log the results at each step because you can back-propagate every single batch/example, or you can just kick back and use integrated callbacks such as with Keras and TensorFlow using `WandbMetricsLogger()`, `WandbModelCheckpoint()`, or `WandbEvalCallback()`, as explained at `https://docs.wandb.ai/guides/integrations/keras`. Integrations do not stop at TensorFlow/Keras; there are suitable commands for common frameworks and packages such as PyTorch, FastAI, scikit-learn, XGBoost, LightGBM, and so on. Just check the integration page for the correct W&B callback to be used: `https://docs.wandb.ai/guides/integrations`.

If you are doing things manually, after finishing each learning step, you can log the measures you need to record and close the run for it to be recorded on W&B:

```
wandb.log({'val_AUC': auc,
           'val_F1_score': f1_score})
run.finish()
```

The `wandb.log()` command is a data collector, gathering all the specified arguments and transmitting them to the W&B instance. This centralized data collection enables seamless access and tracking of these metrics and artifacts within the W&B user interface.

Versioning with artifacts

In addition to tracking experiments, W&B also features a built-in versioning system centered around the concept of **artifacts**. Artifacts are a fundamental tool for versioning datasets, models, and their dependencies. An artifact is essentially a versioned data folder, allowing you to track changes and maintain a history of your data assets. Let's consider versioning a dataset within our current project to exemplify this. To achieve this, we simply need to create an artifact and upload it to W&B. In order to create an artifact, you will need to utilize the `wandb.Artifact()` function, specifying the artifact's name and type.

For instance, to create a dataset artifact named my_dataset, you can use the following code:

```
artifact = wandb.Artifact('my_dataset', type='dataset')
```

In the same way, you can create artifacts relative to models (`type='model'`). Once you've created the artifact, you can add files using the `.add_file()` method. For example, to upload a CSV file named data.csv, you can use the following code:

```
artifact.add_file('data.csv')
```

To finalize the versioning process, you'll need to log the artifact to your W&B run. This can be done using the `.log_artifact()` method:

```
wandb.log_artifact(artifact)
```

With this straightforward process, you can effectively version your datasets, ensuring data integrity and traceability throughout your ML workflow during a Kaggle competition.

Implementing Sweeps optimization

Another W&B feature that plays well in Kaggle competition is the **Sweeps optimization**. You can optimize your hyperparameters with sweeps in three simple steps:

1. **Define the sweep configuration**: Specify the parameters to explore, the search strategy to employ (grid search, random search, or even Bayesian search), and the metric to optimize. This can be done by creating a dictionary-like object containing method, metric, and parameters keys.

2. **Initialize the sweep**: Kickstart the sweep process by calling wandb.sweep(sweep_config), passing in the sweep configuration dictionary. This generates a unique identifier for the sweep. Also, you need to prepare a training function containing the training and evaluation procedure in a way that is not too different from that of Optuna.

3. **Run the sweep agent**: Initiate the actual sweep execution by calling wandb.agent(sweep_ id, function=train). This provides the sweep identifier and the function that defines the model architecture and training process.

You can find a full example of the procedure at https://www.kaggle.com/code/lucamassaron/w-b-sweep-for-xgboost/. This Kaggle Notebook replicates the same optimization as seen in the Optuna example – this time, leveraging W&B Sweeps. Notice how we first define the sweep configuration by providing high and low boundaries for the search space:

```
sweep_config = {
    "method": "bayes",
    "metric": {
      "name": "rmse",
      "goal": "minimize"
    },
    "parameters": {
        'learning_rate': {"max": 1.0, "min": 0.01},
        'reg_lambda': {"max": 100.0, "min": 1e-9},
        'reg_alpha': {"max": 100.0, "min": 1e-9},
        'subsample': {"max": 1.0, "min": 0.1},
        'colsample_bytree': {"max": 1.0, "min": 0.1},
        'max_depth': {"max": 7, "min": 1},
        'min_child_weight': {"max": 7, "min": 1},
        'gamma': {"max": 1.0, "min": 0.1},
    }
}
```

The next step initializes the sweep itself on the W&B platform:

```
sweep_id = wandb.sweep(sweep_config,
                       project="XGBoost-sweeps")
```

We then proceed to define a train function using logging to record the results of each experiment and then launch the optimization:

```
wandb.agent(sweep_id, train, count=100)
```

The results can be conveniently explored using the W&B platform.

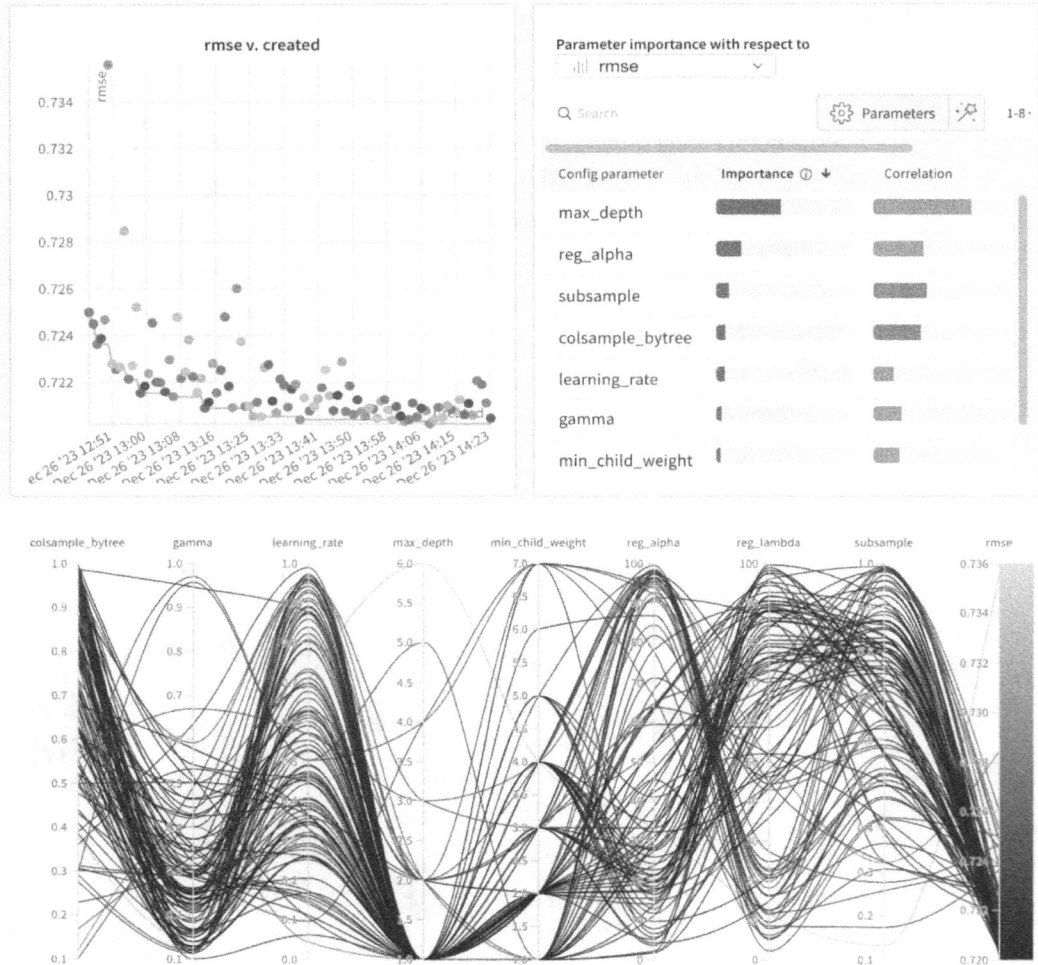

Figure 9.3: Plots of the results of the experiments

The exploration reveals both key optimization parameters and their role in optimization (for instance, the parallel plot reveals how the depth should be at a stump level in this problem). We can also obtain the exact optimal configuration, comprising all used parameters, both optimized and set. Ideally, we could also have saved as an artifact the generated models, thus allowing fully reproducible results useful when winning in some Kaggle competitions to claim your prize and place.

Summary

In this chapter, we discussed hyperparameter optimization at length as a way to increase your model's performance and score higher on the leaderboard. We started by explaining the code functionalities of scikit-learn, such as grid search and random search, as well as the newer halving algorithms.

Then, we progressed to Bayesian optimization and explored scikit-optimize, KerasTuner, and, finally, Optuna. We spent more time discussing the direct modeling of the surrogate function by GPs and how to hack it because it can allow you greater intuition and a more ad hoc solution.

We recognize that, at the moment, Optuna has become a gold standard among Kagglers for tabular competitions and deep neural network ones because of its speedier convergence to optimal parameters in the time allowed in a Kaggle Notebook. However, if you want to stand out among the competition, you should strive to test solutions from other optimizers as well.

In the next chapter, we will move on to discuss another way to improve your performance in Kaggle competitions: ensembling models. By discovering the workings of averaging, blending, and stacking, we will illustrate how you can boost your results beyond what you can obtain by tuning hyperparameters alone.

Join our book's Discord space

Join our community's Discord space for discussions with the authors and other readers:

`https://packt.link/kaggle`

10

Ensembling with Blending and Stacking Solutions

When you start competing on Kaggle, it doesn't take long to realize that you cannot win with a single, well-devised model; you must ensemble multiple models. Next, you will immediately wonder how to set up a working ensemble. While there are a few guides available, much of the understanding still draws from Kaggle's community insights rather than scientific papers.

The point here is that if ensembling is the key to winning in Kaggle competitions, in the real world, ensembling is instead associated with complexity, poor maintainability, difficult reproducibility, and hidden technical costs for little advantage. Often, the small boost that can move you from the lower ranks to the top of the leaderboard doesn't matter for real-life applications because the costs overshadow the advantages. However, that doesn't mean that ensembling is not being used at all in the real world. In a limited way, such as by averaging and mixing a few diverse models, ensembling allows us to create models that can solve many data science problems more effectively and efficiently.

Ensembling in Kaggle is not only a way to gain extra predictive performance, but it is also a teaming strategy. When working with other teammates, putting together everyone's contributions produces a result that often performs better than individual efforts and can help organize the team's work by structuring everyone's efforts toward a clear goal. In fact, collaborative techniques like pair coding are clearly not feasible when work is performed in different time zones and under different constraints for each participant. One team member may be subject to constraints due to office hours, another due to studying and examinations, and so on.

Teams in a competition often don't have the chance to, and do not necessarily have to, synchronize and align all participants on the same tasks. Moreover, the skills within a team may also differ. A good ensembling strategy shared among a team means that individuals can keep working based on their own routines and styles yet still contribute to the group's success. Therefore, even different skills may be advantageous when using ensembling techniques based on diverse predictions.

This chapter will start with the ensembling techniques you already know because they are embedded in algorithms such as random forests and gradient boosting. Then, we will progress to ensembling techniques for multiple models, such as averaging, blending, and stacking. We will provide you with some theory, practice, and code examples you can use as templates when building your solutions on Kaggle.

We will cover these topics:

- A brief introduction to ensemble algorithms
- Averaging models into an ensemble
- Blending models using a meta-model
- Stacking models
- Creating complex stacking and blending solutions

Before leaving you to read this chapter and try all the techniques, we have to mention a great reference on ensembling for us and for all practitioners when competing on Kaggle: the blog post written in 2015 by *Triskelion* (*Hendrik Jacob van Veen*) and by a few collaborators (*Le Nguyen The Dat, Armando Segnini*). The *Kaggle Ensembling Guide* was originally found on the *mlwave* blog (`https://github.com/MLWave/Kaggle-Ensemble-Guide`), which is no longer up, but you can retrieve the contents of the guide from `https://usermanual.wiki/Document/Kaggle20ensembling20gui de.685545114.pdf`. The post arranged most of the implicit and explicit knowledge about ensembling from Kaggle forums at the time.

A brief introduction to ensemble algorithms

The idea that ensembles of models can outperform single ones is not a recent one. We can trace it back to *Sir Francis Galton*, who was alive in Victorian Britain. He figured out that, to guess the weight of an ox at a county fair, it was more useful to take an average from a host of more or less educated estimates from a crowd than having a carefully devised estimate from an expert.

In 1996, *Leo Breiman* formalized the idea of combining multiple models into a more predictive one by illustrating the **bagging** technique (also called the "bootstrap aggregating" procedure) that later led to the development of even more effective **random forest** algorithms. In the following period, other ensembling techniques, such as **gradient boosting** and **stacking**, were also presented, thus completing the range of ensemble methods we use today.

> You can refer to a few articles to figure out how these ensembling algorithms were initially devised:
>
> - For random forests, read Breiman, L. *Bagging predictors.* Machine learning 24.2 − 1996: 123-140.
> - If you want to know more about how boosting originally worked, read Freund, Y. and Schapire, R.E. *Experiments with a new boosting algorithm. icml. Vol. 96 − 1996*, and *Friedman, J. H. Greedy function approximation: a gradient boosting machine. Annals of Statistics* (2001): 1189-1232.
> - As for stacking, refer to Ting, K. M. and Witten, *I. H. Stacking bagged and dagged models*, 1997, for a first formal draft of the technique.

The first basic strategies for ensembling predictors in Kaggle competitions were taken directly from bagging and random forest strategies for classification and regression. They involved making an average of various predictions and were thus named **averaging** techniques. These approaches quickly emerged from the very first Kaggle competitions held over 11 years ago also because of the pre-Kaggle Netflix competition, where strategies based on the average of the results of different models dominated the scene. Given their success, basic ensembling techniques based on averaging set a standard for many competitions to come, and they are still quite valuable and valid even today for scoring more highly on the leaderboard.

Stacking, which is more complex and computationally demanding, emerged a bit later when problems in competitions became more complex and the struggle between participants turned fiercer. Just as the random forest approach has inspired averaging different predictions, boosting heavily inspired stacking approaches. In boosting, by sequentially re-processing information, your learning algorithm can model problems in a better and more complete way. In fact, in gradient boosting, sequential decision trees are built to model the part of data that previous iterations are unable to grasp. This idea is reprised in stacking ensembles, where you stack the results of previous models and re-process them to increase predictive performance.

We spoke to *Rob Mulla*, a 4x Grandmaster and senior devrel at Google. There is a lot we can learn from his experiences about his views on ensembling and what he has learned from Kaggle.

Interview: Rob Mulla

https://www.kaggle.com/robikscube

What's your favorite kind of competition and why? In terms of techniques and solving approaches, what is your specialty on Kaggle?

My favorite type of competitions are ones that involve unique datasets requiring novel solutions that incorporate different types of modeling approaches. I enjoy when a competition isn't just training large models on the dataset, but actually requires understanding the data very well and implementing ideas that leverage architectures specific to the tasks. I don't try to specialize in any particular approach. When I first started Kaggle, I mainly stuck to gradient-boosted models, but in order to be competitive in recent years, I've grown in my understanding of deep learning, computer vision, NLP, and optimization. My favorite competitions require using more than just one technique.

How do you approach a Kaggle competition? How different is this approach to what you do in your day-to-day work?

I approach Kaggle competitions in some ways very similarly to work projects. First comes data understanding. In real-world projects, you may need to work on defining the problem and developing a good metric. In Kaggle, that is already done for you. Next is understanding how the data and metrics relate to each other – and developing and testing modeling techniques that you believe will best solve the problem. The biggest difference in Kaggle compared to real-life data science is the final bit of ensembling and tuning of models to get a slight edge – in many real-world applications, these types of large ensembles are not necessary because the computational expense to performance gain can be small.

Tell us about a particularly challenging competition you entered, and what insights you used to tackle the task.

A very challenging competition that I entered was the NFL Helmet Impact Detection competition. It involved video data, which I had no prior experience with. It also required researching common approaches and reading existing papers on the topic. I had to work on a two-stage approach, which added to the complexity of the solution. A different competition that I found challenging was the Indoor Location Navigation competition. It involved modeling, optimization, and really understanding the data well. I didn't end up doing very well in the competition, but I learned a lot.

Has Kaggle helped you in your career? If so, how?

Yes. Kaggle has played a big part in helping me gain notoriety in the data science space. I've also grown in my knowledge and understanding of new techniques and have met and worked with many brilliant people who have helped me grow in my skills and understanding of machine learning.

My team placed second in the NFL Helmet Impact Detection Competition. I also participated in a number of NFL competitions prior to that competition. The hosts of the competition reached out to me and eventually it helped me land my current role.

In your experience, what do inexperienced Kagglers often overlook? What do you know now that you wish you'd known when you first started?

I think inexperienced Kagglers sometimes worry too much about the ensembling and hyperparameter tuning of models. These are important toward the end of a competition, but they are not important unless you've already built a good base model. I also think that fully understanding the competition metric is extremely important. Many Kagglers overlook how important it is to understand how to optimize your solution to the evaluation metric.

What mistakes have you made in competitions in the past?

A lot. I've overfitted models and spent time working on things that didn't end up being beneficial in the end. However, I feel like this was necessary for me to learn how to better tackle future competitions. The mistakes may have hurt me in the specific competition I was working in, but helped me to be better in later competitions.

Are there any particular tools or libraries that you would recommend using for data analysis or machine learning?

For EDA, know how to manipulate data using NumPy, Pandas, and Matplotlib, or another plotting library. For modeling, know how to set up a proper cross-validation scheme with scikit-learn. The standard models like XGBoost/LightGBM are good to know how to baseline with. Deep learning libraries are mainly TensorFlow/Keras or PyTorch. Getting to know one of the two main deep learning libraries is important.

Averaging models into an ensemble

In order to introduce the averaging ensembling technique better, let's quickly revise all the strategies devised by Leo Breiman for ensembling. His work represented a milestone for ensembling strategies, and what he found out at the time still works fairly well in a wide range of problems.

Breiman explored all these possibilities in order to figure out whether there was a way to reduce the variance of error in powerful models that tended to overfit the training data too much, such as decision trees.

Conceptually, he discovered that ensembling effectiveness was based on three elements: how we deal with the **sampling of training cases**, how we **build the models**, and, finally, how we **combine the different models** obtained.

As for the sampling, the approaches tested and found were:

- **Pasting**, where several models are built using subsamples (sampling without replacements) of the examples (the data rows)
- **Bagging**, where several models are built using random selections of bootstrapped examples (sampling with replacement)
- **Random subspaces**, where a number of models are built using subsamples (sampling without replacements) of the features (the data columns)
- **Random patches**, an approach similar to bagging, except that features are also sampled when each model is selected, as in random subspaces

The reason we sample instead of using the same information is that, by subsampling cases and features, we create models that are all relevant to the same problem while each being different from the others.

This difference also applies to the way each model overfits the sample; we expect all the models to grasp the useful, generalizable information from the data *in the same way* and deal with the noise that is not useful for making predictions *in a different way*. Hence, variation in modeling reduces the variation in predictions because errors tend to cancel each other out.

If this variation is so helpful, then the next step should not just be to modify the *data* the model learns from, but also *the model itself*. We have two main approaches for the models:

- Ensembles of the same type of models
- Ensembles of different models

Interestingly, ensembling in one way or another doesn't help if the models we are putting together are too different in predictive power. The point here is that you get an advantage if you put together models that can correctly guess the same type of predictions, so they can smooth out their errors when averaging the predictions they get wrong. If you are ensembling models with performances that are too different, you will soon find out that there is no point because the net effect will be negative: as you are not smoothing your incorrect predictions, you are also degrading the correct ones.

This is a crucial limit of averaging: it can use a set of different models (for instance, because they are trained using different samples and features) only if they are similar in predictive power. To take an example, linear regression and the *k*-nearest neighbor algorithm have different ways of modeling a problem and capturing signals from data. Thanks to the (distinct) characteristic functional forms at their cores, these algorithms can grasp different predictive nuances from the data and perform better on specific parts of their predictive tasks. Another good example is presented by the Kaggler Owen Zhang in his famous presentation of 2015, *Learn Kaggle techniques from Kaggle #1* (https://www.youtube.com/watch?v=LgLcfZjNF44&ab_channel=NYCDataScienceAcademy), where he suggests averaging gradient boosting models and generalized linear models (glmnets) because they complement each other in a blended average. Still, though working in most cases, you cannot really take advantage of diverse models when using averaging if their performances are too wide apart. In such a case, by contrast, the different ways algorithms have to capture signals is something that another approach (i.e., stacking) actually can leverage because it can take the best results as features from each algorithm.

Grasping different prediction nuances among diverse models is also emphasized when talking about deep learning for tabular problems. In the paper by *Ravid Shwartz-Ziv* and *Amitai Armon, Tabular Data: Deep Learning is Not All You Need* (you can read it at https://arxiv.org/abs/2106.03253), the authors found that when evaluating datasets not featured in the original papers, the deep learning architectures for tabular data exhibited reduced performance compared to a baseline model XGBoost. At this point, they didn't just assert that a gradient-boosting model like XGBoost surpasses deep learning in any problem involving tabular data, but they also conducted additional experiments. Consequently, they introduced the idea of forming an ensemble that combines these deep models with XGBoost, resulting in, in most cases, superior performance on these datasets compared to individual models and also the classical ensemble without deep models made by performing gradient boosting models (XGBoost, LightGBM, CatBoost).

Based on all such evidence coming from experience and different sources, we can summarize that for an ensemble based on averaging (averaging the results of multiple models) to be effective, it should be:

- Built on models that are trained on different samples
- Built on models that use different subsamples from the available features
- Composed of models somewhat similar in predictive power
- Composed of diverse types of models

Technically, this implies that the models' predictions should be as uncorrelated as possible while performing at the same level of accuracy on prediction tasks.

Now that we have discussed the opportunities and limitations of averaging multiple machine learning models, we will finally delve into the technical details. There are three ways to average multiple classification or regression models:

- Majority voting, using the most frequent classification outcome among multiple models (only for classification models): each base model's prediction is considered as a vote towards the final prediction
- Averaging output values or probabilities across the models
- Using a weighted average of the values or probabilities provided by the models

In the following few sections, we will discuss each approach in detail in the context of Kaggle competitions.

Majority voting

Producing different models by varying the examples, features, and models we use in the ensemble (if they are comparable in predictive power, as we discussed before) requires a certain computational effort, but it doesn't require you to build a data processing pipeline that is all that different from what you would set up when using a single model.

In this pipeline, you just need to collect different test predictions, keeping track of the models used, how you sampled examples or features when training, the hyperparameters you used, and the resulting cross-validation performance.

If the competition requires you to predict a class, you can use **majority voting**; that is, for each prediction, you take the class most frequently predicted by your models. Majority voting works for binary and multi-class predictions because it presumes that there are sometimes errors in your models, but they can guess correctly most of the time. Majority voting is used as an "error correction procedure," discarding noise and keeping meaningful signals.

In our first simple example, we demonstrate how majority voting works. We start by creating our example dataset. Using the `make_classification` function from scikit-learn, we generate a *Madelon*-like dataset.

The original Madelon was an artificial dataset containing data points grouped in clusters placed on the vertices of some dimensional hypercube and randomly labeled. It comprises a few informative features mixed with irrelevant and repeated ones (to create multicollinearity between features) and has a certain amount of injected random noise. Ideated by *Isabelle Guyon* (one of the creators of the SVM algorithm) for the *NIPS 2003 Feature Selection Challenge*, the Madelon dataset is the model example of a challenging artificial dataset for a competition. Some Kaggle competitions were even inspired by it: `https://www.kaggle.com/c/overfitting` and the more recent `https://www.kaggle.com/c/dont-overfit-ii`.

We will use this recreation of the Madelon dataset throughout this chapter as a basis for testing ensembling techniques. To get started, follow these steps:

1. In the following code, we start generating a synthetic dataset for a classification task leveraging the `make_classification` function from scikit-learn, which is inspired by the procedure used by the Madelon dataset. The function creates clusters of data points distributed around the corners of a hypercube in an n-dimensional space.

It then introduces interdependence among the features in these clusters, meaning that the features aren't entirely independent of each other (see: https://scikit-learn.org/stable/modules/generated/sklearn.datasets.make_classification.html):

```
from sklearn.datasets import make_classification
from sklearn.model_selection import train_test_split

X, y = make_classification(n_samples=5000, n_features=50,
                           n_informative=10,
                           n_redundant=25, n_repeated=15,
                           n_clusters_per_class=5,
                           flip_y=0.05, class_sep=0.5,
                           random_state=0)

X_train, X_test, y_train, y_test = train_test_split(
    X, y, test_size=0.33, random_state=0)
```

2. After splitting it into a training and a test set, we proceed by instantiating our learning algorithms. We will just use three base algorithms: SVMs, random forests, and k-nearest neighbor classifiers, with default hyperparameters for demonstration purposes. You can try changing them or increasing their number:

```
from sklearn.svm import SVC
from sklearn.ensemble import RandomForestClassifier
from sklearn.neighbors import KNeighborsClassifier
from sklearn.metrics import log_loss, roc_auc_score, accuracy_score

model_1 = SVC(probability=True, random_state=0)
model_2 = RandomForestClassifier(random_state=0)
model_3 = KNeighborsClassifier()
```

3. Next, we train each model on the training set:

```
model_1.fit(X_train, y_train)
model_2.fit(X_train, y_train)
model_3.fit(X_train, y_train)
```

4. At this point, we need to predict the test set for each model and ensemble all these predictions using majority voting. To do this, we will be using the mode function from SciPy:

```
import numpy as np
from scipy.stats import mode
preds = np.stack([model_1.predict(X_test),
                  model_2.predict(X_test),
                  model_3.predict(X_test)]).T
max_voting = np.apply_along_axis(mode, 1, preds)[:,0]
```

5. We can also figure out how many discordant predictions we have (about 24%):

```
discordant = np.sum(np.var(preds, axis=1) > 0) / len(y_test)
print(f"{discordant:0.2f}")
```

6. Then, we check the accuracy for each single model:

```
for i, model in enumerate(['SVC', 'RF ', 'KNN']):
    acc = accuracy_score(y_true=y_test, y_pred=preds[:, i])
    print(f"Accuracy for model {model} is: {acc:0.3f}")
```

We see that the three models have similar performance, around **0.8**.

7. Now, it is time to check the majority voting ensemble:

```
max_voting_accuracy = accuracy_score(y_true=y_test,
                                     y_pred=max_voting)
print(f"Accuracy for majority voting is: {max_voting_accuracy:0.3f}")
```

The voting ensemble is actually more accurate: **0.817**; this is because it managed to put together the correct signals from the majority.

For multilabel problems (when you can predict multiple classes), you can just pick the classes that are predicted above a certain number of times, assuming a relevance threshold separating prediction signals from noise. For instance, if you have five models, you could set this threshold to 3, which means if at least three models predict a class, then the prediction for that class should be considered correct.

In regression problems, as well as when you are predicting probabilities, you cannot actually use majority voting. Majority voting works exclusively with class ownership. Instead, when you have to predict numbers, you must combine the results numerically. In this case, resorting to an **average** or a **weighted average** will provide you with the right way to combine predictions.

Averaging of model predictions

When averaging your predictions from different models in a competition, you can consider all your predictions as having potentially the same predictive power and use the arithmetic mean to derive an average value.

Aside from the arithmetic mean, we have also found it quite effective to use:

- The **geometric mean:** This is where you multiply the n submissions, and then you take the $1/n^{th}$ power of the resulting product.

- The **logarithmic mean:** Analogous to the geometric mean, the logarithmic mean is advantageous when dealing with very large values because it avoids the overflow issues that can arise from multiplying large numbers in the geometric mean calculation. You take the logarithm of your submission (using np.log1p to avoid taking the log of zero if some predictions are zero), average them together, and take the exponentiation of the resulting mean (using np.expm1 if you used np.lopg1p).

- The **harmonic mean:** This is where you take the arithmetic mean of the reciprocals of your submissions, and then you take the reciprocal of the resulting mean.

- The **mean of powers:** This is where you take the average of the nth power of the submissions, then you take the $1/n^{th}$ power of the resulting average.

The simple arithmetic average is always quite effective and is basically a no-brainer that works more often than expected. Sometimes, variants such as the geometric or harmonic mean may work better.

Continuing with the previous example, we will now try to figure out what kind of mean works best when we switch to the **ROC-AUC** score (described in *Chapter 6, Competition Tasks and Metrics*) as our evaluation metric:

1. To begin with, we evaluate the performances of each single model:

```
proba = np.stack([model_1.predict_proba(X_test)[:, 1],
                  model_2.predict_proba(X_test)[:, 1],
                  model_3.predict_proba(X_test)[:, 1]]).T

for i, model in enumerate(['SVC', 'RF ', 'KNN']):
    score = roc_auc_score(y_true=y_test, y_score=proba[:, i])
    print(f"ROC-AUC for model {model} is: {score:0.5f}")
```

The results give us a range from **0.875** to **0.881**.

2. Our first test is performed using the arithmetic mean:

```
arithmetic = proba.mean(axis=1)
score = roc_auc_score(y_true=y_test, y_score=arithmetic)
print(f"Simple averaging ROC-AUC is: {score:0.5f}")
```

The resulting ROC-AUC score is decisively better than the single models' performance: **0.90192**.

3. We also test whether the geometric, harmonic, or logarithmic mean, or the mean of powers, can outperform the arithmetic mean:

```
geometric = proba.prod(axis=1)**(1/3)
score = roc_auc_score(y_true=y_test, y_score=geometric)
print(f"Geometric averaging ROC-AUC is: {score:0.5f}")

harmonic = 1 / np.mean(1. / (proba + 0.00001), axis=1)
score = roc_auc_score(y_true=y_test, y_score=harmonic)
print(f"Harmonic averaging ROC-AUC is: {score:0.5f}")

n = 3
mean_of_powers = np.mean(proba**n, axis=1)**(1/n)
score = roc_auc_score(y_true=y_test, y_score=mean_of_powers)
print(f"Mean of powers averaging ROC-AUC is: {score:0.5f}")

logarithmic = np.expm1(np.mean(np.log1p(proba), axis=1))
score = roc_auc_score(y_true=y_test, y_score=logarithmic)
print(f"Logarithmic averaging ROC-AUC is: {score:0.5f}")
```

Running the code will tell us that none of them can. In this case, the arithmetic mean is the best choice for ensembling. In almost all cases, what works better than the simple mean is putting prior knowledge into how you combine the numbers. This happens when you weigh your models in the mean calculation.

Weighted averages

When weighing your models, you need to find an empirical way to determine the correct weights. A standard method, though very prone to *adaptive overfitting* (a concept discussed in *Chapter 7*, which occurs when a model's hyperparameters or architecture are influenced by the validation or test data), is to test different combinations on the public leaderboard until you find the best combination.

Of course, that won't ensure that you score the same on the private leaderboard. Here, the principle is to weigh what works better. However, as we have discussed at length, very often, the feedback from the public leaderboard cannot be trusted because of important differences with the private test data. Instead, you can use your cross-validation scores or out-of-fold ones (the latter will be discussed along with stacking in a later section). In fact, another viable strategy is to use weights that are **proportional to the models' cross-validation performances**.

Although it might seem a bit counterintuitive, another very effective method is weighting the submissions **inversely proportionally to their covariances or unique variance**. In this way, we are giving more weight to predictions that are less similar to each other, such as those that are less correlated and more diverse. By doing this, we aim to reduce errors by averaging and get results that are more accurate overall, effectively reducing the variance of the estimates.

In the next example, we will first create a **correlation matrix** of our predicted probabilities, and then we proceed by:

1. Removing the one values on the diagonal and replacing them with zeroes.
2. Averaging the correlation matrix by row to obtain a vector.
3. Taking the reciprocal of each row sum.
4. Normalizing their sum to 1.0.
5. Using the resulting weighting vector in a matrix multiplication of our predicted probabilities.

Here is the code for this:

```
cormat = np.corrcoef(proba.T)
np.fill_diagonal(cormat, 0.0)
W = 1 / np.mean(cormat, axis=1)
W = W / sum(W) # normalizing to sum==1.0
weighted = proba.dot(W)
score = roc_auc_score(y_true=y_test, y_score=weighted)
print(f"Weighted averaging ROC-AUC is: {score:0.5f}")
```

The resulting ROC-AUC of **0.90206** is slightly better than the plain average. Giving more importance to more uncorrelated predictions is an ensembling strategy that is often successful. Even if it only provides slight improvements, this could suffice to turn the competition to your advantage.

Averaging in your cross-validation strategy

As we have covered, averaging doesn't require you to build any special complex pipelines, only a certain number of typical data pipelines that create the models you will average, using the same weights for all predictions or some empirically found weights. The only way to test it is to run a submission on the public leaderboard, thus risking adaptive fitting because your evaluation of the averaging will solely be based on the response from Kaggle.

Before testing directly on the leaderboard, though, you may also test at training time by running the averaging operations on the validation fold (the fold you are not using for training your model). This will provide you with less biased feedback than that from the leaderboard. In the following code snippet, you can find an example of how a cross-validation prediction is arranged. Here is a brief explanation of what happens in the code:

1. First, the code initializes a *k*-fold cross-validation strategy (kf in the code) with five splits, shuffling the data before splitting.

2. Then, within each fold, three models (model_1, model_2, model_3) are trained on the data, and their predictions on the test data are stacked horizontally. The arithmetic mean of these predictions is then calculated, providing the ensemble predictions for each sample. ROC-AUC scores are computed for each fold.

3. Finally, the average of the ROC-AUC scores across all folds is calculated, providing an overall evaluation of the ensemble's effectiveness in predicting the target variable:

```
from sklearn.model_selection import KFold

kf = KFold(n_splits=5, shuffle=True, random_state=0)
scores = list()

for k, (train_index, test_index) in enumerate(kf.split(X_train)):
    model_1.fit(X_train[train_index, :], y_train[train_index])
    model_2.fit(X_train[train_index, :], y_train[train_index])
    model_3.fit(X_train[train_index, :], y_train[train_index])

    proba = np.stack(
        [model_1.predict_proba(X_train[test_index, :])[:, 1],
         model_2.predict_proba(X_train[test_index, :])[:, 1],
         model_3.predict_proba(X_train[test_index, :])[:, 1]]).T
```

```
        arithmetic = proba.mean(axis=1)
        score = roc_auc_score(y_true=y_train[test_index],
                              y_score=arithmetic)
        scores.append(score)
        print(f"FOLD {k} Simple averaging ROC-AUC is: {score:0.5f}")

    print(f"CV averaging ROC-AUC is: {np.mean(scores):0.5f}")
```

Relying on the cross-validation results, as in the code above, can help you evaluate which averaging strategy is more promising without testing directly on the public leaderboard.

Correcting averaging for ROC-AUC evaluations

If your task will be evaluated on the ROC-AUC score, simply averaging your results may not suffice. This is because different models may have adopted different optimization strategies, resulting in differences in the range of computed probabilities. A solution is to convert output probabilities into ranks and then average the ranks (or make a weighted average of them – more about weighted averages will follow). The rankdata function from SciPy's stats module helps assign ranks to data, dealing with ties appropriately when necessary. By default, each tied value receives the average of the ranks it would have been assigned, but other options are possible: https://docs.scipy.org/doc/scipy/reference/generated/scipy.stats.rankdata.html. Here is a code example, where rankdata has been used to transform the output probabilities from three different models:

```
from scipy.stats import rankdata

ranks = np.stack(
    [rankdata(model_1.predict_proba(X_test)[:, 1]),
     rankdata(model_2.predict_proba(X_test)[:, 1]),
     rankdata(model_3.predict_proba(X_test)[:, 1])]).T

arithmetic = ranks.mean(axis=1)
score = roc_auc_score(y_true=y_test, y_score=arithmetic)
print(f"Simple rank averaging ROC-AUC is: {score:0.5f}")
```

This approach works perfectly when you are directly handling the test predictions, as in our example.

Blending models using a meta-model

The Netflix competition (which we discussed at length in *Chapter 1*) didn't just demonstrate that averaging would be advantageous for challenging problems in a data science competition; it also brought about the idea that you can use a model to average your models' results more effectively. The winning team, BigChaos, in their paper (Töscher, A., Jahrer, M., and Bell, R.M. *The BigChaos Solution to the Netflix Grand Prize*. Netflix prize documentation – 2009) made many mentions of **blending** and provided many hints about its effectiveness and the way it works.

In a few words, blending is a kind of weighted averaging procedure where the weights used to combine the predictions are estimated by way of a holdout set and a meta-model trained on it. A **meta-model** is simply a machine learning algorithm that learns from the output of other machine learning models. Usually, a meta-model is a linear model (but sometimes, it can also be a non-linear one; more on that in the next section), but you can actually use whatever you want, with some risks we will discuss.

The procedure for obtaining a blending is straightforward:

1. Before building all your models, you randomly extract a holdout sample from the training data (in a team, you should all use the same holdout). Usually, the holdout is about 10% of the available data; however, depending on circumstances (for instance, the number of examples in your training data, stratifications), it could be less as well as more. As always, in sampling, you may enforce stratification to ensure sampling representativeness, and you can test using adversarial validation to confirm that the sample matches the distribution in the rest of the training set.

2. Train all your models on the remaining training data.

3. Predict on the holdout and the test data.

4. Use the holdout predictions as training data in a meta-learner and reuse the meta-learner model to compute the final test predictions using the test predictions from your models. Alternatively, you can use the meta-learner to figure out the selection of predictors and their weights that should be used in a weighted average.

There are quite a few advantages and disadvantages to such a procedure. Let's start with the advantages:

*   It is easy to implement; you just have to figure out what the holdout sample is
*   Using a meta-learning algorithm ensures you will find the best weights without testing on the public leaderboard

Here are some of the weaknesses:

- Sometimes, depending on sample size and the type of models you use, reducing the number of training examples may increase the variance of the predictions of your estimators

- Even if you take great care over how you sample your holdout, you may still fall into adaptive overfitting, that is, finding weights that suit the holdout but are not generalizable, especially if you use a meta-learner that is too complex

- Using a holdout for testing purposes has the same limitations as the training and test split we discussed in the chapter on model validation: you won't have a reliable estimate if the sample size of the holdout is too small or if, for some reason, your sampling is not representative.

Best practices for blending

In blending, the kind of meta-learner you use can make a significant difference. The most common dilemma is to use a linear or a non-linear model. Among linear models, linear or logistic regressions are preferred. Using a regularized model also helps to discard models that are not useful (L1 regularization or Lasso), or reduce the influence of less useful ones (L2 regularization or Ridge), or a combination of the two (Elastic net). One limit to using these kinds of meta-learners is that they may assign some models a negative contribution, as you can see from the coefficient's value in the model. When you encounter this situation, the model is usually overfitting since all models should contribute positively to the ensemble's building (or, at worst, not contribute at all). The most recent versions of scikit-learn allow you to impose only positive weights and remove the intercept. These constraints act as a regularizer and prevent overfitting.

Non-linear models as meta-learners are less common because they tend to overfit in regression and binary classification problems. Still, they often shine in multiclass and multilabel classification problems since they can model the complex relationships between the classes present. They also generally perform better if, aside from the models' predictions, you also provide them with the original features since they can spot any interactions beneficial in helping them correctly select which models to trust more.

In our following example, we first try blending using a linear model (a logistic regression), then a non-linear approach (a random forest):

1. We start by splitting the training set into a training part for the blend elements and a holdout for the meta-learner. Afterward, we fit the models on the trainable part and predict the holdout:

```
from sklearn.preprocessing import StandardScaler

X_blend, X_holdout, y_blend, y_holdout = train_test_split(
    X_train, y_train, test_size=0.25, random_state=0)

model_1.fit(X_blend, y_blend)
model_2.fit(X_blend, y_blend)
model_3.fit(X_blend, y_blend)

proba = np.stack([model_1.predict_proba(X_holdout)[:, 1],
                  model_2.predict_proba(X_holdout)[:, 1],
                  model_3.predict_proba(X_holdout)[:, 1]]).T

scaler = StandardScaler()
proba = scaler.fit_transform(proba)
```

2. We can now train our linear meta-learner using the probabilities predicted on the holdout:

```
from sklearn.linear_model import LogisticRegression
blender = LogisticRegression(solver='liblinear')
blender.fit(proba, y_holdout)
print(blender.coef_)
```

The resulting coefficients are:

```
[[0.78911314 0.47202077 0.75115854]]
```

We can determine which model contributes more to the meta-ensemble by looking at the coefficients. However, remember that coefficients also rescale probabilities when they are not well calibrated, so a larger coefficient for a model may not imply that it is the most important one. If you want to figure out the role of each model in the blend by looking at coefficients, you first have to rescale them by standardization (in our code example, this has been done using scikit-learn's StandardScaler).

Our output shows us that the SVC and k-nearest neighbors models are weighted more in the blend than the random forest one; their coefficients are almost equivalent, and both are larger than the random forest coefficient.

3. Once the meta-model is trained, we just predict our test data and check its performance:

```python
test_proba = np.stack([model_1.predict_proba(X_test)[:, 1],
                       model_2.predict_proba(X_test)[:, 1],
                       model_3.predict_proba(X_test)[:, 1]]).T


blending = blender.predict_proba(test_proba)[:, 1]
score = roc_auc_score(y_true=y_test, y_score=blending)
print(f"ROC-AUC for linear blending is: {score:0.5f}")
```

4. We can try the same thing using a non-linear meta-learner, such as a random forest, for instance:

```python
blender = RandomForestClassifier(random_state=0)
blender.fit(proba, y_holdout)


test_proba = np.stack([model_1.predict_proba(X_test)[:, 1],
                       model_2.predict_proba(X_test)[:, 1],
                       model_3.predict_proba(X_test)[:, 1]]).T


blending = blender.predict_proba(test_proba)[:, 1]
score = roc_auc_score(y_true=y_test, y_score=blending)
print(f"ROC-AUC for non-linear blending is: {score:0.5f}")
```

An alternative to using a linear or non-linear model as a meta-learner is provided by the **ensemble selection** technique formalized by *Caruana, Niculescu-Mizil, Crew*, and *Ksikes* in their famous paper, *Ensemble selection from libraries of models* (Proceedings of the Twenty-First International Conference on Machine Learning, 2004 – http://niculescu-mizil.org/papers/shotgun.icml04.revised.rev2.pdf).

The ensemble selection is actually a weighted average, so it could simply be considered analogous to a linear combination. However, it is a constrained linear combination (because it is part of a hill-climbing optimization) that will also make a selection of models and apply only positive weights to the predictions. All this minimizes the risk of overfitting and ensures a more compact solution because the solution will involve a model selection. From this perspective, ensemble selection is recommended in all problems where the risk of overfitting is high (for instance, because the training cases are few in number or the models are too complex) and in real-life applications because of its simpler yet effective solution.

When using a meta-learner, you are depending on optimizing its own cost function, which may differ from the metric adopted for the competition. Another great advantage of ensemble selection is that it can be optimized to any evaluation function, so it is mostly suggested when the metric for the competition is different from the canon of those typically optimized in machine learning models.

Implementing ensemble selection requires the following steps, as described in the paper mentioned previously:

1. Start with your trained models and a holdout sample.

2. Test all your models on the holdout sample and, based on the evaluation metric, retain the most effective in a selection (the **ensemble selection**).

3. Then, keep testing other models that could be added to the ones in the ensemble selection so that the average of the proposed selection improves over the previous one. You can either do this with replacement or without. Without replacement, you only put a model into the selection ensemble once; in this case, the procedure is just like a simple average after a forward selection. (In a forward selection, you iteratively add to a solution the model that improves the performance the most until adding further models no longer improves the performance.) With replacement, you can put a model into the selection multiple times, thus resembling a weighted average.

4. When you cannot improve any further, stop and use the ensemble selection.

Here is a simple code example of an ensemble selection:

1. We start by deriving a holdout sample and a training selection from our original training data. We fit the models and obtain the predictions on our holdout, as previously seen when blending with a meta-learner:

```
X_blend, X_holdout, y_blend, y_holdout = train_test_split(
    X_train, y_train, test_size=0.5, random_state=0)

model_1.fit(X_blend, y_blend)
model_2.fit(X_blend, y_blend)
model_3.fit(X_blend, y_blend)

proba = np.stack([model_1.predict_proba(X_holdout)[:, 1],
                  model_2.predict_proba(X_holdout)[:, 1],
                  model_3.predict_proba(X_holdout)[:, 1]]).T
```

2. The ensembling is created in the next code snippet through a series of iterations. At each iteration, we try adding all the models in turn to the present ensemble and check if they improve the model. If any of these additions outperforms the previous ensemble on the holdout sample, the ensemble is updated, and the bar is raised to the present level of performance.

    If no addition can improve the ensemble, the loop is stopped, and the composition of the ensemble is reported back:

```python
iterations = 100
proba = np.stack([model_1.predict_proba(X_holdout)[:, 1],
                  model_2.predict_proba(X_holdout)[:, 1],
                  model_3.predict_proba(X_holdout)[:, 1]]).T

baseline = 0.5
print(f"starting baseline is {baseline:0.5f}")
models = []

for i in range(iterations):
    challengers = list()
    for j in range(proba.shape[1]):
        new_proba = np.stack(proba[:, models + [j]])
        score = roc_auc_score(
            y_true=y_holdout,
            y_score=np.mean(new_proba, axis=1)
        )
        challengers.append([score, j])

    challengers = sorted(
        challengers, key=lambda x: x[0], reverse=True
    )
    best_score, best_model = challengers[0]
    if best_score > baseline:
        print(f"Adding model_{best_model+1} to the ensemble",
              end=': ')
        print(f"ROC-AUC increases score to {best_score:0.5f}")
        models.append(best_model)
        baseline = best_score
```

```
        else:
            print("Cannot improve further - Stopping")
```

3. Finally, we count how many times each model has been inserted into the average, and we calculate the weights for our averaging on the test set:

```
from collections import Counter
freqs = Counter(models)
weights = {key: freq/len(models) for key, freq in freqs.items()}
print(weights)
```

The solution provided will display a ROC-AUC score of **0.86779** and weights for the models of **0.2**, **0.4**, and **0.4**, respectively.

You can make the procedure more sophisticated in various ways. Since this approach may overfit, especially at the initial stages, you could start from a randomly initialized ensemble set or, as the authors suggest, you may already be starting with the *n* best-performing models in the set (you decide the value of *n*, as a hyperparameter). Another variation involves applying sampling to the set of models that can enter the selection at each iteration; in other words, you randomly exclude some models from being picked. Not only will this inject randomness into the process, but it will also prevent specific models from dominating the selection.

Stacking models together

Stacking was first mentioned in *David Wolpert*'s paper (*Wolpert, D. H. Stacked generalization.* Neural networks 5.2 – 1992). Still, it took years before the idea became widely accepted and common (only with release 0.22 in December 2019, for instance, has scikit-learn implemented a stacking wrapper). This was due principally to the Netflix competition first and to Kaggle competitions afterward.

In stacking, you always have a meta-learner. This time, however, it is not trained on a holdout but on the entire training set, thanks to the **out-of-fold** (**OOF**) prediction strategy. We already discussed this strategy in *Chapter 7, Designing Good Validation*. In OOF prediction, you start from a replicable *k*-fold cross-validation split. **Replicable** means that, by recording the cases in each training and testing set at each round or by reproducibility assured by a random seed, you can replicate the same validation scheme for each model you need to be part of the stacking ensemble.

In the Netflix competition, stacking and blending were often used interchangeably. However, the actual method devised by Wolpert originally implied leveraging a scheme based on *k*-fold cross-validation, not a holdout set.

In fact, the core idea in stacking is not to reduce the variance, as in averaging; it is mostly to reduce the bias because it is expected that each model involved in the stacking will grasp a part of the information present in the data, to be recomposed in the final meta-learner.

Let's remind ourselves of how OOF predictions on the training data work. When testing a model, at each round of the validation, you train a model on a part of the training data and validate on another part that is held out from the training.

By recording the validation predictions and then reordering them to reconstruct the ordering of the original training cases, you will obtain a prediction of your model on the very same training set that you have used. However, as you have used multiple models and each model has predicted cases it didn't use for training, you should not have any overfitting effects on your training set predictions.

Having obtained OOF predictions for all your models, you can proceed to build a meta-learner that predicts your target based on the OOF predictions (first-level predictions), or you can keep on producing further OOF predictions on top of your previous OOF predictions (second- or higher-level predictions), thus creating multiple stacking layers. This is compatible with an idea presented by Wolpert himself. By using multiple meta-learners, you are actually imitating the structure of a fully connected feedforward neural network without backpropagation, where the weights are optimally calculated to maximize the predictive performance at the level of each layer separately. From a practical point of view, stacking multiple layers has proven very effective and works very well for complex problems where single algorithms cannot obtain the best results.

Moreover, one interesting aspect of stacking is that you don't need models of comparable predictive power, as in averaging and often in blending. In fact, even worse-performing models may be effective as part of a stacking ensemble. A k-nearest neighbors model may not be comparable to a gradient-boosting solution. Still, when you use its OOF predictions for stacking, it may contribute positively and increase the predictive performance of the ensemble.

When you have trained all the stacking layers, it is time to predict. As far as producing the predictions used at various stacking stages, it is important to note that you have two ways to do this. The original Wolpert paper suggests re-training your models on all your training data and then using those re-trained models to predict the test set. However, in practice, many Kagglers don't retrain but directly use the models created for each fold and make multiple predictions on the test set that are averaged at the end.

In our experience, stacking is generally more effective with complete re-training on all available data before predicting on the test set when using a low number of k-folds. In these cases, the sample consistency may really make a difference in the quality of the prediction because training on less data means getting more variance in the estimates. As discussed in *Chapter 7*, when creating OOF predictions, it is always better to use a high number of folds, between 10 and 20. This limits the number of examples that are held out, and without re-training on all the data, you can simply use the average of predictions obtained from the cross-validation trained models for obtaining your prediction on the test set.

Performing stacking

In our next example, for illustrative purposes, we are using a five-fold cross-validation loop, and the results are stacked twice. In the diagram below, you can follow how the data and the models move between different stages of the stacking process:

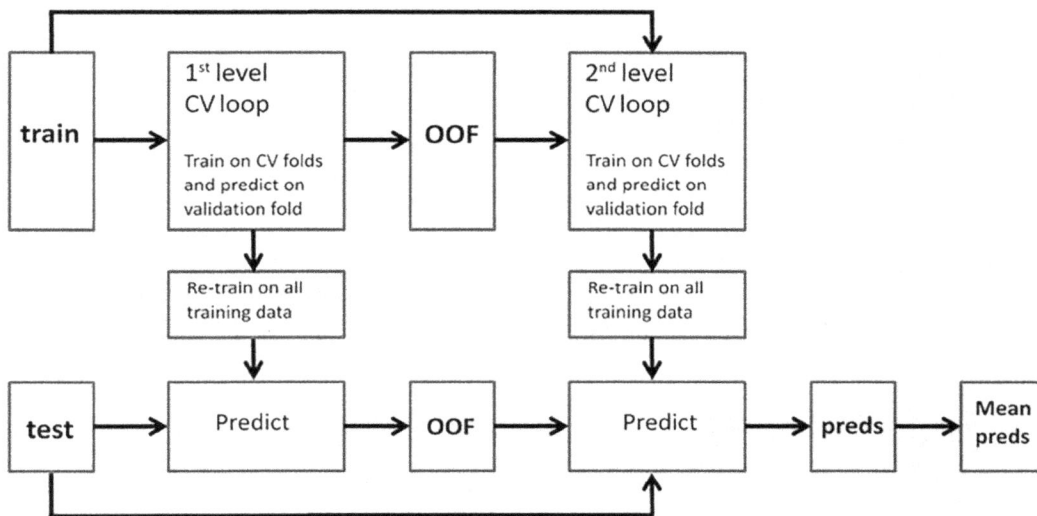

Figure 10.1: Diagram of a two-layer stacking process with final averaging of predictions

Notice that:

- Training data is fed to both levels of the stacking (OOF predictions at the second level of the stacking are joined with the training data)
- After obtaining OOF predictions from the CV loops, models are re-trained on the entire training dataset

- The final predictions are a simple average of all the predictions obtained by the stacked predictors

Let's now take a look at the following code to understand how this diagram translates into Python commands:

1. We start with the first level of training:

```python
from sklearn.model_selection import KFold
kf = KFold(n_splits=5, shuffle=True, random_state=0)
scores = list()
first_lvl_oof = np.zeros((len(X_train), 3))
first_lvl_preds = np.zeros((len(X_test), 3))

for k, (train_index, val_index) in enumerate(kf.split(X_train)):
    model_1.fit(X_train[train_index, :], y_train[train_index])
    first_lvl_oof[val_index, 0] = model_1.predict_proba(
        X_train[val_index, :])[:, 1]

    model_2.fit(X_train[train_index, :], y_train[train_index])
    first_lvl_oof[val_index, 1] = model_2.predict_proba(
        X_train[val_index, :])[:, 1]

    model_3.fit(X_train[train_index, :], y_train[train_index])
    first_lvl_oof[val_index, 2] = model_3.predict_proba(
        X_train[val_index, :])[:, 1]
```

2. After the first layer, we retrain on the full dataset:

```python
model_1.fit(X_train, y_train)
first_lvl_preds[:, 0] = model_1.predict_proba(X_test)[:, 1]

model_2.fit(X_train, y_train)
first_lvl_preds[:, 1] = model_2.predict_proba(X_test)[:, 1]

model_3.fit(X_train, y_train)
first_lvl_preds[:, 2] = model_3.predict_proba(X_test)[:, 1]
```

3. In the second stacking, we will reuse the same models as those in the first layer, adding the stacked OOF predictions to the existing variables. As a good practice, we also compute and store the second-level OOF predictions, which can be useful later for model evaluation, hyperparameter tuning, or further levels of the stacked ensemble:

```
second_lvl_oof = np.zeros((len(X_train), 3))
second_lvl_preds = np.zeros((len(X_test), 3))
skip_X_train = np.hstack([X_train, first_lvl_oof])

for k, (train_index, val_index) in enumerate(kf.split(X_train)):
    model_1.fit(
        skip_X_train[train_index, :], y_train[train_index]
    )
    second_lvl_oof[val_index, 0] = model_1.predict_proba(
        skip_X_train[val_index, :]
    )[:, 1]

    model_2.fit(
        skip_X_train[train_index, :], y_train[train_index]
    )
    second_lvl_oof[val_index, 1] = model_2.predict_proba(
        skip_X_train[val_index, :]
    )[:, 1]

    model_3.fit(
        skip_X_train[train_index, :], y_train[train_index]
    )
    second_lvl_oof[val_index, 2] = model_3.predict_proba(
        skip_X_train[val_index, :]
    )[:, 1]
```

4. Again, we retrain on the full data for the second layer:

```
skip_X_test = np.hstack([X_test, fist_lvl_preds])

model_1.fit(skip_X_train, y_train)
second_lvl_preds[:, 0] = model_1.predict_proba(skip_X_test)[:, 1]

model_2.fit(skip_X_train, y_train)
```

```
second_lvl_preds[:, 1] = model_2.predict_proba(skip_X_test)[:, 1]


model_3.fit(skip_X_train, y_train)
second_lvl_preds[:, 2] = model_3.predict_proba(skip_X_test)[:, 1]
```

5. The stacking is concluded by averaging all the stacked OOF results from the second layer:

```
arithmetic = second_lvl_preds.mean(axis=1)
score = roc_auc_score(y_true=y_test, y_score=arithmetic)
scores.append(score)
print(f"Stacking ROC-AUC is: {score:0.5f}")
```

The resulting ROC-AUC score is about **0.90424**, which is better than previous blending and averaging attempts on the same data and models.

Stacking variations

The main variations on stacking involve changing how test data is processed across the layers, whether to use only stacked OOF predictions or also the original features in all the stacking layers, what model to use as the last one, and various tricks to prevent overfitting.

We discuss some of the most effective here that we have personally experimented with:

- **Optimization may or may not be used**: Some solutions do not care too much about optimizing single models; others optimize only the last layers; others optimize the first layers. Based on our experiences, optimization of single models is important, and we prefer to do it as early as possible in our stacking ensemble.

- **Models can differ at the different stacking layers, or the same sequence of models can be repeated at every stacking layer**: Here, we don't have a general rule, as it really depends on the problem. The kind of models that are more effective may vary according to the problem. As a general suggestion, combining gradient-boosting solutions and neural networks has never disappointed us.

- **At the first level of the stacking procedure, just create as many models as possible**: For instance, you can try a regression model if your problem is a classification, and vice versa. You can also use different models with different hyperparameter settings, thus avoiding too much extensive optimization because the stacking will decide for you. If you are using neural networks, just changing the random initialization seed could suffice to create a diverse bag of models.

You can also try models using different feature engineering techniques and even use unsupervised learning (like *Mike Kim* did when he used *t-SNE* dimensions in a solution of his: `https://www.kaggle.com/c/otto-group-product-classification-challenge/discussion/14295`). The idea is to select all such contributions during the second stacking level. This means that, at such a point, you do not have to experiment any further; you just need to focus on a narrower set of better-performing models. By applying stacking, you can re-use all your experiments and let the stacking decide for you to what degree you should use something in your modeling pipeline.

- Some stacking implementations take on all the features or a selection of them to further stages, reminiscent of skip layers in neural networks. We have noticed that bringing in features at later stages in the stacking can improve your results, but be careful: it also brings in more noise and risk of overfitting.

- Ideally, your OOF predictions should be made from cross-validation schemes with a high number of folds, in other words, between 10 and 20, but we have also seen solutions working with a lower number, such as five folds.

- For each fold, bagging the data (resampling with repetition) multiple times for the same model and then averaging all the results from the model (OOF predictions and test predictions) helps to avoid overfitting and produces better results in the end.

- **Beware of early stopping in stacking**: Using it directly on the validation fold may cause a degree of overfitting, which may or may not be mitigated by the stacking procedure. We suggest you play it safe and always apply early stopping based on a validation sample from your training folds, not your validation one.

The possibilities are endless. Once you have grasped the basic concept of this ensembling technique, all you need is to apply your creativity to the problem at hand. We will discuss this key concept in the final section of this chapter, where we will look at a stacking solution for a Kaggle competition.

Creating complex stacking and blending solutions

At this point in the chapter, you may wonder to what extent you should apply the techniques we have discussed. In theory, you could use all the ensembling techniques we have presented in any competition on Kaggle, not just tabular ones, but you have to consider a few limiting factors:

- Sometimes, datasets are massive, and training a single model takes a long time.
- In image recognition competitions, you are limited to using deep learning methods.

- Even if you can manage to stack models in a deep learning competition, you have a limited choice for stacking different models. Since you are restricted to deep learning solutions, you can only vary small design aspects of the networks and some hyperparameters (or sometimes just the initialization seed) without degrading the performance. In the end, given the same type of models and more similarities than differences in the architectures, the predictions will tend to be too similar and more correlated than they should be, limiting the effectiveness of ensembling.

Under these conditions, complex stacking regimes are usually not feasible. By contrast, averaging and blending are usually possible with large datasets.

In earlier competitions and all recent tabular competitions, complex stacking and blending solutions ruled the day. To give you an idea of the complexity and creativity that needs to be put into stacking for a competition, in this last section, we will discuss the solution provided by *Gilberto Titericz* (`https://www.kaggle.com/titericz`) and *Stanislav Semenov* (`https://www.kaggle.com/stasg7`) to the *Otto Group Product Classification Challenge* (`https://www.kaggle.com/c/otto-group-product-classification-challenge`). The competition was held in 2015, and its task required classifying over 200,000 products into nine distinct classes based on 93 features.

The solution proposed by Gilberto and Stanislav comprised three levels:

1. On the first level, there were 33 models. All the models used quite different algorithms, apart from a cluster of *k*-nearest neighbors where only the *k* parameter varied. They also used unsupervised t-SNE. In addition, they engineered eight features based on dimensionality manipulation (computations performed on distances from nearest neighbors and clusters) and row statistics (the number of non-zero elements in each row). All the OOF predictions and features were passed to the second level.

2. On the second level, they started optimizing hyperparameters and doing model selection and bagging (they created multiple versions of the same model by resampling and averaging the results for each model). In the end, they had only three models that they re-trained on all the data: XGBoost, AdaBoost, and a neural network.

3. On the third level, they prepared a weighted average of the results by first doing a geometric mean of XGBoost and the neural network and then averaging it with AdaBoost.

We can learn a lot from this solution, and not just limited to this competition. Aside from the complexity (on the second level, the number of times they resampled was in the order of hundreds for each model), it is noticeable that there are multiple variations on the schemes we discussed in this chapter. Creativity and trial and error clearly dominate the solution. This is quite typical of many Kaggle competitions, where the problems are seldom the same from one competition to another, and each solution is unique and not easily repeatable.

Many AutoML engines, such as **AutoGluon**, more or less explicitly try to take inspiration from such procedures in order to offer a predefined series of automated steps that can ensure you a top result by stacking and blending.

> See `https://arxiv.org/abs/2003.06505` for a list of the algorithms used by AutoGluon to build its stacked models. The list is quite long, and you will find many ideas for your own stacking solutions.

However, despite implementing some of the best practices around, their results are always subpar compared to what can be achieved by a good team of Kagglers, because creativity in the way you experiment and compose the ensemble is the key to success. The same goes for this chapter of ours. We have shown you the best practices for ensembling; take them as a starting point and create your own by mixing ideas and innovating based on the Kaggle competition or the real-life problem you are dealing with.

Finally, we have to mention that complex ensembling is not just the domain of tabular data problems, but you can find many examples in Kaggle competitions on images and text where blending (stacking is too complex and computationally unfeasible) the solutions from different deep learning methods achieve better results. For a good example, you can have a look at the *CommonLit Readability Prize* competition (`https://www.kaggle.com/competitions/commonlitreadabilityprize`), where the top places in the competitions were taken by teams using blends of different large language models instead of a single well fine-tuned single model.

To conclude the chapter, we caught up with *Xavier Conort*. Xavier is a Competitions Grandmaster who ranked #1 in 2012-2013. An inspiration for many Kagglers at the beginning of Kaggle history, he is now co-founder of FeatureByte. He spoke to us about his experiences with Kaggle, his career, and more.

Interview: Xavier Conort

`https://www.kaggle.com/xavierconort`

What's your favorite kind of competition and why? In terms of techniques and solving approaches, what is your specialty on Kaggle?

I really enjoyed competitions where feature engineering from multiple tables was required to get good results. I liked to mine for good features, especially for business problems that were new to me. This gave me a lot of confidence in my capacity to tackle new problems. In addition to good feature engineering, stacking helped me get good results. I used it to blend multiple models or transform text or high categorical variables into numeric features. My favorite algorithm was GBM, but I tested many other algorithms to add diversity to my blends.

How do you approach a Kaggle competition? How different is this approach to what you do in your day-to-day work?

My primary goal was to learn as much as possible from each competition. Before entering a competition, I tried to assess which skills I would develop. I was not afraid to go beyond my comfort zone. Thanks to the leaderboard feedback, I knew I could learn rapidly from my mistakes. Day-to-day work rarely offers this opportunity. It is difficult to assess the actual quality of the solution we are working on. So, we just play safe and tend to repeat past recipes. I don't think I could have learned as much as I did without Kaggle.

Tell us about a particularly challenging competition you entered, and what insights you used to tackle the task.

My favorite competition is GE Flight Quest, a competition organized by GE where competitors had to predict the arrival times of domestic flights in the US. I especially liked the way the competition's private leaderboard was designed. It tested our capacity to predict future events by scoring our predictions on flights that happened after the competition deadline.

As we had only a few months of history (3 or 4, if my memory is correct), I knew there was a strong risk of overfitting. To mitigate this risk, I decided to build only features that had an obvious causal relation with flight delays, such as features measuring weather conditions and traffic. And I was very careful to exclude the name of the airport from my primary feature lists. Indeed, some airports hadn't experienced bad weather conditions during the few months of history. So, I was very concerned that my favorite ML algorithm, GBM, would use the name of the airport as a proxy for good weather and then fail to predict well for those airports in the private leaderboard. To capture the fact that some airports are better managed than others and improve my leaderboard score slightly, I eventually did use the name of the airport, but as a residual effect only. It was a feature of my second layer of models that used the predictions of my first layer of models as an offset. This approach can be considered a two-step boosting, where you censor some information during the first step. I learned it from actuaries applying this approach in insurance to capture geospatial residual effects.

Has Kaggle helped you in your career? If so, how?

It definitely helped me in my career as a data scientist. Before converting to data science, I was an actuary in the insurance industry, didn't know anything about machine learning, and didn't know any data scientists. Thanks to Kaggle's diversity of competitions, I boosted my learning curve. Thanks to my good results, I could show a track record and convince employers that a 39-year-old actuary could successfully develop new skills on his own. And thanks to Kaggle's community, I connected with many passionate data scientists across the world. I first had a lot of fun competing with or against them. Finally, I had the chance to work with some of them. Jeremy Achin and Tom De Godoy, the DataRobot founders, were my competition teammates before they asked me to join DataRobot. Without Kaggle's help, I think I would still be working as an actuary in the insurance industry.

Have you ever used something you have done in Kaggle competitions in order to build your portfolio to show to potential employers?

I have to confess that I did enter a few competitions with the goal of impressing my employer or potential clients. It worked well, but it was much less fun and much more pressure.

In your experience, what do inexperienced Kagglers often overlook? What do you know now that you wish you'd known when you first started?

I would advise inexperienced Kagglers not to look at the solutions posted during the competition but to try to find good solutions on their own. I am happy that competitors didn't share code during the early days of Kaggle. It forced me to learn the hard way.

What mistakes have you made in competitions in the past?

One mistake is to keep on competing in competitions that are badly designed with leaks. It is just a waste of time. You don't learn much from those competitions.

Are there any particular tools or libraries that you would recommend using for data analysis or machine learning?

Gradient Boosting Machine is my favorite algorithm. I first used R's gbm, then scikit-learn's GBM, then XGBoost, and finally LightGBM. Most of the time, it has been the principal ingredient of my winning solution. To get some insight into what GBM learns, I would recommend the SHAP package.

What's the most important thing someone should keep in mind or do when they're entering a competition?

Compete to learn. Compete to connect with other passionate data scientists. Don't compete only to win.

Summary

In this chapter, we discussed how ensembling multiple solutions works and proposed some basic code examples you can use to start building your own solutions. We started from the ideas that power model ensembles, such as random forests and gradient boosting. Then, we explored the different ensembling approaches, from the simple averaging of test submissions to meta-modeling across multiple layers of stacked models.

As we discussed at the end, ensembling is more of an art form based on some shared common practices. When we explored a successful complex stacking regime that won a Kaggle competition, we were amazed by how the combinations were tailored to the data and the problem itself. You cannot just take a stacking, replicate it on another problem, and hope it will be the best solution. You can only follow guidelines and find the best solution consisting of averaging/stacking/blending diverse models yourself through lots of experimentation and computational effort.

The next chapter will delve into deep learning competitions, beginning with computer vision ones for classification and segmentation tasks.

Join our book's Discord space

Join our community's Discord space for discussions with the authors and other readers:

https://packt.link/kaggle

11

Modeling for Computer Vision

Computer vision tasks are among the most popular problems in the practical applications of machine learning; they were the gateway into deep learning for many Kagglers, including the authors of this book. Over the last few years, there has been tremendous progress in the field, and new **State Of The Art** (**SOTA**) libraries continue to be released. In this chapter, we will give you an overview of the most popular competition types in computer vision.

Here is an walkthrough of the topics covered in this chapter:

- **Introduction to Computer Vision Modeling** with an overview of the most popular competition types in computer vision and a look at essential augmentation strategies.
- **End-to-End Pipelines for Core Tasks** comprising:
    - **Image Classification**: A complete walkthrough from data preparation to model setup and visualization.
    - **Object Detection**: A guide to identifying and locating objects within an image, featuring the YOLOv5 model.
    - **Image Segmentation**: Techniques for classifying each pixel in an image to identify areas of interest, using the Detectron2 library.
- **Capstone Case Study**: The CryoET Object Identification Competition, where you adapt 2D computer vision knowledge to complex 3D volumetric data, and explore advanced techniques required for success in a cutting-edge, real-world challenge.

Augmentation strategies

While deep learning techniques have been highly successful in computer vision tasks such as image recognition, segmentation, or object detection, the underlying algorithms are typically highly data-intensive, requiring large amounts of data to avoid overfitting. However, not all domains of interest satisfy that requirement, which is where **data augmentation** comes in. This refers to a group of image processing techniques that create modified versions of images, thereby enhancing the size and quality of training datasets, which, in turn, leads to improved performance of deep learning models. The augmented data will typically represent a more comprehensive set of possible data points, thereby minimizing the distance between the training and validation set, as well as any future test sets.

In this section, we will review some of the more common augmentation techniques, along with choices for their software implementations. The most frequently used transformations include:

- **Flipping**: Flipping the image (along the horizontal or vertical axis)
- **Rotation**: Rotating the image by a given angle (clockwise or anti-clockwise)
- **Cropping**: Selecting a random subsection of the image
- **Brightness**: Modifying the brightness of the image
- **Scaling**: Increasing or decreasing the image to a higher (outward) or lower (inward) size

Below, we demonstrate how those transformations work in practice using the image of an American acting legend and comedian, Betty White:

Figure 11.1: Betty White image

We can flip the image along the vertical or horizontal axis:

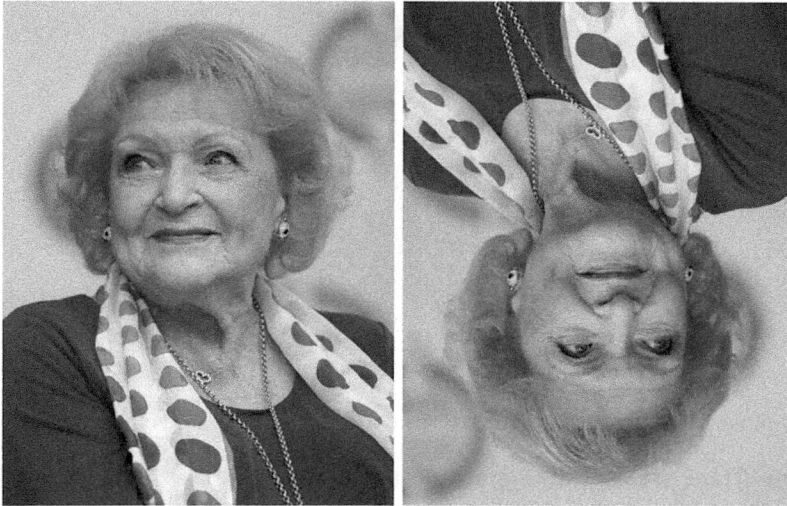

Figure 11.2: Betty White image – flipped vertically (left) and horizontally (right)

Rotations are pretty self-explanatory; notice the automatic padding of the image in the background:

Figure 11.3: Betty White image – rotated clockwise

We can also crop an image to the region of interest:

Figure 11.4: Betty White image – cropped

On a high level, we can say that augmentations can be applied in one of two ways:

- **Offline**: These are typically applied to smaller datasets (fewer images or smaller sizes), although the definition of "small" depends on the available hardware. The idea is to generate modified versions of the original images as a preprocessing step for your dataset, and then use those alongside the "original" ones.

- **Online**: These are used for bigger datasets. The augmented images are not saved on disk; the augmentations are applied in mini-batches and fed to the model.

In the following sections, we will provide an overview of two of the most common methods for augmenting your image dataset: the built-in Keras functionality and the `Albumentations` package. There are several other options available out there (`skimage`, `OpenCV`, `imgaug`, Augmentor, and **SOLT - Streaming Over Lightweight data Transformations**, see `https://github.com/imedslab/solt`), but we will focus on the most popular ones.

> The methods discussed in this chapter focus on image analysis powered by a GPU. The use of **tensor processing units (TPUs)** is an emerging but still somewhat niche application. For those of you interested in image augmentation in combination with TPU-powered analysis, you are encouraged to check out the excellent work of *Chris Deotte* (**@cdeotte**):
>
> `https://www.kaggle.com/code/cdeotte/triple-stratified-kfold-with-tfrecords`
>
> Chris is a quadruple Kaggle Grandmaster and a fantastic educator, thanks to the notebooks he creates and the discussions he participates in; overall, a person definitely worth following for any Kaggle user, regardless of their level of experience.

We will be using data from the *Cassava Leaf Disease Classification* competition (https://www.kaggle.com/c/cassava-leaf-disease-classification). As usual, we begin with the groundwork:

1. First, we load the necessary packages:

```
import os
import glob
import numpy as np
import scipy as sp
import pandas as pd

import cv2
from skimage.io import imshow, imread, imsave

# imgaug
import imageio
import imgaug as ia
import imgaug.augmenters as iaa

# Albumentations
import albumentations as A

# Visualization
import matplotlib.pyplot as plt
import matplotlib.image as mpimg
%matplotlib inline
import seaborn as sns
from IPython.display import HTML, Image

# Warnings
import warnings
warnings.filterwarnings("ignore")
```

2. Next, we define some helper functions that will streamline the presentation later. We need a way to load images into arrays:

```
def load_image(image_id):
    file_path = image_id
    image = imread(Image_Data_Path + file_path)
    return image
```

3. We would like to display multiple images in a gallery style, so we create a function that takes as input an array containing the images along with the desired number of columns, and outputs the array reshaped into a grid with a given number of columns:

```python
def gallery(array, ncols=3):

    nindex, height, width, intensity = array.shape
    nrows = nindex//ncols
    assert nindex == nrows*ncols
    result = (array.reshape(nrows, ncols, height, width, intensity)
              .swapaxes(1,2)
              .reshape(height*nrows, width*ncols, intensity))
    return result
```

4. With the boilerplate code taken care of, we can load the images for augmentation:

```python
data_dir = '../input/cassava-leaf-disease-classification/'
Image_Data_Path = data_dir + '/train_images/'
train_data = pd.read_csv(data_dir + '/train.csv')

# We load and store the first 10 images in memory for faster access
train_images = train_data["image_id"][:10].apply(load_image)
```

5. Let's load a single image so we know what our reference is:

```python
curr_img = train_images[7]
plt.figure(figsize = (15,15))
plt.imshow(curr_img)
plt.axis('off')
```

Here is the resulting image:

Figure 11.5: Reference image

In the following sections, we will demonstrate how to generate augmented images from this reference image using both built-in Keras functionality and the albumentations library.

Keras built-in augmentations

The Keras library has built-in functionality for augmentations. While not as extensive as dedicated packages, it has the advantage of easy integration with your code. We do not need a separate code block for defining the augmentation transformations, but can incorporate them inside ImageDataGenerator, a functionality we are likely to use anyway.

The first Keras approach we examine is based on the ImageDataGenerator class. As the name suggests, it can be used to generate batches of image data with real-time data augmentations.

The ImageDataGenerator approach

We begin by instantiating an object of the ImageDataGenerator class in the following manner:

```
import tensorflow as tf
from tensorflow.keras.preprocessing.image import (
    ImageDataGenerator,
    array_to_img,
    img_to_array,
    load_img
)
datagen = ImageDataGenerator(
    rotation_range=40,
    shear_range=0.2,
    zoom_range=0.2,
    horizontal_flip=True,
    brightness_range=(0.5, 1.5)
)
```

We define the desired augmentations as arguments to ImageDataGenerator. The official documentation does not seem to address the topic, but practical results indicate that the augmentations are applied in the order in which they are defined as arguments.

> In the above example, we utilize only a limited subset of possible options; for a complete list, you are encouraged to consult the official documentation: https://keras.io/api/preprocessing/image/.

Next, we iterate through the images with the `.flow` method of the `ImageDataGenerator` object. The class provides three different functions to load the image dataset into memory and generate batches of augmented data:

- `flow`
- `flow_from_directory`
- `flow_from_dataframe`

They all achieve the same objective, but differ in the way the locations of the files are specified. In our example, the images are already in memory, so we can iterate using the `.flow()` method:

```
i = 0
for batch in datagen.flow(
    curr_img_array,
    batch_size=1,
    save_to_dir='.',
    save_prefix='Augmented_image',
    save_format='jpeg'
):
    i += 1
    # A hard-coded stop to avoid entering an infinite loop
    if i > 9:
        break
```

We can examine the augmented images using the helper functions we defined earlier:

```
aug_images = []
for img_path in glob.glob("*.jpeg"):
    aug_images.append(mpimg.imread(img_path))
plt.figure(figsize=(20, 20))
plt.axis('off')
plt.imshow(
    gallery(np.array(aug_images[0:9]), ncols=3)
)
plt.title('Augmentation examples')
```

Here is the result:

Augmentation examples

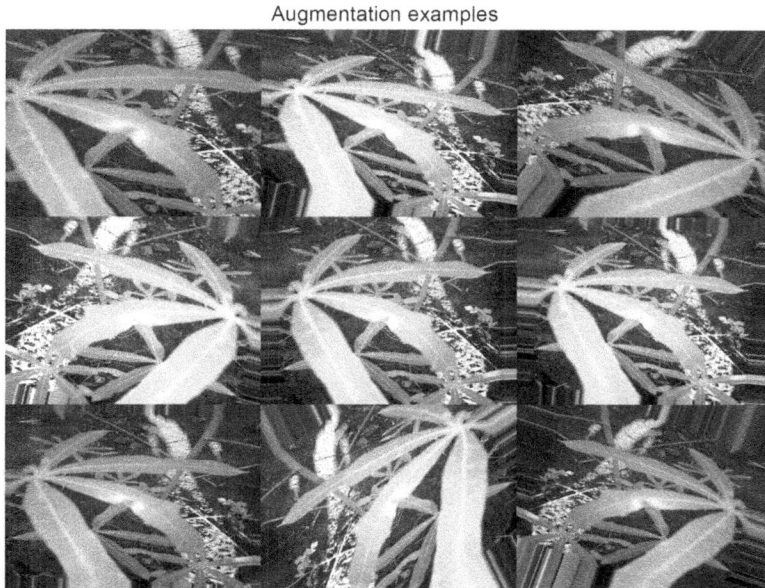

Figure 11.6: A collection of augmented images

Augmentations are a handy tool, but using them efficiently requires a judgment call. First, it is a good idea to visualize them to get a sense of the impact on the data. On the one hand, we want to introduce some variation in the data to increase the model's generalization. On the other hand, if we change the images too radically, the input data will be less informative, and the model's performance is likely to suffer. In addition, the choice of augmentations to use can also be problem-specific, as evident in comparing different competitions.

Look at *Figure 11.6* above (the reference image from the *Cassava Leaf Disease Classification* competition). The leaves on which we are supposed to identify the disease can vary in size, orientation, and other characteristics, due to both the shapes of the plants and differences in how the images are captured. This means transformations such as vertical or horizontal flips, cropping, and rotations all make sense in this context.

By contrast, we can look at a sample image from the *Severstal: Steel Defect Detection* competition (`https://www.kaggle.com/c/severstal-steel-defect-detection`) as shown in *Figure 11.7*. In this competition, participants had to localize and classify defects on a steel sheet. All the images had the same size and orientation, which means that rotations or crops would have produced unrealistic photos, adding to the noise and harming the generalization capabilities of an algorithm.

28a3f7928.jpg has defect 2

5a8203514.jpg has defect 3

acc4098ce.jpg has defect 1

6e7bace88.jpg has defect 3

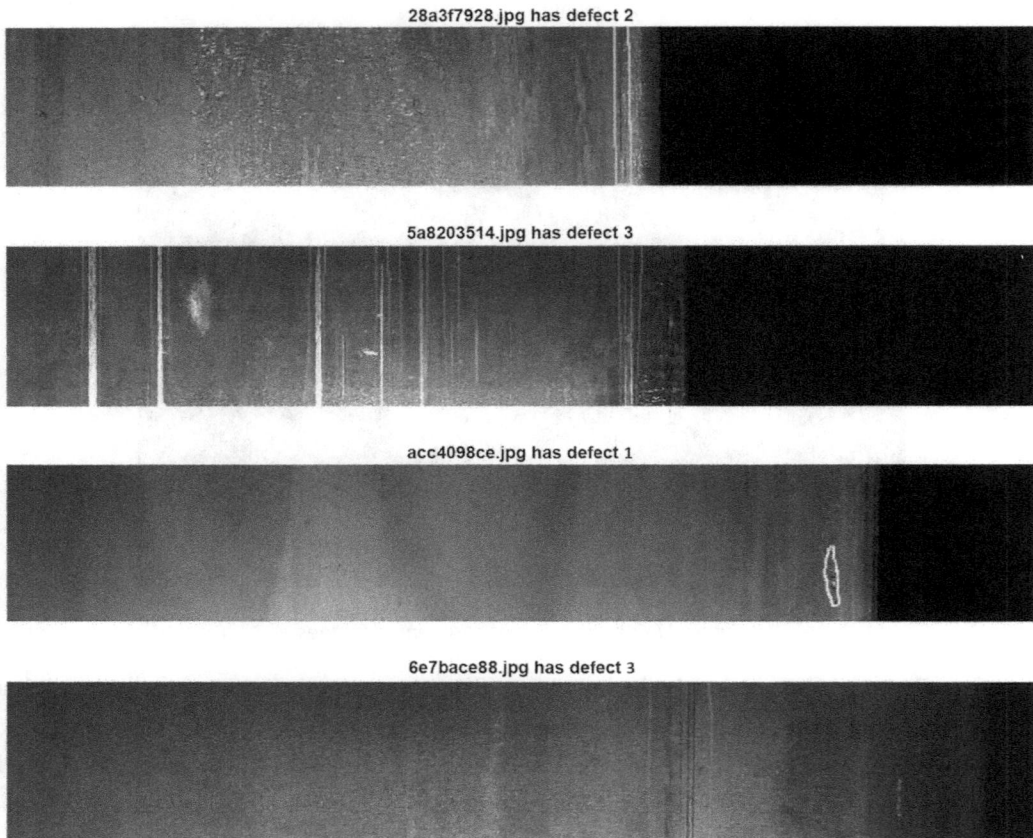

Figure 11.7: Sample images from the Severstal competition

Preprocessing layers

An alternative approach to data augmentation as a preprocessing step in a native Keras manner is to use the preprocessing layers API. The functionality is remarkably flexible: these pipelines can be used either in combination with Keras models or independently, in a manner similar to ImageDataGenerator.

Below, we show briefly how a preprocessing layer can be set up. First, the imports:

```
import tensorflow as tf
from tensorflow.keras import layers
```

We load a pretrained model in the standard Keras manner:

```
pretrained_base = tf.keras.applications.VGG16(
    weights='imagenet',      # Use pre-trained ImageNet weights
    include_top=False,       # Drop the final Dense layers
                             #(so you can add your own)
    input_shape=(224, 224, 3)  # Standard input shape
                               #(adjust if your images are different)
)
pretrained_base.trainable = False
```

The preprocessing layers can be used in the same way as other layers are used inside the Sequential constructor; the only requirement is that they need to be specified before any others, at the beginning of our model definition:

```
model = tf.keras.Sequential([
    layers.InputLayer(shape=(224, 224, 3)),

    # Preprocessing
    layers.RandomFlip('horizontal'),
    layers.RandomContrast(0.5),

    # VGG16 requires specific preprocessing
    # (it expects pixels 0-255, centered)
    layers.Lambda(tf.keras.applications.vgg16.preprocess_input),

    # Base Model
    pretrained_base,

    # Model Head
    layers.Flatten(),
    layers.Dense(6, activation='relu'),
    layers.Dense(1, activation='sigmoid'),
])

model.summary()
```

A deeper dive: the albumentations package

The `albumentations` package is a fast image augmentation library built as a wrapper around other libraries.

> The package is the result of intensive coding in quite a few Kaggle competitions (see `https://medium.com/@iglovikov/the-birth-of-albumentations-fe38c1411cb3`), and claims among its core developers and contributors quite a few notable Kagglers, including *Eugene Khvedchenya* (`https://www.kaggle.com/bloodaxe`), *Vladimir Iglovikov* (`https://www.kaggle.com/iglovikov`), *Alex Parinov* (`https://www.kaggle.com/creafz`), and *ZFTurbo* (`https://www.kaggle.com/zfturbo`).
>
> The full documentation can be found at `https://albumentations.readthedocs.io/en/latest/`.

Below, we list the important characteristics:

- A unified API for different data types
- Support for all common computer vision tasks
- Integration with both TensorFlow and PyTorch

Using the `albumentations` functionality to transform an image is straightforward:

1. We begin by initializing the required transformations:

```
import albumentations as A
horizontal_flip = A.HorizontalFlip(p=1)
rotate = A.ShiftScaleRotate(p=1)
gaus_noise = A.GaussNoise()
bright_contrast = A.RandomBrightnessContrast(p=1)
gamma = A.RandomGamma(p=1)
blur = A.Blur()
```

2. Next, we apply the transformations to our reference image:

```
img_flip = horizontal_flip(image = curr_img)
img_gaus = gaus_noise(image = curr_img)
img_rotate = rotate(image = curr_img)
img_bc = bright_contrast(image = curr_img)
img_gamma = gamma(image = curr_img)
img_blur = blur(image = curr_img)
```

We can access the augmented images with the `'image'` key and visualize the results:

```python
img_list = [
    img_flip['image'],img_gaus['image'],
    img_rotate['image'], img_bc['image'],
    img_gamma['image'], img_blur['image']
]

plt.figure(figsize=(20,20))
plt.axis('off')
plt.imshow(
    gallery(np.array(img_list), ncols = 3)
)
plt.title('Augmentation examples')
```

Here are our results:

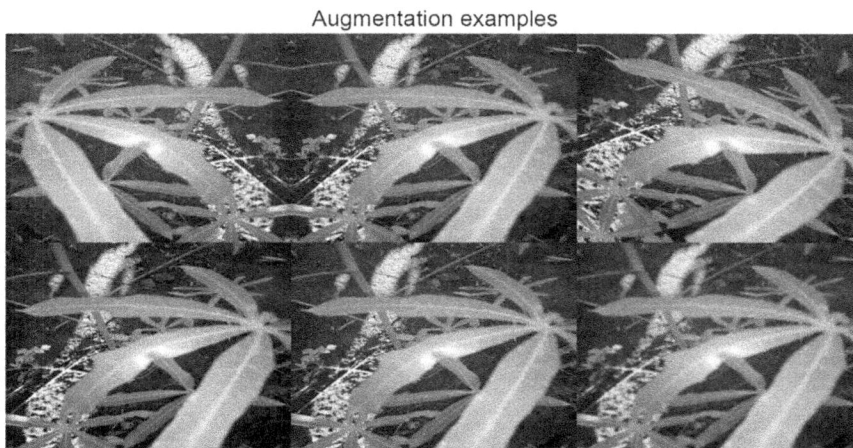

Figure 11.8: Image augmented using the albumentations library

Having discussed augmentation as a crucial preprocessing step in approaching a computer vision problem, we are now in a position to apply this knowledge in the following sections, beginning with a very common task: image classification.

Before we proceed, let's look at a brief conversation we had with Chris Deotte, whom we've mentioned quite a few times in this book (including earlier in this chapter), and for good reason. He is a quadruple Kaggle Grandmaster and senior data scientist and researcher at NVIDIA, who joined Kaggle in 2019.

Interview: Chris Deotte

`https://www.kaggle.com/cdeotte`

What's your favorite kind of competition and why? In terms of techniques and solving approaches, what is your specialty on Kaggle?

I enjoy competitions with fascinating data and competitions that require building creative novel models. My specialty is analyzing trained models to determine their strengths and weaknesses. Afterward, I enjoy improving the models and/or developing post-processing to boost CV LB.

How do you approach a Kaggle competition? How different is this approach from what you do in your day-to-day work?

I begin each competition by performing exploratory data analysis (EDA), creating a local validation, building some simple models, and submitting to Kaggle for leaderboard scores. This fosters an intuition of what needs to be done in order to build an accurate and competitive model.

Tell us about a particularly challenging competition you entered, and what insights you used to tackle the task.

Kaggle's Shopee – Price Match Guarantee was a challenging competition that required both image models and natural language models. A key insight was extracting embeddings from the two types of models and then determining how to use both image and language information together to find product matches.

Has Kaggle helped you in your career? If so, how?

Yes. Kaggle helped me become a senior data scientist at NVIDIA by improving my skills and boosting my resume's marketability.

Many employers peruse the work on Kaggle to find employees with specific skills to help solve their specific projects. In this way, I have been solicited about many job opportunities.

In your experience, what do inexperienced Kagglers often overlook? What do you know now that you wish you'd known when you first started?

In my opinion, inexperienced Kagglers often overlook the importance of local validation. Seeing your name on the leaderboard is exciting. And it's easy to focus on improving our leaderboard scores instead of our cross-validation scores.

What mistakes have you made in competitions in the past?

Many times, I have made the mistake of trusting my leaderboard score over my cross-validation score and selecting the wrong final submission.

Are there any particular tools or libraries that you would recommend using for data analysis or machine learning?

Absolutely. Feature engineering and quick experimentation are important when optimizing tabular data models. In order to accelerate the cycle of experimentation and validation, using NVIDIA RAPIDS cuDF and cuML on GPUs is essential.

What's the most important thing someone should keep in mind or do when they're entering a competition?

The most important thing is to have fun and learn. Don't worry about your final placement. If you focus on learning and having fun, then over time, your final placements will become better and better.

Do you use other competition platforms? How do they compare to Kaggle?

Yes, I have entered competitions outside of Kaggle. Individual companies like Booking.com or Twitter.com (now X) will occasionally host a competition. These competitions are fun and involve high-quality, real-life data.

Image classification

In this section, we will present an end-to-end pipeline that can serve as a template for handling image classification problems. We will walk through the necessary steps, from data preparation to model setup and estimation to results visualization. Apart from being informative (and cool), this last step can also be very useful if you need to examine your code in-depth to get a better understanding of the performance.

We will continue using the data from the *Cassava Leaf Disease Classification* contest (https://www.kaggle.com/c/cassava-leaf-disease-classification).

As usual, we begin by loading the necessary libraries:

```
import numpy as np
import pandas as pd
import matplotlib.pyplot as plt
import datetime

from sklearn.model_selection import train_test_split
from sklearn.metrics import accuracy_score

import tensorflow as tf

from tensorflow.keras import models, layers
from tensorflow.keras.preprocessing import image
from tensorflow.keras.preprocessing.image import ImageDataGenerator
from tensorflow.keras.callbacks import (
    ModelCheckpoint,
    EarlyStopping,
    ReduceLROnPlateau
)
from tensorflow.keras.applications import EfficientNetB0
from tensorflow.keras.optimizers import Adam

import os, cv2, json
from PIL import Image
```

It is usually a good idea to define a few helper functions; it makes for code that is easier to both read and debug. If you are approaching a general image classification problem, a good starting point can be provided by a model from the EfficientNet family, introduced in 2019 in a paper from the Google Research Brain Team (https://arxiv.org/abs/1905.11946). The basic idea is to balance network depth, width, and resolution to enable more efficient scaling across all dimensions and, subsequently, better performance. For our solution, we will use the simplest member of the family, EfficientNetB0, as our convolutional backbone. When used as a feature extractor without its final classification layers, this highly efficient architecture has approximately **4 million trainable parameters**, making it an excellent starting point that balances performance and computational cost.

For an appropriately detailed explanation of the EfficientNet networks, you are encouraged to explore `https://research.google/blog/efficientnet-improving-accuracy-and-efficiency-through-automl-and-model-scaling/` as a starting point.

We construct our model with B0 as the basis, followed by a pooling layer for improved translation invariance and a dense layer with an activation function suitable for our multiclass classification problem:

```python
class CFG:
    # config
    WORK_DIR = '../input/cassava-leaf-disease-classification'
    BATCH_SIZE = 8
    EPOCHS = 5
    TARGET_SIZE = 512


def create_model():
    conv_base = EfficientNetB0(
        include_top=False,
        weights=None,
        input_shape=(CFG.TARGET_SIZE, CFG.TARGET_SIZE, 3)
    )

    model = conv_base.output
    model = layers.GlobalAveragePooling2D()(model)
    model = layers.Dense(5, activation="softmax")(model)
    model = models.Model(conv_base.input, model)
    model.compile(
        optimizer=Adam(learning_rate=0.001),
        loss="sparse_categorical_crossentropy",
        metrics=["acc"]
    )
    return model
```

Some brief remarks on the parameters we pass to the `EfficientNetB0` function are as follows:

- The `include_top` parameter allows you to decide whether to include the final dense layers. As we want to use the pre-trained model as a feature extractor, a default strategy would be to skip them and then define the head ourselves.

- weights can be set to None if we want to train the model from scratch, or to imagenet or noisy-student if we instead prefer to utilize the weights pre-trained on large image collections.

To better understand what our model has learned, we can visualize its *activations*. An activation, or *feature map*, is the output of a layer in the network. Visualizing these helps us peek inside the black box and see which features the model is identifying, from simple edges in early layers to more complex patterns in deeper ones. Our helper function creates a "restricted" model that outputs the activations from each layer up to a specified point. By passing an image through this model, we can capture and plot these feature maps. This is an invaluable technique for debugging and gaining intuition about a model's behavior. For a deeper dive into this topic, we recommend exploring resources like Jason Brownlee's excellent guide on visualizing filters and feature maps: https://machinelearningmastery.com/how-to-visualize-filters-and-feature-maps-in-convolutional-neural-networks/.

The following helper function enables us to visualize the activation layer, allowing us to examine the network's performance from a visual perspective. This is frequently helpful in developing an intuition in a field notorious for its opacity:

```python
def activation_layer_vis(img, activation_layer=0, layers=10):
    layer_outputs = [
        layer.output for layer in model.layers[:layers]
    ]
    activation_model = models.Model(
        inputs=model.input,
        outputs=layer_outputs
    )
    activations = activation_model.predict(img)

    rows = int(activations[activation_layer].shape[3] / 3)
    cols = int(activations[activation_layer].shape[3] / rows)

    fig, axes = plt.subplots(
        rows,
        cols,
        figsize=(15, 15 * cols)
    )
    axes = axes.flatten()
```

```
for i, ax in zip(
    range(activations[activation_layer].shape[3]),
    axes
):
    ax.matshow(
        activations[activation_layer][0, :, :, i],
        cmap='viridis'
    )
    ax.axis('off')

plt.tight_layout()
plt.show()
```

We generate the activations by creating predictions for a given model based on a restricted model – in other words, using the entire architecture up until the penultimate layer; this is the code up to the `activations` variable.

The rest of the function ensures that we show the right layout of activations, corresponding to the shape of the filter in the appropriate convolution layer.

Next, we process the labels and set up the validation scheme; there is no special structure in the data (for example, a time dimension or overlap across classes), so we can use a simple random split:

```
train_labels = pd.read_csv(os.path.join(CFG.WORK_DIR, "train.csv"))

STEPS_PER_EPOCH = int(len(train_labels) * 0.8 / CFG.BATCH_SIZE)
VALIDATION_STEPS = int(len(train_labels) * 0.2 / CFG.BATCH_SIZE)
```

> For a refresher on more elaborate validation schemes, refer to *Chapter 6, Designing Good Validation.*

We are now able to set up the data generators, which are necessary for our TF-based algorithm to loop through the image data.

First, we instantiate two `ImageDataGenerator` objects; this is when we incorporate the image augmentations. For the purpose of this demonstration, we will go with the Keras built-in ones. After that, we create the generator using the `flow_from_dataframe()` method, which is used to generate batches of tensor image data with real-time data augmentation:

```python
train_labels.label = train_labels.label.astype('str')

train_datagen = ImageDataGenerator(
    validation_split=0.2,
    preprocessing_function=None,
    rotation_range=45,
    zoom_range=0.2,
    horizontal_flip=True,
    vertical_flip=True,
    fill_mode='nearest',
    shear_range=0.1,
    height_shift_range=0.1,
    width_shift_range=0.1
)

train_generator = train_datagen.flow_from_dataframe(
    train_labels,
    directory=os.path.join(CFG.WORK_DIR, "train_images"),
    subset="training",
    x_col="image_id",
    y_col="label",
    target_size=(CFG.TARGET_SIZE, CFG.TARGET_SIZE),
    batch_size=CFG.BATCH_SIZE,
    class_mode="sparse"
)

validation_datagen = ImageDataGenerator(validation_split=0.2 )

validation_generator = validation_datagen.flow_from_dataframe(
    train_labels,
    directory=os.path.join(CFG.WORK_DIR, "train_images"),
    subset="validation",
    x_col="image_id",
    y_col="label",
    target_size=(CFG.TARGET_SIZE, CFG.TARGET_SIZE),
    batch_size=CFG.BATCH_SIZE,
    class_mode="sparse"
)
```

With the data structures specified, we can create the model:

```
model = create_model()
model.summary()
```

Once our model is created, we can quickly examine a summary. This is mostly useful for sanity checks, because unless you have a photographic memory, chances are you are not going to remember the layer composition batches of a sophisticated model like **EfficientNet-B0 (EffNetB0)**. In practice, you can use the summary to verify whether the dimensions of output filters are correct and whether the parameter counts (trainable and non-trainable) align with expectations. For the sake of compactness, we only demonstrate the first few lines of the output below; inspecting the architecture diagram for B0 will give you an idea of how long the complete output would be.

```
Model: "functional_1"


                                        Output Shape            Param # Connected to
=======================================================================================


rescaling (Rescaling)                   (None, 512, 512, 3)  0        input_1[0][0]


normalization (Normalization) (None, 512, 512, 3)  7        rescaling[0][0]


stem_conv_pad (ZeroPadding2D) (None, 513, 513, 3)  0        normalization[0]
- -
```

```
stem_conv (Conv2D)                      (None, 256, 256, 32) 864      stem_conv_


stem_bn (BatchNormalization)            (None, 256, 256, 32) 128      stem_conv[0][0]


stem_activation (Activation)            (None, 256, 256, 32) 0        stem_bn[0][0]


block1a_dwconv (DepthwiseConv2D (None, 256, 256, 32) 288      stem_

activation[0][0]
```

Figure 11.9: Summary output for the EfficientNetB0 model, showing the initial layers, output shapes, and parameter counts

With the above steps taken care of, we can proceed to fitting the model. In this step, we can also very conveniently define callbacks. The first one is `ModelCheckpoint`. The model can be found from here: `https://www.kaggle.com/datasets/maksymshkliarevskyi/cassava-leaf-disease-models`:

```
model_save = ModelCheckpoint(
    filepath = './EffNetB0_512_8_best_weights.weights.h5',
    save_best_only = True,
    save_weights_only = True,
    monitor = 'val_loss',
    mode = 'min',
    verbose = 1
)
```

The checkpoint uses a few parameters worth elaborating on:

- We can preserve the best set of model weights by setting `save_best_only = True`.
- We reduce the size of the model by only keeping the weights, instead of the complete set of optimizer state.
- We decide on which model is optimal by locating a minimum for validation loss.

Next, we use one of the popular methods for preventing overfitting, **early stopping**. We monitor the performance of the model on the holdout set and stop the algorithm if the metric stops improving for a given number of epochs – in this case, 5:

```
early_stop = EarlyStopping(
    monitor = 'val_loss',
    min_delta = 0.001,
    patience = 5,
    mode = 'min',
    verbose = 1,
    restore_best_weights = True
)
```

The ReduceLROnPlateau callback monitors the loss on the holdout set, and if no improvement is seen for a patience number of epochs, the learning rate is reduced, in this case, by a factor of 0.3. While not a universal solution, it can frequently help with convergence:

```
reduce_lr = ReduceLROnPlateau(
    monitor = 'val_loss',
    factor = 0.3,
    patience = 2,
    min_delta = 0.001,
    mode = 'min',
    verbose = 1
)
```

We are now ready to fit the model:

```
history = model.fit(
    train_generator,
    steps_per_epoch = STEPS_PER_EPOCH,
    epochs = CFG.EPOCHS,
    validation_data = validation_generator,
    validation_steps = VALIDATION_STEPS,
    callbacks = [model_save, early_stop, reduce_lr]
)
```

We will briefly explain the two parameters we have not encountered before:

- The training generator yields steps_per_epoch batches per training epoch.
- When the epoch is finished, the validation generator produces validation_steps batches. An example output after calling model.fit() is given here:

```
Epoch 00001: val_loss improved from inf to 0.57514, saving model to
./ EffNetB0_512_8_best_weights.h5
```

Once a model is fitted, we can examine the activations on a sample image using the helper function we wrote at the start. While this is not necessary for successful model execution, it can help determine what sort of features our model is extracting before applying the classification layer at the top:

```
activation_layer_vis(img_tensor, 0)
```

Here is what we might see:

Figure 11.10: Sample activations from a fitted model

We can generate the predictions with `model.predict()`:

```python
ss = pd.read_csv(
    os.path.join(CFG.WORK_DIR, "sample_submission.csv")
)
preds = []

for image_id in ss.image_id:
    image = Image.open(
        os.path.join(CFG.WORK_DIR, "test_images", image_id)
    )
    image = image.resize((CFG.TARGET_SIZE, CFG.TARGET_SIZE))
    image = np.expand_dims(image, axis = 0)
    preds.append(np.argmax(model.predict(image)))

ss['label'] = preds
```

We build the predictions by iterating through the list of images. For each of them, we reshape the image to the required dimensions and pick the channel with the strongest signal (the model predicts class probabilities, of which we pick the largest one with argmax). The final predictions are class numbers, in line with the metric utilized in the competition.

We have now demonstrated a minimal end-to-end pipeline for image classification. Numerous improvements are, of course, possible – for instance, more augmentations, bigger architecture, and callback customization – but the basic underlying template should provide you with a good starting point going forward.

We will now move on to a second popular problem in computer vision: object detection.

Object detection

Object detection is a computer vision/image processing task where we need to identify instances of semantic objects of a certain class in an image or video. In classification problems such as those discussed in the previous section, we simply need to assign a class to each image, whereas in object detection tasks, we want to draw a **bounding box** around an object of interest to locate it within an image.

In this section, we will use data from the *Global Wheat Detection* competition (https://www.kaggle.com/competitions/global-wheat-detection). In this competition, participants had to detect wheat heads, which are spikes on top of plants containing grain. Detection of these in plant images is used to estimate the size and density of wheat heads across crop varieties. We will demonstrate how to train a model for solving this using **YOLOv5**, a well-established model in object detection, which was state-of-the-art until late 2021, when it was, based on preliminary results, surpassed by the YOLOX architecture. YOLOv5 gave rise to extremely competitive results in the competition, and although it was eventually disallowed by the organizers due to licensing issues, it is very well suited for the purpose of this demonstration.

Figure 11.11: Sample image visualizations of detected wheat heads

An important point worth mentioning before we begin is the different formats for bounding box annotations; there are different (but mathematically equivalent) ways of describing the coordinates of a rectangle.

The most common types are COCO, VOC-Pascal, and YOLO. The differences between them are clear from the figure below:

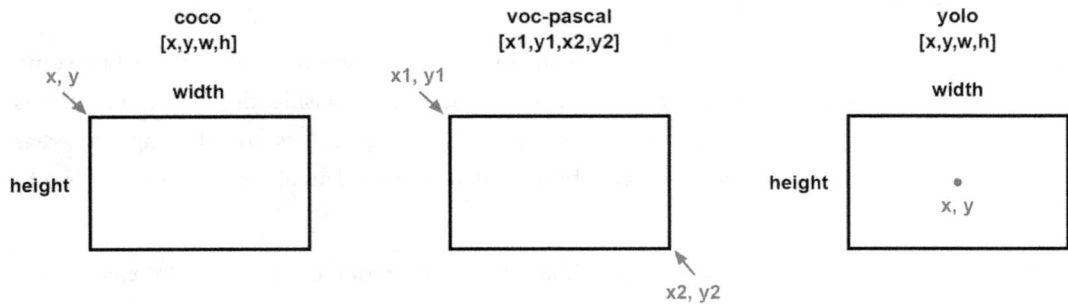

Figure 11.12: Annotation formats for bounding boxes

One more part we need to define is the grid structure: YOLO detects objects by placing a grid over an image and checking for the presence of an object of interest (wheat head, in our case) in any of the cells. The bounding boxes are reshaped to be offset within the relevant cells of the image, and the (x, y, w, h) parameters are scaled to the unit interval:

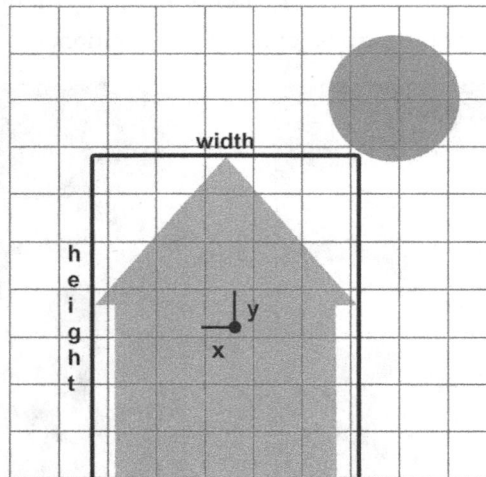

Figure 11.13: YOLO annotation positioning

We start by loading the annotations for our training data:

```
df = pd.read_csv('../input/global-wheat-detection/train.csv')
df.head(3)
```

Let's inspect a few:

	image_id	width	height	bbox	source
0	b6ab77fd7	1024	1024	[834.0, 222.0, 56.0, 36.0]	usask_1
1	b6ab77fd7	1024	1024	[226.0, 548.0, 130.0, 58.0]	usask_1
2	b6ab77fd7	1024	1024	[377.0, 504.0, 74.0, 160.0]	usask_1

Figure 11.14: Training data with annotations

We extract the actual coordinates of the bounding boxes from the bbox column:

```
bboxs = np.stack(df['bbox'].apply(
    lambda x: np.fromstring(x[1:-1], sep=',')))
bboxs
```

Let's look at the array:

```
array([[834., 222., 56., 36.],
       [226., 548., 130., 58.],
       [377., 504., 74., 160.],
       ...,
       [134., 228., 141., 71.],
       [430., 13., 184., 79.],
       [875., 740., 94., 61.]])
```

The next step is to extract the coordinates in YOLO format into separate columns:

```
for i, column in enumerate(['x', 'y', 'w', 'h']):
    df[column] = bboxs[:,i]

df.drop(columns=['bbox'], inplace=True)
df['x_center'] = df['x'] + df['w']/2
df['y_center'] = df['y'] + df['h']/2
df['classes'] = 0

df = df[[
    'image_id','x', 'y', 'w', 'h',
    'x_center','y_center','classes'
]]
df.head(3)
```

The implementation from Ultralytics has some requirements on the structure of the dataset, specifically, where the annotations are stored and the folders for training/validation data.

The creation of the folders in the code below is fairly straightforward, but I encourage anyone curious to consult the official documentation at https://github.com/ultralytics/yolov5/wiki/Train-Custom-Data:

```python
# stratify on source
source = 'train'

# Pick a single fold for demonstration's sake
fold = 0

val_index = set(df[df['fold'] == fold]['image_id'])

# Loop through the bounding boxes per image
for name, mini in tqdm(df.groupby('image_id')):
    # Where to save the files
    if name in val_index:
        path2save = 'valid/'
    else:
        path2save = 'train/'

    # Storage path for labels
    if not os.path.exists(
        'convertor/fold{}/labels/'.format(fold) + path2save
    ):
        os.makedirs(
            'convertor/fold{}/labels/'.format(fold) + path2save
        )

    with open(
        'convertor/fold{}/labels/'.format(fold) +
        path2save + name + ".txt", 'w+'
    ) as f:
        # Normalize the coordinates in accordance with
        # the Yolo format requirements
        row = mini[
            ['classes', 'x_center', 'y_center', 'w', 'h']
        ].astype(float).values
```

```
            row = row / 1024
            row = row.astype(str)
            for j in range(len(row)):
                text = ' '.join(row[j])
                f.write(text)
                f.write("\n")

        if not os.path.exists(
            'convertor/fold{}/images/{}'.format(fold, path2save)
        ):
            os.makedirs('convertor/fold{}/images/{}'.format(fold, path2save))

        # No preprocessing needed for images => copy them as a batch
        sh.copy(
            "../input/global-wheat-detection/{}/{}.jpg".format(source, name),
            'convertor/fold{}/images/{}/{}.jpg'.format(fold, path2save, name)
        )
```

The next thing we do is install the YOLO package itself. If you are running this in a Kaggle notebook or Colab, make sure to double-check GPU is enabled; Yolo installation will actually work without it, but you are likely to run into all sorts of timeouts and memory issues due to CPU versus GPU performance differences:

```
!git clone https://github.com/ultralytics/yolov5 && cd yolov5 && pip
install -r requirements.txt
```

We omit the output, as it is rather extensive. The last bit of preparation needed is the YAML configuration file, where we specify the training and validation data locations and the number of classes. We are only interested in detecting wheat heads and not distinguishing between different types, so we have one class (its name is only provided for notational consistency and can be an arbitrary string in this instance):

```
yaml_text = """train: /kaggle/working/convertor/fold0/images/train/ val: /
kaggle/working/convertor/fold0/images/valid/
nc: 1
names: ['wheat']"""

with open("wheat.yaml", 'w') as f:
    f.write(yaml_text)
%cat wheat.yaml
```

With that, we can start training our model:

```
!python ./yolov5/train.py --img 512 --batch 2 --epochs 3 --workers 2
--data wheat.yaml --cfg "./yolov5/models/yolov5s.yaml" --name yolov5x_
fold0 --cache
```

Unless you are used to launching things from the command line, the incantation above is positively cryptic, so let's discuss its composition in some detail:

- `train.py` is the workhorse script for training a YoloV5 model, starting from pre-trained weights.
- `--img 512` means we want the original images (which, as you can see, we did not pre-process in any way) to be rescaled to 512x512. For a competitive result, you should use a higher resolution, but this code was executed in a Kaggle notebook, which has certain limitations on resources.
- `--batch` refers to the batch size in the training process.
- `--epochs 3` means we want to train the model for three epochs.
- `--workers 2` specifies the number of workers in the data loader. Increasing this number might help performance, but there is a known bug in version 6.0 (the most recent one available in the Kaggle Docker image, as of the time of this writing) when the number of workers is too high, even on a machine where more might be available.
- `--data wheat.yaml` is the file pointing to our data specification YAML file, defined above.
- `--cfg "./yolov5/models/yolov5s.yaml"` specifies the model architecture and the corresponding set of weights to be used for initialization. You can use the ones provided with the installation (check the official documentation for details), or you can customize your own and keep them in the same `.yaml` format.
- `--name` specifies where the resulting model is to be stored.

We break down the output of the training command below. First, the groundwork:

```
Downloading the pretrained weights, setting up Weights&Biases https://
wandb.ai/site integration, GitHub sanity check.
Downloading https://ultralytics.com/assets/Arial.ttf to /root/.config/
Ultralytics/Arial.ttf...
wandb: (1) Create a W&B account
wandb: (2) Use an existing W&B account
wandb: (3) Don't visualize my results
wandb: Enter your choice: (30 second timeout)
```

```
wandb: W&B disabled due to login timeout.
train: weights=yolov5/yolov5s.pt, cfg=./yolov5/models/yolov5s.yaml,
data=wheat.yaml, hyp=yolov5/data/hyps/hyp.scratch-low.yaml, epochs=3,
batch_size=2, imgsz=512, rect=False, resume=False, nosave=False,
noval=False, noautoanchor=False, evolve=None, bucket=, cache=ram,
image_weights=False, device=, multi_scale=False, single_cls=False,
optimizer=SGD, sync_bn=False, workers=2, project=yolov5/runs/train,
name=yolov5x_fold0, exist_ok=False, quad=False, cos_lr=False, label_
smoothing=0.0, patience=100, freeze=[0], save_period=-1, local_rank=-1,
entity=None, upload_dataset=False, bbox_interval=-1, artifact_alias=latest

github: up to date with https://github.com/ultralytics/yolov5 ✅ YOLOv5 🚀
v6.1-76-gc94736a torch 1.9.1 CUDA:0 (Tesla P100-PCIE-16GB, 16281MiB)

hyperparameters: lr0=0.01, lrf=0.01, momentum=0.937, weight_decay=0.0005,
warmup_epochs=3.0, warmup_momentum=0.8, warmup_bias_lr=0.1, box=0.05,
cls=0.5, cls_pw=1.0, obj=1.0, obj_pw=1.0, iou_t=0.2, anchor_t=4.0, fl_
gamma=0.0, hsv_h=0.015, hsv_s=0.7, hsv_v=0.4, degrees=0.0, translate=0.1,
scale=0.5, shear=0.0, perspective=0.0, flipud=0.0, fliplr=0.5, mosaic=1.0,
mixup=0.0, copy_paste=0.0

Weights & Biases: run 'pip install wandb' to automatically track and
visualize YOLOv5 🚀 runs (RECOMMENDED)

TensorBoard: Start with 'tensorboard --logdir yolov5/runs/train', view at
http://localhost:6006/

Downloading https://github.com/ultralytics/yolov5/releases/download/v6.1/
yolov5s.pt to yolov5/yolov5s.pt...
100%|████████████████████████████| 14.1M/14.1M [00:00<00:00,
40.7MB/s]
```

Then comes the model. We see a summary of the architecture, the optimizer setup, and the augmentations used:

```
Overriding model.yaml nc=80 with nc=1
                from  n    params  module
arguments
  0             -1  1     3520  models.common.Conv
[3, 32, 6, 2, 2]
```

```
  1                 -1  1      18560  models.common.Conv
[32, 64, 3, 2]
  2                 -1  1      18816  models.common.C3
[64, 64, 1]
  3                 -1  1      73984  models.common.Conv
[64, 128, 3, 2]
  4                 -1  2     115712  models.common.C3
[128, 128, 2]
  5                 -1  1     295424  models.common.Conv
[128, 256, 3, 2]
  6                 -1  3     625152  models.common.C3
[256, 256, 3]
  7                 -1  1    1180672  models.common.Conv
[256, 512, 3, 2]
  8                 -1  1    1182720  models.common.C3
[512, 512, 1]
  9                 -1  1     656896  models.common.SPPF
[512, 512, 5]
 10                 -1  1     131584  models.common.Conv
[512, 256, 1, 1]
 11                 -1  1          0  torch.nn.modules.upsampling.Upsample
[None, 2, 'nearest']
 12            [-1, 6]  1          0  models.common.Concat
[1]
 13                 -1  1     361984  models.common.C3
[512, 256, 1, False]
 14                 -1  1      33024  models.common.Conv
[256, 128, 1, 1]
 15                 -1  1          0  torch.nn.modules.upsampling.Upsample
[None, 2, 'nearest']
 16            [-1, 4]  1          0  models.common.Concat
[1]
 17                 -1  1      90880  models.common.C3
[256, 128, 1, False]
 18                 -1  1     147712  models.common.Conv
[128, 128, 3, 2]
 19           [-1, 14]  1          0  models.common.Concat
[1]
 20                 -1  1     296448  models.common.C3
```

```
[256, 256, 1, False]
 21                 -1  1     590336  models.common.Conv
[256, 256, 3, 2]
 22          [-1, 10]  1          0  models.common.Concat
[1]
 23                 -1  1    1182720  models.common.C3
[512, 512, 1, False]
 24      [17, 20, 23]  1      16182  models.yolo.Detect
[1, [[10, 13, 16, 30, 33, 23], [30, 61, 62, 45, 59, 119], [116, 90, 156,
198, 373, 326]], [128, 256, 512]]
YOLOv5s summary: 270 layers, 7022326 parameters, 7022326 gradients, 15.8
GFLOPs


Transferred 342/349 items from yolov5/yolov5s.pt
Scaled weight_decay = 0.0005
optimizer: SGD with parameter groups 57 weight (no decay), 60 weight, 60
bias
albumentations: Blur(always_apply=False, p=0.01, blur_limit=(3, 7)),
MedianBlur(always_apply=False, p=0.01, blur_limit=(3, 7)), ToGray(always_
apply=False, p=0.01), CLAHE(always_apply=False, p=0.01, clip_limit=(1,
4.0), tile_grid_size=(8, 8))
train: Scanning '/kaggle/working/convertor/fold0/labels/train' images and
labels
train: New cache created: /kaggle/working/convertor/fold0/labels/train.
cache
train: Caching images (0.0GB ram): 100%|██████████| 51/51 [00:00&lt;00:00,
76.00it/
val: Scanning '/kaggle/working/convertor/fold0/labels/valid' images and
labels..
val: New cache created: /kaggle/working/convertor/fold0/labels/valid.cache
val: Caching images (2.6GB ram): 100%|██████████| 3322/3322
[00:47&lt;00:00, 70.51i
Plotting labels to yolov5/runs/train/yolov5x_fold0/labels.jpg...


AutoAnchor: 6.00 anchors/target, 0.997 Best Possible Recall (BPR). Current
anchors are a good fit to dataset ✔
Image sizes 512 train, 512 val
Using 2 dataloader workers
```

This is followed by the actual training log:

```
Starting training for 3 epochs...


     Epoch   gpu_mem        box        obj        cls     labels   img_size
       0/2    0.371G     0.1196    0.05478          0         14        512:
100%|     |
               Class     Images     Labels          P          R
mAP@.5 mAP@WARNING: NMS time limit 0.120s exceeded
               Class     Images     Labels          P          R
mAP@.5 mAP@
                 all       3322     147409    0.00774     0.0523
0.00437    0.000952
     Epoch   gpu_mem        box        obj        cls     labels   img_size
       1/2    0.474G     0.1176    0.05625          0          5        512:
100%|     |
               Class     Images     Labels          P          R
mAP@.5 mAP@WARNING: NMS time limit 0.120s exceeded
               Class     Images     Labels          P          R
mAP@.5 mAP@WARNING: NMS time limit 0.120s exceeded
               Class     Images     Labels          P          R
mAP@.5 mAP@
                 all       3322     147409    0.00914     0.0618
0.00493    0.00108
     Epoch   gpu_mem        box        obj        cls     labels   img_size
       2/2    0.474G     0.1146    0.06308          0         12        512:
100%|     |
               Class     Images     Labels          P          R
mAP@.5 mAP@
                 all       3322     147409    0.00997     0.0674
0.00558    0.00123


3 epochs completed in 0.073 hours.
Optimizer stripped from yolov5/runs/train/yolov5x_fold0/weights/last.pt,
14.4MB
```

```
Optimizer stripped from yolov5/runs/train/yolov5x_fold0/weights/best.pt,
14.4MB
Validating yolov5/runs/train/yolov5x_fold0/weights/best.pt...
Fusing layers...
YOLOv5s summary: 213 layers, 7012822 parameters, 0 gradients, 15.8 GFLOPs
                Class      Images      Labels          P          R
mAP@.5 mAP@WARNING: NMS time limit 0.120s exceeded
                Class      Images      Labels          P          R
mAP@.5 mAP@WARNING: NMS time limit 0.120s exceeded
                Class      Images      Labels          P          R
mAP@.5 mAP@WARNING: NMS time limit 0.120s exceeded
                Class      Images      Labels          P          R
mAP@.5 mAP@WARNING: NMS time limit 0.120s exceeded
                Class      Images      Labels          P          R
mAP@.5 mAP@WARNING: NMS time limit 0.120s exceeded
                Class      Images      Labels          P          R
mAP@.5 mAP@WARNING: NMS time limit 0.120s exceeded
                Class      Images      Labels          P          R
mAP@.5 mAP@WARNING: NMS time limit 0.120s exceeded
                Class      Images      Labels          P          R
mAP@.5 mAP@WARNING: NMS time limit 0.120s exceeded
                Class      Images      Labels          P          R
mAP@.5 mAP@WARNING: NMS time limit 0.120s exceeded
                Class      Images      Labels          P          R
mAP@.5 mAP@WARNING: NMS time limit 0.120s exceeded
                Class      Images      Labels          P          R
mAP@.5 mAP@
                  all        3322      147409     0.00997     0.0673
0.00556     0.00122
Results saved to yolov5/runs/train/yolov5x_fold0
```

The results from both training and validation stages can be examined; they are stored in the yolov5 folder under ./yolov5/runs/train/yolov5x_fold0:

Figure 11.15: Validation data with annotations

Once we have trained the model, we can use the weights from the best-performing model (Yolov5 has a neat functionality of automatically keeping both the best and last epoch model, storing them as best.pt and last.pt) to generate predictions on the test data:

```
!python ./yolov5/detect.py --weights ./yolov5/runs/train/yolov5x_fold0/
weights/best.pt --img 512 --conf 0.1 --source /kaggle/input/global-wheat-
detection/test --save-txt --save-conf --exist-ok
```

We will discuss the parameters that are specific to the inference stage:

- --weights points to the location of the best weights from our model trained above.
- --conf 0.1 specifies which candidate bounding boxes generated by the model should be kept. As usual, it is a compromise between precision and recall (too low a threshold gives a high number of false positives, while moving the threshold too high means we might not find any wheat heads at all).
- --source is the location of the test data.

The labels created for our test images can be inspected locally:

```
!ls ./yolov5/runs/detect/exp/labels/
```

This is what we might see:

```
2fd875eaa.txt 53f253011.txt aac893a91.txt f5a1f0358.txt 348a992bb.txt
796707dd7.txt cc3532ff6.txt
```

Let's look at an individual prediction:

```
!cat 2fd875eaa.txt
```

It has the following format:

```
0 0.527832 0.580566 0.202148 0.838867 0.101574
0 0.894531 0.587891 0.210938 0.316406 0.113519
```

This means that in image 2fd875eaa, our trained model detected two bounding boxes (their coordinates are entries 2–5 in the row), with confidence scores above 0.1 given at the end of the row.

How do we go about combining the predictions into a submission in the required format? We start by defining a helper function that helps us convert the coordinates from the yolo format to COCO (as required in this competition): it is a matter of simple rearrangement of the order and normalizing to the original range of values by multiplying the fractions by the image size:

```
def convert(s):
    x = int(1024 * (s[1] - s[3]/2))
    y = int(1024 * (s[2] - s[4]/2))
    w = int(1024 * s[3])
    h = int(1024 * s[4])

    return(
        str(s[5]) + ' ' + str(x) + ' ' + str(y) + ' ' +
        str(w) + ' ' + str(h)
    )
```

We then proceed to generate a submission file:

1. We loop over the files listed above.

2. For each file, all rows are converted into strings in the required format (one row represents one bounding box detected).

3. The rows are then concatenated into a single string corresponding to this file. The code is as follows:

```python
with open('submission.csv', 'w') as myfile:

    # Prepare submission
    wfolder = './yolov5/runs/detect/exp/labels/'
    for f in os.listdir(wfolder):
        fname = wfolder + f
        xdat = pd.read_csv(fname, sep = ' ', header = None)
        outline = f[:-4] + ' ' + ' '.join(
            list(xdat.apply(lambda s: convert(s), axis = 1))
        )
        myfile.write(outline + '\n')

    myfile.close()
```

Let's see what it looks like:

```
!cat submission.csv
```

This is how the result appears:

```
53f253011 0.100472 61 669 961 57 0.106223 0 125 234 183 0.1082 96 696 928
126 0.108863 515 393 86 161 0.11459 31 0 167 209 0.120246 517 466 89 147
aac893a91 0.108037 376 435 325 188
796707dd7 0.235373 684 128 234 113
cc3532ff6 0.100443 406 752 144 108 0.102479 405 87 4 89 0.107173 576 537
138 94 0.113459 256 498 179 211 0.114847 836 618 186 65 0.121121 154 544
248 115 0.125105 40 567 483 199
2fd875eaa 0.101398 439 163 204 860 0.112546 807 440 216 323
348a992bb 0.100572 0 10 440 298 0.101236 344 445 401 211
f5a1f0358 0.102549 398 424 295 96
```

The generated `submission.csv` file completes our pipeline.

In this section, we have demonstrated how to use YOLOv5 to solve the object detection problem, including handling annotations in various formats, customizing a model for a specific task, training it, and evaluating the results.

Based on this knowledge, you should be able to start working with object detection problems. We now move on to the third popular class of computer vision tasks: semantic segmentation.

Semantic segmentation

The easiest way to think about **segmentation** is that it classifies each pixel in an image, assigning it to a corresponding class; combined, those pixels form areas of interest, such as regions with disease on an organ in medical images. By contrast, object detection (discussed in the previous section) classifies patches of an image into different object classes and creates bounding boxes around them.

We will demonstrate the modeling approach using data from the *Sartorius – Cell Instance Segmentation* competition (`https://www.kaggle.com/c/sartorius-cell-instance-segmentation`). In this task, the participants were tasked with training models for instance segmentation of neural cells using a set of microscopy images.

Our solution will be built around Detectron2, a library created by Facebook AI Research that supports multiple detection and segmentation algorithms.

> Detectron2 is a successor to the original Detectron library (`https://github.com/facebookresearch/Detectron/`) and the Mask R-CNN project (`https://github.com/facebookresearch/maskrcnn-benchmark/`).
>
> Installing Detectron2 can be challenging, especially on native Windows. We recommend using a Linux environment, Windows Subsystem for Linux (WSL), or Google Colab. Please refer to the official Detectron2 installation guide for detailed instructions:
>
> `https://detectron2.readthedocs.io/en/latest/tutorials/install.html`

We begin by installing the extra packages:

```
!pip install pycocotools
!pip install 'git+https://github.com/facebookresearch/detectron2.git'
```

We install pycocotools (`https://github.com/cocodataset/cocoapi/tree/master/PythonAPI/pycocotools`), which we will use to format the annotations, and Detectron2 (`https://github.com/facebookresearch/detectron2`), our primary tool for this task.

Before we can train our model, we need some preparation: the annotations must be converted from the **run-length encoding (RLE)** format provided by the organizers to the COCO format required as input for Detectron2. The basic idea behind RLE is saving space: creating a segmentation means marking a group of pixels in a specific manner. Since an image can be thought of as an array, this area can be denoted by a series of straight lines (row- or column-wise).

You can encode each of those lines by listing the indices, or by specifying a starting position and the length of the subsequent contiguous block. A visual example is given below:

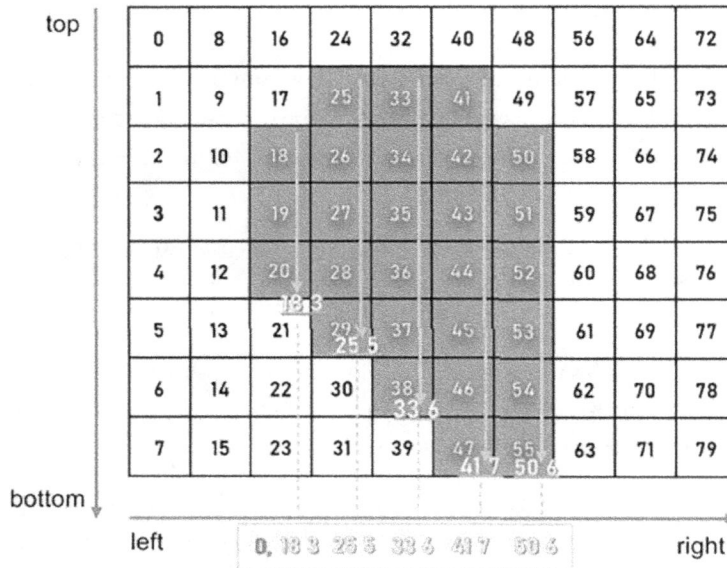

Figure 11.16: Visual representation of RLE

Microsoft's **Common Objects in Context (COCO)** format is a specific JSON structure dictating how labels and metadata are saved for an image dataset. Below, we demonstrate how to convert RLE to COCO and combine it with a *k*-fold validation split, resulting in the required train/validation pairs of JSON files for each fold.

Let's begin:

```
# from pycocotools.coco import COCO
import skimage.io as io
```

```python
import matplotlib.pyplot as plt
from pathlib import Path
from PIL import Image

import pandas as pd
import numpy as np
from tqdm.notebook import tqdm
import json
import itertools
from sklearn.model_selection import GroupKFold

# Config
class CFG:
    data_path = ('../input/sartorius-cell-instance-segmentation/')
    nfolds = 5
```

We need three functions to go from RLE to COCO:

1. First, we need to convert from RLE to a binary mask:

    ```python
    # From https://www.kaggle.com/stainsby/fast-tested-rle
    def rle_decode(mask_rle, shape):
        '''
        mask_rle: run-length as string formatted (start length)
        shape: (height,width) of array to return
        Returns numpy array, 1 - mask, 0 - background
        '''
        s = mask_rle.split()
        starts, lengths = [
            np.asarray(x, dtype=int)
            for x in (s[0:][::2], s[1:][::2])
        ]
        starts -= 1
        ends = starts + lengths

        img = np.zeros(shape[0] * shape[1], dtype=np.uint8)
        for lo, hi in zip(starts, ends):
            img[lo:hi] = 1
        return img.reshape(shape)  # Needed to align to RLE direction
    ```

2. The second one converts a binary mask to RLE:

```python
# From https://stackoverflow.com/questions/49494337/encode-numpy-
array-using-uncompressed-rle-for-coco-dataset
def binary_mask_to_rle(binary_mask):
    rle = {'counts': [], 'size': list(binary_mask.shape)}
    counts = rle.get('counts')
    for i, (value, elements) in enumerate(
        itertools.groupby(binary_mask.ravel(order='F'))
    ):
        if i == 0 and value == 1:
            counts.append(0)
        counts.append(len(list(elements)))
    return rle
```

3. Finally, we combine the two in order to produce the COCO output:

```python
def coco_structure(train_df):
    cat_ids = {
        name: id+1
        for id, name in enumerate( train_df.cell_type.unique())
    }

    cats = [
        {'name': name, 'id': id}
        for name, id in cat_ids.items()
    ]

    images = [
        {
            'id': id,
            'width': row.width,
            'height': row.height,
            'file_name':f'train/{id}.png'
        }
        for id, row in train_df.groupby('id')
        .agg('first').iterrows()
    ]

    annotations = []
```

```
        for idx, row in tqdm(train_df.iterrows()):
            mk = rle_decode(row.annotation, (row.height, row.width))
            ys, xs = np.where(mk)
            x1, x2 = min(xs), max(xs)
            y1, y2 = min(ys), max(ys)
            enc =binary_mask_to_rle(mk)
            seg = {
                'segmentation':enc,
                'bbox': [int(x1), int(y1), int(x2-x1+1), int(y2-y1+1)],
                'area': int(np.sum(mk)),
                'image_id':row.id,
                'category_id':cat_ids[row.cell_type],
                'iscrowd':0,
                'id':idx
            }
            annotations.append(seg)
    return {
        'categories':cats,
        'images':images,
        'annotations':annotations
    }
```

We split our data into non-overlapping folds:

```
train_df = pd.read_csv(CFG.data_path + 'train.csv')


gkf = GroupKFold(n_splits = CFG.nfolds)


train_df["fold"] = -1
y = train_df.width.values


for f, (t_, v_) in enumerate(
    gkf.split(X=train_df, y=y, groups=train_df.id.values)
):
    train_df.loc[v_, "fold"] = f


fold_id = train_df.fold.copy()
```

4. We can now loop over the folds:

```
all_ids = train_df.id.unique()

# For fold in range(CFG.nfolds):
for fold in range(4,5):
    train_sample = train_df.loc[fold_id != fold]
    root = coco_structure(train_sample)

    with open(
        'annotations_train_f' + str(fold) + '.json',
        'w', encoding='utf-8'
    ) as f:
        json.dump(root, f, ensure_ascii=True, indent=4)

    valid_sample = train_df.loc[fold_id == fold]
    print('fold ' + str(fold) + ': produced')

for fold in range(4,5):
    train_sample = train_df.loc[fold_id == fold]
    root = coco_structure(train_sample)

    with open(
        'annotations_valid_f' + str(fold) + '.json',
        'w', encoding='utf-8'
    ) as f:
        json.dump(root, f, ensure_ascii=True, indent=4)

    valid_sample = train_df.loc[fold_id == fold]
    print('fold ' + str(fold) + ': produced')
```

The reason the loop has to be executed in pieces is the size limit of the Kaggle environment: the maximum size of notebook output is limited to 20 GB. Five folds, each with two files (training/validation), would result in a total of 10 JSON files, exceeding this limit.

Such practical considerations are worth keeping in mind when running code in a Kaggle notebook. However, for such "preparatory" work, you can, of course, produce the results elsewhere and upload them as Kaggle Datasets afterward.

With the splits produced, we can move toward training a Detectron2 model for our dataset. As usual, we start by loading the necessary packages:

```
from datetime import datetime
import os

import pandas as pd
import numpy as np
import pycocotools.mask as mask_util
import detectron2
from pathlib import Path
import random, cv2, os
import matplotlib.pyplot as plt

# Import some common detectron2 utilities
from detectron2 import model_zoo
from detectron2.engine import DefaultPredictor, DefaultTrainer
from detectron2.config import get_cfg
from detectron2.utils.visualizer import Visualizer, ColorMode
from detectron2.data import MetadataCatalog, DatasetCatalog
from detectron2.data.datasets import register_coco_instances
from detectron2.utils.logger import setup_logger
from detectron2.evaluation.evaluator import DatasetEvaluator
from detectron2.engine import BestCheckpointer
from detectron2.checkpoint import DetectionCheckpointer

setup_logger()

import torch
```

While the number of imports from Detectron2 can seem intimidating at first, their function will become clear as we progress with the task definition; we start by specifying paths to the input data folder, annotations folder, and a YAML file defining our preferred model architecture:

```
class CFG:
    wfold = 4
    data_folder = '../input/sartorius-cell-instance-segmentation/'
    anno_folder = '../input/sartoriusannotations/'
    model_arch = 'mask_rcnn_R_50_FPN_3x.yaml'
```

```
nof_iters = 10000
seed = 45
```

One point worth mentioning here is the iterations parameter (`nof_iters` above). Usually, model training is parametrized in terms of the number of epochs – in other words, complete passes through the training data. Detectron2 is engineered differently: one iteration refers to one mini-batch, and different mini-batch sizes are used in various parts of the model.

In order to ensure the results are reproducible, we fix the random seeds used by different components of the model:

```
def seed_everything(seed):
    random.seed(seed)
    os.environ['PYTHONHASHSEED'] = str(seed)
    np.random.seed(seed)
    torch.manual_seed(seed)
    torch.cuda.manual_seed(seed)
    torch.backends.cudnn.deterministic = True


seed_everything(CFG.seed)
```

The competition metric was the mean average precision at different **intersection over union (IoU)** thresholds. As a refresher from *Chapter 5, Competition Tasks and Metrics*, the IoU of a proposed set of object pixels and a set of true object pixels is calculated as:

$$IoU(A, B) = |A \cap B| / |A \cup B|$$

The metric sweeps over a range of IoU thresholds, calculating an average precision value at each point. The threshold values range from 0.5 to 0.95, with increments of 0.05.

At each threshold value, a precision value is calculated based on the number of **true positives (TPs)**, **false negatives (FNs)**, and **false positives (FPs)** resulting from comparing the predicted object with all ground truth objects. Lastly, the score returned by the competition metric is the mean taken over the individual average precisions of each image in the test dataset.

Below, we define the functions necessary to calculate the metric and use it directly inside the model as the objective function:

```
# Taken from https://www.kaggle.com/theoviel/competition-metric-map-iou
def precision_at(threshold, iou):
    matches = iou > threshold
```

```
        true_positives = np.sum(matches, axis=1) == 1 # Correct objects
        false_positives = np.sum(matches, axis=0) == 0 # Missed objects
        false_negatives = np.sum(matches, axis=1) == 0 # Extra objects
        return (
            np.sum(true_positives),
            np.sum(false_positives),
            np.sum(false_negatives)
        )
def score(pred, targ):
    pred_masks = pred['instances'].pred_masks.cpu().numpy()
    enc_preds = [
        mask_util.encode(np.asarray(p, order='F'))
        for p in pred_masks
    ]
    enc_targs = list(map(lambda x:x['segmentation'], targ))
    ious = mask_util.iou(enc_preds, enc_targs, [0]*len(enc_targs))
    prec = []
    for t in np.arange(0.5, 1.0, 0.05):
        tp, fp, fn = precision_at(t, ious)
        p = tp / (tp + fp + fn)
        prec.append(p)
    return np.mean(prec)
```

With the metric defined, we can use it in the model:

```
class MAPIOUEvaluator(DatasetEvaluator):
    def __init__ (self, dataset_name):
    dataset_dicts = DatasetCatalog.get(dataset_name)
    self.annotations_cache = {
        item['image_id']:item['annotations']
        for item in dataset_dicts
    }

    def reset(self):
        self.scores = []

    def process(self, inputs, outputs):
        for inp, out in zip(inputs, outputs):
```

```
            if len(out['instances']) == 0:
                self.scores.append(0)
            else:
                targ = self.annotations_cache[inp['image_id']]
                self.scores.append(score(out, targ))


    def evaluate(self):
        return {"MaP IoU": np.mean(self.scores)}
```

This gives us the basis for creating a `Trainer` object, which is the workhorse of our solution built around Detectron2:

```
class Trainer(DefaultTrainer):
    @classmethod
    def build_evaluator(cls, cfg, dataset_name, output_folder=None):
        return MAPIOUEvaluator(dataset_name)


    def build_hooks(self):
        # copy of cfg
        cfg = self.cfg.clone()

        # build the original model hooks
        hooks = super().build_hooks()

        # add the best checkpointer hook
        hooks.insert(
            -1,
            BestCheckpointer(
                cfg.TEST.EVAL_PERIOD,
                DetectionCheckpointer(self.model, cfg.OUTPUT_DIR),
                "MaP IoU",
                "max",
            )
        )
        return hooks
```

We now proceed to load the training/validation data in Detectron2 style:

```
from pathlib import Path

dataDir=Path(CFG.data_folder)
register_coco_instances(
    'sartorius_train',{}, CFG.anno_folder +
    'annotations_train_f' + str(CFG.wfold) + '.json', dataDir)
register_coco_instances(
    'sartorius_val',{}, CFG.anno_folder +
    'annotations_valid_f' + str(CFG.wfold) + '.json', dataDir)

metadata = MetadataCatalog.get('sartorius_train')

train_ds = DatasetCatalog.get('sartorius_train')
```

Before we instantiate a Detectron2 model, we need to take care of configuring it. Most of the values can be left at the default values (at least, in a first pass); if you decide to tinker a bit more, start with BATCH_SIZE_PER_IMAGE (for increased generalization performance) and SCORE_THRESH_TEST (to limit false negatives):

```
cfg = get_cfg()
cfg.INPUT.MASK_FORMAT='bitmask'
cfg.merge_from_file(
    model_zoo.get_config_file(
        'COCO-InstanceSegmentation/' + CFG.model_arch
    )
)
cfg.DATASETS.TRAIN = ("sartorius_train",)
cfg.DATASETS.TEST = ("sartorius_val",)
cfg.DATALOADER.NUM_WORKERS = 2
cfg.MODEL.WEIGHTS = model_zoo.get_checkpoint_url(
    'COCO-InstanceSegmentation/' + CFG.model_arch
)
cfg.SOLVER.IMS_PER_BATCH = 2
cfg.SOLVER.BASE_LR = 0.001
cfg.SOLVER.MAX_ITER = CFG.nof_iters
cfg.SOLVER.STEPS = []
cfg.MODEL.ROI_HEADS.BATCH_SIZE_PER_IMAGE = 512
```

```
cfg.MODEL.ROI_HEADS.NUM_CLASSES = 3
cfg.MODEL.ROI_HEADS.SCORE_THRESH_TEST = .4
cfg.TEST.EVAL_PERIOD = len(DatasetCatalog.get('sartorius_train'))
// cfg.SOLVER.IMS_PER_BATCH
```

Training a model is straightforward:

```
os.makedirs(cfg.OUTPUT_DIR, exist_ok=True)


trainer = Trainer(cfg)
trainer.resume_or_load(resume=False)
trainer.train()
```

You will notice that the output during training is rich in information about the progress of the procedure:

```
[01/06 22:26:36 d2.data.datasets.coco]: Loading ../input/sartorius-annotations/annotations_t
rain_f4.json takes 1.16 seconds.
[01/06 22:26:36 d2.data.datasets.coco]: Loaded 485 images in COCO format from ../input/sarto
rius-annotations/annotations_train_f4.json
[01/06 22:26:38 d2.data.build]: Removed 0 images with no usable annotations. 485 images lef
t.
[01/06 22:26:38 d2.data.build]: Distribution of instances among all 3 categories:
|  category  | #instances  |  category  | #instances  |  category  | #instances  |
|:----------:|:-----------:|:----------:|:-----------:|:----------:|:-----------:|
|   shsy5y   |    41952    |   astro    |    8360     |    cort    |    8556     |
|            |             |            |             |            |             |
|   total    |    58868    |            |             |            |             |
[01/06 22:26:38 d2.data.dataset_mapper]: [DatasetMapper] Augmentations used in training: [Re
sizeShortestEdge(short_edge_length=(640, 672, 704, 736, 768, 800), max_size=1333, sample_sty
le='choice'), RandomFlip()]
[01/06 22:26:38 d2.data.build]: Using training sampler TrainingSampler
[01/06 22:26:38 d2.data.common]: Serializing 485 elements to byte tensors and concatenating
them all ...
[01/06 22:26:38 d2.data.common]: Serialized dataset takes 6.79 MiB

model_final_f10217.pkl: 178MB [00:04, 35.8MB/s]
```

Figure 11.17: Training output from Detectron2

Once the model is trained, we can save the weights and use them for inference (potentially in a separate notebook – see the discussion earlier in this chapter) and submission preparation. We start by adding new parameters that allow us to regularize the prediction, setting confidence thresholds and minimal mask sizes:

```
THRESHOLDS = [.18, .35, .58]
MIN_PIXELS = [75, 150, 75]
```

We need a helper function for encoding a single mask into RLE format:

```
def rle_encode(img):
    '''
    img: numpy array, 1 - mask, 0 - background
    Returns run length as string formatted
    '''
    pixels = img.flatten()
    pixels = np.concatenate([[0], pixels, [0]])
    runs = np.where(pixels[1:] != pixels[:-1])[0] + 1
    runs[1::2] -= runs[::2]
    return ' '.join(str(x) for x in runs)
```

Below is the main function for producing all masks per image, filtering out the dubious ones (with confidence scores below THRESHOLDS) with small areas (containing fewer pixels than MIN_PIXELS):

```
def get_masks(fn, predictor):
    im = cv2.imread(str(fn))
    pred = predictor(im)

    pred_class = torch.mode(
        pred['instances'].pred_classes
    )[0]

    take = pred['instances'].scores >= THRESHOLDS[pred_class]

    pred_masks = pred['instances'].pred_masks[take]
    pred_masks = pred_masks.cpu().numpy()

    res = []
    used = np.zeros(im.shape[:2], dtype=int)

    for mask in pred_masks:
        mask = mask * (1-used)
        # Skip predictions with small area
        if mask.sum() >= MIN_PIXELS[pred_class]:
            used += mask
```

```
            res.append(rle_encode(mask))

    return res
```

We then prepare the lists where image IDs and masks will be stored:

```
from pathlib import Path

dataDir=Path(CFG.data_folder)

ids, masks=[],[]
test_names = list((dataDir/'test').iterdir())
```

Competitions with large image sets – like the ones discussed in this section – often require training models for longer than 9 hours, which is the time limit imposed in Code competitions (see https://www.kaggle.com/docs/competitions). This means that training a model and running inference within the same notebook becomes impossible. A typical workaround is to run a training notebook/script first as a standalone notebook in Kaggle, Google Colab, GCP, or locally. The output of this first notebook (the trained weights) is used as input to the second one – in other words, to define the model used for predictions.

We proceed in that manner by loading the weights of our trained model:

```
cfg = get_cfg()
cfg.merge_from_file(
    model_zoo.get_config_file(
        "COCO-InstanceSegmentation/"+ CFG.arch+".yaml"
    )
)

cfg.INPUT.MASK_FORMAT = 'bitmask'
cfg.MODEL.ROI_HEADS.NUM_CLASSES = 3

cfg.MODEL.WEIGHTS = (
    CFG.model_folder + 'output/model_best.pth'
)

cfg.MODEL.ROI_HEADS.SCORE_THRESH_TEST = 0.5
cfg.TEST.DETECTIONS_PER_IMAGE = 1000

predictor = DefaultPredictor(cfg)
```

We can visualize some of the predictions:

```
encoded_masks = get_masks(test_names[0], predictor)


_, axs = plt.subplots(1,2, figsize = (40, 15)) axs[1].imshow(cv2.
imread(str(test_names[0])))


for enc in encoded_masks:
    dec = rle_decode(enc)
    axs[0].imshow(np.ma.masked_where(dec == 0, dec))
```

Here is an example:

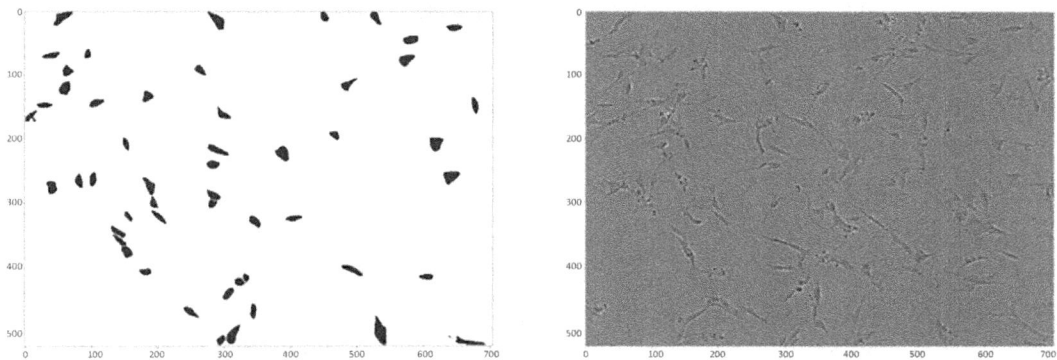

Figure 11.18: Visualizing a sample prediction from Detectron2 alongside the source image

With the helper functions defined above, producing the masks in RLE format for submission is straightforward:

```
for fn in test_names:
    encoded_masks = get_masks(fn, predictor)
    for enc in encoded_masks:
        ids.append(fn.stem)
        masks.append(enc)


pd.DataFrame({
    'id':ids,
    'predicted':masks
}).to_csv('submission.csv', index=False)


pd.read_csv('submission.csv').head()
```

Here are the first few rows of the final submission:

	id	predicted
0	7ae19de7bc2a	139541 4 140244 7 140948 8 141652 8 142356 9 1...
1	7ae19de7bc2a	96418 4 97121 6 97825 7 98529 8 99233 8 99937...
2	7ae19de7bc2a	26627 14 27329 17 28031 19 28733 21 29435 23 3...
3	7ae19de7bc2a	148230 2 148931 6 149633 9 150336 11 151039 13...
4	7ae19de7bc2a	224918 2 225620 7 226324 9 227027 12 227731 13...

Figure 11.19: Formatted submission from a trained Detectron2 model

We have reached the end of the section. The pipeline above demonstrates how to set up a semantic segmentation model and train it. We have used a small number of iterations, but in order to achieve competitive results, longer training is necessary.

To wrap up this chapter, let's see what Kaggler Laura Fink has to say about her time on the platform. As well as being a Notebooks Grandmaster and producing many masterful notebooks, she is also a senior data scientist at H2O.ai.

Interview: Laura Fink

```
https://www.kaggle.com/allunia
```

What's your favorite kind of competition and why? In terms of techniques and solving approaches, what is your specialty on Kaggle?

My favorite competitions are those that aim to yield something good to humanity. I especially like all healthcare-related challenges. Nonetheless, each competition feels like an adventure for me with its own puzzles to be solved. I really enjoy learning new skills and exploring new kinds of datasets or problems. Consequently, I'm not focused on specific techniques but rather on learning something new. I think I'm known for my strengths in EDA.

How do you approach a Kaggle competition? How different is this approach from what you do in your day-to-day work?

When entering a competition, I start by reading the problem statement and the data description. After browsing through the forum and public notebooks for collecting ideas, I usually start by developing my own solutions. In the initial phase, I spend some time on EDA to search for hidden groups and get some intuition. This helps quite a lot in setting up a proper validation strategy, which I believe is the foundation of all remaining steps. Then, I start to iterate through different parts of the machine learning pipeline, like feature engineering or preprocessing, improving the model architecture, asking questions about the data collection, searching for leakages, doing more EDA, or building ensembles. I try to improve my solution in a greedy fashion. Kaggle competitions are very dynamic, and one needs to try out diverse ideas and different solutions to survive in the end.

This is definitely different from my day-to-day work, where the focus is more on gaining insights from data and finding simple but effective solutions to improve business processes. Here, the task is often more complex than the models used. The problem to be solved has to be defined very clearly, which means that one has to discuss with experts of different backgrounds which goals should be reached, which processes are involved, and how the data needs to be collected or fused. Compared to Kaggle competitions, my daily work needs much more communication than machine learning skills.

Tell us about a particularly challenging competition you entered, and what insights you used to tackle the task.

The G2Net Gravitational Wave Detection competition was one of my favorites. The goal was to detect simulated gravitational wave signals that were hidden in noise originating from detector components and terrestrial forces. An important insight during this competition was that you should have a critical look at standard ways to analyze data and try out your own ideas. In the papers I read, the data was prepared mainly by using the Fourier or Constant-Q transform after whitening the data and applying a bandpass filter.

It came out very quickly that whitening was not helpful, as it used spline interpolation of the power spectral density, which was itself very noisy. Fitting polynomials to small subsets of noisy data adds another source of errors because of overfitting.

After dropping the whitening, I tried out different hyperparameters of the Constant-Q transform, which turned out to be the leading method in the forum and public notebooks for a long time. As there were two sources of gravitational waves that can be covered by different ranges of Q-values, I tried out an ensemble of models that differed in these hyperparameters. This turned out to be helpful in improving my score, but then I reached a limit. The Constant-Q transform applies a series of filters to time series and transforms them into the frequency domain. I started to ask myself if there was a method that does these filtering tasks in a better, more flexible way. It was at the same time that the idea of using 1-dimensional CNNs came up in the community, and I loved it. We all know that filters of 2-dimensional CNNs are able to detect edges, lines, and textures in given image data. The same could be done with "classical" filters like the Laplace or Sobel filter. For this reason, I asked myself: can't we use the 1-dimensional CNN to learn the most important filters on its own, instead of applying transformations that are already fixed somehow?

I was not able to get my 1-dimensional CNN solution to work, but it turned out that many top teams managed it well. The G2Net competition was one of my favorites, even though I missed out on the goal of winning a medal. However, the knowledge I gained along the way and the lessons I learned about so-called standard approaches were very valuable.

Has Kaggle helped you in your career? If so, how?

I started my first job after university as a Java software developer, even though I already had my first contact with machine learning during my master's thesis. I was interested in doing more data analytics, but at that time, there were almost no data science jobs, or they were not named this way. When I heard about Kaggle for the first time, I was trapped right from the start. Since then, I often found myself on Kaggle during the evenings to have some fun. It was not my intent to change my position at that time, but then a research project came up that needed machine learning skills. I was able to show that I was a suitable candidate for this project because of the knowledge I gained by participating on Kaggle. This turned out to be the entry point for my data science career.

Kaggle has always been a great place for me to try out ideas, learn new methods and tools, and gain practical experience. The skills I obtained this way have been quite helpful for data science projects at work. It's like a boost of knowledge, as Kaggle provides a sandbox for you to try out different ideas and to be creative without risk. Failing in a competition means that there was at least one lesson to learn, but failing in a project can have a huge negative impact on yourself and other people.

Besides taking part in competitions, another great way to build up your portfolio is to write notebooks. In doing so, you can show the world how you approach problems and how to communicate insights and conclusions. The latter is very important when you have to work with management, clients, and experts from different backgrounds.

In your experience, what do inexperienced Kagglers often overlook? What do you know now that you wish you'd known when you first started?

I think many beginners who enter competitions are seduced by the public leaderboard and build their models without having a good validation strategy. While measuring their success on the leaderboard, they are likely overfitting to the public test data. After the end of the competition, their models are not able to generalize to the unseen private test data, and they often fall hundreds of places. I still remember how frustrated I was during the Mercedes-Benz Greener Manufacturing competition, as I was not able to climb up the public leaderboard. But when the final standings came out, it was a big surprise how many people were shuffled up and down the leaderboard. Since then, I have always kept in mind that a proper validation scheme is very important for managing the challenges of under- and overfitting.

What mistakes have you made in competitions in the past?

My biggest mistake so far was spending too much time and effort on the details of my solution at the beginning of a competition. Indeed, it's much better to iterate fast through diverse and different ideas after building a proper validation strategy. That way, it's easier and faster to find promising directions for improvements, and the danger of getting stuck somewhere is much smaller.

Are there any particular tools or libraries that you would recommend using for data analysis or machine learning?

There are a lot of common tools and libraries you can learn and practice when becoming active in the Kaggle community, and I can only recommend them all. It's important to stay flexible and to learn about their advantages and disadvantages. This way, your solutions don't depend on your tools, but rather on your ideas and creativity.

What's the most important thing someone should keep in mind or do when they're entering a competition?

Data science is not about building models, but rather about understanding the data and the way it was collected. Many competitions I have entered so far have shown leakages or had hidden groups in the test data that one could find with EDA.

Exploring a capstone case study: CZII – CryoET Object Identification competition

Having established a foundation with 2D classification, detection, and segmentation in the previous sections of this chapter, we now turn to a domain where these ideas must be extended to **volumetric** data, stricter evaluation regimes, and more stringent computational constraints. In this final section, the recent CZII – CryoET Object Identification competition (`https://www.kaggle.com/competitions/czii-cryo-et-object-identification`) serves as a capstone case study, introducing more advanced techniques (3D U-Nets, patch-wise training/inference, recall-weighted metrics, and robust ensembling) to demonstrate how to adapt our toolkit to substantially more complex scientific imagery. Due to the significant computational and data storage requirements for processing 3D tomograms, this section focuses on the strategic and conceptual approaches used by top competitors. A fully runnable code example is beyond the scope of a standard notebook environment. However, you are encouraged to explore the official competition notebooks (linked here: `https://www.kaggle.com/competitions/czii-cryo-et-object-identification/code`) to see these principles in action.

This capstone transitions from familiar 2D computer vision pipelines to volumetric (3D) imaging, where inputs are tomograms, positives are extremely sparse, memory is the primary limiting factor, and the metric skews decisions toward recall. We will reuse the ideas you practiced for classification, detection, and segmentation, then adapt them to 3D with patch-wise training/inference, recall-weighted losses, and careful post-processing that converts masks into point detections.

Cryo-electron tomography (CryoET) produces 3D microscopic images (tomograms) of cellular samples at near-atomic resolution, revealing how protein complexes are arranged inside cells. The **CZII – CryoET Object Identification** competition in 2024 challenged participants to develop machine learning models to automatically annotate five classes of protein complexes in these 3D tomograms. In other words, the task was to detect the centers of small biological structures (specific protein complexes) within large 3D volumes of CryoET data. Successful solutions would help accelerate biological discovery by automating the labor-intensive process of particle picking (identifying protein locations) in CryoET images.

With the problem framed at a high level, we first make the data concrete: what the tomograms contain, how classes differ in size and frequency, and where simulated vs. experimental data introduce a domain gap we'll have to bridge.

The key characteristics of the dataset

The competition provided a curated real-world CryoET dataset comprising *hundreds of tomograms* along with ground truth annotations for protein complexes. Notably, the training data included two subsets: (1) *experimental tomograms* with manually annotated protein centers (the result of months of expert particle picking), and (2) *simulated tomograms* ("phantoms") designed to mimic cellular CryoET data with known particle positions. Each tomogram is a 3D array (volume), typically on the order of ~500^3 voxels or larger. The target particles belonged to six protein complex types – Apo-ferritin, β-amylase, β-galactosidase, cytosolic ribosome (80S), thyroglobulin, and a **virus-like particle** (**VLP**). Of these, five classes were scored for the competition (β-amylase was present in the data but excluded from evaluation, serving as a distractor/non-scored class). Each protein type has a characteristic size and shape, ranging from small enzymes to large ribosomes, which adds complexity because the detection method needed to handle particles of varying scales and densities within the volume.

Modeling implications

Expect severe class imbalance (tiny foreground vs. vast background) and multi-scale targets; both motivate class-aware sampling of 3D patches, augmentations that respect volumetric structure (axis flips and right-angle rotations), and sometimes class-specific label radii to teach the network the right footprint for each protein type. The mix of experimental and simulated tomograms is a built-in domain shift, so many strong solutions pretrain or cotrain on simulated data and finetune on experimental data to transfer robustness.

Next, we pin down how inputs and labels are represented so we can choose sensible training targets and submission outputs.

Talking about the data format

Input data was provided as 3D image arrays (tomograms) along with corresponding label information. For training, participants had access to ground truth annotations, typically in the form of either segmentation masks marking particle locations or lists of particle center coordinates. For the **test phase**, the models were given a list of 3D volumes (each referred to as an "experiment") with no annotations. The expected output was a single CSV file listing the predicted particle locations for all five scored particle types in each volume. Each prediction in the CSV needed to include the *experiment (volume) identifier*, the *particle type/class*, and the (x, y, z) *coordinates* of the predicted particle's center. In essence, this was a **3D object detection** task where each object is identified by a point (center) in 3D space and a class label.

Unlike 2D detection with boxes, the competition's pointbased targets and CSV submissions favor two practical routes: (1) segmentation → centroid (train a 3D segmenter, then reduce connected components to particle centers), or (2) direct coordinate prediction (heatmaps/peaks or point-detection losses). Either way, memory pressure in 3D means tiling volumes into overlapping patches at training and testing time and stitching predictions back together.

Now that we understand the data's structure, let's analyze the evaluation metric and its implications. With inputs/labels fixed, we examine how you're scored and what makes the task hard – both drive architectural choices, loss weighting, thresholds, and ensembling later.

Evaluation strategy and challenges

The competition's evaluation used an F-beta score (F-score) with $\beta = 4$ as the primary metric. This F4-score heavily prioritized recall over precision – in fact, missing a true particle (false negative) was significantly worse than reporting an extra false positive. In formula terms, $F_\beta = \frac{(1+\beta^2)\cdot\text{Precision}\cdot\text{Recall}}{\beta^2\cdot\text{Precision}+\text{Recall}}$, so with $\beta = 4$, recall is weighted $4^2 = 16$ times more than precision. This choice reflects the scientific importance of identifying as many true particles as possible (since undetected proteins could lead to missed biological insights) while tolerating some false alarms. Practically, a model that finds most particles but also some extras could score higher than one that misses many particles, even if it has few false detections.

Matching criteria

To compute the F4-score, a predicted point was considered a correct detection (true positive) if it lay within a certain distance of a ground truth particle center. The allowed distance was typically tied to the particle's size – for example, one solution used **half the particle's radius as the matching threshold**. This means a prediction had to fall relatively close to the true center to count as a hit. Detections outside this radius are counted as false positives, and any ground truth not matched is counted as a miss (false negative). After matching, the counts of **true positives (TPs)**, **false negatives (FNs)**, and **false positives (FPs)** across all volumes and classes were plugged into the F4 formula to yield the final score (around 0.78 for top teams, indicating a balance of high recall and reasonable precision).

Key challenges

This competition was notably difficult due to several factors:

- **High dimensionality:** Each sample was a large 3D image (voxel grid). Processing 3D data is memory-intensive and computationally heavy. Models had to handle volumes far larger than typical 2D images, often requiring creative memory management (such as patch-based processing) to fit into GPU RAM.

- **Class imbalance and sparsity:** The number of particles in a tomogram is relatively small compared to the total number of voxels. Most voxels are background (no particle). This extreme class imbalance (tiny foreground vs. huge background) makes model training tricky – a naive model could predict "no particle anywhere" and be correct for the vast majority of voxels, but entirely useless. Moreover, some protein types were rarer than others in the dataset. Competitors addressed this via techniques like oversampling of patches containing particles, and using special loss functions (Dice, Tversky, etc.) or class weighting to prevent the model from being overwhelmed by negatives.

- **Noise and variability:** CryoET images are inherently noisy and have varying signal-to-noise ratios. Denoising and normalization steps were important to make particle signals stand out. Also, differences between the simulated data and real experimental data (a domain gap) meant models needed to generalize well. Some top teams explicitly handled this by multi-stage training – e.g., first training on abundant simulated data, then fine-tuning on real data to adapt to real noise characteristics.

- **Varying particle sizes:** The five target protein complexes varied significantly in size (for instance, a ribosome is much larger than an enzyme). A fixed-size annotation or detection method might not work equally well for all. The ground truth annotations were often encoded as spherical blobs of a certain radius around each particle center. Competitors sometimes customized the annotation radius per class (for example, using a larger "target" radius for bigger particles and a smaller one for others) to help their model localize each class more accurately. This required incorporating domain knowledge about particle dimensions into the training process.

- **3D augmentations and patch sampling**: Effective data augmentation was crucial given the limited number of annotated volumes. Competitors used 3D variants of flips and rotations (e.g., random 90° rotations about axes, and flips along axes) to augment the training data. They also employed patch-based training: extracting many small sub-volumes (patches) from each tomogram for training, rather than feeding entire volumes at once. A common strategy was to sample patches of size $96 \times 96 \times 96$ voxels that contained at least one particle (using MONAI's RandCropByLabelClassesd transform to ensure class-balanced patch sampling). This way, the model sees balanced examples of particles vs. background. Patch-based training not only alleviates memory usage but also implicitly increases the dataset size by providing many training examples per volume.

Before committing to a model architecture, we will use code to explore and understand the data: how tomograms are stored, how class labels are represented, and how to sample informative 3D patches efficiently.

Data exploration overview

The code in this section is based on the official overview notebook for the CZII CryoET competition: `https://github.com/czimaginginstitute/2024_czii_mlchallenge_notebooks/blob/main/overview.ipynb`

The dataset is packaged around the Copick API, which presents each tomogram as an OMEZarr volume with associated annotations. The same API works whether you mirror the competition data locally or access it through the CZ CryoET Data Portal, and the examples below mirror the structure used in the notebook.

The notebook begins with a minimal environment setup so that the data model and helper routines are available in the kernel. On Kaggle, this is typically done with a single cell that pulls in Copick and common numerics/plotting libraries:

```
!pip -q install "copick[all]" copick-utils zarr matplotlib torch
```

The core concept in Copick is a project that contains runs, voxel spacings, tomograms, and their overlays (e.g., dense segmentations). You can either point to a local configuration JSON file (if you've synced the competition data) or open a project directly from the Data Portal. The notebook demonstrates both entry points. Below, we show the idiomatic pattern used in that overview:

```
# Option A: open a locally-synced competition mirror
import copick
root = copick.from_file("/path/to/copick_config.json")


# Option B: open directly from the CZ CryoET Data Portal (dataset id
10440)
# This mirrors what the overview uses when exploring portal-hosted data.
root = copick.from_czcdp_datasets([10440], overlay_root="/tmp/overlay")
```

Either way, you get an object that exposes runs (each run corresponds to one tomogram acquisition), the available voxel spacings, and the class catalog. This mirrors the "print everything" cells in the notebook:

```
# Introspection cells from the overview
print("Pickable objects (name → label id):")
for o in root.pickable_objects:
    print(f"  {o.name:>22s}  →  {o.label}")


print("\nFirst few runs in this project:")
```

```
for r in root.runs[:5]:
    print("  ", r.name)
```

The project's pickable objects are the protein classes you will detect. In the competition's phantom dataset, the examples and tutorial material enumerate six classes: apo-ferritin, beta-amylase, beta-galactosidase, ribosome, thyroglobulin, and virus-like particles, which are also expressed in the Copick configuration and propagated into the Zarr overlays. The sync tutorial includes a concise, reproducible recipe for mirroring the portal data into a local, Kaggle-compatible project directory with exactly these class names.

With the project open, the notebook executes a single run and reads a denoised tomogram at a 10 Å voxel spacing. The Copick API intentionally abstracts the Zarr layout, but the examples/snippets page shows the exact code the notebook uses to pull a NumPy array for plotting:

```
import numpy as np, zarr
run = root.get_run(root.runs[0].name) # e.g., "TS_5_4"
vs = run.get_voxel_spacing(10.0)        # 10 Å (≈ competition scale)
tomo = vs.get_tomogram("denoised")      # commonly "wbp" in raw form

# Zarr stores multiple scales; "0" is unbinned, "1" is bin-2, etc.
tomo_vol = np.array(zarr.open(tomo.zarr())["0"]) # shape: (Z, Y, X)
print("Tomogram volume:", tomo_vol.shape)
```

Similarly, dense segmentation overlays are loaded for visualization and to build patch-sampling probability maps. Again, the snippets API call is what the notebook uses under the hood:

```
# Read a dense segmentation (semantic labels) into a NumPy array
# If multiple users/algorithms exist, choose the one you want to inspect.
# 'object_name=None' ⇒ all-classes mask
seg = run.get_segmentations()[0]
# same (Z, Y, X) layout as tomo
seg_vol = np.array(zarr.open(seg.zarr())["0"])
print("Segmentation volume:", seg_vol.shape, seg_vol.dtype)
```

Sometimes only particle picks are present; the CZII tools provide a tested conversion from picks to dense masks by rasterizing spherical regions of class-specific diameters. The competition documentation is explicit that the reference masks are created in this way. That detail matters when you compare patchlevel label densities or reproduce the training inputs from raw picks.

With a tomogram and its labels in hand, the notebook spends time simply examining the data: a single Z-slice for visual texture, along with the corresponding semantic mask. The following excerpt recreates the "slice and labels" plot shown in the overview:

```python
import matplotlib.pyplot as plt

z = tomo_vol.shape[0] // 2    # middle slice
fig, (ax0, ax1) = plt.subplots(1, 2, figsize=(10, 5))
ax0.imshow(tomo_vol[z], cmap="gray")
ax0.set_title("Tomogram Slice")
ax0.axis("off")

ax1.imshow(seg_vol[z], interpolation="nearest")
ax1.set_title("Segmentation Mask")
ax1.axis("off")
plt.tight_layout()
```

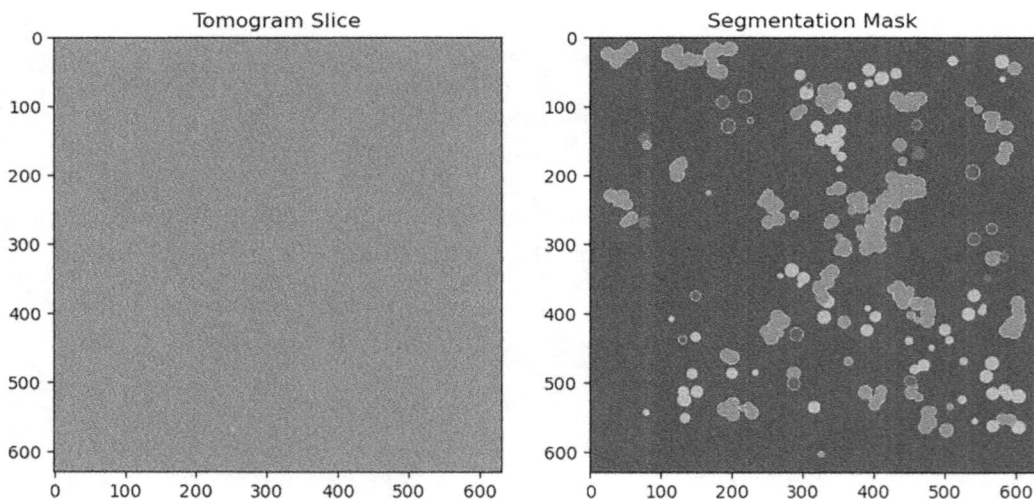

Figure 11.20: A central slice from a 3D tomogram (left) and its corresponding segmentation mask (right)

The most practically useful part of the notebook is its treatment of patch sampling. Full volumes are too large to fit as single inputs; instead, you train on 3D crops. The competition tooling supports density-weighted sampling, so crops fall more often on informative regions and less often on empty ice, which the organizers expose through their PyTorch data utilities.

The high-level idea is to collapse or blur the 3D label volume into a probability field and then sample crop centers according to that field. The notebook's visualization makes this concrete by drawing the crop rectangles on a heatmap-like projection of the underlying label density. The preprint accompanying the challenge describes this as "*patchwisesampling through the morphospaces module*," which is precisely what the overview notebook demonstrates.

The following distilled cell mirrors that logic in pure NumPy/Matplotlib for clarity.

```python
# Build a simple 2D sampling field by summing across z
# (For 3D patching you keep it volumetric; here we visualize the
footprint.)
density_2d = (seg_vol > 0).sum(axis=0).astype(np.float32)
density_2d = density_2d / (density_2d.max() + 1e-6)  # normalize to [0,1]

# Sample patch centers with probability ∝ density
rng = np.random.default_rng(0)
H, W = density_2d.shape
ps = 96   # patch size (example)
n = 30    # how many rectangles to draw
flat_indices = rng.choice(H * W, size=n, p=np.ravel(density_2d))
ys, xs = np.unravel_index(flat_indices, (H, W))

# Visualize the sampling plan
fig, ax = plt.subplots(figsize=(5, 5))
ax.imshow(density_2d, cmap="jet")
for y, x in zip(ys, xs):
    r0, c0 = max(0, y-ps//2), max(0, x-ps//2)
    rect = plt.Rectangle((c0, r0), ps, ps, linewidth=1.5,
                         edgecolor="white", facecolor="none", alpha=0.7)
    ax.add_patch(rect)
ax.set_title("Patches")
ax.axis("off")
plt.tight_layout()
```

It produces the same kind of "where the sampler spends its time" figure that appears in the overview:

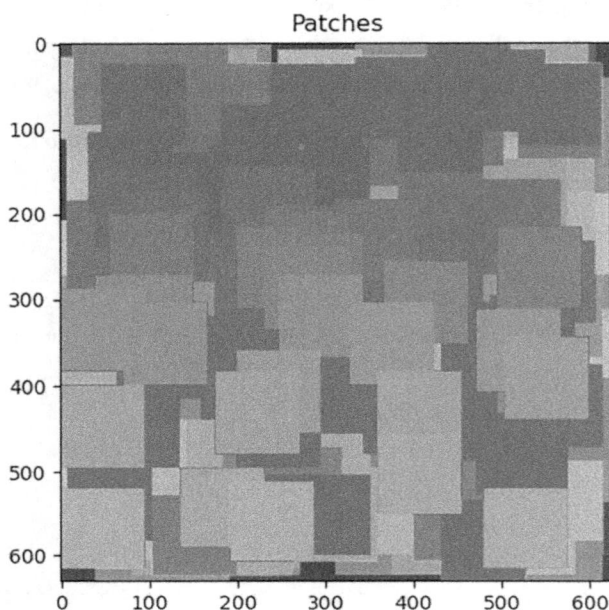

Figure 11.21: A density-weighted patch sampling

In the CZII tooling, this sampling is provided as part of the dataset/data loader, rather than being hand-rolled using NumPy. However, the principle remains identical: build a probability map from the segmentation, draw crop centers from that distribution, and yield 3D crops of the tomogram along with crops of the label volume. The notebook's sequence makes two training consequences explicit. First, weighting toward positive regions dramatically reduces wasted batches, which is crucial when small targets are sparse relative to the volume size. Second, the sampler is configurable (stride, patch shape, and number of samples per volume) so that class balance and context size can be tuned per model. The competition paper and repository README mention this sampler as part of the open-source stack released with the challenge.

Finally, to minimize friction when transitioning from exploration to training, the notebook reiterates the steps for creating a local mirror of the exact data layout used in the examples. The "Syncing CZII Kaggle Challenge Dataset" tutorial provides a concise, reproducible script that renames portal objects to match the Kaggle class names and generates a `copick_config.json` file that can be committed to your project. Once that file exists, all of the `root = copick.from_file(...)` cells above work unchanged on your machine or on Kaggle.

We will now proceed to a straightforward and reproducible baseline: 3D U-Net segmentation, combined with centroid extraction, which directly maps from the 2D segmentation workflow you used earlier to its 3D counterpart.

A first baseline solution, the 3D U-Net segmentation approach

As a starting point, organizers provided example notebooks demonstrating baseline approaches. One simple yet effective baseline was to cast the problem as **3D semantic segmentation** on a voxel grid and then post-process to get particle centers. In this approach, a neural network predicts a 3D mask where each voxel is classified into one of the particle classes or background. Predicted connected regions in the mask are then reduced to points (centroids) to produce the final detections. This essentially turns object detection into a segmentation problem, which is easier to optimize with dense voxel-wise labels.

Baseline model

The baseline used a small **3D U-Net** architecture implemented with the MONAI library (a medical-imaging focused toolkit). The U-Net took a 96^3 voxel patch as input (single-channel density values) and output a $96^3 \times 7$ mask (7 channels = 6 particle types + background). MONAI's built-in U-Net was configured with modest depth and width: for example, four levels with channel sizes of 48, 64, 80, and 80, and downsampling strides of 2, 2, 1 (i.e., the last two layers operate at the same resolution). A snippet of the model definition is shown below:

```python
import lightning.pytorch as pl
from monai.networks.nets import UNet
from monai.losses import TverskyLoss
from monai.metrics import DiceMetric


class Model(pl.LightningModule):
    def __init__(self, spatial_dims=3, in_channels=1, out_channels=7):
        super().__init__()
        # 3D U-Net with residual units
        self.model = UNet(
            spatial_dims=spatial_dims,
            in_channels=in_channels,
            out_channels=out_channels,
            channels=(48, 64, 80, 80),
            strides=(2, 2, 1),
            num_res_units=1
        )
```

```python
        # Tversky Loss (alpha,beta tuned via include_background)
        self.loss_fn = TverskyLoss(
            include_background=True,
            to_onehot_y=True,
            softmax=True
        )
        # Dice score metric for validation
        self.metric_fn = DiceMetric(
            include_background=False,
            reduction="mean",
            ignore_empty=True
        )

    def forward(self, x):
        return self.model(x)

    def training_step(self, batch, batch_idx):
        x, y = batch['image'], batch['label'] # 3D patch & label mask
        y_pred = self(x)                       # forward pass
        loss = self.loss_fn(y_pred, y)         # compute Tversky loss
        return loss

    def validation_step(self, batch, batch_idx):
        x, y = batch['image'], batch['label']
        y_pred = self(x)
        # Convert predictions & labels to one-hot for Dice metric
        pred_onehot = [
            AsDiscrete(argmax=True, to_onehot=7)(p)
            for p in decollate_batch(y_pred)
        ]
        label_onehot = [
            AsDiscrete(to_onehot=7)(t)
            for t in decollate_batch(y)
        ]
        self.metric_fn(
            y_pred=pred_onehot, y=label_onehot)  # accumulate Dice
```

Starting from a Kaggle baseline

In this baseline, **Tversky loss** was used for training – this is a generalized Dice loss that was found to be robust in the heavily imbalanced scenario (it can be tuned to weight false negatives more than false positives, similar to F_β behavior). The model outputs a probability for each class at each voxel (with softmax=True in loss). During validation, the code computes a **Dice score** for feedback by converting the softmax outputs to one-hot masks and comparing them to one-hot ground truth.

Regarding the training loop, the baseline was implemented using PyTorch Lightning, which abstracts away much of the boilerplate code associated with the training loop. Essentially, the training loop fed random 96^3 patches (with augmentations like flips and rotations applied) to the model. The training_step instance computed the loss on each batch, and Lightning handled the backward pass and optimizer step. They trained for a small number of epochs just to demonstrate the pipeline (the baseline wasn't fully optimized for the highest score, achieving around ~0.53 F_4 as a proof of concept).

As for the inference and post-processing phase, it is essential to note that, at test time, the trained model was applied to each full tomogram to predict the locations of particles. Since a full volume might be too large to run in one go, the volume was first tiled into patches (96^3, overlapping as needed), and the model was run on each patch. The output patch predictions were then stitched back together to form a segmentation of the entire volume. From this 3D mask, connected component analysis was used to identify distinct blobs for each class. Each connected component's centroid gave the (x, y, z) location of a detected particle, and the component's class label gave the particle type. The baseline utilized the cc3d library to identify connected components in the 3D mask and calculated the centroids and sizes of each blob. Very small blobs (volume < 500 voxels in the mask) were filtered out as likely noise. Finally, all the detected coordinates were output to the submission CSV. The following pseudocode (simplified for clarity) illustrates the inference process:

```
# Pseudocode for model inference on a tomogram

volume = load_volume(experiment_id) # Load 3D numpy array for the tomogram

patches, coords = extract_3d_patches(
    volume, patch_size=96
) # tile the volume
```

```python
mask_preds = np.zeros_like(volume, dtype=np.int8) # initialize empty mask

for patch, coord in zip(patches, coords):
    patch_tensor = torch.tensor(
        patch[None, None, ...]
    ).to(device)  # shape (1,1,96,96,96)

    output = model(patch_tensor) # output shape (1,7,96,96,96)
    probs = torch.softmax(
        output[0], dim=0)  # 7-channel probability map for the patch

    # convert to hard labels:
    # pick class with prob > 0.5 (else 0 = background)

    # thresholding yields a 7-channel boolean mask
    mask_patch = (probs > 0.5).int()

    # take the argmax across classes for each voxel
    _, pred_class = mask_patch.max(dim=0)

    # place patch prediction into full volume mask
    mask_preds[coord] = pred_class.cpu().numpy()

# Now perform connected components on each class in the full mask:
detected_particles = []

for class_label in [1,2,3,4,5,6]: # skip 0 (background)
    binary_mask = (mask_preds == class_label)
    cc = cc3d.connected_components(binary_mask) # label connected regions
    stats = cc3d.statistics(cc)

    for label_val in range(1, stats['num_objects']+1):
        # (z, y, x) in voxel indices
        centroid_vox = stats['centroids'][label_val]
        size = stats['voxel_counts'][label_val]

        if size < BLOB_MIN_SIZE:
            continue  # skip tiny noise
```

```
# Convert voxel coordinates to physical (x,y,z) and
# record detection

centroid_xyz = voxel_to_world(
    centroid_vox) # e.g. multiply by voxel size
detected_particles.append(
    (experiment_id, class_label, *centroid_xyz)
)
```

In summary, the baseline treated each particle as a small segmented blob and reduced the segmentation to points. This approach of segmentation-to-centroid was the foundation for many top solutions as well, though they significantly improved on it with more advanced models and ensembling.

The baseline provides a solid foundation. To see how top competitors built upon this, we will now analyze the second-place team's winning strategy. In the following solution notes, focus on how teams trade off recall versus precision under the Fβ metric, control GPU memory with patch sizes, and combine diverse architectures (3D U-Nets, VoxRes-style models, hybrid 2.5D, or direct detectors) via **Test-Time Augmentation** (**TTA** - an inference technique that improves a model's predictive performance by averaging the predictions made on multiple augmented versions of the same input data) and ensembling.

Examining the second-place solution: an ensemble of lightweight 3D models

The **second-place team (LuoZiqian&Lion)** achieved a remarkable balance of accuracy and speed by ensembling multiple efficient 3D segmentation models (https://cryoetdataportal. czscience.com/depositions/10320). Their approach can be summarized as a "many-model averaging" strategy, where each model is relatively lightweight (fewer parameters), but together they produce a strong consensus.

Model architectures

Instead of relying on one extensive network, they trained an ensemble of *diverse 3D CNN architectures*. The ensemble included classic designs, such as **3D U-Net**, as well as variants specifically designed for volumetric data, including **VoxResNet, VoxHRNet, SegResNet, DenseVNet**, and a so-called **UNet2E3D** model. In total, the models' parameter counts ranged from as low as 0.87 million to about 14.2 million, which is modest by modern standards – this helped ensure that each model could be trained on the limited data and used within inference time constraints. A few of these architectures are worth noting:

- *VoxResNet* and *DenseVNet* are inspired by medical imaging networks (volumetric ResNets and DenseNets) known for good performance on 3D segmentation tasks.

- *VoxHRNet* presumably adapts the high-resolution net concept to 3D, maintaining detailed feature maps at multiple scales – a feature useful for locating small particles.

- *UNet2E3D* likely refers to a U-Net that uses 2D pre-trained encoders expanded to 3D (a clever trick some competitors used by taking a 2D segmentation model and "inflating" it to 3D, leveraging 2D ImageNet pre-training for better initial weights).

By combining these different architectures, the team aimed to capture a variety of feature extraction styles. Each model might make slightly different errors, and the ensemble could average out individual weaknesses.

Training strategy

Each model in the ensemble was trained on the competition data using segmentation labels. The team put careful thought into *preprocessing and loss functions* to handle the challenges:

- They normalized and scaled the input data consistently and opted for **instance normalization (InstanceNorm3d)** layers in their networks, rather than batch normalization. InstanceNorm is often favored in medical images as it normalizes each 3D sample independently, making training more stable when batch sizes are small or intensity varies between volumes.

- They used **PReLU** activations (parametric ReLU), which can adapt the negative slope, potentially offering slight gains over standard ReLU in retaining signal for sparse features. This likely helped, given the high sparsity of particle signals.

- To address class imbalance, the team experimented with loss functions known to work well for segmentation: **Tversky loss, Dice loss, and cross-entropy loss**. Different models may employ different losses or a combination of them (for example, some models are trained with a Dice loss, which directly optimizes the overlap of predicted versus true masks, while others use cross-entropy for stable probabilistic predictions). Tversky, as noted, allows weighting false negatives higher, which aligns with the recall-heavy metric.

- **Customized mask radii per particle type:** Uniquely, the second-place team mentioned in their solution write-up that they used customized mask radii for each particle class. This suggests that when preparing training labels, they did not treat all particles the same – instead, they may have dilated or eroded the ground truth marker for each class differently.

For instance, a large particle like a ribosome might be annotated with a larger radius blob in the training mask, whereas a small protein might use a smaller radius. By doing so, each model can learn an appropriate "footprint" for each class (ensuring that the segmentation of each particle is neither too small to miss the object nor so large that multiple nearby objects blend). This is a form of domain-informed label augmentation.

They likely trained each model on random patches (such as 96^3), similarly to the baseline, possibly using cross-validation or different random seeds to obtain slightly varied models. The mention of **7 to 10 models** in their ensemble implies that they trained a considerable number of networks; some may have been variants of the same architecture with different initializations or data folds, while others were completely different architectures.

How to perform ensembling and inference

At inference time, the second-place team averaged the outputs of their models to get a more robust prediction. Concretely, this means for each test volume (after tiling it into patches and running each model on the patches), they would take the probability outputs from all models and average them voxel-wise. The result is an **ensemble probability volume** for each class. Averaging in probability space tends to smooth out outlier predictions – a voxel must consistently be predicted as a certain class by most models to end up with a high probability of being classified as such. They also applied **TTA**: running the inference multiple times per volume with different transformations (e.g., flipping the volume or rotating it) and averaging those predictions as well. TTA helps because a model might be sensitive to orientation or specific noise patterns; by averaging predictions over transforms, one can reduce variance.

After obtaining the final probability maps, the post-processing to obtain particle centers was similar to the baseline: thresholding the probabilities to obtain binary segmentations, finding connected components, and computing centroids. The second-place solution carefully tuned this step by using the **voxel count statistics** of connected components. They would have known from the training data roughly how many voxels each particle's mask should occupy (since they set the mask radii). For example, if they expect an Apo-ferritin particle to appear as ~300 voxels of connected volume, they can ignore any connected blob far off that size (too small likely noise; too large might be merged particles or artifact). By filtering out implausible detections using size and possibly shape cues, they improved precision without sacrificing recall of real particles. Finally, the centroids of the remaining components were output as detections for the CSV.

Overall, the second-place approach demonstrated the power of ensemble learning in this 3D domain: instead of one very complex model, a committee of simpler models (each specializing in the segmentation task) produced a superior result. This ensemble achieved a private leaderboard F_4 score of 0.7838, narrowly missing first place.

Looking at other top solutions and insights

The CryoET competition attracted a range of creative approaches. While a full breakdown of every solution is beyond the scope of this book, we will highlight a few notable strategies from the winning and other top teams and extract common trends:

- **First place (Team "Daddies") – two-headed monster:** The winners employed an ensemble that combined segmentation and direct object detection models. They trained *two kinds of models*: (1) several 3D U-Nets (with ResNet and EfficientNet-B3 backbones) producing dense segmentations, and (2) object detection models using architectures like SegResNet and DynUNet (from MONAI) that directly output particle coordinates. The segmentation models focused on voxel-wise labeling (with a heavily weighted cross-entropy loss – they gave a 256:1 weight ratio to positive vs. negative voxels to force the model to learn the sparse particles). In contrast, the detection models employed a point-based loss inspired by YOLO (specifically, a modified PaddlePaddle PP-YOLO loss) that computes IoU-like metrics on detected against true points. Both types were trained on 96^3 patches for manageable memory. During inference, this team's trick was to merge the outputs of the segmentation and detection pipelines using a *novel feature map scaling technique* – essentially, aligning the probability maps from segmentation with the confidence scores from detection so they could be combined. This yielded final predictions that benefited from the strengths of both approaches (segmentation for recall, and detection for precision). They also put a heavy effort into optimization, converting models to TensorRT and running inference on dual GPUs in parallel, which achieves a 3× faster inference speed. This was likely crucial to meet the competition's runtime limits, given the large test volume size. The first-place solution's emphasis on an ensemble of different paradigms (segmentation and detection) and extreme recall-bias (256:1 loss weighting, $\beta = 4$ metric) shows how important it was to not miss any particles.

- **Third place (Team "ONCE UPON A MOON") – large U-Net ensemble with EMA:** The third-place solution stuck to a pure segmentation approach, but with larger models and careful training. They used an ensemble of 3D U-Nets with a deep ResNet-101 encoder backbone – a very powerful network for feature extraction. To deal with memory, they trained on smaller patches (64×128×128) but exploited a clever trick: during inference, they could use larger patches (64×256×256) since the batch size can be 1 and more memory can be allocated then. This means a higher resolution context at prediction time, improving accuracy. They also utilized the **Exponential Moving Average (EMA)** of model weights (decay 0.995) during training – EMA smooths out training noise and typically yields a more generalized model at test time. They trained a 7-fold cross-validation and then took an ensemble of 4 models for submission (selecting the best 4), and applied extensive TTA (flips along all axes and 90° rotations). This yielded a score nearly equal to second place (0.7835). The takeaway here is the use of heavy ensembling and stabilization (EMA and cross-validation) to eke out performance in a high-variance setting.

- **Fourth place (Team "yu4u & tattaka") – 2.5D hybrid approach:** One interesting approach came from the fourth-place team, which actually published their method in detail (https://arxiv.org/abs/2502.13484). They built two different models and ensembled them: "yu4u's model" was a ConvNeXt-based 3D CNN that produces a heatmap where each particle center is a peak (trained with a custom MSE loss, weighting the positive peaks); "tattaka's model" was a lightweight 2.5D U-Net with a ResNetRS50 encoder. The 2.5D U-Net worked by reducing the depth dimension early (via strided pooling) to compress 3D data into a more 2D-like representation, then using joint upsampling to recover the full volume. This effectively cut down computation while still leveraging 3D context. Both models output heatmaps (representing the probability of a particle center at each voxel), which were then combined. They addressed class imbalance by tailoring loss functions: e.g., splitting the MSE loss into separate positive and negative components with different weights. This approach demonstrates how domain-specific architecture tweaks (treating the problem as predicting heatmaps and reducing 3D to 2.5D to conserve resources) can be effective. It's a reminder that not all top solutions used out-of-the-box U-Nets; some designed custom networks suited to this task.

These additional ideas generalize beyond CryoET, including curriculum or two-stage training from simulated to real-world settings, leveraging pretrained 2D encoders inflated to 3D, and robust post-processing that filters implausible blobs based on expected size.

- **Other techniques:** A number of different teams introduced noteworthy ideas. For example, one competitor pretrained their model on simulated data, then fine-tuned on real data by *freezing* certain layers in a two-stage training – this kind of transfer learning helped adapt to the real data domain. Others explored using **pre-trained 2D networks** (like EfficientNet backbones) by replicating filters across the depth dimension (the essence of the UNet2E3D mentioned in second place). Nearly all top teams agreed on **extensive data augmentation** (random flips, rotations, and even small intensity shifts) to make models more robust to orientation and noise differences. And universally, model ensembling and test-time augmentation were employed – no single model, no matter how complex, was as good as a combination, given the diverse challenge of detecting multiple particle types in varying conditions.

Stepping back, the patterns you've seen echo earlier 2D chapters—only amplified by 3D constraints: validation first, augmentation with intent, loss/metric alignment, and ensembles for stability.

Overall trends emerged in the competition

The competition underlined a few key lessons for machine learning in 3D vision domains:

- **Ensembles and diversity:** Combining multiple models was a dominant strategy. Whether it was mixing segmentation with detection (first place), ensembling many small segmenters (second place), or cross-validation folds (third place), consensus improved reliability. Diverse architectural choices (CNN vs. transformer vs. 2.5D, etc.) can complement each other.

- **Focus on recall:** Due to the evaluation metric, solutions were tuned to be overly sensitive rather than excessively specific. It's better to grab an extra blob (and maybe filter it later) than to miss a true particle. Loss functions and training sampling were biased accordingly (huge class weights, Tversky loss, etc.). Post-processing thresholds were set to favor inclusion of potential particles. This is a good example of matching the optimization target to the metric – a common theme in competitive machine learning.

- **Efficient 3D processing:** Handling full 3D data pushed teams to use patch-wise processing, multi-GPU computing, and optimized inference. The top teams all mention patch sizes (such as 96^3), indicating a de facto standard tile size that balances context and memory. Some utilized libraries like MONAI for 3D transforms and architectures, which accelerated the development of strong models. There was also a clever use of pre-trained models (when possible) to extract more signal from limited data, as well as strategies such as EMA to stabilize training on such a small dataset.

- **Domain knowledge integration:** Little touches inspired by domain knowledge made differences, such as using known particle size (radii) to set detection thresholds, or training including an unscored class to force models to distinguish it (third place included the non-scored β – amylase in their labels, not to misclassify it as something else). Understanding the physics of CryoET (such as anisotropic resolution or typical noise) could guide data preprocessing (one could apply filtering or crop out known artifact regions, though not explicitly described in top solutions). In short, top solutions did not treat it as just any computer vision task – they leveraged the *structure of the problem*, whether through custom losses or post-processing rules.

Taken together, these techniques demonstrate how to **port and extend** your 2D computer vision toolbox to volumetric science data: start from a clean segmentation baseline, respect memory with patches, align training with the **recall-heavy** objective, and conclude with disciplined post-processing and ensembles.

In conclusion, the CZII CryoET Object Identification competition demonstrated the application of modern deep learning techniques to cutting-edge scientific data. Through a combination of 3D convolutional networks, creative training regimes, and ensemble strategies, participants successfully developed algorithms that can automatically identify particles in 3D microscopic images with high accuracy. This represents a significant step toward automating analysis in structural biology, demonstrating the power of combining domain expertise with advanced machine learning. The approaches developed – from 3D U-Nets to hybrid detection models – and the tricks to handle class imbalance and large-volume inference provide a valuable case study in adapting computer vision methods to complex 3D data. As CryoET and similar volumetric imaging techniques become more common, these solutions pave the way for generalizable pipelines for object detection in 3D space. The competition not only advanced research in this specific problem but also contributed insights and open-source code to the broader community, bridging the gap between AI and science.

Summary

In this chapter, we gave you an overview of the most important topics related to computer vision from a Kaggle competition angle. We introduced augmentations, an essential class of techniques used to extend the generalization capabilities of an algorithm, and demonstrated end-to-end pipelines for three of the most common problems: image classification, object detection, and semantic segmentation. We then provided a few examples of how these techniques can be concretely applied in the context of Kaggle competitions.

In the next chapter, we switch our attention to natural language processing, another extremely broad and popular category of problems.

Get This Book's PDF Version and Exclusive Extras

UNLOCK NOW

Scan the QR code (or go to packtpub.com/unlock). Search for this book by name, confirm the edition, and then follow the steps on the page.

Note: Keep your invoice handy. Purchases made directly from Packt don't require an invoice.

Join our book's Discord space

Join our community's Discord space for discussions with the authors and other readers:

https://packt.link/kaggle

12
Modeling for NLP

Natural language processing (NLP) is a field operating at the intersection of linguistics, computer science, and AI. Its primary focus is on algorithms to process and analyze large amounts of natural language data. Over the last few years, it has become an increasingly popular topic of Kaggle competitions. While the domain itself is vast and encompasses very popular topics, such as chatbots and machine translation, this chapter will focus on specific subsets that Kaggle contests frequently address.

Sentiment analysis, as a simple classification problem, is extremely popular and widely discussed, so we'll begin with a somewhat more interesting variation on the issue: identifying sentiment-supporting phrases in a tweet. We'll proceed to describe an example solution to the problem of open-domain question-answering and conclude with a section on augmentation for NLP problems, a topic that receives significantly less attention than its computer vision counterpart.

To summarize, we will cover:

- Sentiment analysis
- Toxic comments classification
- Open domain Q&A
- Text augmentation strategies

Sentiment analysis

X is one of the most popular social media platforms and an important communication tool for many individuals and companies alike.

Capturing sentiment in language is particularly important in the latter context: a positive tweet can go viral and spread the word, while a particularly negative one can be harmful. Since human language is complicated, it is important not only to decide on the sentiment but also to investigate *how* it was determined: which words actually led to the sentiment description?

We will demonstrate an approach to this problem by using data from the *Tweet Sentiment Extraction* competition (https://www.kaggle.com/c/tweet-sentiment-extraction). For brevity, we have omitted the imports from the following code, but you can find them in the corresponding notebook in the GitHub repository for this chapter.

To get a better feel for the problem, let's start by looking at the data:

```
df = pd.read_csv('/kaggle/input/tweet-sentiment-extraction/train.csv')
df.head()
```

Here are the first few rows:

	textID	text	selected_text	sentiment
0	cb774db0d1	I'd have responded, if I were going	I'd have responded, if I were going	neutral
1	549e992a42	Sooo SAD I will miss you here in San Diego!!!	Sooo SAD	negative
2	088c60f138	my boss is bullying me...	bullying me	negative
3	9642c003ef	what interview! leave me alone	leave me alone	negative
4	358bd9e861	Sons of ****, why couldn't they put them on t...	Sons of ****,	negative

Figure 12.1: Sample rows from the training data

The actual tweets are stored in the text column. Each of them has an associated sentiment, along with the support phrase stored in the selected_text column (the part of the tweet that was the basis for the decision on sentiment assignment).

We start by defining basic cleanup functions. First, we want to get rid of website URLs and non-characters and replace the stars people use in place of swear words with a single token, swear. We use some regular expressions to help us do this:

```
def basic_cleaning(text):
    text=re.sub(r'https?://www\.\S+\.com','',text)
    text=re.sub(r'[^A-Za-z|\s]','',text)
    text=re.sub(r'\*+','swear',text) # Capture swear words that are ****
out
    return text
```

Next, we remove HTML from the content of the tweets, as well as emojis:

```
def remove_html(text):
    html=re.compile(r'<.*?>')
    return html.sub(r'',text)
def remove_emoji(text):
    emoji_pattern = re.compile(
        "["
        u"\U0001F600-\U0001F64F" #emoticons
        u"\U0001F300-\U0001F5FF" #symbols & pictographs
        u"\U0001F680-\U0001F6FF" #transport & map symbols
        u"\U0001F1E0-\U0001F1FF" #flags (iOS)
        u"\U00002702-\U000027B0"
        u"\U000024C2-\U0001F251"
        "]+",
        flags=re.UNICODE
    )
    return emoji_pattern.sub(r'', text)
```

Lastly, we want to be able to remove repeated characters (for example, so we have "way" instead of "waaaayyyyy"):

```
def remove_multiplechars(text):
    text = re.sub(r'(.)\1+',r'\1', text)
    return text
```

For convenience, we combine the four functions into a single cleanup function:

```
def clean(df):
    for col in ['text']:#,'selected_text']:
        df[col]=df[col].astype(str).apply(lambda x:basic_cleaning(x))
        df[col]=df[col].astype(str).apply(lambda x:remove_emoji(x))
        df[col]=df[col].astype(str).apply(lambda x:remove_html(x))
        df[col]=df[col].astype(str).apply(
            lambda x:remove_multiplechars(x))
    return df
```

The last bit of preparation involves writing functions for creating the embeddings based on a pretrained model (the `tokenizer` argument):

```python
def fast_encode(texts, tokenizer, chunk_size=256, maxlen=128):
    tokenizer.enable_truncation(max_length=maxlen)
    tokenizer.enable_padding(length=maxlen)
    all_ids = []

    for i in range(0, len(texts), chunk_size):
        text_chunk = texts[i:i+chunk_size].tolist()
        encs = tokenizer.encode_batch(text_chunk)
        all_ids.extend([enc.ids for enc in encs])

    return np.array(all_ids)
```

Next, we create a preprocessing function enabling us to work with the entire corpus:

```python
def preprocess_news(df,stop=stop,n=1,col='text'):
    '''Function to preprocess and create corpus'''
    new_corpus=[]
    stem=PorterStemmer()
    lem=WordNetLemmatizer()
    for text in df[col]:
        words=[w for w in word_tokenize(text) if (w not in stop)]
        words=[lem.lemmatize(w) for w in words if(len(w)>n)]
        new_corpus.append(words)

    new_corpus=[word for l in new_corpus for word in l]
    return new_corpus
```

Using our previously prepared functions, we can clean and prepare the training data. The sentiment column is our target, and we convert it to dummy variables (one-hot encoding) to be compatible with the model's loss function:

```python
df.dropna(inplace=True)
df_clean = clean(df)
df_clean_selection = df_clean.sample(frac=1)
X = df_clean_selection.text.values
y = pd.get_dummies(df_clean_selection.sentiment)
```

A common next step is tokenization. For a traditional LSTM model built with Keras, we can use the legacy preprocessing functions such as Tokenizer (which require installation with pip install keras-preprocessing), as shown below, as well as conversion into sequences (along with padding, to ensure equal lengths across dataset):

```
tokenizer = text.Tokenizer(num_words=20000)
tokenizer.fit_on_texts(list(X))
list_tokenized_train = tokenizer.texts_to_sequences(X)
X_t = sequence.pad_sequences(list_tokenized_train, maxlen=128)
```

DistilBERT is a lightweight version of BERT, with a tradeoff of 3% performance loss for 40% fewer parameters. We could train the embedding layer and gain performance—at the cost of massively increased training time. However, for modern Transformer-based models like DistilBERT, a specific tokenizer from the Hugging Face library is required. We will now create the embeddings for our model using this tokenizer and utilize them as is without further training.

```
tokenizer = transformers.AutoTokenizer.from_pretrained("distilbert-base-
uncased")

# Save the loaded tokenizer locally
save_path = '/kaggle/working/distilbert_base_uncased/'
if not os.path.exists(save_path):
    os.makedirs(save_path)
tokenizer.save_pretrained(save_path)

# Reload it with the huggingface tokenizers library
fast_tokenizer = BertWordPieceTokenizer(
    'distilbert_base_uncased/vocab.txt', lowercase=True)
fast_tokenizer
```

We can use the previously defined fast_encode function, along with the fast_tokenizer defined above, to encode the tweets:

```
X = fast_encode(df_clean_selection.text.astype(str),
                fast_tokenizer,
                maxlen=128)
```

With the data prepared, we can construct the model. For the sake of this demonstration, we will go with a fairly standard architecture for these applications: a combination of LSTM layers, normalized by global pooling and dropout, and a dense layer on top. In order to achieve a truly competitive solution, some tweaking of the architecture would be needed: a "heavier" model, bigger embeddings, more units in the LSTM layers, and so on.

```python
transformer_layer = transformers.TFDistilBertModel.from_
pretrained('distilbert-base-uncased')
embedding_size = 128
inp = Input(shape=(128, ))
embedding_matrix=transformer_layer.weights[0].numpy()
x = Embedding(embedding_matrix.shape[0],
              embedding_matrix.shape[1],
              embeddings_initializer=Constant(embedding_matrix),
              trainable=False)(inp)
x = Bidirectional(LSTM(50, return_sequences=True))(x)
x = Bidirectional(LSTM(25, return_sequences=True))(x)
x = GlobalMaxPool1D()(x)
x = Dropout(0.5)(x)
x = Dense(50, activation='relu', kernel_regularizer='L1L2')(x)
x = Dropout(0.5)(x)
x = Dense(3, activation='softmax')(x)
model_DistilBert = Model(inputs=[inp], outputs=x)
model_DistilBert.compile(loss='categorical_crossentropy',
                         optimizer='adam',
                         metrics=['accuracy'])
```

There is no special need to pay attention to a temporal dimension of the data, so we are fine with a random split into training and validation, which can be achieved inside a call to the `fit` method:

```python
model_DistilBert.fit(X,y,batch_size=32,epochs=10,validation_split=0.1)
```

Below is some sample output:

```
Epoch 1/10
27480/27480 [==============================] - 480s 17ms/step - loss:
0.5100 - accuracy: 0.7994
Epoch 2/10
27480/27480 [==============================] - 479s 17ms/step - loss:
0.4956 - accuracy: 0.8100
```

```
Epoch 3/10
27480/27480 [==============================] - 475s 17ms/step - loss:
0.4740 - accuracy: 0.8158
Epoch 4/10
27480/27480 [==============================] - 475s 17ms/step - loss:
0.4528 - accuracy: 0.8275
Epoch 5/10
27480/27480 [==============================] - 475s 17ms/step - loss:
0.4318 - accuracy: 0.8364
Epoch 6/10
27480/27480 [==============================] - 475s 17ms/step - loss:
0.4069 - accuracy: 0.8441
Epoch 7/10
27480/27480 [==============================] - 477s 17ms/step - loss:
0.3839 - accuracy: 0.8572
```

Generating a prediction from the fitted model proceeds in a straightforward manner. In order to utilize all the available data, we begin by re-training our model on all available data (so no validation):

```
df_clean_final = df_clean.sample(frac=1)
X_train = fast_encode(df_clean_selection.text.astype(str),
                      fast_tokenizer,
                      maxlen=128)
y_train = y
```

We refit the model on the entire dataset before generating the predictions:

```
optimizer = Adam(learning_rate=0.001)
model_DistilBert.compile(loss='categorical_crossentropy',
                         optimizer=optimizer, metrics=['accuracy'])
history = model_DistilBert.fit(X_train, y_train,
                               batch_size=32, epochs=10)
```

Our next step is to process the test data into the same format we are using for training data fed into the model:

```
df_test = pd.read_csv('/kaggle/input/tweet-sentiment-extraction/test.csv')
df_test.dropna(inplace=True)
df_clean_test = clean(df_test)
X_test = fast_encode(df_clean_test.text.values.astype(str),
```

```
                              fast_tokenizer,
                              maxlen=128)
y_test = df_clean_test.sentiment
```

Finally, we generate the predictions:

```
y_preds = model_DistilBert.predict(X_test)
y_predictions = pd.DataFrame(y_preds,
                             columns=['negative','neutral','positive'])
y_predictions_final = y_predictions.idxmax(axis=1)
accuracy = accuracy_score(y_test,y_predictions_final)
print(f"The final model shows {accuracy:.2f} accuracy on the test set.")
```

The final model shows 0.74 accuracy on the test set. Below we show a sample of what the output looks like; as you can see already from these few rows, there are some instances where the sentiment is obvious to a human reader, but the model fails to capture it:

	textID	text	sentiment	predicted_negative	predicted_neutral	predicted_positive
0	f87dea47db	Last session of the day httptwitpiccomezh	neutral	0.022949	0.967165	0.009886
1	96d74cb729	Shanghai is also really exciting precisely s...	positive	0.000075	0.012165	0.987760
2	eee518ae67	Recession hit Veronique Branquinho she has to ...	negative	0.993622	0.006364	0.000014
3	01082688c6	happy bday	positive	0.000020	0.005859	0.994122
4	33987a8ee5	httptwitpiccomwp I like it	positive	0.006184	0.119946	0.873870
5	726e501993	thats great weee visitors	positive	0.000165	0.019434	0.980401
6	261932614e	I THINK EVERYONE HATES ME ON HERE lol	negative	0.916203	0.081649	0.002148
7	afa11da83f	so wish i could but im in school and myspace ...	negative	0.877504	0.116624	0.005871
8	e64208b4ef	and within a short time of the last clue all ...	neutral	0.116272	0.859304	0.024424
9	37bcad24ca	What did you get My day is alright havent do...	neutral	0.223977	0.756474	0.019550

Figure 12.2: Example rows from the predicted results

We have now demonstrated a sample pipeline for solving sentiment attribution problems using a Transformer's embeddings with an LSTM model. There are some improvements that can be made if you want to achieve competitive performance, given below in order of likely impact:

- **Larger embeddings:** This allows us to capture more information already at the (processed) input data level
- **Bigger model:** More units in the LSTM layers
- **Longer training:** In other words, more epochs

While the improvements listed above will undoubtedly boost the performance of the model, the core elements of our pipeline are reusable:

- Data cleaning and preprocessing
- Creating text embeddings
- Incorporating recurrent layers and regularization in the target model architecture

We'll now move on to a discussion of open domain question-answering, a frequent problem encountered in NLP competitions.

Interview: Abhishek Thakur

`https://www.kaggle.com/abhishek`

We caught up with *Abhishek Thakur*, the world's first quadruple Kaggle Grandmaster. He previously worked at Hugging Face, where he built AutoNLP; he also wrote pretty much the only book on Kaggle in English (aside from this one!), Approaching (Almost) Any Machine Learning Problem.

What's your specialty on Kaggle?

None. Every competition is different and there is so much to learn from each one of them. If I were to have a specialty, I would win all competitions in that domain.

How do you approach a Kaggle competition? How different is this approach from what you do in your day-to-day work?

The first thing I do is take a look at the data and try to understand it a bit. If I'm late to the competition, I take the help of public EDA kernels.

The first thing I do when approaching a problem on (or off) Kaggle is build a benchmark. Building a benchmark is very important as it provides you with a baseline you can compare your future models to. If I'm late to the game, for building the baseline, I try not to take the help of public notebooks. If we do that, we think only in a single direction. At least, that's what I feel.

When I am done with a benchmark, I try to squeeze as much as possible out of it without doing anything complicated like stacking or blending. Then I go over the data and models again and try to improve on the baseline, one step at a time.

Day-to-day work sometimes has a lot of similarities. Most of the time there is a benchmark and then you have to come up with techniques, features, and models that beat the benchmark.

What was the most interesting competition you entered? Did you have any special insights?

Every competition is interesting.

Has Kaggle helped you in your career?

Sure, it has helped. In the last few years, Kaggle has gained a very good reputation when it comes to hiring data scientists and machine learning engineers. Kaggle rank and experience with many datasets are things that surely help in the industry in one way or another. The more experienced you are with approaching different types of problems, the faster you will be able to iterate. And that's something very useful in industries. No one wants to spend several months doing something that doesn't bring any value to the business.

In your experience, what do inexperienced Kagglers often overlook? What do you know now that you wish you'd known when you first started?

Most beginners give up quite easily. It's very easy to join a Kaggle competition and get intimidated by top scorers. If beginners want to succeed on Kaggle, they have to have perseverance. In my opinion, perseverance is key. Many beginners also fail to start on their own and stick to public kernels. This makes them think like the authors of public kernels. My advice would be to start with competitions on your own, look at data, build features, build models, and then dive into kernels and discussions to see what others might be doing differently. Then incorporate what you have learned into your own solution.

Open domain Q&A

In this section, we will be looking at the *Google QUEST Q&A Labeling* competition (https://www.kaggle.com/c/google-quest-challenge/overview/description). In this competition, question-answer pairs were evaluated by human raters on a diverse set of criteria, such as "question conversational," "question fact-seeking," or "answer helpful." The task was to predict a numeric value for each of the target columns (corresponding to the criteria); since the labels were aggregated across multiple raters, the objective was effectively a multivariate regression output, with target columns normalized to the unit range.

Before engaging in modeling with advanced techniques (like transformer-based models for NLP), it is frequently a good idea to establish a baseline with simpler methods. As with the previous section, we will omit the imports for brevity, but you can find them in the notebook in the GitHub repo.

We begin by defining several helper functions, which can help us extract different aspects of the text. First is a function that will output a word count given a string:

```
def word_count(xstring):
    return xstring.split().str.len()
```

The metric used in the competition was **Spearman correlation** (linear correlation computed on ranks: https://en.wikipedia.org/wiki/Spearman%27s_rank_correlation_coefficient).

Since we intend to build a scikit-learn pipeline, it is useful to define the metric as a scorer (the make_scorer method is a wrapper in scikit-learn that takes a scoring function—like accuracy or MSE—and returns a callable that scores an output of the estimator):

```
def spearman_corr(y_true, y_pred):
    if np.ndim(y_pred) == 2:
        corr = np.mean([
            stats.spearmanr(y_true[:, i], y_pred[:, i])[0]
            for i in range(y_true.shape[1])
        ])
    else:
        corr = stats.spearmanr(y_true, y_pred)[0]
    return corr


custom_scorer = make_scorer(spearman_corr, greater_is_better=True)
```

Next is a small helper function to extract successive chunks of size n from 1. This will help us later with generating embeddings for our body of text without running into memory problems:

```
def chunks(l, n):
    for i in range(0, len(l), n):
        yield l[i:i + n]
```

Part of the feature set we will use is embeddings from pretrained models. Recall that the idea of this section is the construction of a baseline without training elaborate models, but this need not prevent us from using existing ones.

We begin by importing the tokenizer and model, and then we process the corpus in chunks, encoding each question/answer into a fixed-size embedding:

```
def fetch_vectors(string_list, batch_size=64):
    # Inspired by
    # https://jalammar.github.io/a-visual-guide-to-using-bert- for-the-
first-time/
    DEVICE = torch.device("cuda")
    tokenizer = transformers.DistilBertTokenizer.from_pretrained(
        "../input/distilbertbaseuncased/")
    model = transformers.DistilBertModel.from_pretrained(
        "../input/distilbertbaseuncased/")
    model.to(DEVICE)
    fin_features = []
    for data in chunks(string_list, batch_size):
        tokenized = []
        for x in data:
            x = " ".join(x.strip().split()[:300])
            tok = tokenizer.encode(x, add_special_tokens=True)
            tokenized.append(tok[:512])
        max_len = 512
        padded = np.array([i + [0] * (max_len - len(i)) for i in
tokenized])
        attention_mask = np.where(padded != 0, 1, 0)
        input_ids = torch.tensor(padded).to(DEVICE)
        attention_mask = torch.tensor(attention_mask).to(DEVICE)
        with torch.no_grad():
            last_hidden_states = model(input_ids,
```

```
                                              attention_mask=attention_mask)
        features = last_hidden_states[0][:, 0, :].cpu().numpy()
        fin_features.append(features)
    fin_features = np.vstack(fin_features)
    return fin_features
```

We can now proceed to load the data:

```
xtrain = pd.read_csv(data_dir + 'train.csv')
xtest = pd.read_csv(data_dir + 'test.csv')
xtrain.head(4)
```

Here are the first few rows:

question_title	question_body	question_user_name	question_user_page	answer
What am I losing when using extension tubes in...	After playing around with macro photography on...	ysap	https://photo.stackexchange.com/users/1024	I just got extension tubes, so here's the skin...
What is the distinction between a city and a s...	I am trying to underetand what kinds of places...	russellpierce	https://rpg.stackexchange.com/users/8774	It might be helpful to look into the definitio...
Maximum protusion length for through-hole comp...	I'm working on a PCB that has through-hole com...	Joe Baker	https://electronics.stackexchange.com/users/10157	Do you even need grooves? We make several pro...
Can an affidavit be used in Beit Din?	An affidavit, from what i understand, is basic...	Scimonster	https://judaism.stackexchange.com/users/5151	Sending an "affidavit" it is a dispute between...

Figure 12.3: Sample rows from the training data

We specify our 30 target columns of interest:

```
target_cols = ['question_asker_intent_understanding',
               'question_body_critical',
               'question_conversational',
               'question_expect_short_answer',
               'question_fact_seeking',
               'question_has_commonly_accepted_answer',
               'question_interestingness_others',
```

```
                    'question_interestingness_self',
                    'question_multi_intent',
                    'question_not_really_a_question',
                    'question_opinion_seeking', 'question_type_choice',
                    'question_type_compare', 'question_type_consequence',
                    'question_type_definition', 'question_type_entity',
                    'question_type_instructions',
                    'question_type_procedure',
                    'question_type_reason_explanation',
                    'question_type_spelling',
                    'question_well_written', 'answer_helpful',
                    'answer_level_of_information', 'answer_plausible',
                    'answer_relevance', 'answer_satisfaction',
                    'answer_type_instructions', 'answer_type_procedure',
                    'answer_type_reason_explanation',
                    'answer_well_written']
```

For a discussion of their meaning and interpretation, we refer you to the competition's Data page, at https://www.kaggle.com/c/google-quest-challenge/data.

Next, we proceed with **feature engineering**. We start by counting the words in the title and body of the question, as well as the answer. This is a simple yet surprisingly useful feature in many applications:

```
for colname in ['question_title', 'question_body', 'answer']:
    newname = colname + '_word_len'
    xtrain[newname] = xtrain[colname].str.split().str.len()
    xtest[newname] = xtest[colname].str.split().str.len()
```

The next feature we create is **lexical diversity**, counting the proportion of unique words in a chunk of text:

```
colname = 'answer'
xtrain[colname+'_div'] = xtrain[colname].apply(
    lambda s: len(set(s.split())) / len(s.split()) )
xtest[colname+'_div'] = xtest[colname].apply(
    lambda s: len(set(s.split())) / len(s.split()) )
```

When dealing with information sourced online, we can extract potentially informative features by examining the components of a website address (where we define components as elements of the address separated by dots); we count the number of components, and store individual ones as features:

```
for df in [xtrain, xtest]:
    df['domcom'] = df['question_user_page'].apply(
        lambda s: s.split('://')[1].split('/')[0].split('.'))
    # Count components
    df['dom_cnt'] = df['domcom'].apply(lambda s: len(s))
    # Pad the length in case some domains have fewer components in the
name
    df['domcom'] = df['domcom'].apply(lambda s: s + ['none', 'none'])
    # Components
    for ii in range(0,4):
        df['dom_'+str(ii)] = df['domcom'].apply(lambda s: s[ii])
```

Numerous target columns deal with how relevant the answer is for a given question. One possible way of quantifying this relationship is by evaluating **shared words** within a pair of strings:

```
# Shared elements
for df in [xtrain, xtest]:
    df['q_words'] = df['question_body'].apply(
        lambda s: [f for f in s.split() if f not in eng_stopwords] )
    df['a_words'] = df['answer'].apply(
        lambda s: [f for f in s.split() if f not in eng_stopwords] )
    df['qa_word_overlap'] = df.apply(
        lambda s: len(np.intersect1d(s['q_words'], s['a_words'])), axis=1)
    df['qa_word_overlap_norm1'] = df.apply(
        lambda s: s['qa_word_overlap']/(1 + len(s['a_words'])), axis = 1)
    df['qa_word_overlap_norm2'] = df.apply(
        lambda s: s['qa_word_overlap']/(1 + len(s['q_words'])), axis = 1)
    df.drop(['q_words', 'a_words'], axis = 1, inplace = True)
```

Stopwords and punctuation occurrence patterns can tell us something about the style and intent:

```
for df in [xtrain, xtest]:

    ## Number of characters in the text ##
    df["question_title_num_chars"] = df["question_title"].apply(
```

```
        lambda x: len(str(x)))
df["question_body_num_chars"] = df["question_body"].apply(
        lambda x: len(str(x)))
df["answer_num_chars"] = df["answer"].apply(
        lambda x: len(str(x)))


## Number of stopwords in the text ##
df["question_title_num_stopwords"] = df["question_title"].apply(
        lambda x: len([w for w in str(x).lower().split()
                        if w in eng_stopwords]))
df["question_body_num_stopwords"] = df["question_body"].apply(
        lambda x: len([w for w in str(x).lower().split()
                        if w in eng_stopwords]))
df["answer_num_stopwords"] = df["answer"].apply(
        lambda x: len([w for w in str(x).lower().split()
                        if w in eng_stopwords]))


## Number of punctuations in the text ##
df["question_title_num_punctuations"] =df['question_title'].apply(
        lambda x: len([c for c in str(x) if c in string.punctuation]) )
df["question_body_num_punctuations"] =df['question_body'].apply(
        lambda x: len([c for c in str(x) if c in string.punctuation]) )
df["answer_num_punctuations"] =df['answer'].apply(
        lambda x: len([c for c in str(x) if c in string.punctuation]) )


## Number of title case words in the text ##
df["question_title_num_words_upper"] = df["question_title"].apply(
        lambda x: len([w for w in str(x).split() if w.isupper()]))
df["question_body_num_words_upper"] = df["question_body"].apply(
        lambda x: len([w for w in str(x).split() if w.isupper()]))
df["answer_num_words_upper"] = df["answer"].apply(
        lambda x: len([w for w in str(x).split() if w.isupper()]))
```

With the "vintage" features prepared, where our focus is on simple summary statistics of the text, without paying heed to semantic structure, we can move on to creating **embeddings** for the questions and answers. We could theoretically train a separate word2vec-type model on our data (or fine-tune an existing one), but for the sake of this demonstration, we will use a pretrained model as is. A practical choice is the **Universal Sentence Encoder** from Google (https://www.kaggle.com/ models/google/universal-sentence-encoder/tensorFlow2/universal-sentence-encoder/2). This model is trained on a variety of data sources. It takes as input a piece of English text and outputs a 512-dimensional vector.

```
module_url = "https://www.kaggle.com/models/google/universal-sentence-
encoder/TensorFlow2/universal-sentence-encoder/2"
embed = hub.load(module_url)
```

The code for turning the text fields into embeddings is presented below: we loop through the entries in the training/test sets in batches, embed each batch (for memory efficiency reasons), and then append them to the original list.

The final DataFrames are constructed by stacking each list of batch-level embeddings vertically:

```
embeddings_train = {}
embeddings_test = {}
for text in ['question_title', 'question_body', 'answer']:
    train_text = (
        xtrain[text].str.replace('?', '.').str.replace('!', '.').tolist()
    )
    test_text = (
        xtest[text].str.replace('?', '.').str.replace('!', '.').tolist()
    )
    curr_train_emb = []
    curr_test_emb = []
    batch_size = 4

    ind = 0
    while ind*batch_size < len(train_text):
        curr_train_emb.append(embed(train_text[
            ind*batch_size: (ind + 1)*batch_size]).numpy())
        ind += 1

    ind = 0
```

```
    while ind*batch_size < len(test_text):
        curr_test_emb.append(embed(test_text[
            ind*batch_size: (ind + 1)*batch_size]).numpy())
        ind += 1


    embeddings_train[text + '_embedding'] = np.vstack(curr_train_emb)
    embeddings_test[text + '_embedding'] = np.vstack(curr_test_emb)
    print(text)
```

Given the vector representations for both questions and answers, we can calculate the semantic similarity between the fields by using different distance metrics on the pairs of vectors. The idea behind trying different metrics is the desire to capture diverse types of characteristics; an analogy in the context of classification would be to use both accuracy and entropy to get a complete picture of the situation:

```
l2_dist = lambda x, y: np.power(x - y, 2).sum(axis=1)
cos_sim = lambda x, y: (x*y).sum(axis=1)


dist_features_train = np.array([
    l2_dist(embeddings_train['question_title_embedding'],
            embeddings_train['answer_embedding']),
    l2_dist(embeddings_train['question_body_embedding'],
            embeddings_train['answer_embedding']),
    l2_dist(embeddings_train['question_body_embedding'],
            embeddings_train['question_title_embedding']),
    cos_sim(embeddings_train['question_title_embedding'],
            embeddings_train['answer_embedding']),
    cos_sim(embeddings_train['question_body_embedding'],
            embeddings_train['answer_embedding']),
    cos_sim(embeddings_train['question_body_embedding'],
            embeddings_train['question_title_embedding'])
]).T


dist_features_test = np.array([
    l2_dist(embeddings_test['question_title_embedding'],
            embeddings_test['answer_embedding']),
    l2_dist(embeddings_test['question_body_embedding'],
            embeddings_test['answer_embedding']),
```

```
        l2_dist(embeddings_test['question_body_embedding'],
                embeddings_test['question_title_embedding']),
        cos_sim(embeddings_test['question_title_embedding'],
                embeddings_test['answer_embedding']),
        cos_sim(embeddings_test['question_body_embedding'],
                embeddings_test['answer_embedding']),
        cos_sim(embeddings_test['question_body_embedding'],
                embeddings_test['question_title_embedding'])
    ]).T
```

Let's gather the distance features in separate columns:

```
for ii in range(0,6):
    xtrain['dist'+str(ii)] = dist_features_train[:,ii]
    xtest['dist'+str(ii)] = dist_features_test[:,ii]
```

Finally, we can also create **TF-IDF** representations of the text fields; the general idea is to create multiple features based on diverse transformations of the input text, and then feed them to a relatively simple model.

This way, we can capture the characteristics of the data without the need to fit a sophisticated deep learning model.

We can achieve it by analyzing the text at the word and character levels. To limit the memory consumption, we put an upper bound on the maximum number of both kinds of features (your mileage might vary; with more memory, these limits can be upped):

```
limit_char = 5000
limit_word = 25000
```

We instantiate character- and word-level vectorizers. The setup of our problem lends itself to a convenient usage of the `Pipeline` functionality from scikit-learn, allowing a combination of multiple steps in the model-fitting procedure. We begin by creating two separate transformers for the title column (word- and character-level):

```
title_col = 'question_title'
title_transformer = Pipeline([
    ('tfidf', TfidfVectorizer(
        lowercase = False, max_df = 0.3, min_df = 1,
        binary = False, use_idf = True, smooth_idf = False,
        ngram_range = (1,2), stop_words = 'english',
```

```
        token_pattern = '(?u)\\b\\w+\\b' , max_features = limit_word
    ))
])
title_transformer2 = Pipeline([
    ('tfidf2',  TfidfVectorizer(sublinear_tf=True,
        strip_accents='unicode', analyzer='char',
        stop_words='english', ngram_range=(1, 4),
        max_features= limit_char
    ))
])
```

We use the same logic (two different pipelined transformers) for the body:

```
body_col = 'question_body'
body_transformer = Pipeline([
    ('tfidf',TfidfVectorizer(
        lowercase = False, max_df = 0.3, min_df = 1,
        binary = False, use_idf = True, smooth_idf = False,
        ngram_range = (1,2), stop_words = 'english',
        token_pattern = '(?u)\\b\\w+\\b' , max_features = limit_word
    ))
])
body_transformer2 = Pipeline([
    ('tfidf2',  TfidfVectorizer( sublinear_tf=True,
        strip_accents='unicode', analyzer='char',
        stop_words='english', ngram_range=(1, 4),
        max_features= limit_char
    ))
])
```

And finally, we have the following for the answer column:

```
answer_col = 'answer'
answer_transformer = Pipeline([
    ('tfidf', TfidfVectorizer(
        lowercase = False, max_df = 0.3, min_df = 1,
        binary = False, use_idf = True, smooth_idf = False,
        ngram_range = (1,2), stop_words = 'english',
        token_pattern = '(?u)\\b\\w+\\b' , max_features = limit_word
    ))
```

```
    ])
answer_transformer2 = Pipeline([
    ('tfidf2',  TfidfVectorizer( sublinear_tf=True,
        strip_accents='unicode', analyzer='char',
        stop_words='english', ngram_range=(1, 4),
        max_features= limit_char
    ))
])
```

We wrap up the feature-engineering part by processing the numerical features. We use simple methods only: missing value imputation to take care of N/A values and a power transformer to stabilize the distribution and make it closer to Gaussian (which is frequently helpful if you are using a numerical feature inside a neural network):

```
num_cols = [
    'question_title_word_len', 'question_body_word_len',
    'answer_word_len', 'answer_div',
    'question_title_num_chars','question_body_num_chars',
    'answer_num_chars',
    'question_title_num_stopwords','question_body_num_stopwords',
    'answer_num_stopwords',
    'question_title_num_punctuations',
    'question_body_num_punctuations','answer_num_punctuations',
    'question_title_num_words_upper',
    'question_body_num_words_upper','answer_num_words_upper',
    'dist0', 'dist1', 'dist2', 'dist3', 'dist4', 'dist5'
]
num_transformer = Pipeline([
    ('impute', SimpleImputer(strategy='constant', fill_value=0)),
    ('scale', PowerTransformer(method='yeo-johnson'))
])
```

A useful feature of pipelines is that they can be combined and nested. Next, we add functionality to handle categorical variables, and then put it all together in a ColumnTransformer object to streamline the data preprocessing and feature engineering logic. Each part of the input can be handled in its own appropriate manner:

```
cat_cols = [
    'dom_0',  'dom_1', 'dom_2',
```

```
        'dom_3', 'category','is_question_no_name_user',
        'is_answer_no_name_user','dom_cnt'
]
cat_transformer = Pipeline([
    ('impute', SimpleImputer(strategy='constant', fill_value='')),
    ('encode', OneHotEncoder(handle_unknown='ignore'))
])
preprocessor = ColumnTransformer(
    transformers = [
        ('title', title_transformer, title_col),
        ('title2', title_transformer2, title_col),
        ('body', body_transformer, body_col),
        ('body2', body_transformer2, body_col),
        ('answer', answer_transformer, answer_col),
        ('answer2', answer_transformer2, answer_col),
        ('num', num_transformer, num_cols),
        ('cat', cat_transformer, cat_cols)
    ]
)
```

Finally, we are ready to use a `Pipeline` object combining preprocessing and model fitting:

```
pipeline = Pipeline([
    ('preprocessor', preprocessor),
    ('estimator',Ridge(random_state=RANDOM_STATE))
])
```

It is always a good idea to evaluate the performance of your model out of sample: a convenient way to go about this is to create out-of-fold predictions, which we discussed in *Chapter 6*. The procedure involves the following steps:

1. Split the data into folds. In our case, we use `GroupKFold`, since one question can have multiple answers (in separate rows of the data frame). In order to prevent information leakage, we want to ensure each question only appears in one fold.

2. For each fold, train the model using the data in the other folds, and generate the predictions for the fold of choice, as well as the test set.

3. Average the predictions on the test set.

We start by preparing the "storage" matrices in which we will store the predictions. mvalid will contain the out-of-fold predictions, while mfull is a placeholder for the predictions on the entire test set, averaged across folds. Since several questions contain more than one candidate answer, we stratify our KFold split on question_body:

```
nfolds = 5
mvalid = np.zeros((xtrain.shape[0], len(target_cols)))
mfull = np.zeros((xtest.shape[0], len(target_cols)))
kf = GroupKFold(n_splits= nfolds).split(X=xtrain.question_body,
                                   groups=xtrain.question_body)
```

We loop through the folds and build the separate models:

```
for ind, (train_index, test_index) in enumerate(kf):

    # Split the data into training and validation
    x0, x1 = xtrain.loc[train_index], xtrain.loc[test_index]
    y0, y1 = ytrain.loc[train_index], ytrain.loc[test_index]
    for ii in range(0, ytrain.shape[1]):
        # Fit model
        be = clone(pipeline)
        be.fit(x0, np.array(y0)[:,ii])
        filename = 'ridge_f' + str(ind) + '_c' + str(ii) + '.pkl'
        pickle.dump(be, open(filename, 'wb'))

        # Storage matrices for the OOF and test predictions, respectively
        mvalid[test_index, ii] = be.predict(x1)
        mfull[:,ii] += be.predict(xtest)/nfolds

    print('---')
```

Once the fitting part is done, we can evaluate the performance in accordance with the metric specified in the competition:

```
corvec = np.zeros((ytrain.shape[1],1))
for ii in range(0, ytrain.shape[1]):
    mvalid[:,ii] = rankdata(mvalid[:,ii])/mvalid.shape[0]
    mfull[:,ii] = rankdata(mfull[:,ii])/mfull.shape[0]

    corvec[ii] = stats.spearmanr(ytrain[ytrain.columns[ii]], mvalid[:,ii])[0]

print(corvec.mean())
```

The final score is 0.34, which is fairly acceptable as a starting point.

In this section, we have demonstrated how to build descriptive features on a body of text. While this is not a winning formula for an NLP competition (the score is OK, but not a guarantee for landing in the medal zone), it is a useful tool to keep in your toolbox. We close this chapter with a section providing an overview of text augmentation techniques.

Interview: Shotaro Ishihara

`https://www.kaggle.com/sishihara`

Our second interview of the chapter is with *Shotaro Ishihara*, aka u++, a Competitions and Notebooks Master who was a member of the winning team in the PetFinder.my Adoption Prediction competition. He is currently a senior data scientist and researcher at the Japanese news media company Nikkei, and has also published various books in Japanese on Kaggle, including a translation of Abhishek Thakur's book. He maintains a weekly newsletter in Japanese on Kaggle initiatives.

Where can we find the Kaggle books you've written/translated?

`https://www.kspub.co.jp/book/detail/5190067.html` is where you can find a Kaggle primer for beginners based on the Titanic GettingStarted competition.

`https://book.mynavi.jp/ec/products/detail/id=123641` is where you can find the Japanese translation of Abhishek Thakur's Approaching (Almost) Any Machine Learning Problem.

What's your favorite kind of competition and why? In terms of techniques and solving approaches, what is your specialty on Kaggle?

In Kaggle, I love joining competitions with tabular or text datasets. These types of datasets are familiar to me because they are widely used in news media companies. I have a good knowledge of the approaches used to handle these datasets.

How do you approach a Kaggle competition? How different is this approach from what you do in your day-to-day work?

The first process is the same: thinking about how to tackle the problem through exploratory data analysis. Kaggle assumes the use of advanced machine learning, but this is not the case in business. In practice, I try to find ways to avoid using machine learning. Even when I do use it, I prefer working with classical methods such as TF-IDF and linear regression rather than advanced methods such as BERT.

We are interested in learning more about how to avoid using machine learning in real-world problems. Can you give us some examples?

When working on automated article summaries at work, we adopt a more straightforward extractive approach (`https://www.jstage.jst.go.jp/article/pjsai/JSAI2021/0/ JSAI2021_1D20S3a03/_article/-char/en`) rather than a neural network-based method (`https://www.jstage.jst.go.jp/article/pjsai/JSAI2021/0/JSAI2021_1D40S3c02/_ article/-char/en`).

It is difficult to guarantee 100% performance with machine learning, and simple methods that are easy for humans to understand and engage with are sometimes preferred.

Tell us about a particularly challenging competition you entered, and what insights you used to tackle the task.

In the PetFinder.my Adoption Prediction competition, a multi-modal dataset was provided. Many participants tried to explore and use all types of data, and the main approach was to extract features from images and texts, concatenate them, and train LightGBM. I also employed the same approach. Surprisingly, one of my teammates, Takuoko (`https://www.kaggle.com/takuok`), developed a great neural network that handles all datasets from end to end. Well-designed neural networks have the potential to outperform LightGBM in multi-modal competitions. This is a lesson I learned in 2019.

Is that lesson still valid today? Has Kaggle helped you in your career? If so, how?

I think the answer is yes. Compared to 2019, neural networks are getting better and better at handling multimodal data.

Yes. Kaggle gave me a lot of experience in data analysis. The machine learning knowledge I've gained from Kaggle has significantly helped me to work more successfully. My achievements in Kaggle and business work were one of the main reasons why I received the 30 Under 30 Awards and Grand Prize in 2020 from the International News Media Association. Kaggle has also allowed me to get to know a lot of people. These relationships have definitely contributed to my career development.

How have you built up your portfolio thanks to Kaggle?

Learned skills, achieved competition results, and published notebooks, books, newsletters, and so on.

How do you promote your publishing?

I have various communication channels and I use the appropriate tools for promotion. For example, Twitter, personal blogs, and YouTube.

In your experience, what do inexperienced Kagglers often overlook? What do you know now that you wish you'd known when you first started?

The importance of exploratory data analysis. In the field of machine learning, there is a concept of No Free Lunch. We should not only learn algorithms, but also learn how to address challenges. The No Free Lunch theorem is a statement that there is no universal model that performs well on all problems.

In machine learning competitions, it is essential to find a model that is appropriate to the characteristics of the dataset and the task in order to improve your score.

What mistakes have you made in competitions in the past?

Overfitting to the public leaderboard. In the LANL Earthquake Prediction competition, I scored pretty well on the public leaderboard and finished the competition at the rank of fifth. However, my final ranking was 211st, which means I believed too much in a limited dataset. Overfitting is a very popular concept in machine learning, and I realized the importance of this with pain through Kaggle.

Do you suggest any particular way to avoid overfitting?

It is important to observe carefully how the training and evaluation datasets are divided. I try to build a validation set that reproduces this partitioning.

Are there any particular tools or libraries that you would recommend using for data analysis or machine learning?

I love pandas, which is an essential library for handling tabular datasets. I use it for exploratory data analysis by extracting, aggregating, and visualizing.

What do you suggest readers do to master pandas?

You can look at some community tutorials. Kaggle also provides some learning tutorial courses on pandas and feature engineering.

Do you use other competition platforms? How do they compare to Kaggle?

I sometimes use Japanese platforms like Signate, Nishika, etc. (`https://upura.github.io/projects/data_science_competitions/`). These are obviously inferior to Kaggle in terms of functionality and UX/UI, but it's interesting to see familiar subjects like the Japanese language.

Toxic comments classification

The first "Toxic Comments" competition was one of the first NLP competitions that one of us (Bojan) spent a significant amount of time on and seriously participated in. It was also one of the last few Kaggle NLP competitions before Transformers, LLMs, and similar technologies became widely available and popular. In that regard, it serves as a good testbed for comparing "classical" NLP techniques with more modern ones. Our team finished third in that competition, and some of the special tricks and techniques we used have stood the test of time, such as bespoke text preprocessing and unorthodox text augmentation techniques.

To get a better feel for the problem, let's start by looking at the data:

```
train = pd.read_csv('../input/train.csv.zip').fillna(' ')
train.head()
```

The output of the above code is the following:

	id	comment_text	toxic	severe_toxic	obscene	threat	insult	identity_hate
0	0000997932d777bf	Explanation\nWhy the edits made under my usern...	0	0	0	0	0	0
1	000103f0d9cfb60f	D'aww! He matches this background colour I'm s...	0	0	0	0	0	0
2	000113f07ec002fd	Hey man, I'm really not trying to edit war. It...	0	0	0	0	0	0
3	0001b41b1c6bb37e	"\nMore\nI can't make any real suggestions on ...	0	0	0	0	0	0
4	0001d958c54c6e35	You, sir, are my hero. Any chance you remember...	0	0	0	0	0	0

Figure 12.4: Heading rows from the training data

The output shows the first few rows of the dataset. Each row contains a unique `id` (a hash value), the raw `comment_text`, which is what we use as a predictor, and six binary columns (`toxic`, `severe_toxic`, `obscene`, `threat`, `insult`, and `identity_hate`) that serve as our target labels, indicating which toxicity categories the comment belongs to.

Text classification with TF-IDF and logistic regression

Behind all machine learning algorithms lies the problem of how to represent the data we have in numerical terms. For NLP problems, this reduces to converting a corpus of texts into a numerical representation. One of the conceptually simplest ways of accomplishing this is by turning each separate word into a vector space of the same dimension as the total vocabulary. In that case, each data point becomes just a collection of word vectors, each weighed by the number of times it appears in the given data point. We call this a **bag-of-words (BoW)** encoding, because we are literally treating each individual piece of text (a sentence, paragraph, etc.) as an unordered collection of words. The order of words in the sentence doesn't matter. This seems like an incredibly simplistic reduction, but for many real-world problems, it works remarkably well, especially when we have a large, representative, and labeled dataset. (Considering the philosophical implications of what this means about our communication with each other is left as an exercise for you.)

The limitations of BoW are the following:

- **Sparsity**: If the vocabulary is large, the resulting vectors are often sparse (many 0s).
- **No semantic information**: BoW doesn't capture the meaning or context of words.
- **Word order loss**: Important relationships between words are ignored.

For our toxic comments classification, we'll use a variation of the BoW encoding called TF-IDF. It's an improvement over BoW that adjusts word counts based on their frequency of appearance across all documents. It is implemented in the scikit-learn package and is fast and easy to use. We'll build a machine learning pipeline for multi-label text classification using logistic regression. It is designed to predict various types of toxic comments (such as "toxic," "obscene," "insult," etc.) from text data. The approach involves extracting features from the text using TF-IDF for both word-level and character-level n-grams and applying logistic regression to make predictions.

Importing the necessary libraries

Before we begin processing data or training models, we need to import the necessary Python libraries and frameworks that will assist us in the machine learning process. The code imports NumPy and pandas for general numerical and data manipulation tasks. It also utilizes scikit-learn (sklearn) for its machine learning utilities, specifically a logistic regression classifier and a cross-validation scoring function. Additionally, `scipy.sparse` is used to efficiently store and combine sparse matrix representations, which are crucial for large-scale text data. The following code is used to import the relevant libraries:

```
import numpy as np
import pandas as pd


from sklearn.feature_extraction.text import TfidfVectorizer
from sklearn.linear_model import LogisticRegression
from sklearn.model_selection import cross_val_score
from scipy.sparse import hstack
```

Defining the target labels and loading the dataset

In text classification problems, the labels are categories into which each text instance is to be classified. In this case, we have six different categories of toxicity: toxic, severe_toxic, obscene, threat, insult, and identity_hate. These are stored in a list, allowing us to easily iterate over them later. Next, the code reads in the training and testing data from CSV files. Any missing values are filled with whitespace to avoid empty strings that could disrupt text processing.

By separating train_text and test_text, and also combining them into all_text, we set ourselves up to fit text vectorizers on all available text at once. This ensures that the training and testing sets are represented in the same feature space. A key preprocessing step is to ensure that both sets share the same vocabulary and TF-IDF transformations, which can be done by fitting the vectorizer on the concatenated text and then transforming each subset:

```
class_names = ['toxic', 'severe_toxic', 'obscene', 'threat',
               'insult', 'identity_hate']
train = pd.read_csv('../input/train.csv').fillna(' ')
test = pd.read_csv('../input/test.csv').fillna(' ')

train_text = train['comment_text']
test_text = test['comment_text']
all_text = pd.concat([train_text, test_text])
```

Word-level TF-IDF vectorization

The code utilizes two distinct TF-IDF vectorizers: one at the word level and one at the character level. The reasoning behind using both is that subtle differences in how language is used (including character n-grams) may help identify nuances of toxicity that single words alone do not capture. For example, misspellings or character-level patterns that may signify harassment can be detected via character n-grams.

As shown in the following code, the first vectorizer focuses on words. It uses a sublinear TF scaling, strips Unicode accents, ignores English stop words, and considers only words with a length of at least one character. It also sets a maximum number of features to limit the dimensionality of the vector space. After fitting this vectorizer on `all_text`, the code transforms the `train_text` and `test_text` into numeric matrices:

```python
word_vectorizer = TfidfVectorizer(
    sublinear_tf=True,
    strip_accents='unicode',
    analyzer='word',
    token_pattern=r'\w{1,}',
    stop_words='english',
    ngram_range=(1, 1),
    max_features=10000
)
word_vectorizer.fit(all_text)
train_word_features = word_vectorizer.transform(train_text)
test_word_features = word_vectorizer.transform(test_text)
```

Character-level TF-IDF vectorization

Similarly, the character-level TF-IDF vectorizer captures sequences of characters (2 to 6 characters long) rather than whole words. This can help detect patterns such as elongated words, special punctuation sequences, or character-level insults that do not form coherent words. As before, we fit on the concatenated text and then transform the training and test sets:

```python
char_vectorizer = TfidfVectorizer(
    sublinear_tf=True,
    strip_accents='unicode',
    analyzer='char',
    ngram_range=(2, 6),
    max_features=50000
)
char_vectorizer.fit(all_text)
train_char_features = char_vectorizer.transform(train_text)
test_char_features = char_vectorizer.transform(test_text)
```

Combining word and character features

We now have two sets of features for both training and test data: character-level features and word-level features. We aim to combine them into a single set of features for each instance, as both representations may capture complementary information. The hstack function from scipy. sparse allows us to efficiently combine sparse matrices horizontally (side by side), producing a single sparse feature matrix that can be passed to the classifier:

```
train_features = hstack([train_char_features, train_word_features])
test_features = hstack([test_char_features, test_word_features])
```

Training the logistic regression model and cross-validation

For each of the class names (the toxicity categories), we will train a separate logistic regression classifier. Logistic regression is often a strong baseline method for text classification problems, and it is relatively fast to train. We use C=0.1 to control regularization strength and solver='sag' for optimization. Before training, we evaluate the model using cross-validation. Cross-validation splits the training data into multiple folds, trains the model on some of the folds, and tests it on the remaining fold, then repeats this process. The average result (cv_score) provides a more reliable estimate of how well the model will perform on unseen data.

After computing the cross-validation scores, the code prints these scores for each class. It then fits the classifier on the entire training set and uses the trained model to produce predictions for the test set. Notice that instead of simple class predictions, we use predict_proba to obtain a probability of belonging to each class. These probabilities can be very useful for ranking or for threshold-based decision-making later on.

```
scores = []
submission = pd.DataFrame.from_dict({'id': test['id']})
for class_name in class_names:
    train_target = train[class_name]
    classifier = LogisticRegression(C=0.1, solver='sag')

    cv_score = np.mean(cross_val_score(classifier, train_features,
                                       train_target, cv=3,
                                       scoring='roc_auc'))
    scores.append(cv_score)
    print('CV score for class {} is {}'.format(class_name, cv_score))

    classifier.fit(train_features, train_target)
    submission[class_name] = classifier.predict_proba(test_features)[:, 1]
```

Calculating the total cross-validation score

After finishing the loop for all classes, the code prints out the average cross-validation score across all toxicity categories, giving a single number that summarizes model quality:

```
print('Total CV score is {}'.format(np.mean(scores)))
```

Saving the predictions to a CSV file

Finally, the predictions are saved into a `submission.csv` file:

```
submission.to_csv('submission.csv', index=False)
```

This code implements a standard text classification pipeline: loading data, creating TF-IDF features at both the word and character level, combining these features, training a logistic regression model for each target class, evaluating it via cross-validation, and finally, generating predictions on the test set. This approach is typical in text classification tasks and provides a strong baseline for more complex methods, such as neural networks or ensemble models.

After saving the predictions, our pipeline is complete. This code implements a standard text classification pipeline, and we can discuss how to refine our approach and obtain better results.

Text preprocessing and cleanup

Even though most modern NLP algorithms and embeddings are powerful enough to get very good classification results right out of the box, thoughtful preprocessing, especially when the specific context of the machine learning problem is taken into account, can still have a significant impact on the quality of the final models. This is particularly the case when every little bit of improvement counts, such as in Kaggle competitions.

Below, we will walk through a step-by-step guide to a data preprocessing and normalization script used before feeding text data into machine learning models for classification tasks, such as toxicity detection. This code reads in raw training and test data, normalizes and cleans the text, and prepares it for further modeling. We will first explore the primary purpose of each library import and the constants defined, then delve into the dictionaries and patterns used to clean text, and finally, examine the normalization pipeline that transforms raw user-generated content into cleaner, tokenized input for a neural model or traditional machine learning classifier.

Importing libraries, loading files, and setting up global variables

We start by importing standard Python modules for file handling, string operations, and more specialized libraries for text processing, vectorization, and model splitting. Also, basic hyperparameters and constants such as the names of text columns, label classes, and thresholds for text length are defined. These values guide the downstream operations in the normalization and modeling steps. We also read a sample submission file that might be used as a template for final predictions. The train_filepath and test_filepath are set to locate the raw training and testing CSV files. Additional dictionary files such as hyphenation, misspellings, merged corrections, and lists of known toxic words are located in corresponding binary files. These serve as references during text normalization.

First, let's import all the relevant libraries and set a few general parameters for our environment:

```
import os, math, operator, csv, random, pickle,re

import gc

from nltk.tokenize import TweetTokenizer
#from spacy.symbols import nsubj, VERB, dobj
import spacy
import en_core_web_sm
from unidecode import unidecode
from sklearn.model_selection import KFold, train_test_split

import numpy as np
import pandas as pd

from sklearn.model_selection import train_test_split

import sys
sys.setrecursionlimit(1500)
```

Next, we'll set a few constants and parameters and load the data. Many of these datasets have been preprocessed in advance and serve the purpose of using them for text preprocessing and cleaning:

```
TEXT_COLUMN = 'comment_text'
list_classes = ["toxic", "severe_toxic", "obscene",
                "threat", "insult", "identity_hate"]
CHARS_TO_REMOVE =
"""!"#$%&()*+,-./:;<=>?@[\]^_`{|}~\t\n"""'∞θ ÷ α • à − βØ³π'₹'°£€\× ™√² —"""
submission = pd.read_csv("../input/sample_submission.csv.zip")

categories = ["toxic", "severe_toxic", "obscene",
              "threat", "insult", "identity_hate"]

data_folder = "../input/"
pretrained_folder = "../input/"
train_filepath = data_folder + "train.csv.zip"
test_filepath = data_folder + "test.csv.zip"

submission_path = data_folder + "submission.csv"

# ALL the bin files can be found at https://www.kaggle.com/datasets/
tunguz/cleaning-dictionaries/
hyphens_filepath = (
    "../input/cleaning-dictionaries/hyphens_dictionary.bin")
misspellings_filepath = (
    "../input/cleaning-dictionaries/misspellings_all_dictionary.bin")
merged_filepath = (
    "../input/cleaning-dictionaries/merged_all_dictionary.bin")
toxic_words_filepath = (
    "../input/cleaning-dictionaries/toxic_words.bin")
asterisk_words_filepath = (
    "../input/cleaning-dictionaries/asterisk_words.bin")
fasttext_filepath = (
    "../input/cleaning-dictionaries/merged_all_dictionary.bin")
```

The next part utilizes the file path definitions for loading the pretrained dictionary file.

Loading pretrained dictionaries

Text normalization often involves mapping commonly misspelled or variant forms of words back to a standard dictionary form. Here, we load multiple dictionaries to address issues such as hyphenation, misspellings, and known toxic words.

These dictionaries are stored in binary (.bin) format and loaded using the pickle module. By reading them into memory, the preprocessing functions can quickly reference these mappings.

After loading these dictionaries, the code prints out their lengths, giving a quick sense of how comprehensive each dictionary is. This is useful for debugging or confirming that the dictionaries have been loaded correctly.

```
with open(hyphens_filepath, mode='rb') as file:
    hyphens_dict = pickle.load(file)
with open(misspellings_filepath, mode='rb') as file:
    misspellings_dict = pickle.load(file)
with open(merged_filepath, mode='rb') as file:
    merged_dict = pickle.load(file)
with open(toxic_words_filepath, mode='rb') as file:
    toxic_words = pickle.load(file)
with open(asterisk_words_filepath, mode='rb') as file:
    asterisk_words = pickle.load(file)
with open(fasttext_filepath, mode='rb') as file:
    fasttext_misspelings = pickle.load(file)

print(len(hyphens_dict))
print(len(misspellings_dict))
print(len(merged_dict))
print(len(toxic_words))
print(len(asterisk_words))
print(len(fasttext_misspelings))
```

Setting preprocessing parameters

Before processing the text, it is common to define parameters that guide the normalization process. Below, limits such as length_threshold (the maximum length of a comment before truncation) and word_count_threshold (the maximum number of words allowed) help ensure that extremely long comments do not overwhelm memory or downstream models.

Furthermore, `words_limit` may be used in certain vectorization steps to limit the number of words considered globally. Additionally, the code defines sets of valid characters, which helps clean out unwanted symbols and focus on core alphanumeric and punctuation characters.

```
training_samples_count = 149571
validation_samples_count = 10000

length_threshold = 20000 # Truncate comments longer than this character
length
word_count_threshold = 900 # Truncate comments with more than this many
words
words_limit = 310000

valid_characters = (
    " " + "@$" + "'!?-" + "abcdefghijklmnopqrstuvwxyz"
    + "abcdefghijklmnopqrstuvwxyz".upper()
)
valid_characters_ext = valid_characters + "abcdefghijklmnopqrstuvwxyz".
upper()
valid_set = set(x for x in valid_characters)
valid_set_ext = set(x for x in valid_characters_ext)
```

Defining contraction patterns

Many English texts contain contractions like "don't" or "I'm." Since machine learning models often benefit from a more standardized input form, these patterns are defined to expand contractions into their fully spelled-out forms. The code uses regular expressions to identify and replace such patterns, ensuring more consistent tokenization and ultimately better model performance. Here, patterns is a list of compiled regular expressions ready for quick substitution operations during normalization.

```
cont_patterns = [
    (r'(W|w)on\'t', r'will not'),
    (r'(C|c)an\'t', r'can not'),
    (r'(I|i)\'m', r'i am'),
    (r'(A|a)in\'t', r'is not'),
    (r'(\w+)\'ll', r'\g<1> will'),
    (r'(\w+)n\'t', r'\g<1> not'),
    (r'(\w+)\'ve', r'\g<1> have'),
```

```
      (r'(\w+)\'s', r'\g<1> is'),
      (r'(\w+)\'re', r'\g<1> are'),
      (r'(\w+)\'d', r'\g<1> would'),
    ]
    patterns = [(re.compile(regex), repl) for (regex, repl) in cont_patterns]
```

Splitting toxic words

A key part of this pipeline is identifying and isolating known "toxic" words. The function split_
word() checks if a given word contains any known toxic sub-words from the toxic_words dictionary. If such sub-words are found, the code splits the word into cleaner segments, effectively
segmenting a longer token that might contain inappropriate content into more meaningful pieces.
This allows models to more accurately represent and handle harmful language. This recursive
strategy ensures that multiple toxic words within a single token are fully isolated.

```
    def split_word(word, toxic_words):
        if word == "":
            return ""

        lower = word.lower()
        for toxic_word in toxic_words:
            start = lower.find(toxic_word)
            if start >= 0:
                end = start + len(toxic_word)
                result = " ".join([word[0:start], word[start:end],
                                   split_word(word[end:], toxic_words)])
                return result.replace("  ", " ").strip()
        return word
```

Tokenizing with TweetTokenizer

The TweetTokenizer from NLTK is well suited for handling informal text, often seen in social
media or online discussion forums. It manages issues like repeated characters and special tokens
commonly found in user-generated content. The *word_tokenize()* function below also enacts some
simple replacements (e.g., $ to s, @ to a) and inserts spaces around punctuation. Ultimately, this
leads to a more consistent list of tokens ready for further normalization.

```
    tknzr = TweetTokenizer(strip_handles=False, reduce_len=True)
    def word_tokenize(sentence):
        sentence = sentence.replace("$", "s")
```

```
sentence = sentence.replace("@", "a")
sentence = sentence.replace("!", " ! ")
sentence = sentence.replace("?", " ? ")


return tknzr.tokenize(sentence)
```

URL replacement

URLs in text often contain no useful semantic information for certain classification tasks and can add noise. The replace_url() function identifies words that look like URLs and replaces them with an empty string, effectively removing them from the token stream.

```
def replace_url(word):
    if ("http://" in word or "www." in word or "https://" in word
        or "wikipedia.org" in word):
        return ""
    return word
```

Normalizing by dictionary

The normalization process includes dictionary lookups to correct misspellings or variations of words. The normalize_by_dictionary() function takes a previously normalized word, splits it, and checks each sub-word against a given dictionary. If a match is found, it replaces the word with the corrected form from the dictionary. Applying multiple dictionaries sequentially refines the input text step by step. Here, uppercase words are preserved in uppercase if matched, acknowledging that capitalization might matter in certain contexts.

```
def normalize_by_dictionary(normalized_word, dictionary):
    result = []
    for word in normalized_word.split():
        if word == word.upper():
            if word.lower() in dictionary:
                result.append(dictionary[word.lower()].upper())
            else:
                result.append(word)
        else:
            if word.lower() in dictionary:
                result.append(dictionary[word.lower()])
            else:
                result.append(word)

    return " ".join(result)
```

Loading a spaCy model

The script loads a spaCy model (en_core_web_sm) for potential lemmatization, part-of-speech tagging, or named entity recognition. Though not fully shown in use here, spaCy's pipeline is a powerful tool for more advanced text preprocessing steps.

```
nlp = en_core_web_sm.load()
```

The main normalization function

The normalize_comment() function ties together all the steps described above. It first converts text to ASCII with unidecode, applies truncation thresholds, and then looks for patterns of asterisk words to replace. After tokenization, each word is cleaned: URLs are removed, punctuation is filtered, and toxic sub-words are split out. Subsequent dictionary lookups standardize spelling, hyphenation, and other textual quirks. Finally, the comment text is converted to lowercase unless a word is entirely uppercase (often meaning it might be an acronym or intentionally emphasized), and certain known multi-word terms, like sock puppet are standardized.

```python
def normalize_comment(comment):
    comment = unidecode(comment)
    comment = comment[:length_threshold]

    # Replace known asterisk patterns
    for w in asterisk_words:
        if w[0] in comment:
            comment = comment.replace(w[0], w[1])
        if w[0].upper() in comment:
            comment = comment.replace(w[0].upper(), w[1].upper())

    normalized_words = []
    for word in word_tokenize(comment):
        word = replace_url(word)
        if word.count(".") == 1:
            word = word.replace(".", " ")
        filtered_word = "".join([x for x in word if x in valid_set])

        # Split toxic words inside larger tokens
        normalized_word = split_word(filtered_word, toxic_words)
```

```
        # Apply multiple dictionary normalizations
        normalized_word = normalize_by_dictionary(
            normalized_word, hyphens_dict)
        normalized_word = normalize_by_dictionary(
            normalized_word, merged_dict)
        normalized_word = normalize_by_dictionary(
            normalized_word, misspellings_dict)
        normalized_word = normalize_by_dictionary(
            normalized_word, fasttext_misspelings)

        normalized_words.append(normalized_word)

    # Convert words to lowercase unless fully uppercase
    normalized_comment = " ".join(normalized_words)
    result = []
    for word in normalized_comment.split():
        if word.upper() == word:
            result.append(word)
        else:
            result.append(word.lower())

    result = " ".join(result)
    # Merge certain specific words
    if "sock puppet" in result:
        result = result.replace("sock puppet", "sockpuppet")
    if "SOCK PUPPET" in result:
        result = result.replace("SOCK PUPPET", "SOCKPUPPET")

    return result
```

Reading and normalizing data

The function read_data_files() reads in the training and testing CSV files. It extracts both the raw text and their labels (in the training set) and applies the normalize_comment() function to transform the data. The result is a pair of arrays ready for modeling. By using np.vectorize() to apply normalization, the code quickly processes large volumes of text. Once cleaned, the data arrays are printed to confirm their shapes and then returned for saving or further use.

```
def read_data_files(train_filepath, test_filepath):
    # read train data
    train = pd.read_csv(train_filepath)
    labels = train[categories].values

    # read test data
    test = pd.read_csv(test_filepath)
    test_comments = test["comment_text"].fillna("_na_").values

    # normalize comments
    np_normalize = np.vectorize(normalize_comment)
    comments = train["comment_text"].fillna("_na_").values
    normalized_comments = np_normalize(comments)
    del comments
    gc.collect()

    comments = test["comment_text"].fillna("_na_").values
    normalized_test_comments = np_normalize(test_comments)
    del comments
    gc.collect()

    print('Shape of data tensor:', normalized_comments.shape)
    print('Shape of label tensor:', labels.shape)
    print('Shape of test data tensor:', normalized_test_comments.shape)

    return (labels, normalized_comments, normalized_test_comments)

labels, x_train, x_test = read_data_files(train_filepath, test_filepath)
```

Saving the processed data

Finally, once the labels and normalized comment arrays are prepared, they are saved as .npy files. These NumPy arrays can be quickly loaded later, allowing model training and experimentation without needing to repeat the lengthy normalization process.

```
np.save("../cleaned_data/lables", labels)
np.save("../cleaned_data/x_train", x_train)
np.save("../cleaned_data/x_test", x_test)
```

In summary, this section has demonstrated how a textual dataset is transformed from raw input to a sanitized and standardized form suitable for model training. Steps included removing unwanted characters, expanding contractions, fixing misspellings, splitting toxic words, and leveraging multiple dictionaries to ensure the input text is as clean and semantically consistent as possible. This thorough preprocessing often leads to improved performance and better generalization in machine learning models, particularly those dealing with user-generated content and toxicity classification.

Text classification with RNNs

For many years, **recurrent neural networks (RNNs)**, such as GRU and LSTM, have been the primary workhorses in deep learning applications for NLP. For the past six-plus years, however, they have been overshadowed by transformers and LLMs. Yet, for most everyday use cases, they remain close to being optimal. When it comes to NLP applications, RNNs need to be used in conjunction with classical Word2Vec embeddings. One of the key strategies our team employed in this competition was to utilize **dual** embeddings within the same RNN architecture. In the following sections, we will show how that was done.

Imports and environment setup

Before we begin building and training our model, we need to import several dependencies and packages that provide the necessary functionality for data manipulation, model building, and evaluation. Here, you will see a variety of imports that support tasks such as loading, cleaning, and tokenizing the data; building, training, and evaluating deep neural networks; and utilizing common utilities like regular expressions and file operations. The libraries include powerful frameworks like TensorFlow and scikit-learn, as well as convenient text processing packages such as nltk and unidecode.

```
import os, math, operator, csv, random, pickle, re
import pandas as pd
import numpy as np
import gc

import tensorflow as tf
print(tf.__version__)
tf.test.is_gpu_available(
    cuda_only=False,
    min_cuda_compute_capability=None
)
```

```python
from tensorflow.keras.models import Model
from tensorflow.keras.layers import (
    MaxPooling1D, BatchNormalization, Permute, Lambda, Activation, Conv1D,
    GlobalAveragePooling1D, GlobalMaxPooling1D, Dense, Embedding, Dropout,
    Input, Flatten, TimeDistributed, concatenate, SpatialDropout1D,
    Bidirectional, LSTM, GRU, add
)
from tensorflow.keras.callbacks import LearningRateScheduler
from tensorflow.keras.preprocessing.text import Tokenizer
from tensorflow.keras.preprocessing.sequence import pad_sequences
from tensorflow.keras import backend as K

from nltk.tokenize import TweetTokenizer

from unidecode import unidecode

from sklearn.model_selection import KFold, train_test_split
from sklearn.metrics import roc_auc_score
from sklearn.model_selection import train_test_split
```

Loading preprocessed data

At this stage, we assume the data has already been preprocessed and saved in a suitable format. We load preprocessed labels and tokenized text sequences from .npy files. Storing data in NumPy arrays allows for efficient loading and memory usage, which is crucial for large-scale text datasets.

We also load the trained tokenizer object, which has been previously fit on the training data. This tokenizer is crucial for converting the raw text into sequences of integer indices, allowing the model to treat words as discrete numeric tokens. After loading the tokenizer, we convert our text data into these tokenized sequences and pad them to ensure all sequences are of equal length—a common practice required by most RNNs and other sequence-based models.

```python
# Load all the preprocessed data as numpy text arrays.
labels = np.load('../input/labels.npy')
x_train = np.load('../input/x_train.npy')
x_test = np.load('../input/x_test.npy')

fileObject = open('../dictionaries/tokenizer','rb')
tokenizer = pickle.load(fileObject)
```

```
x_train = tokenizer.texts_to_sequences(x_train)
x_test = tokenizer.texts_to_sequences(x_test)
x_train = sequence.pad_sequences(x_train, maxlen=MAX_LEN)
x_test = sequence.pad_sequences(x_test, maxlen=MAX_LEN)
```

Loading embeddings

Pretrained embeddings, such as GloVe and fastText, provide dense vector representations of words that capture semantic relationships. Incorporating these embeddings typically improves model performance on text classification tasks. Here, we load a precomputed embedding matrix that maps each token ID to a corresponding embedding vector. By using this matrix as the weights for an embedding layer, our model can leverage rich semantic information from the start.

```
# Load the dual embeddings matrix:
embedding_matrix = np.load('../embeddings/embedding_matrix_big.npy')
```

Splitting the datasets

To properly measure how well our model generalizes, we need to hold back a portion of the training data as a validation set. The `train_test_split` function from scikit-learn splits our labeled examples into training and validation sets. The validation set helps us tune hyperparameters, avoid overfitting, and select the best model version without over-optimizing on the training data.

```
# split the train data into the train and validation sets
x_train, x_valid, y_train, y_valid = train_test_split(
    x_train, labels, test_size = 0.1)
```

Building the Keras model

Here we define the neural network architecture using Keras' functional API. We start with an Input layer that will receive integer-encoded words. Next, we include a non-trainable Embedding layer initialized with our pretrained embeddings, followed by a SpatialDropout1D to reduce overfitting by randomly dropping entire 1D feature maps.

We then stack recurrent layers, bidirectional GRU and bidirectional LSTM to capture both the forward and backward context in the text. The output of these recurrent layers is pooled using both `GlobalMaxPooling1D` and `GlobalAveragePooling1D` to condense the sequence information into fixed-length vectors. Finally, we concatenate these pooled features and feed them through a couple of dense layers with residual connections to form a richer representation before outputting predictions through a sigmoid activation for a multi-label classification scenario.

```python
# Define the Keras model
def build_model(embedding_matrix):
    words = Input(shape=(None,))
    x = Embedding(*embedding_matrix.shape,
                  weights=[embedding_matrix],
                  trainable=False)(words)
    x = SpatialDropout1D(0.2)(x)
    x = Bidirectional(GRU(LSTM_UNITS, return_sequences=True))(x)
    x = Bidirectional(LSTM(LSTM_UNITS, return_sequences=True))(x)

    hidden = concatenate([
        GlobalMaxPooling1D()(x),
        GlobalAveragePooling1D()(x),
    ])
    hidden = add([
        hidden,
        Dense(DENSE_HIDDEN_UNITS, activation='relu')(hidden)
    ])
    hidden = add([
        hidden,
        Dense(DENSE_HIDDEN_UNITS, activation='relu')(hidden)
    ])
    result = Dense(6, activation='sigmoid')(hidden)

    model = Model(inputs=words, outputs=result)
    model.compile(loss='binary_crossentropy', optimizer='adam')

    return model
```

Training and averaging multiple seeds

Training deep neural networks can be sensitive to initial random weight initialization. To mitigate this variability and potentially improve overall results, it's common to train the model multiple times with different random seeds and then average the predictions. Here, we train the model for a few epochs, adjusting the learning rate each time using a scheduler to stabilize training and improve convergence.

For each seed, we evaluate the model on the validation set and compute the ROC-AUC score, which is a robust metric for multi-label classification. After training multiple models, we average their predictions on the test set to produce a more stable final prediction.

```python
# Train the model and make predictions on the test set.
# In order to improve performance we use a 10 seed average.

EPOCHS = 5
SEEDS = 10

pred = 0

for ii in range(SEEDS):
    model = build_model(embedding_matrix)
    for global_epoch in range(EPOCHS):
        print(global_epoch)
        model.fit(
            x_train,
            y_train,
            validation_data = (x_valid, y_valid),
            batch_size=128,
            epochs=1,
            verbose=2,
            callbacks=[LearningRateScheduler(
                lambda _: 1e-3 * (0.55 ** global_epoch))
            ]
        )
        val_preds = model.predict(x_valid)
        AUC = 0
        for i in range(6):
            AUC += roc_auc_score(y_valid[:,i], val_preds[:,i])/6.
        print(AUC)

    pred += model.predict(x_test, batch_size = 1024, verbose = 1)/SEEDS
```

Creating a submission file

Here, we load a sample submission file, replace the placeholder predictions with our averaged model predictions, and then save it. This final step completes our pipeline, allowing us to submit our predictions to a leaderboard or deploy them in a production environment.

```
# We create the submission file

list_classes = ["toxic", "severe_toxic", "obscene",
                "threat", "insult", "identity_hate"]
submission = pd.read_csv('../input/sample_submission.csv')
submission[list_classes] = pred
submission.to_csv('../submissions/submission.csv', index=False)
submission.head()


# This model scores 0.98644 on the Private Leaderboard, and 0.98653 on the
public leaderboard
```

In short, this workflow demonstrates that carefully engineered RNNs, augmented with high-quality, dual-pretrained embeddings, judicious regularization, and a modest ensemble of random seeds, can still deliver near-state-of-the-art results on real-world text classification tasks. By combining bidirectional GRU + LSTM layers with both max and average pooling, we effectively capture local and longrange context while controlling model size and training time. Although transformers now dominate many NLP leaderboards, this chapter demonstrates that a well-tuned RNN pipeline remains a robust, resourceefficient choice—especially when compute or latency constraints make larger models impractical.

Text classification with DistilBERT

Modern NLP applications are built around transformers. BERT was one of the earliest widely used transformers, and it was later "distilled" into a more compact version called DistilBERT. The code in this section, added ahead, has been developed by Sebastian Raschka and fine-tuned by Bojan to obtain a score that's on par with some of the better RNNs.

Setting up the environment and dependencies

In the code below, we import both fundamental Python libraries and machine learning frameworks that will aid in creating and training our model. We also define key constants that determine aspects like sequence length, batch sizes, and learning rates. This part of the code ensures we have the right tools at hand and a working environment ready for model development.

```python
import pandas as pd
import torch
from tqdm import tqdm
from torch.utils.data import Dataset, DataLoader, RandomSampler,
SequentialSampler
from transformers import DistilBertTokenizer, DistilBertModel


MAX_LEN = 320
TRAIN_BATCH_SIZE = 32
VALID_BATCH_SIZE = 32
EPOCHS = 2
LEARNING_RATE = 1e-05
DEVICE = 'cuda:0' if torch.cuda.is_available() else 'cpu'
print(DEVICE)
```

Here, we load `pandas` for data manipulation, `torch` and related classes for model training, and the `transformers` library from Hugging Face, which provides the pretrained DistilBERT model and tokenizer. Key hyperparameters such as `MAX_LEN` (the maximum token length for BERT), `TRAIN_BATCH_SIZE`, and `EPOCHS` are defined upfront so that we can easily reference and adjust them later. We also detect whether a GPU is available and choose to run on cuda:0 if so, otherwise defaulting to CPU.

Loading and preparing the training data

In supervised learning tasks, it's critical to load and prepare your data carefully. The dataset we're working with includes labels for toxicity classification, and we want to merge these multiple labels into a single list per training example. By doing this, we create a suitable target format for multi-label classification using DistilBERT.

```python
train_data = pd.read_csv('../input/train.csv.zip')

label_columns = ["toxic", "severe_toxic", "obscene",
                 "threat", "insult", "identity_hate"]
```

```
train_data['labels'] = train_data[label_columns].apply(lambda x: list(x),
axis=1)

train_data.drop(['id'], inplace=True, axis=1)
train_data.drop(label_columns, inplace=True, axis=1)

print(train_data.head())
```

We start by reading the training data from a CSV file. Next, we create a new column, labels, that stores all six target values—toxic, severe_toxic, obscene, threat, insult, and identity_hate—as a list for each row. This makes it easier to feed the targets into our model later. We then remove unnecessary columns like the original id and the individual label columns, since we've condensed them into a single labels column, keeping the data tidy and directly usable for the training process.

Creating a custom Dataset class for multi-label classification

Deep learning frameworks, such as PyTorch, expect datasets to be organized in a specific way. By creating a custom Dataset class, we standardize how we will tokenize, encode, and ultimately feed our data into the model.

```
class MultiLabelDataset(Dataset):

    def __init__(self, dataframe, tokenizer, max_len, new_data=False):
        self.tokenizer = tokenizer
        self.data = dataframe
        self.text = dataframe.comment_text
        self.new_data = new_data

        if not new_data:
            self.targets = self.data.labels
        self.max_len = max_len

    def __len__(self):
        return len(self.text)

    def __getitem__(self, index):
        text = str(self.text[index])
        text = " ".join(text.split())
```

```
        inputs = self.tokenizer.encode_plus(
            text,
            None,
            add_special_tokens=True,
            max_length=self.max_len,
            pad_to_max_length=True,
            return_token_type_ids=True
        )
        ids = inputs['input_ids']
        mask = inputs['attention_mask']
        token_type_ids = inputs["token_type_ids"]

        out = {
            'ids': torch.tensor(ids, dtype=torch.long),
            'mask': torch.tensor(mask, dtype=torch.long),
            'token_type_ids': torch.tensor(token_type_ids,
                                        dtype=torch.long),
        }
        if not self.new_data:
            out['targets'] = torch.tensor(self.targets[index],
                                        dtype=torch.float)
        return out
```

This class takes a DataFrame and the `tokenizer` as inputs. The tokenizer from Hugging Face's Transformers library converts raw text into input IDs, attention masks, and token type IDs—formats expected by the DistilBERT model. The __getitem__ method prepares each text for input to the model by splitting and encoding it and, if it's training data (not new/unlabeled data), attaches the corresponding targets. For test or inference data, we only return inputs without targets (`new_data=True`).

Splitting the data into training and validation sets

To ensure that our model generalizes well, we typically split our dataset into training and validation sets. The training set is used to adjust the model's parameters, while the validation set helps us monitor the model's performance during training and avoid overfitting.

```
train_size = 1.0


train_df = train_data.sample(frac=train_size, random_state=123)
```

```
val_df = train_data.drop(train_df.index).reset_index(drop=True)
train_df = train_df.reset_index(drop=True)

print("Orig Dataset: {}".format(train_data.shape))
print("Training Dataset: {}".format(train_df.shape))
print("Validation Dataset: {}".format(val_df.shape))
```

Here, for demonstration purposes, we chose train_size = 1.0, meaning we're not actually leaving any samples for validation. However, the code is structured to allow a fraction of data to be reserved as a validation set. In a real scenario, train_size would be less than 1.0, for example, 0.9, so that we have a dedicated portion of the data for validation.

Initializing the tokenizer and creating data loaders

Once we have defined our dataset class and prepared our training/validation splits, we need to instantiate the tokenizer and create PyTorch DataLoader objects. The DataLoader handles batching of samples and shuffling, making training more efficient and less prone to errors.

```
tokenizer = DistilBertTokenizer.from_pretrained(
    'distilbert-base-uncased', truncation=True, do_lower_case=True)
training_set = MultiLabelDataset(train_df, tokenizer, MAX_LEN)
val_set = MultiLabelDataset(val_df, tokenizer, MAX_LEN)

train_params = {'batch_size': TRAIN_BATCH_SIZE,
                'shuffle': True,
                'num_workers': 8}

val_params = {'batch_size': VALID_BATCH_SIZE,
              'shuffle': False,
              'num_workers': 8}

training_loader = DataLoader(training_set, **train_params)
#val_loader = DataLoader(val_set, **val_params)
```

We download a DistilBERT tokenizer that matches the pretrained DistilBERT model we'll use. Then, by passing train_df and val_df to MultiLabelDataset, we get dataset objects that can be passed into DataLoader instances. The DataLoader parameters, such as batch_size, shuffle, and num_workers, are chosen to ensure efficient loading and possibly parallel tokenization.

Defining the model architecture

At the heart of our pipeline is the DistilBERT-based classification model. We load the pretrained DistilBERT model and add a classification head on top of it. The classification head consists of linear layers and a dropout for regularization, transforming DistilBERT's hidden states into our desired set of output classes.

```python
class DistilBERTClass(torch.nn.Module):
    def __init__(self):
        super(DistilBERTClass, self).__init__()
        self.bert = DistilBertModel.from_pretrained(
            "distilbert-base-uncased")
        self.classifier = torch.nn.Sequential(
            torch.nn.Linear(768, 768),
            torch.nn.ReLU(),
            torch.nn.Dropout(0.1),
            torch.nn.Linear(768, 6)
        )

    def forward(self, input_ids, attention_mask, token_type_ids):
        output_1 = self.bert(input_ids=input_ids,
                             attention_mask=attention_mask)
        hidden_state = output_1[0]
        out = hidden_state[:, 0]  # [CLS] token representation
        out = self.classifier(out)
        return out
```

DistilBERT outputs a sequence of hidden states for each token. We take the hidden state corresponding to the first token ([CLS]) as a representation of the entire sentence. Our classifier layers then produce logits for each of the six labels. The entire network is trainable, allowing the pretrained DistilBERT weights to adjust slightly to the new classification task.

Preparing the model and optimizer for training

Now we move our model to the chosen device (GPU if available, else CPU) and set up an optimizer. The optimizer updates the model's weights based on the gradients computed from the loss function.

```python
model = DistilBERTClass()
model.to(DEVICE)
```

```
optimizer = torch.optim.Adam(params=model.parameters(), lr=LEARNING_RATE)
```

By using the Adam optimizer with a low learning rate, we give the model room to adjust its pretrained weights to the new task without making overly aggressive updates. This fine-tuning approach typically yields better results than training from scratch.

Training loop

Training the model involves iterating over the dataset multiple times (epochs), calculating the loss on each batch, and updating model weights accordingly. During training, we use a binary cross-entropy loss with logits, appropriate for multi-label classification tasks, and we track the loss to ensure the model is learning properly.

```python
def train(epoch):
    model.train()

    for _, data in tqdm(enumerate(training_loader, 0)):
        ids = data['ids'].to(DEVICE, dtype=torch.long)
        mask = data['mask'].to(DEVICE, dtype=torch.long)
        token_type_ids = data[
            'token_type_ids'].to(DEVICE, dtype=torch.long)
        targets = data['targets'].to(DEVICE, dtype=torch.float)

        outputs = model(ids, mask, token_type_ids)

        optimizer.zero_grad()
        loss = torch.nn.functional.binary_cross_entropy_with_logits(
            outputs, targets)

        if _ % 5000 == 0:
            print(f'Epoch: {epoch}, Loss:  {loss.item()}')

        loss.backward()
        optimizer.step()

for epoch in range(EPOCHS):
    train(epoch)
```

Within each epoch, we:

- Put the model in training mode.

- Iterate over batches from `training_loader`.

- Move the inputs and targets to the correct device.

- Run a forward pass through the model and compute the loss.

- Zero out previous gradients, backpropagate the new gradients, and step the optimizer to update weights.

- Optionally print progress every 5,000 batches, giving insight into training progress.

Preparing and processing the test data

After training, we often need to run the model on new, unseen test data. We follow a similar process to the training set: load the test data, create a dataset, and feed it into a `DataLoader`. Since there are no targets for the test data, the Dataset class will return only the inputs.

```
test_data = pd.read_csv('../input/test.csv.zip')
print(test_data.head())


test_set = MultiLabelDataset(test_data, tokenizer, MAX_LEN, new_data=True)
test_loader = DataLoader(test_set, **val_params)


all_test_pred = []
```

We define a `new_data=True` flag when creating `test_set`, indicating that no target labels will be returned, and the dataset is only for inference.

Inference on the test data

For prediction (inference), we put the model in evaluation mode. In this mode, the model doesn't update its weights; it only processes the inputs and returns predictions. We also wrap our inference logic in a with `torch.inference_mode()` block to improve efficiency and reduce unnecessary computations.

```
def test(epoch):
    model.eval()

    with torch.inference_mode():
        for _, data in tqdm(enumerate(test_loader, 0)):
            ids = data['ids'].to(DEVICE, dtype=torch.long)
```

```
                mask = data['mask'].to(DEVICE, dtype=torch.long)
                token_type_ids = data[
                    'token_type_ids'].to(DEVICE, dtype=torch.long)
                outputs = model(ids, mask, token_type_ids)
                probas = torch.sigmoid(outputs)

                all_test_pred.append(probas)

        return probas

    probas = test(model)
```

We apply the sigmoid function to the raw logits to obtain probabilities for each label. These probabilities indicate how likely the comment text belongs to each toxic category. We accumulate all predictions into all_test_pred.

Formatting and saving the predictions

Finally, we take our predictions, align them with the test dataset, and output them in a CSV file. This final step is crucial for submitting results to benchmarks or competitions.

```
    all_test_pred = torch.cat(all_test_pred)

    submit_df = test_data.copy()
    submit_df.drop("comment_text", inplace=True, axis=1)

    label_columns = ["toxic", "severe_toxic", "obscene",
                     "threat", "insult", "identity_hate"]

    for i,name in enumerate(label_columns):
        submit_df[name] = all_test_pred[:, i].cpu()

    submit_df.to_csv('../submissions/distilbert_0.csv', index=False)

    submit_df.head()
```

Here, we restore the original test data indices and associate each sample with its predicted probabilities. We remove the comment_text column since submissions usually require only the IDs and the predicted scores. Finally, we write the predictions to a CSV file, which can then be evaluated on a leaderboard or used in a downstream application.

This walkthrough demonstrated how a distilled Transformer, such as DistilBERT, can be fine-tuned from end to end for multi-label toxicity detection with only a few dozen lines of PyTorch code. By pairing a lightweight tokenizer, a compact classification head, and a carefully chosen learning rate, we retain much of BERT's accuracy while cutting memory footprint and training time, making deployment on resource-constrained hardware far more practical. Once the model is trained, a simple inference loop and CSV export transform raw comment text into leaderboard-ready probability scores, completing a pipeline that can be adapted to virtually any text classification task with just a dataset swap and minimal hyperparameter tuning.

Text classification with AutoTrain

Automated machine learning (AutoML) refers to the practice of using algorithms and tools to streamline and automate the typically labor-intensive process of developing machine learning models. Traditional machine learning workflows often require substantial human effort in areas like data preprocessing, feature engineering, model selection, hyperparameter tuning, and performance evaluation. AutoML systems aim to reduce this burden by programmatically searching through model configurations, architectures, and parameter values to identify optimal solutions with minimal human intervention. By leveraging techniques such as reinforcement learning, evolutionary computation, Bayesian optimization, and neural architecture search, these systems can produce highly competitive models that rival or surpass those crafted by expert data scientists. In doing so, AutoML democratizes machine learning by enabling domain experts, who may not have deep machine learning expertise, to harness the power of predictive modeling and advanced analytics more efficiently and at scale. In this section, we will utilize an AutoML solution, Hugging Face's AutoTrain to develop a deep learning transformer-based model for text classification in the Toxic Comments competition.

Setting up the environment and dependencies

Before we can start working with our data and models, we need to import the necessary libraries and modules. This typically involves loading fundamental packages for data manipulation and building machine learning models. We use the Kaggle UserSecretsClient to securely retrieve a token for Hugging Face's Hub, ensuring safe authentication rather than hard-coding credentials. After this, we import modules and functions from the Hugging Face AutoTrain library, as well as PyTorch, scikit-learn, and the Hugging Face Transformers library. These imports prepare the environment for text classification tasks, tokenization of textual inputs, and training the model.

```
import numpy as np # linear algebra
import pandas as pd # data processing, CSV file I/O (e.g. pd.read_csv)
import os

from kaggle_secrets import UserSecretsClient
user_secrets = UserSecretsClient()

HF_USERNAME = "tunguz"
HF_TOKEN = user_secrets.get_secret("HUGGING_FACE_HUB_TOKEN")

from autotrain.params import TextClassificationParams
from autotrain.project import AutoTrainProject

import torch
from sklearn import model_selection, metrics
from transformers import (
    AutoModelForSequenceClassification,
    AutoTokenizer, TrainingArguments, Trainer
)
```

We now read in our training and testing data.

```
train = pd.read_csv('../input/toxic-train/train.csv')
test = pd.read_csv('../input/toxic-train/test.csv')

print(train.head())
```

Setting up AutoTrain parameters

Here we define TextClassificationParams, a configuration object provided by autotrain. These parameters help us specify model details and training hyperparameters without manually coding them all. We choose a model (in this case, "google-bert/bert-base-uncased"), which is a pretrained BERT model from the Hugging Face Hub. We also inform Autotrain where our data is located, which columns represent text and labels, and how the data should be split into training and validation sets.

Beyond this, we specify training parameters such as epochs, `batch_size`, `max_seq_length`, and the learning rate (lr). We also set up optimization details (e.g., using the `adamw_torch` optimizer with a linear learning rate scheduler) and any other advanced settings, such as `mixed_precision`, for faster training. We can integrate logging and decide whether or not to push the final trained model to the Hugging Face Hub. All of these parameters together streamline the subsequent training and deployment steps.

```
params = TextClassificationParams(
    model="google-bert/bert-base-uncased",
    data_path="../input/toxic-train/",
    text_column="comment_text",
    target_column="toxic",
    train_split="train",
    valid_split=None,
    epochs=3,
    batch_size=8,
    max_seq_length=512,
    lr=1e-5,
    optimizer="adamw_torch",
    scheduler="linear",
    gradient_accumulation=1,
    mixed_precision="fp16",
    project_name="autotrain-model",
    log="tensorboard",
    push_to_hub=False,
    username=HF_USERNAME,
    token=HF_TOKEN,
)
```

Initializing and creating the Autotrain project

With parameters defined, we instantiate an AutoTrainProject. By specifying `backend="local"`, the training will occur locally in the current environment, rather than on a remote server. Calling `project.create()` triggers the setup process. At this stage, the code prepares all necessary components to initiate the training run, including tokenizers, model directories, and data processors.

```
project = AutoTrainProject(params=params, backend="local", process=True)
project.create()
```

Loading a pretrained model and tokenizer

After preparing the project, we load the tokenizer and the model. A tokenizer is responsible for converting raw text into a sequence of token IDs that the model can understand. The AutoTokenizer automatically downloads and configures the correct tokenizer based on the model we specify. Similarly, AutoModelForSequenceClassification retrieves a model pretrained on general text tasks, which we will fine-tune for our toxicity classification objective.

Here, the model and tokenizer are loaded from a previously trained directory, ../input/toxic-autotrain-toxic/autotrain-model, which presumably contains the results from a previous training run or experiment.

```
tokenizer = AutoTokenizer.from_pretrained(
    "../input/toxic-autotrain-toxic/autotrain-model", use_fast=True)
model = AutoModelForSequenceClassification.from_pretrained(
    "../input/toxic-autotrain-toxic/autotrain-model")
```

Preparing the test data for inference

The test dataset might have multiple label columns for various toxicity categories. Since we are focusing on a single binary toxicity label, we reset all toxicity flags to zero. This ensures that during inference, we can run predictions on a known baseline, thereby validating the model's performance. We do not, however, actually need these label columns at test time; they may be a relic of the training process or a preparation step for final submission formats.

```
test.loc[:, "toxic"] = 0
test.loc[:, "severe_toxic"] = 0
test.loc[:, "obscene"] = 0
test.loc[:, "threat"] = 0
test.loc[:, "insult"] = 0
test.loc[:, "identity_hate"] = 0
print(test.head())
```

Creating a custom Dataset class

To feed data into a PyTorch-based model in a transformers-style workflow, we often create a custom dataset class. The ClassificationDataset defined here takes the data and the tokenizer as input. In its __getitem__ method, it retrieves the text comment and label for a specific item, then uses the tokenizer to convert the raw text into input IDs and attention masks. The returned dictionary has all the necessary keys (input_ids, attention_mask, and labels) that the model and trainer need for processing.

The dataset class encapsulates logic for indexing into the data and running on-the-fly tokenization. This makes the training loop cleaner and more flexible.

```python
class ClassificationDataset:
    def __init__(self, data, tokenizer):
        self.data = data
        self.tokenizer = tokenizer

    def __len__(self):
        return len(self.data)

    def __getitem__(self, item):
        text = str(self.data["comment_text"].values[item])
        target = int(self.data["toxic"].values[item])
        inputs = self.tokenizer(
            text,
            max_length=512,
            padding="max_length",
            truncation=True
        )

        ids = inputs["input_ids"]
        mask = inputs["attention_mask"]

        return {
            "input_ids": torch.tensor(ids, dtype=torch.long),
            "attention_mask": torch.tensor(mask, dtype=torch.long),
            "labels": torch.tensor(target, dtype=torch.long),
        }
```

Running predictions with the trainer

Now we want to generate predictions on the test dataset. We instantiate our `ClassificationDataset` for the test set and then use the Hugging Face Trainer to run inference (predictions) without explicitly training. By passing the dataset to `trainer.predict()`, we obtain raw model outputs known as **logits**.

Logits are the unnormalized scores that the model outputs before applying an activation function, such as softmax. For classification tasks, we typically convert these logits to probabilities (e.g., using softmax) and then make a decision on the predicted class. The comment notes this step at the end: we must transform the logits into probabilities before submitting the final result.

```
dataset = ClassificationDataset(test, tokenizer)
trainer = Trainer(model)
preds = trainer.predict(dataset).predictions

# Preds will be in the form of logits,
# and need to be converted into probabilities before submission.
```

In this walkthrough, you saw how Hugging Face AutoTrain collapses an end-to-end text classification pipeline from data ingestion and splitting to hyperparameter tuning, model training, and local inference into only a handful of declarative steps. Once the returned logits are transformed into probabilities and thresholded to determine the final label, the predictions are immediately ready for submission to a competition or integration into a downstream application. The same template can be reused for any classification problem simply by pointing AutoTrain at a new dataset and tweaking a few parameter values, making it easy to iterate on ideas or scale solutions across projects without rewriting boilerplate code. In short, AutoTrain demonstrates how modern AutoML tooling can bring state-of-the-art transformer models within the reach of practitioners who prefer to focus on problem framing and evaluation rather than low-level engineering.

Text classification with LLM embeddings and logistic regression

Getting simple vector embeddings of various textual inputs is often the most effective solution for many classes of NLP problems. Embeddings from some of the best models can be almost as good for text classification as some of the most advanced fine-tuned deep learning models. In this section, we'll show how to create two such embeddings: one from the latest large proprietary OpenAI model, and another from an open-source NVIDIA model.

OpenAI embeddings

For the machine learning and language modeling parts, the key component is the openai library, which provides the interface to OpenAI's large language models and embedding services.

```
import pandas as pd
import numpy as np
```

```
import os

import openai
from openai import OpenAI
```

Initializing the OpenAI client

To interact with the OpenAI API, we first initialize a client. The OpenAI() class encapsulates the configuration and authentication details, so that we can simply call its methods to generate embeddings later on. Typically, you need to set environment variables or API keys before proceeding to this step. However, we assume that this has already been taken care of.

```
client = OpenAI()
```

Once the client is initialized, it can be passed along to functions that request embeddings or other outputs from the OpenAI models. This abstraction helps keep the code organized and makes it easy to switch models or endpoints if needed.

Defining a helper function for embeddings

In this section, we define a helper function, get_embedding, that takes in a piece of text and a specified embedding model. It uses the client object to request embeddings from OpenAI's API. Before sending the text to the embedding model, we perform a small preprocessing step by replacing newline characters (\ n) with spaces to ensure the input is formatted in a way that the model expects.

```
def get_embedding(text, model="text-embedding-3-large"):
    text = text.replace("\n", " ")
    return client.embeddings.create(
        input = [text], model=model
    ).data[0].embedding
```

This function returns a vector representation (an embedding) of the text. Embeddings are numeric representations of text where semantically similar texts are mapped close to each other in a high-dimensional vector space. Here, the model name "text-embedding-3-large" is used as a default, but you can change it to another model if needed. The returned embedding is essentially a list of floats representing the semantic content of the input text.

Loading and cleaning the data

The next step involves loading training and test datasets from CSV files. The `.fillna(' ')` method ensures that any missing values in the comment_text column are replaced with a space character, preventing embedding failures due to empty or NaN values. We then select only the 'comment_text' column from each dataset, assuming that is the text field of interest.

```
train = pd.read_csv('../input/train.csv.zip').fillna(' ')[['comment_
text']]
test = pd.read_csv('../input/test.csv.zip').fillna(' ')[['comment_text']]
```

Handling specific data anomalies

Sometimes, certain data entries may cause issues or need special handling. In the example code, three entries in the test set are overwritten with asterisks, '*'. Perhaps these lines were empty, corrupted, or contained characters that were problematic for the embedding model. By explicitly setting them to '*', we are providing a placeholder text rather than leaving them as is. This ensures that even these problematic entries can still be processed by the embedding function.

```
test.at[9932, 'comment_text'] = '*'
test.at[55331, 'comment_text'] = '*'
test.at[97708, 'comment_text'] = '*'
```

Overwriting data points like this is a practical, albeit somewhat ad hoc, data cleaning strategy. The key is to ensure that the model receives a consistent and safe input. An asterisk, '*', may represent a neutral placeholder, ensuring the embedding call will succeed.

Generating embeddings for the data

We now apply the embedding function we defined earlier to both the training and test sets. Using the pandas `apply` method, we run `get_embedding` on each entry of the comment_text column. This line effectively transforms each text comment into a high-dimensional numeric vector and stores these vectors in new columns named embedding_3_large.

```
train['embedding_3_large'] = train.comment_text.apply(
    lambda x: get_embedding(x, model='text-embedding-3-large'))
test['embedding_3_large'] = test.comment_text.apply(
    lambda x: get_embedding(x, model='text-embedding-3-large'))
```

Converting embeddings to NumPy arrays

While working with pandas is convenient for data manipulation, machine learning frameworks often expect input data in NumPy array format. Thus, we convert the list of embeddings into arrays of float values. This step involves taking each row from the embedding_3_large column, which is currently a list, and transforming it into a NumPy array. Then we stack these arrays together to form a two-dimensional array of shape (num_samples, embedding_dimension).

```
train_embeds = np.array(
    [np.array(i) for i in train.embedding_3_large.values])
test_embeds = np.array(
    [np.array(i) for i in test.embedding_3_large.values])
```

The resulting arrays, train_embeds and test_embeds, provide a fixed-length vector representation for every comment. This is ideal for direct ingestion into downstream machine learning models, enabling tasks like classification, clustering, similarity search, or visualization through dimensionality reduction techniques.

Saving the embeddings for later use

Finally, after generating embeddings, it is common practice to save them to disk. This is because embedding generation can be time-consuming and expensive, especially if we have a very large dataset. By saving the embeddings as .npy files, we can quickly reload them later without needing to call the OpenAI API again. This makes experiments reproducible and efficient.

```
np.save('../input/test_embs_3_large', test_embeds)
np.save('../input/train_embs_3_large', train_embeds)
```

NVIDIA embeddings

The code below imports AutoModel from the transformers library, pandas, and Numpy. It then uses AutoModel.from_pretrained() to initialize a model called 'nvidia/NV-Embed-v2'. The trust_remote_code=True argument allows the loading of custom code that might be embedded in the model repository—a useful feature if the model relies on specific methods or tokenizers not present in the standard Transformers codebase.

```
from transformers import AutoModel
import pandas as pd
import numpy as np

model = AutoModel.from_pretrained('nvidia/NV-Embed-v2', trust_remote_
code=True)
```

Defining a function to get embeddings

Once you have a model, you need a consistent way to obtain vector embeddings from raw text. The get_embedding function does just that. Before this step, it's assumed you have some predefined variables, such as passage_prefix and max_length, which guide the model's encoding process. The presence of a passage_prefix might be a prompt template that the model expects, while max_length sets the upper bound on the number of tokens processed.

```
def get_embedding(text):
    text = text.replace("\n", " ")
    return model.encode([text], instruction=passage_prefix,
                        max_length=max_length)[0]
```

The rest of the code follows the same pattern as that for the OpenAI embeddings.

Next, we'll train a logistic regression on the Nvidia embeddings. The code uses cross-validation and a particular metric (ROC AUC) to train and evaluate these models.

Setting the stage: Data and dependencies

The following lines import essential packages: NumPy, pandas, and selected utilities from scikit-learn. LogisticRegression will be our classification model of choice. For evaluation, we use the roc_auc_score.

```
import numpy as np
import pandas as pd


from sklearn.linear_model import LogisticRegression
from sklearn.model_selection import KFold
from sklearn.metrics import roc_auc_score
```

Next, we load the target columns and the data from the previously saved numpy embedding files:

```
class_names = ['toxic', 'severe_toxic', 'obscene',
               'threat', 'insult', 'identity_hate']
target = pd.read_csv(
    '../input/train.csv.zip').fillna(' ')[class_names].values
train_features = np.load('../input/train_embs_NV_2.npy')
test_features = np.load('../input/test_embs_NV_2.npy')
sample_submission = pd.read_csv('../input/sample_submission.csv.zip')
```

We train a separate logistic regression model for each class and then combine their predictions. To store these predictions, we initialize a NumPy array of zeros:

```
preds = np.zeros((test_features.shape[0], target.shape[1]))
```

We also define a list of complexity parameters (C) for each classifier. The C parameter in logistic regression controls the strength of the regularization. A higher C value indicates weaker regularization and can lead to more complex models, whereas a lower C value indicates stronger regularization and simpler models. By fine-tuning C for each target class, we aim to achieve better performance.

```
Cs = [4, 1, 4, 3, 2, 2]
```

Cross-validation and training

To assess the predictive performance of the models locally, we train a K-fold cross-validation.

```
errors = []
train_oof = np.zeros(target.shape)
kf = KFold(n_splits=5, random_state=137, shuffle=True)
```

Here, kf is a 5-fold cross-validator. The train_oof array (out-of-fold predictions) will store the predictions made on the validation splits. This is useful for meta-modeling or stacking methods, where predictions from one layer become features in another layer.

Iterating over each target

Since we have multiple classes, we loop through each of them and train a separate model. For each target class, ii, we go through the 5-fold cross-validation splits:

```
for ii in range(6):
    print("Fitting target", ii+1)
    for jj, (
        train_index, val_index) in enumerate(kf.split(train_features)
    ):
        print("Fitting fold", jj+1)
        train_x = train_features[train_index]
        val_x = train_features[val_index]
        train_target = target[train_index, ii]

        classifier = LogisticRegression(
            C=Cs[ii], solver='sag', max_iter=10)
        classifier.fit(train_x, train_target)
```

We split the training data into a train_x set (features for training), a val_x set (features for validation), and corresponding train_target labels. The logistic regression model is initialized with the specified C value and a 'sag' solver, which is efficient for large datasets. The max_iter=10 parameter sets the maximum number of iterations for convergence. Although this is a small number, it might be sufficient if the data is well conditioned and the features are well prepared.

Making predictions and recording performance

After fitting the model, we use it to predict probabilities on the validation set and the test set:

```
train_oof[val_index, ii] = classifier.predict_proba(val_x)[:,1]
preds[:, ii] += classifier.predict_proba(test_features)[:,1]/5
train_target = target[train_index, ii]
```

For each validation split, the probabilities are stored in train_oof. This gives us an unbiased estimate of how well the model performs on unseen data. We also average the test set predictions over all folds, producing a more stable final prediction.

Once we finish training all folds for a particular target, we measure performance using the ROC AUC score:

```
print(roc_auc_score(target[:,ii], train_oof[:,ii]))
errors.append(roc_auc_score(target[:,ii], train_oof[:,ii]))
```

This helps us track how well each classifier is performing. The ROC AUC score is particularly suitable when dealing with imbalanced classes, as it remains invariant to changes in class distribution.

Preparing the submission

Finally, once all classes have been trained and predictions generated, we place the test predictions in the submission DataFrame and save it:

```
sample_submission[class_names] = preds
sample_submission.to_csv('../input/NV_2_LR.csv', index=False)
```

This creates a CSV file ready for submission to a competition platform. This submission achieves a public leaderboard score of 0.98318 and a private leaderboard score of 0.98357.

In practice, large-scale text embeddings give us a powerful "universal feature" layer: once each text is reduced to a dense vector, even a simple linear model such as logistic regression can rival far more elaborate neural architectures on many classification tasks.

In this chapter, we showed that this holds for both proprietary services (OpenAI's textembedding3large) and fully opensource options (NVIDIA's NVEmbedv2); with either source, the workflow with clean text, generate embeddings once, cache them to disk, and sweep a handful of regularization strengths remains the same. The resulting models train in seconds and can be deployed in resource-constrained environments because all the heavy lifting was done during the one-off embedding step. While state-of-the-art results will still demand task-specific fine-tuning for some problems, these experiments demonstrate that "embeddings + logistic regression" should always be a good baseline to check out, and often, it might be all you need.

Text augmentation strategies

We discussed augmentation strategies for computer vision problems extensively in the previous chapter. By contrast, similar approaches for textual data are a less well-explored topic (as evidenced by the fact that there is no single package like albumentations). In this section, we demonstrate some of the possible approaches to addressing the problem.

Basic techniques

As usual, it is informative to examine the basic approaches first, focusing on random changes and synonym handling. A systematic study of the basic approaches is provided in *Wei* and *Zou* (2019) at https://arxiv.org/abs/1901.11196.

We begin with **synonym replacement**. Replacing certain words with their synonyms produces text that is close in meaning to the original, but slightly perturbed (see the project page at https://wordnet.princeton.edu/ if you are interested in more details, like where the synonyms are actually coming from):

```python
def get_synonyms(word):

    synonyms = set()

    for syn in wordnet.synsets(word):
        for l in syn.lemmas():
            synonym = l.name().replace("_", " ").replace("-", " ").lower()
            synonym = "".join([char for char in synonym
                               if char in ' qwertyuiopasdfghjklzxcvbnm'])
            synonyms.add(synonym)
    if word in synonyms:
        synonyms.remove(word)
```

```
        return list(synonyms)
```

We create a simple wrapper around the workhorse function defined above, specifying a chunk of text (a string containing multiple words) and replacing at most *n* of the words:

```
def synonym_replacement(words, n):
    words = words.split()
    new_words = words.copy()
    random_word_list = list(set([word for word in words
                                 if word not in stop_words]))
    random.shuffle(random_word_list)
    num_replaced = 0

    for random_word in random_word_list:
        synonyms = get_synonyms(random_word)

        if len(synonyms) >= 1:
            synonym = random.choice(list(synonyms))
            new_words = [synonym if word == random_word else word
                         for word in new_words]
            num_replaced += 1

        if num_replaced >= n: # Only replace up to n words
            break
    sentence = ' '.join(new_words)
    return sentence
```

Let's see how the function works in practice:

```
print(f" Example of Synonym Replacement: {synonym_replacement"
      f"('The quick brown fox jumps over the lazy dog',4)}")
```

This is the output:

```
Example of Synonym Replacement: The spry brown university fox jumpstart
over the lazy detent
```

Not quite what you would call Shakespearean. As the example shows, although naive synonym replacement can result in grammatically correct but semantically awkward sentences, it still conveys the same message while markedly changing the style and wording. We can extend this approach by creating multiple new sentences per tweet:

```
trial_sent = data['text'][25]
print(trial_sent)
>>
the free fillin' app on my ipod is fun, im addicted


for n in range(3):
    print(f" Example of Synonym Replacement: "
          f"{synonym_replacement(trial_sent,n)}")
```

This is the output:

```
Example of Synonym Replacement: the free fillin' app on my ipod is fun, im
addict
Example of Synonym Replacement: the innocent fillin' app on my ipod is
fun, im addicted
Example of Synonym Replacement: the relinquish fillin' app on my ipod is
fun, im addict
```

As you can see, generating variations of a text chunk using synonyms is relatively straightforward.

Next, **swapping** is a simple and efficient method; we create a modified sentence by randomly swapping the order of words in the text.

Carefully applied, this can be viewed as a potentially useful form of **regularization**, as it disturbs the sequential nature of the data that models like LSTM rely on. The first step is to define a function that swaps words:

```
def swap_word(new_words):
    random_idx_1 = random.randint(0, len(new_words)-1)
    random_idx_2 = random_idx_1
    counter = 0
    while random_idx_2 == random_idx_1:
        random_idx_2 = random.randint(0, len(new_words)-1)
        counter += 1
        if counter > 3:
```

```
                return new_words

    new_words[random_idx_1], new_words[random_idx_2] = \
        new_words[random_idx_2], new_words[random_idx_1]
    return new_words
```

Then, we write a wrapper around this function:

```
# n is the number of words to be swapped
def random_swap(words, n):
    words = words.split()
    new_words = words.copy()

    for _ in range(n):
        new_words = swap_word(new_words)

    sentence = ' '.join(new_words)
    return sentence
```

Synonyms and swapping do not affect the length of the sentence we are modifying. If in a given application it is useful to modify an attribute in a given application, we can remove or add words to the sentence.

The most common way to implement the former is to delete words at random:

```
def random_deletion(words, p):
    words = words.split()

    # Obviously, if there's only one word, don't delete it
    if len(words) == 1:
        return words
    # Randomly delete words with probability p
    new_words = []
    for word in words:
        r = random.uniform(0, 1)
        if r > p:
            new_words.append(word)
    # If you end up deleting all words, just return a random word
    if len(new_words) == 0:
        rand_int = random.randint(0, len(words)-1)
```

```
            return [words[rand_int]]
    sentence = ' '.join(new_words)

    return sentence
```

Let's look at some examples:

```
print(random_deletion(trial_sent,0.2))
print(random_deletion(trial_sent,0.3))
print(random_deletion(trial_sent,0.4))
```

This is the output:

```
the free fillin' app on my is fun, addicted
free fillin' app on my ipod is im addicted
the free on my ipod is fun, im
```

If we can remove, we can also add, of course. Random insertion of words to a sentence can be viewed as the NLP equivalent of adding noise or blur to an image:

```
def random_insertion(words, n):
    words = words.split()
    new_words = words.copy()
    for _ in range(n):
        add_word(new_words)
    sentence = ' '.join(new_words)
    return sentence
def add_word(new_words):
    synonyms = []
    counter = 0

    while len(synonyms) < 1:
        random_word = new_words[random.randint(0, len(new_words)-1)]
        synonyms = get_synonyms(random_word)
        counter += 1
        if counter >= 10:
            return
    random_synonym = synonyms[0]
    random_idx = random.randint(0, len(new_words)-1)
    new_words.insert(random_idx, random_synonym)
```

Here is the function in action:

```
print(random_insertion(trial_sent,1))
print(random_insertion(trial_sent,2))
print(random_insertion(trial_sent,3))
```

This is the output:

```
the free fillin' app on my addict ipod is fun, im addicted
the complimentary free fillin' app on my ipod along is fun, im addicted
the free along fillin' app addict on my ipod along is fun, im addicted
```

We can combine all the transformations discussed above into a single function, producing four variants of the same sentence:

```
def aug(sent,n,p):
    print(f" Original Sentence : {sent}")
    print(f" SR Augmented Sentence : {synonym_replacement(sent,n)}")
    print(f" RD Augmented Sentence : {random_deletion(sent,p)}")
    print(f" RS Augmented Sentence : {random_swap(sent,n)}")
    print(f" RI Augmented Sentence : {random_insertion(sent,n)}")
aug(trial_sent,4,0.3)
```

This is the output:

```
Original Sentence : the free fillin' app on my ipod is fun, im addicted
SR Augmented Sentence : the disembarrass fillin' app on my ipod is fun, im
hook
RD Augmented Sentence : the free app on my ipod fun, im addicted
RS Augmented Sentence : on free fillin' ipod is my the app fun, im
addicted
RI Augmented Sentence : the free fillin' app on gratis addict my ipod is
complimentary make up fun, im addicted
```

The augmentation methods discussed above do not exploit the structure of text data—to give one example, even analyzing a simple characteristic like "part of speech" can help us construct more useful transformations of the original text. This is the approach we will now focus on.

Text augmentation with back-and-forth translation

The idea for this text augmentation technique originated with Pavel Ostyakov in the Toxic Comments competition. The idea is pretty simple: translate text into some intermediate language and then translate it back to English. The original implementation of this technique was done with the TextBlob library, which works using the Google Translate API. Since then, there have been other implementations, such as ones using the googletrans library. The one that we'll use in this example is deep-translator, which, as of this writing, seems to be the most reliable one.

We first import the necessary libraries:

```
import os
import pandas as pd
from deep_translator import GoogleTranslator
```

We then provide a brief example of how the back-and-forth translations work. In this case, we use French as the intermediate language. This is a slightly modified example, which was originally made by John Miller.

```
train = pd.read_csv('../input/train.csv')


translator = GoogleTranslator()
for i,t in enumerate(train.comment_text[19:22]):
        encoded = translator.translate(t, dest='fr').text
        decoded = translator.translate(encoded, dest='en').text
        print(f"\nSet {i}\n"
              f"Original: {t}\n\n"
              f"Recoded: {decoded}\n")
```

The output of the above code will be:

```
Set 0
Original: Don't mean to bother you


I see that you're writing something regarding removing anything posted
here and if you do oh well but if not and you can acctually discuss this
with me then even better.


I'd like to ask you to take a closer look at the Premature wrestling
deaths catagory and the men listed in it, surely these men belong together
in some catagory. Is there anything that you think we can do with the
```

```
catagory besides delting it?

Recoded: I do not want to bother you

I see that you are writing something regarding the removal of everything
that is posted here and if you do it right, but if not and you can discuss
it with me, then even better.

I would ask you to take a closer look at the category of premature
wrestling deaths and the men listed there, these men surely belong
together in one category. Is there anything you think we can do with the
category in addition to destroying it?


Set 1
Original: "

 Regarding your recent edits

Once again, please read WP:FILMPLOT before editing any more film articles.
Your edits are simply not good, with entirely too many unnecessary
details and very bad writing.  Please stop before you do further damage.
-'''''The '45 "

Recoded: "

Regarding your recent changes

Again, please read WP: FILMPLOT before editing other movie articles. Your
changes are just not good, with too much unnecessary detail and very poor
writing. Please stop before doing other damage. - '' '' '' The '45 "


Set 2
Original: "
Good to know. About me, yeah, I'm studying now.(Deepu) "

Recoded: "
Good to know. About me, yes, I'm studying now. (Deepu) "
```

As you can see, the recoded texts are usually subtly different from the original ones, but they generally carry the same meaning. This augmentation technique has helped the top team in the Toxic Comments competition win first place.

nlpaug

We conclude this section by demonstrating the capabilities provided by the `nlpaug` package (`https://github.com/makcedward/nlpaug`). It aggregates different methods for text augmentation and is designed to be lightweight and easy to incorporate into a workflow. We demonstrate some examples of the functionality contained therein below.

```
! pip install nlpaug
```

We import the character- and word-level augmenters, which we will use to plug in specific methods:

```
import nlpaug.augmenter.char as nac
import nlpaug.augmenter.word as naw
test_sentence = "I genuinely have no idea what the output of this sequence
of words will be - it will be interesting to find out what nlpaug can do
with this!"
```

What happens when we apply a **simulated typo** to our test sentence? This transformation can be parametrized in several ways; for a comprehensive list of parameters and their explanations, the reader is encouraged to consult the official documentation: `https://nlpaug.readthedocs.io/en/latest/augmenter/char/keyboard.html`.

```
aug = nac.KeyboardAug(name='Keyboard_Aug', aug_char_min=1,
                      aug_char_max=10, aug_char_p=0.3, aug_word_p=0.3,
                      aug_word_min=1, aug_word_max=10, stopwords=None,
                      tokenizer=None, reverse_tokenizer=None,
                      include_special_char=True, include_numeric=True,
                      include_upper_case=True, lang='en', verbose=0,
                      stopwords_regex=None, model_path=None, min_char=4)
test_sentence_aug = aug.augment(test_sentence)
print(test_sentence)
print(test_sentence_aug)
```

This is the output:

```
I genuinely have no idea what the output of this sequence of words will be
- it will be interesting to find out what nlpaug can do with this!
```

```
I geb&ine:y have no kdeZ qhQt the 8uYput of tTid sequsnDr of aorVs will be
- it wi,k be jnterewtlHg to find out what nlpaug can do with this!
```

We can simulate an **OCR error** creeping into our input:

```
aug = nac.OcrAug(name='OCR_Aug', aug_char_min=1, aug_char_max=10,
                 aug_char_p=0.3, aug_word_p=0.3, aug_word_min=1,
                 aug_word_max=10, stopwords=None, tokenizer=None,
                 reverse_tokenizer=None, verbose=0,
                 stopwords_regex=None, min_char=1)
test_sentence_aug = aug.augment(test_sentence)
print(test_sentence)
print(test_sentence_aug)
```

We get:

```
I genuinely have no idea what the output of this sequence of words will be
- it will be interesting to find out what nlpaug can do with this!
I 9enoine1y have no idea what the ootpot of this sequence of wokd8 will be
- it will be inteke8tin9 to find out what nlpaug can du with this!
```

While useful, character-level transformations have a limited scope when it comes to creative changes in the data. Let us examine what possibilities nlpaug offers when it comes to word-level modifications. Our first example is replacing a fixed percentage of words with their antonyms:

```
aug = naw.AntonymAug(name='Antonym_Aug', aug_min=1, aug_max=10, aug_p=0.3,
                     lang='eng', stopwords=None, tokenizer=None,
                     reverse_tokenizer=None, stopwords_regex=None,
                     verbose=0)
test_sentence_aug = aug.augment(test_sentence)
print(test_sentence)
print(test_sentence_aug)
```

We get:

```
I genuinely have no idea what the output of this sequence of words will be
- it will be interesting to find out what nlpaug can do with this!
I genuinely lack no idea what the output of this sequence of words will
differ - it will differ uninteresting to lose out what nlpaug can unmake
with this!
```

nlpaug also offers us a possibility of, for example, replacing synonyms; such transformations can also be achieved with the more basic techniques discussed above. For completeness's sake, we demonstrate a small sample below, which uses a BERT architecture under the hood:

```
aug = naw.ContextualWordEmbsAug(model_path='bert-base-uncased',
                                model_type='', action='substitute',
                                # temperature=1.0,
                                top_k=100,
                                # top_p=None,
                                name='ContextualWordEmbs_Aug', aug_min=1,
                                aug_max=10, aug_p=0.3,
                                stopwords=None, device='cpu',
                                force_reload=False,
                                # optimize=None,
                                stopwords_regex=None,
                                verbose=0, silence=True)
test_sentence_aug = aug.augment(test_sentence)
print(test_sentence)
print(test_sentence_aug)
```

Here is the result:

```
I genuinely have no idea what the output of this sequence of words will be
- it will be interesting to find out what nlpaug can do with this!
i genuinely have no clue what his rest of this series of words will say -
its will seemed impossible to find just what we can do with this!
```

As you can see, nlpaug offers a broad range of options for modifying your textual input to generate augmentations. Which ones should actually be chosen is very much context-dependent, and the decision requires a little bit of domain knowledge, suited to a particular application.

Some places for further exploration would be beginner competitions such as *Natural Language Processing with Disaster Tweets* (https://www.kaggle.com/c/nlp-getting-started), as well as more intermediate or advanced ones like *Jigsaw Rate Severity of Toxic Comments* (https://www.kaggle.com/c/jigsaw-toxic-severity-rating) or *Google QUEST Q&A Labeling* (https://www.kaggle.com/c/google-quest-challenge). In all of these cases, nlpaug has been widely used— including in the winning solutions.

Summary

In this chapter, we discussed modeling for NLP competitions. We demonstrated both vintage and state-of-the-art methods applicable to a diverse range of problems appearing in Kaggle competitions. In addition, we touched upon the frequently ignored topic of text augmentation.

In the next chapter, we will discuss simulation competitions, a new class of contests that has been gaining popularity over the last few years.

Join our book's Discord space

Join our community's Discord space for discussions with the authors and other readers:

`https://packt.link/kaggle`

13

Generative AI in Kaggle Competitions

Generative AI—particularly **Large Language Models (LLMs)**—has quickly become a transformative force in data science workflows. Unlike traditional models that predict values or labels, LLMs can generate entirely new data: from synthesized text and code to feature suggestions and documentation. This capability has opened up new possibilities across the entire ML pipeline, from data preparation to model development.

In this chapter, you will learn how the transformative power of generative AI, particularly LLMs, is being practically applied in the competitive data science landscape of Kaggle. We will move beyond theoretical concepts to explore the specific strategies and techniques that are giving competitors a substantial edge. You will discover how to leverage LLMs across the entire machine learning pipeline, from cleaning and augmenting data to generating code and summarizing complex information. We will dissect crucial skills such as prompt engineering, fine-tuning open-source models like Google's Gemma, and implementing advanced methods like **Retrieval-Augmented Generation (RAG)** to build specialized AI assistants. Each technique will be grounded in real-world examples from recent competitions, illustrating how to navigate challenges like data scarcity and platform resource constraints.

The end goal of this chapter is to equip you with a practical and strategic understanding of how to integrate generative AI into your own data science workflows effectively. By analyzing the winning solutions from three distinct and high-stakes Kaggle competitions, you will gain a blueprint for tackling a new class of data problems. You will not only learn the "how" through code snippets and detailed breakdowns, but also the "why" behind specific architectural choices made by top Kagglers. By the end, you will be prepared to harness these powerful models to drive innovation and efficiency in your own projects, whether on Kaggle or in real-world applications.

This knowledge is critically important because generative AI represents a fundamental shift in the practice of data science. What was once a niche area of research has rapidly become an indispensable tool for analysis, development, and problem-solving. As Kaggle competitions increasingly feature LLMs, both as a tool and as the subject of the challenge itself, proficiency in this domain is no longer optional for staying competitive. The skills covered here—fine-tuning models for specific languages, reverse-engineering AI behavior, and building custom data assistants—reflect the evolving demands on data scientists in the industry. Mastering them will not only improve your performance in competitions but also position you at the forefront of this technological wave.

Here are the main topics we will cover in this chapter:

- Understanding generative AI and LLMs
- Unlocking global communication with Gemma: fine-tuning LLMs for new languages
- LLM prompt recovery
- AI assistants for data tasks with Gemma

Each case study will include an overview of the competition and an analysis of one of its top solutions. Along the way, we'll encounter code snippets that illustrate techniques such as fine-tuning LLMs, utilizing LLMs for knowledge retrieval, and crafting effective prompts. Generative AI is a fast-moving field—but as of this writing, it's clear that Kaggle competitors who harness these models wisely can gain a substantial edge in both innovation and efficiency.

Understanding generative AI and LLMs

Generative AI refers to models that can create new content—such as text, images, or audio—resembling the data they were trained on. In the context of Kaggle and data science, the most prominent generative models are LLMs. Formally, *an LLM is a machine learning model designed for natural language processing tasks, especially language generation.* These models typically have hundreds of millions to billions of parameters and are trained on massive text corpora via self-supervised learning (e.g., predicting the next word in a sentence).

Kaggle competitions have evolved to reflect this shift. Some recent challenges directly focus on LLMs—either as tools embedded within the solution process or as the subject of prediction themselves. Even in more conventional competitions, participants now routinely use generative AI to clean and standardize text, suggest imputations, summarize large datasets, or write reusable code snippets.

Tools like ChatGPT are commonly used for brainstorming, debugging, or explaining obscure errors, while pre-trained models are being embedded directly into Kaggle notebooks for tasks like text augmentation or documentation generation. Kaggle has also begun integrating LLMs into its platform offerings—for instance, by making Google's open-source Gemma models available in select environments.

As a result, prompt engineering—the art of writing effective instructions for LLMs—has become an essential new skill. A well-crafted prompt can guide an LLM to perform complex tasks such as extracting structured data, cleaning messy inputs, or generating exploratory insights. For example:

"You are a data-cleaning assistant. Clean the following survey responses, standardizing yes/no answers and fixing typos..."

This kind of prompt can yield model-ready output with minimal manual effort. Kagglers often experiment with different phrasings and techniques—such as few-shot prompting—to achieve the desired results. With API access and local LLM integrations, these iterative workflows can now be embedded directly into the modeling pipeline.

Notably, generative AI isn't limited to text—it spans image generation, audio synthesis, and more—but in Kaggle competitions, text-based LLMs have so far been the primary focus.

The working of LLMs

Most state-of-the-art LLMs are based on the Transformer architecture, which introduced the concept of *self-attention* to effectively capture long-range dependencies in text. Unlike older recurrent networks, Transformers process words in parallel and use attention mechanisms to decide which other words (or tokens) in the input are most relevant to a given token. The model consists of stacked layers of self-attention and feed-forward networks (plus optional encoder/decoder structure, depending on the model). For example, the GPT family of models uses a stack of decoder-only Transformer blocks to predict text autoregressively. *Figure 13.1* illustrates a high-level view of a GPT-style Transformer architecture, where each input token is first converted to an embedding (plus positional encoding) and then processed through multiple Transformer blocks. The output is a probability distribution over the vocabulary for the next token. By iteratively sampling the next token and feeding it back in, the model can generate sequences of tokens comprising text.

Output
Probabilities

Softmax

Generator

Linear

Layer Norm

Decoder Layer N

Decoder

...

Decoder Layer 2

Decoder Layer 1

(+)

Positional Encoding

Embedding

Input

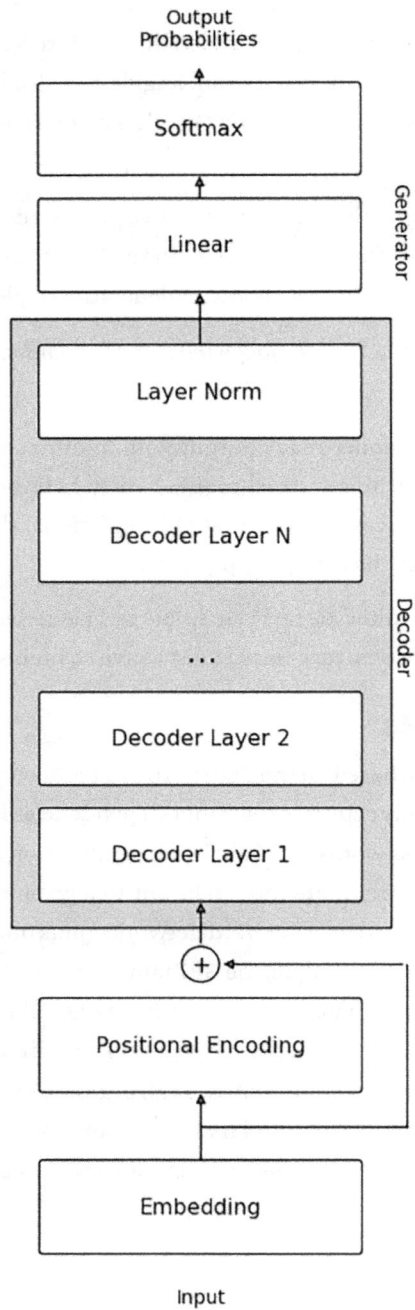

Figure 13.1: High-level architecture of a GPT-style Transformer model

Tokens are encoded into embeddings with positional information, pass through **N** layers of self-attention and feed-forward network (decoder blocks), then are transformed via a linear layer and softmax to predict the next token's probability distribution.

Training such a model from scratch requires enormous data and compute. The model learns statistical patterns of language, developing an implicit understanding of syntax, semantics, and even some world knowledge from its training corpus. The result is an LLM that can generate coherent paragraphs of text, translate between languages, answer questions, write code, and more. Indeed, modern LLMs have demonstrated *emergent abilities*—unexpected skills that arise from scale, such as performing summarization or reasoning tasks that were not explicitly trained.

However, an *out-of-the-box* LLM is not always aligned with what a user wants. This is where **fine-tuning and instruction following** come in. Many LLMs undergo a second phase of training called *instruction tuning*, where the model is fine-tuned on datasets of question-answer pairs or tasks with human-written instructions and responses. This teaches the model to better follow explicit prompts. Additionally, techniques like **Reinforcement Learning from Human Feedback** (**RLHF**) are used to further align model outputs with human preferences. In RLHF, human annotators first rank or label model outputs based on their quality, and a reward model is trained to predict these human preferences. The base LLM is then fine-tuned (often via reinforcement learning algorithms) to **maximize the reward** (i.e., produce outputs that humans would rate highly), thereby making its responses more helpful, honest, and harmless.

LLMs are versatile and can be applied to a wide range of tasks. Some of the typical use cases include:

- **Text generation and completion**: Producing coherent continuations of a prompt or even full articles/stories. For example, given a few sentences, an LLM can continue with a plausible next paragraph.

- **Summarization**: Condensing a long document or article into a concise summary. Large models have shown strong performance at summarizing while preserving key information.

- **Question answering**: Answering factual or reading-comprehension questions by either directly retrieving embedded knowledge or by generating answers from context.

- **Translation and style transformation**: Converting text from one language to another or altering the style/tone of text (e.g., making text more formal, simpler, or in the style of a specific author).

- **Classification and extraction**: While built for generation, LLMs can perform classification by outputting category labels or extract structured information from unstructured text, especially when instruction-tuned.

- **Code generation and debugging**: A particularly impactful use case—models like OpenAI's Codex or Google's Gemini Code Assistant can generate code snippets or even entire functions given a description. They can also assist in finding bugs or explaining code.

- **Generative AI in Kaggle Workflows**: For Kagglers, generative models offer new tools in the toolkit. Even in traditional competitions not centered on NLP, one can use LLMs to generate synthetic data (e.g., augmenting a dataset with additional textual samples), to automate feature engineering (by parsing and summarizing text fields), or to assist in writing analysis reports and documentation. Some Kagglers use chatbots like ChatGPT to help brainstorm modeling strategies or to debug code. It's not uncommon now to see participants discuss how an LLM helped them refactor their code or generate insights from raw data. Given this growing influence, it was only a matter of time before Kaggle competitions explicitly focused on **building and fine-tuning LLMs themselves**.

To better understand how competition works, let's quickly demonstrate how one might interact with an LLM in code. Thanks to high-level libraries like Hugging Face Transformers, as well as less popular but effective options like Keras (which is particularly suitable for simpler tasks), using a pre-trained LLM for text generation can be accomplished in just a few lines. In our example, we utilize the recent KerasHub, as KerasNLP (which contains Keras functions specialized for text processing) and KerasCV (which contains Keras functions suitable for vision tasks) have been consolidated into this new, unified library: `https://github.com/keras-team/keras-hub`. Existing code that utilizes `keras-nlp` and `keras-cv` will continue to function as expected. You can still install and import them as you did before. However, moving forward, new features and models for both NLP and CV will be incorporated into KerasHub.

For example, here's how you could load a text-generation pipeline and produce a continuation of a prompt:

```
import keras_hub

# Load the GPT-2 model from KerasHub
model = keras_hub.models.GPT2CausalLM.from_preset("gpt2_base_en")

# Define the prompt
prompt = "Kaggle is an online platform for"

# Generate text
result = model.generate(prompt, max_length=100)
```

```
# Print the result
print(result)
```

This code produces the following output:

```
Kaggle is an online platform for developers and designers that aims to
provide a platform to share their work, collaborate, and create content in
collaboration with others.

Kaggle is a community based platform where you can create, share, and
collaborate on projects, create new features, collaborate and create code
for your own projects, and collaborate on other projects.

We've been building Kaggle for more than a year and have been able to
bring you a lot of great features
```

As you can see, the output of this prompt is nonsensical and has nothing to do with what Kaggle is about. Fortunately, larger GPT-2 models can get us better results, with the following output when we set the model to be gpt2_extra_large_en:

```
Kaggle is an online platform for machine learning competitions, where
users submit their machine learning models to a competition, which is then
ranked and judged. It has become the go-to place for many machine learning
researchers and companies to share and showcase their work.

Kaggle has become a popular destination for researchers who are looking to
publish their work. It is one of the few places where you can publish your
work in a timely manner.
```

When these models were first trained and introduced, they were quite demanding in terms of resources. However, these days they can comfortably run on even a very moderately powerful laptop, and even more powerful "small" LLMs have been gaining ground over the years.

Kaggle has introduced competitions where *the objective revolves around generative AI*, whether it's adapting an LLM to a specific domain or solving a problem that requires prompt engineering and model understanding. These competitions often fall into the "Analytics" or "Research" category, meaning the submission is often a Jupyter notebook (with code, discussion, and results) rather than just predictions on a test set. Submissions may be evaluated using a combination of automated metrics and human judgment. We will now explore three such competitions in detail, each highlighting different aspects of working with LLMs on Kaggle.

This shift toward lighter, laptop-friendly LLMs has democratized experimentation—setting the stage for community-driven efforts like Google's late2024 Kaggle challenge to fine-tune Gemma 2 for underrepresented languages.

Unlocking global communication with Gemma: fine-tuning LLMs for new languages

One of the landmark generative AI competitions on Kaggle was **Google's Unlocking Global Communication with Gemma**, an analytics competition with a $150,000 prize pool. Announced in late 2024, this competition invited participants to fine-tune Google's **Gemma 2** LLM for a specific language or cultural domain. The backdrop here is that Gemma is a family of open-source language models (built with the same underlying tech as Google's Gemini models) that Google released to foster a community-driven ecosystem of language-specific models. By the time of the competition, a "Gemmaverse" of developers had already begun adapting Gemma to languages ranging from Arabic to Zulu. The competition's goal was to accelerate this trend: each team would pick one of many under-represented languages (or a unique cultural niche of language use) and fine-tune Gemma 2 to excel at it. Competitors documented their approach in Kaggle notebooks, demonstrating improvements in tasks like the model's **language fluency**, ability to handle **literary traditions or historical texts** of that language, and other culturally relevant capabilities.

Competition format and data

As an *analytics* competition, this wasn't about submitting a model for automated scoring on a hidden test set. Instead, participants shared notebooks with their fine-tuned Gemma variants and qualitative/quantitative evaluations. Judges (including the Gemma model developers) assessed the submissions on criteria such as innovation, performance improvements, and the insightfulness of the approach. Google provided baseline resources: the base Gemma 2 model (with variants of 2B parameters that could be fine-tuned on Kaggle's GPUs) and a list of ~70 eligible languages that were considered "under-resourced" in the LLM context. Participants often had to assemble their own fine-tuning datasets for the chosen language—drawing from public text sources or creating synthetic data—since by definition, many of these languages had limited ready-made datasets.

Top solutions overview

The winners of this competition delivered some impressive and instructive solutions. Many leveraged **synthetic data generation** to overcome data scarcity, creating large corpora of question-answer pairs or translated sentences using existing LLMs.

For example, one of the winning teams focused on Italian (a moderately resourced language, but they aimed to push Gemma's abilities in Italian to a new level). Their approach, as described by team member `Stefano Fiorucci`, was a "cheap recipe" that combined multiple techniques: *Synthetic data generation (with LLM-as-a-judge), Supervised Fine-Tuning (SFT), Direct Preference Optimization, and efficient training with a method called Spectrum.* In simpler terms, they first used a large model to generate Italian text data (and employed an LLM to judge/filter the quality of this synthetic data), then fine-tuned Gemma on this data (SFT). Next, they applied **Direct Preference Optimization (DPO)**—a technique related to RLHF—to further align the model's outputs with human preferences by using a smaller reward model and optimizing the LLM against it. Finally, they used *Spectrum*, which is a memory-efficient fine-tuning strategy that selects which parts of the model to train, allowing them to fine-tune the 2B parameter model on Kaggle's limited hardware. This multifaceted pipeline yielded an improved **Gemma-It** model that performed better in Italian language tasks than the original Gemma. You can find the Kaggle notebook for this solution here: `https://www.kaggle.com/code/anakin87/post-training-gemma-for-italian-and-beyond`.

Another top team, headed by `Justin Yang` (which placed 2nd in the competition), tackled the task for Traditional Chinese and produced a notable open-source project called **Kyara**. In their write-up, the team explains that they generated *over one million* synthetic QA pairs for fine-tuning, plus an additional 150k prompts for preference optimization (**Direct Preference Optimization DPO**). This massive dataset was created by a method they dub "Retrieve, Rewrite, and Reformulate," which uses a retrieval-augmented approach to generate question-answer pairs covering a wide range of knowledge. They also translated and paraphrased existing datasets from other languages to enrich the Chinese training data. By the end, they had essentially built a Chinese-centric instruction-following model on top of Gemma. The results, according to their report, showed *strong performance compared to the original Gemma models*—in other words, their fine-tuned **Kyara** model could understand and generate Chinese with greater accuracy and cultural relevance than baseline Gemma. Impressively, they released the Kyara model and the huge dataset on Hugging Face for the community, contributing back to the open-source ecosystem. You can find the Kyara model here: `https://huggingface.co/datasets/zake7749/kyara-zh-sample-1M`. Justin's Kaggle solution notebook can be found here: `https://www.kaggle.com/code/zake7749/kyara-retrieval-augmentation-for-llm-fine-tuning`.

Fine-tuning Gemma in practice: example

How does one actually fine-tune and use an LLM like Gemma in a Kaggle notebook? A common workflow is:

1. Prepare a fine-tuning dataset of prompts and outputs.

2.   Train or fine-tune the model on this data (often using Hugging Face's Transformers library or Google's JAX/TPU tools if provided).

3.   After fine-tuning, load the new model and test it on some examples. Many top teams used PyTorch with Hugging Face Transformers for this process, as it's a straightforward way to implement custom training loops or use Trainer APIs.

Key techniques used in fine-tuning

The top solutions to the Gemma competition highlighted a few recurring themes that are instructive for any generative AI project on Kaggle:

- **Synthetic data generation with LLMs**: When real training data is scarce, use a larger LLM (or multiple LLMs) to generate additional data. For example, you might prompt GPT-4 to produce Q&A pairs in the target language, or to translate and rephrase English text into the target language. One team mentioned using an "LLM-as-a-judge" approach—they generated candidate outputs and then used another model (or heuristic) to judge which outputs were high-quality, ensuring that the fine-tuning data was clean and relevant.

- **Supervised Fine-Tuning (SFT)**: This is the standard next step—take the base model and fine-tune on the supervised dataset (prompt → ideal response pairs). This aligns the model with the task. In practice, teams often had to train for multiple epochs and monitor an evaluation set (if available) to avoid overfitting. Libraries like HuggingFace Transformers make this easier via the `Trainer` class or `peft.Lora` for parameter-efficient fine-tuning. However, interestingly, one team reported that they attempted full LLM fine-tuning with LoRA but abandoned it due to poor validation correlation—indicating that not all fine-tuning attempts guarantee success, especially if the evaluation metrics are tricky (more on this in the *Prompt Recovery* section).

- **Direct Preference Optimization (DPO) or RLHF**: Several top entrants went beyond SFT and performed a second stage of tuning to better align the model's outputs with human preferences. DPO is an approach where, instead of doing full reinforcement learning (which can be complex to implement), one can fine-tune the model to maximize a reward score (like a proxy for human preference) in a simpler, more direct way. To do this, you typically need a reward model—sometimes participants trained a smaller model to act as a judge between outputs, or they reused an existing one. By optimizing Gemma with DPO, teams were essentially making the model's tone and style more user-friendly and its answers more "helpful" or correct where possible. Google's Gemma being open meant that participants could experiment with these cutting-edge alignment techniques right on Kaggle.

- **Efficient training and deployment**: Handling a multi-billion-parameter model within Kaggle's constraints (limited GPU RAM and runtime) is non-trivial. Teams used tricks like 4-bit quantization of model weights to reduce memory usage, gradient checkpointing to trade compute for memory, and transferring parts of the training to faster hardware off-platform when allowed. The Italian team's use of Spectrum and the general use of **BitsAndBytes** (for 4-bit quantization) are examples of how winners squeezed more performance out of the available resources. In practice, Kaggle kernels with A100 GPUs (if provided) can handle fine-tuning a 2B model, but doing RLHF or DPO on top might require careful memory management. Participants often had to innovate on the engineering side as much as the modeling side.

By the end of the competition, Google and the community gained a suite of fine-tuned Gemma models in many languages. This showcased a path towards truly multilingual AI that is not dominated by only high-resource languages. The winning notebooks illustrated how thoughtful data curation and novel training strategies can localize a large model to perform impressively well on niche languages or dialects. For example, one project fine-tuned Gemma for a dialect from Korea's Jeju Island to help preserve that dialect—something far outside the reach of most commercial models. The competition underscored the **practical insights** that:

- LLMs can be fine-tuned with surprisingly good results using public tools and a bit of creativity
- Even without going into low-level model architecture changes, one can achieve significant performance boosts through data and training strategy alone.

  As Google's official blog post put it, this collaborative effort helps *"build a future where AI transcends language barriers,"* and it exemplifies how Kaggle participants are contributing to the frontier of generative AI.

While the Gemma competition demonstrated the power of fine-tuning open-source LLMs for multilingual and culturally aware performance, the next challenge shifted the focus inward—toward understanding the behavior of LLMs themselves. Instead of asking what these models can produce, the LLM Prompt Recovery competition asked: *Can we reverse-engineer the very instructions that shaped those outputs?* Let's now explore how this innovative Kaggle competition reframed the prompt as the central object of prediction.

LLM prompt recovery

Imagine you have an original piece of text, and then you have a second piece of text that is a **transformed version of the first**—for example, maybe the second text is a summary of the first, or it's the first text translated into Shakespearean English, or perhaps it's the first text with all numerical information removed. In modern NLP, such transformations can be done by prompting an LLM. For instance, you might prompt an LLM: *Translate the following passage to French* or *Rewrite this paragraph in a polite tone*. The LLM takes the original and produces the transformed version.

The LLM prompt recovery competition asked the reverse: given the original text and the transformed text, recover the prompt that was used to generate the transformation. In other words, participants had to figure out what instruction an LLM had been given. This challenge was a Kaggle Featured Competition held in early 2024, with a hefty $200,000 prize pool and over 2,000 teams participating. It was one of the first competitions to explicitly focus on *prompt engineering* and *LLM behavior* as the core problem.

Let us go into the details of this competition in the following sections.

Competition overview

The task in LLM Prompt Recovery can be viewed as an *inverse problem*. Normally, we craft prompts to get outputs from an LLM. Here, we have the outputs (and the inputs) and must deduce the prompt. For example, if the original text is a verbose passage and the transformed text is a short bulleted list, the hidden prompt might have been *Summarize the following text in bullet points*. If the transformed text is in Spanish, the prompt likely was *Translate to Spanish*. If the transformed text is a playful version of the original, maybe *Rewrite this text in a lighthearted, humorous tone*. The challenge is that we are not told what kinds of transformations exist; participants had to infer patterns from data.

Examining the data

The competition dataset consisted of many pairs of texts: (original, transformed), and the goal was to predict the exact text of the prompt that was used to go from original to transformed. Importantly, the prompt was applied using a specific LLM (likely a Gemma model or similar) in a controlled environment. The competition description hinted that Google's Gemma models were used to generate the transformed text given the prompt. So the transformations were realistic and fluent, not rule-based modifications. Participants were given some training examples with known prompts to learn from, and a test set where they had to predict prompts.

One complication: prompts could be phrased in multiple ways (e.g., "Translate to French" vs "Translate this text into French language"). To evaluate predictions objectively, Kaggle defined a special metric called **LLM Nerd-Off Sharpened Cosine Similarity**. Despite the whimsical name, this metric was essentially a measure of semantic similarity between the predicted prompt and the true prompt, using embeddings. Likely, they embedded both prompts with a language model and computed cosine similarity, then perhaps raised it to a power or scaled it ("sharpened") to emphasize differences. In short, participants didn't need to match the prompt exactly word for word; they needed to capture the same meaning. A submission earned a high score if its prompts were semantically very close to the actual prompts used in generation.

This evaluation method meant that the task was about **capturing the essence of the prompt**. If the true prompt was *Summarize the article briefly*, and a prediction was *Give a short summary of the above passage*, that should score very high (as it's semantically equivalent). If a prediction missed the mark (e.g., *Translate to French* when the prompt was actually *Simplify the language*), the cosine similarity would be low.

This competition encouraged participants to think about *how different instructions manifest in text*. They had to practically build a system that reads an original and its transformed version, and then outputs a plausible instruction. It's a bit like playing detective with an AI's behavior.

Evaluating the challenges of the competition

There are potentially **hundreds of possible prompt types**. Without additional structure, this is a challenging NLP problem—essentially, natural language understanding combined with some creativity. A straightforward approach would be to fine-tune a sequence-to-sequence model that takes the original and transformed text as input and tries to output the prompt. But doing that naïvely might be tough with limited training data. Also, some prompt types might be extremely rare or ambiguous.

To succeed, competitors incorporated external knowledge and clever strategies. They suspected that certain transformation categories existed (such as translation, summarization, tone change, and information extraction). Likely, the competition forum discussions (and perhaps an initial "analysis" notebook by organizers) indicated the scope of tasks.

Participants, therefore, had to leverage multiple techniques:

- **NLP heuristics**: For example, if the transformed text is much shorter than the original and covers similar content, it's probably a summarization prompt. If the transformed text is in parentheses or brackets, it's possible that the prompt asked for something to be extracted.

- **Embedding-based similarity**: You can embed the original and transformed text to examine the differences. For instance, if the embedding of the transformed text is close to that of a known French translation of the original, that signals a translation prompt. Kagglers might cluster or classify pairs using embeddings to identify prompt categories.

- **LLMs themselves**: Ironically, one could use an LLM to solve this LLM problem. For example, one could feed the original and transformed text into GPT-4 with a prompt like: *Given the above original text and its modified version, what instruction was likely given to the model?* This might give a very good answer much of the time.

- **Fine-tuning custom models**: Many top teams fine-tuned their own smaller LLMs or sequence models on the training data, along with a substantial amount of **synthetic data**. They simulated the process: take a large amount of text, apply various known prompts using an API (such as OpenAI or a local model) to generate transformed text, and thereby create a large dataset of (original, prompt, transformed) text. This synthetic corpus could then be used to train a model for prompt inversion.

The **creativity** factor was high. In fact, the highest-performing strategies did some non-intuitive things, as we'll see with the highlighted solution.

To read from here (`https://www.kaggle.com/suicaokhoailang`) exploited the metric by formulating prompts that maximized similarity without necessarily being exact matches (an "adversarial" approach to the scoring metric). Khoi observed that predicting just the first half of the true prompt often yields a higher similarity score than the full prompt. This clever hack suggests that he might generate prompts that cover the key words and phrasing common to many actual prompts, thereby ensuring high embedding overlap. In Khoi's words, "I think the special token pulls a sentence to some focal point in the embedding space, maybe the center of it. If the sentences are far enough from each other, it's more likely that the new distance (from point B to point A </s>) is shorter than the old one (base of the triangle). But if the two sentences are already very close to each other, this can hurt performance."

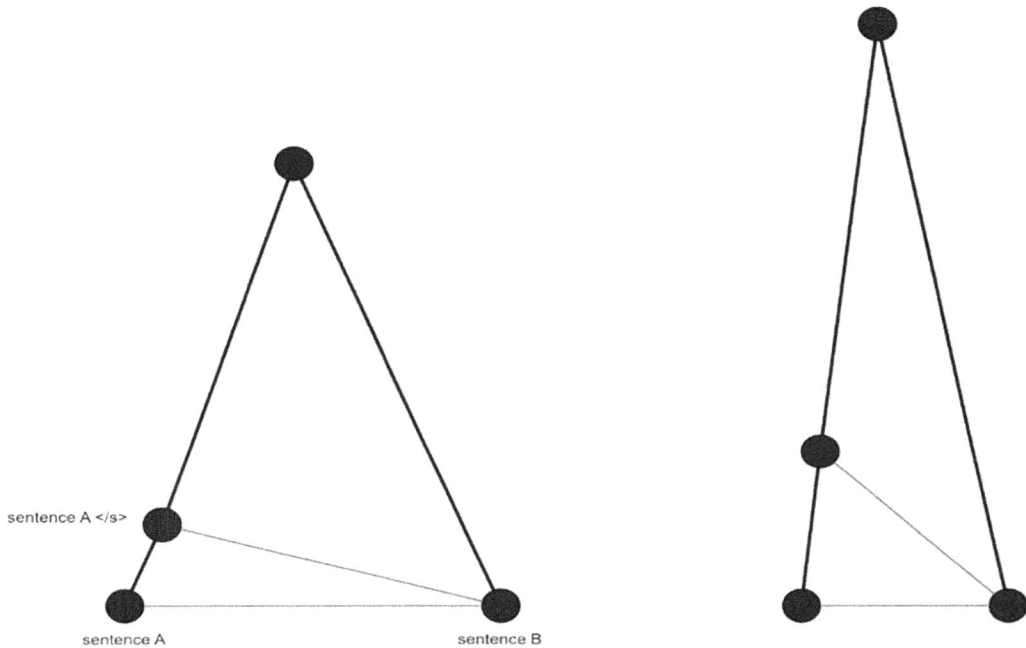

Figure 13.2: Visualization of various distances between prompts and sentences

Khoi used a mix of Mistral 7b and Gemma-7b-1.1-it, and trained them on different datasets. He concatenated the predictions of both models, and the final mixture helped him achieve a winning score. However, this approach, while maximizing the score, might produce prompts that sound incomplete or odd to a human.

Third-place solution (team prompt = "don't say anything")

The third-place team tackled the prompt recovery task with a **hybrid strategy** combining a robust *"mean prompt"* baseline with several fine-tuned models (https://www.kaggle.com/competitions/llm-prompt-recovery/discussion/494621). In essence, they constructed a fixed template prompt (a kind of *universal prompt*) and then augmented it with dynamic content predicted by models. By blending this optimized static prompt with learned predictions, the team could closely mimic the style and content of the true prompts. All of this was done in a neutral, third-person manner in their final assembled prompts, as required by the competition's rules.

The solution can be viewed as a **hybrid ensemble of a prompt template and model predictions**. It consists of five main components working in concert:

- **Mean prompt template**: A fixed template string (discovered through optimization) that serves as a baseline prompt, into which model-generated phrases are inserted.

- **Full prompt prediction model**: A fine-tuned language model (based on Mistral-7B) that attempts to predict the full rewrite prompt for a given sample.

- **Gate classifier**: A filtering model that checks whether the predicted prompt from the previous component is credible (i.e., consistent with the given original and rewritten text) and should be used.

- **Tags prediction model**: Another language model that predicts auxiliary tags (keywords describing style or instructions) relevant to the rewrite prompt.

- **Clustering mechanism**: A strategy that groups test samples into clusters of similar examples and selects the most suitable prompt template for each cluster.

In the final inference pipeline, these components interact as follows. For a given test sample, the **cluster assignment** is first determined based on the sample's characteristics (more on clustering below). A corresponding **prompt template** is chosen for that cluster, or a global default template is used if the cluster is uncertain. Then the **full prompt model** generates a candidate rewrite prompt for the sample, and the **tags model** produces supplementary keywords (such as a target style or tone). The **gate model** evaluates the candidate prompt against the original and rewritten text; if the prompt seems clearly incorrect (irrelevant or inconsistent), the system discards it. If it passes the gate, the final submission prompt is constructed by **starting with the mean prompt template** and **inserting the additional predicted words** (the tags and/or the full prompt prediction) at a specific position within the template. Notably, they only insert words that are *not already present* in the template to avoid redundancy. Through experimentation, the team found that the optimal insertion point was after the third word of the mean prompt template (this placement yielded the best validation performance). The result is a single coherent prompt string that includes the general template phrasing plus any unique details predicted by the models. This approach proved very effective, as using the first four components (template and models without clustering) already achieved an SCS of approximately **0.71** on validation. The clustering step added a small further boost (roughly +0.005)—not enough to reach 0.72, but every bit helped on the leaderboard.

The final solution was the mean prompt + tags + full prompt (if passes the gate). The schematic of the solution can be found below:

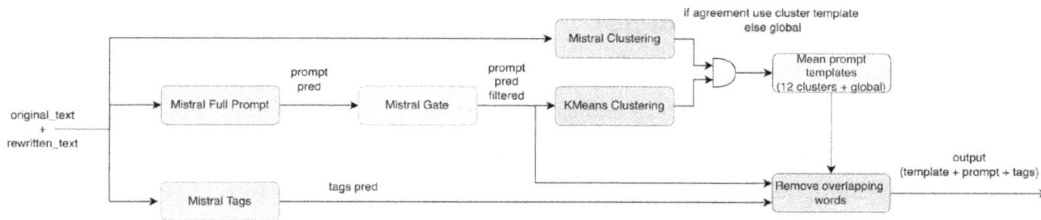

Figure 13.3: Third-place solution schematic

Below, we take a closer look at each component in detail and how they were developed and tuned.

Origins of the mean prompt

Long before any model was trained, the team explored whether a single fixed instruction could, on average, resemble many hidden prompts. They began by producing tens of thousands of candidate instructions using strong external LLMs, such as GPT-3.5 and Gemini. Each candidate was applied to an LLM-generated passage to fabricate a plausible rewritten_text instance. Because the public leaderboard exposed similarity scores for half the test set, the team could estimate how well any individual instruction aligned with real hidden prompts. By iteratively sampling subsets of their synthetic data whose score distribution mirrored that of the public leaderboard, they obtained a surrogate development set. Then, they executed a token-level beam search across a few thousand word candidates. The search objective was simply to maximize average SCS on that surrogate set.

The resulting sentence—obscure, repetitive, and studded with the Romanian word *lucrarea*— would look nonsensical to a human reader, yet it aligned strikingly well with the vectors used in the competition metric. Including *lucrarea* several times exploited an accidental quirk of the embedding space. This seemingly random token sat near the centroid of many true prompts, so multiplying it within the instruction nudged predictions closer in cosine space. With no modeling at all, the "mean prompt" alone achieved a score of 0.70 SCS and provided a strong backbone for later components. Here is the final mean prompt sentence:

improve phrasing text {here we format full prompt prediction + tags} lucrarea tone lucrarea rewrite this creatively formalize discours involving lucrarea anyone emulate lucrarea description send casual perspective information alter it lucrarea ss plotline speaker recommend doing if elegy tone lucrarea more com n paraphrase ss forward this st text redesign poem above etc possible llm clear lucrarea

Building a full-prompt predictor

Although the universal instruction covered broad stylistic ground, it could not capture sample-specific details such as *"translate to French"* or *"summarize in bullet points."* For that, the team fine-tuned Mistral-7B Instruct v0.2 under **Low-Rank Adaptation (LoRA)**. They discovered that careful dataset design, rather than exotic training techniques, drove performance.

The data pipeline ran backward from the goal. First, external LLMs generated diverse rewrite instructions, each enriched with stylistic hints so that a future model could infer the prompt's intended effect. Next, the same LLMs paraphrased every instruction several times, yielding clusters of semantically identical yet lexically distinct prompts. For each prompt variant, the team asked an LLM to draft an `original_text` whose content harmonized with the instruction—ensuring, for instance, that a "formalize this casual email" prompt indeed paired with a casual email. To create the target `rewritten_text`, they let a quantized Gemma-2B model, accelerated with Unsloth, apply the prompt to the original passage. The outcome was a synthetic triple: original, prompt, rewritten.

Because some prompt families now outnumber others, the team embedded every prompt with a Sentence-T5 model, clustered them via HDBSCAN, and subsampled each cluster to an equal size. This balanced curriculum prevented the model from memorizing only the most frequent rewrite patterns. After stripping boilerplate phrases such as "Sure, here is your text," the dataset contained hundreds of thousands of clean examples. Fine-tuned on this mixture, the Mistral LoRA model achieved approximately 0.62 SCS when asked, zero-shot, to guess the hidden prompt from a genuine competition pair. That raw score was modest but crucial: when its predictions were appended to the mean template, overall similarity jumped markedly.

Safeguarding with a gate classifier

Even a well-trained predictor occasionally hallucinated. A single rogue instruction—"Translate to Klingon," say—could drag the composite prompt far from the ground truth and erase gains accrued by the mean template. To defend against such outliers, the team equipped another Mistral instance with a sequence-classification head. Its input concatenated the original passage, a candidate prompt, and the rewritten passage, framed inside a short meta-instruction asking whether the rewriting matched the candidate's prompt.

Positive examples comprised 40 percent of the classifier's training: genuine triples drawn from the synthetic corpus. Another 20 percent were easy negatives, formed by pairing a random prompt with an unrelated rewritten text. The remaining 40 percent were hard negatives: prompts retrieved from the nearest neighbors of the correct instruction in embedding space, therefore superficially similar yet semantically misaligned.

Through this curriculum, the gate learned nuanced distinctions. During inference, it scored each predicted prompt; only those above a conservative threshold graduated to the final composition, while the rest were discarded, leaving the mean template unperturbed.

Isolating attributes with a tag model

Certain facets of a rewrite—such as musicality, rhyme, or an imperative to summarize—proved hard for the full-prompt model to express reliably. The team addressed this by training a second causal-LM head, again atop Mistral-7B, whose sole aim was to emit a comma-separated list of keywords describing the transformation. The tag model digested the same original-rewritten pair and produced elements such as "summarize," "formal tone," "poem," or "third-person perspective." After inference, the system filtered out any tag already present in either the universal template or the accepted full-prompt prediction, then spliced the remaining tokens into the template at a fixed slot—immediately after the third word—found via validation sweep to be optimal. The tag model's brevity reduced its error surface, allowing it to contribute correct micro-details that nudged SCS upwards.

Riding the cluster spectrum

The final refinement sought a middle road between a single template and one-prompt-per-sample prediction. By embedding ground-truth prompts with Sentence-T5 and applying K-Means, the team observed 12 coherent clusters in their local validation set. For each cluster, they re-ran the beam-search optimization to craft a specialized mean prompt. Upon validation, the hypothetical score obtained by always selecting the ideal cluster template approached 0.76 SCS, far exceeding any published leaderboard result, revealing untapped headroom.

Realizing that perfect cluster identification was impossible, the team trained yet another multi-class Mistral classifier to label unseen pairs with cluster IDs. They reinforced that decision with an independent heuristic: the predicted prompt from the full-prompt model was embedded and assigned to a cluster by the earlier K-Means centroids. Only when both the classifier and heuristic agreed did the pipeline adopt the cluster-specific template; otherwise, it reverted to the global mean prompt. This cautious protocol trimmed risk yet still delivered roughly five thousandths of a point on cross-validation, enough to matter in a contest decided on the third decimal place.

End-to-end inference flow

At test time, each sample passed through a deterministic cascade. Sentence-T5 embeddings were computed to map the pair into a provisional cluster. In parallel, the full-prompt Mistral proposed an instruction, the tag model suggested keywords, and the gate classifier vetted the instruction's plausibility.

The system next chose a base template—cluster-specific when confidently classified, otherwise the global version. Into this template, it stitched, in order, unique tokens from the vetted full prompt and the filtered tag list. Because the universal instruction already contained many high-yield generic phrases, additive tokens were often scarce, resulting in concise yet customized prompts. The composite string then stood as the team's prediction.

Quantitative outcome and qualitative lessons

The five-stage assembly scored just above 0.71 SCS on the hidden private set, securing third place. Though marginally shy of the winning mark, it demonstrated several principles valuable beyond this singular task. First, metric-aware prompt engineering—especially exploiting embedding artifacts such as lucrarea—can convert a daunting search space into a hill-climb around a sturdy baseline. Second, synthetic data, if diverse and cluster-balanced, can teach an LLM to infer latent instructions even when no labelled ground truth exists. Third, small specialist heads (gate and tag models) can prune and polish the output of a larger generator, providing robustness and incremental gains. Finally, intermediate granularity—here, a dozen clusters—yields a pragmatic compromise between universality and per-sample overfitting.

All code, from data generation through LoRA fine-tuning to inference, relied heavily on open-source scaffolding. Yet, the decisive advantage arose less from tooling than from a mindset of relentless empirical probing. The team iterated through thousands of candidate tokens, balanced dozens of clusters, and rejected any enhancement that could not demonstrate reproducible uplift on a public subset that faithfully mirrored the test distribution. By combining a rule-based backbone with discriminative and generative neural modules, they reconstructed hidden prompts with a degree of fidelity previously thought unattainable in a zero-label setting.

Although the competition's idiosyncratic metric may never reappear, the architecture of this solution—static template plus gated generative refinement, augmented by attribute tagging and distribution-aware clustering—offers a portable recipe for tasks where the goal is to reverse-engineer or approximate opaque human instructions. In that sense, the third-place team delivered not merely a leaderboard score but a blueprint for marrying prompt engineering and fine-tuned language models under extreme supervision scarcity.

Having explored the complexities of recovering prompts from transformed text using LLMs, we now turn to a different but equally innovative challenge: designing AI assistants that proactively help users with data tasks.

AI assistants for data tasks with Gemma

As generative AI became mainstream, many began to dream of personal "data science assistants"—AI agents that could help with the grunt work of analysis, answer technical questions, or educate users on complex topics. Google and Kaggle tapped into this idea with the "Google – AI Assistants for Data Tasks with Gemma" competition, launched in early 2024 (`https://www.kaggle.com/competitions/data-assistants-with-gemma`). This was an Analytics competition with a $50,000 prize pool, where instead of predicting values, participants created interactive notebooks showcasing an AI assistant that uses the Gemma LLM to help with some aspect of data science or Kaggle workflows.

Competition overview

The goal of this competition was open-ended yet targeted: *"Create the best notebook demonstrating how to use Gemma to accomplish a data science–oriented task."* More specifically, the competition prompt listed **five possible tasks/use cases** that entries could focus on:

- **Explain or teach basic data science concepts**: For example, an assistant that can answer "What is cross-validation?" or "Explain gradient boosting in simple terms."

- **Answer common questions about the Python programming language**: Essentially a Python help chatbot: "How do I open a file in Python?", "What does the `zip()` function do?"

- **Summarize Kaggle competition solution write-ups**: An assistant that can take a long winning solution post from a Kaggle competition and produce a succinct summary.

- **Explain concepts from Kaggle competition solution write-ups**: Similar to #3, but specifically to teach the ML concepts used in solutions. For instance, if a solution used XGBoost with out-of-fold blending, the assistant could explain those concepts in detail.

- **Answer common questions about the Kaggle platform**: A "Kaggle FAQ" assistant: "How do I make a submission?", "What does the Gold medal threshold mean in Kaggle?", etc.

Participants could choose one (or multiple) of these themes and build a notebook around it. The emphasis was on showcasing techniques with Gemma—meaning they should demonstrate prompt engineering, fine-tuning, retrieval augmentation, or other advanced methods to make the assistant effective. Creativity in presentation was also key, as it's a notebook competition; the engaging and informative nature of the notebook would influence the judging.

Because it was a showcase competition, the evaluation was subjective. A panel of judges scored notebooks on criteria such as the usefulness of the assistant, originality, technical soundness (whether they properly fine-tuned or used RAG, etc.), and clarity of explanation in the notebook. In addition, there was a mid-point prize where the organizers highlighted top notebooks halfway through (to encourage sharing and improvement). For example, Kaggle GM Luca Massaron's notebook was one of the five outstanding public notebooks in the mid-point announcement: `https://www.kaggle.com/code/lucamassaron/data-science-ai-assistant-with-gemma-2b-it`—this indicates early favorites.

The competition offered a valuable opportunity for participants to apply their knowledge of LLMs *in a practical setting*. It wasn't just theory—you had to build a working system (within Kaggle notebook constraints, which meant possibly using smaller Gemma models, such as the 2B variant, due to limited GPU memory). It also required integrating external knowledge, as tasks such as answering Python questions or Kaggle FAQs require accurate information. Here are the common techniques used in solutions that highlight a large scope of techniques and different data being used:

- **Retrieval-Augmented Generation (RAG):** Many assistants employed a retrieval step. For instance, a Python assistant might have a knowledge base of Python documentation or Stack Overflow Q&As. When the user asks a question, the system retrieves relevant information and feeds it to Gemma to generate the answer. This ensures factual accuracy and depth, mitigating hallucinations.

- **Fine-tuning or LoRA:** Some contestants fine-tuned Gemma on domain-specific **question-and-answer (Q&A)** pairs (such as a set of Q&As about Python) to give it an edge in that domain. Techniques like LoRA allow fine-tuning a 2B model on Kaggle's GPUs in a reasonable time. Fine-tuning would help the model internalize jargon or structured response styles (like always providing code examples for programming questions).

- **Prompt engineering and chaining:** Instead of one-shot answers, an assistant might break the task into steps (e.g., first use Gemma to outline an answer, then use another prompt to refine it or add examples). Some users may use multiple prompts to achieve improved results (this can be done synchronously in the notebook).

- **Interactive widgets or UI elements:** Kaggle allows limited interactivity (like toggling examples). Solutions might include an interface where you type a question and see the model's answer (within the notebook environment).

- **Multi-turn dialogue handling:** some assistants maintained context over multiple user questions. That requires managing conversation history and feeding it into prompts in an appropriate manner.

Given the broad scope, a variety of entries were made. Let's spotlight one top solution, which won one of the categories: an assistant for answering Python questions, called "PyGEM," based on an entry by David Troxell (`https://www.kaggle.com/davidtroxellucla`), whose notebook indeed was titled **PyGEM: A Chatbot for Python Questions using Gemma** (`https://www.kaggle.com/code/davidtroxellucla/pygem-a-chatbot-for-python-questions-using-gemma`), and which won the Python Q&A category.

Top solution: "PyGEM" — a Python programming chatbot

This notebook took first place in the "common Python questions" category, earning a $10,000 prize. The approach exemplified how to combine Gemma's capabilities with external knowledge to create a genuinely helpful assistant.

Here is the solution summary expectation: PyGEM is a conversational agent that can answer questions about Python, ranging from *What does list.append do?* to *How can I read a CSV file using pandas?*. Under the hood, the solution uses a **RAG** pipeline and fine-tuning to achieve high-quality answers.

- **Knowledge base creation:** The author prepared a knowledge base of Python information. This included conceptual Q&A pairs and possibly documentation. In the provided files, we see **Python_Conceptual_Questions.jsonl** and **python_code_questions.json**, indicating he compiled (or wrote) many Q&A examples covering Python's concepts and common code questions. This serves as a reference corpus.

- **Embedding and retrieval:** To answer a new question, the system first retrieves relevant information from the knowledge base. He embedded all Q&As using an embedding model (for instance, using Gemma's embedding or a smaller model like MiniLM) and built an index. When a user question is submitted, it's embedded, and the top-k similar Q&As from the knowledge base are retrieved. This ensures the assistant has factual content to draw from, rather than relying purely on Gemma's memory (which might be incomplete or hallucinated).

- **Gemma LLM generation:** The retrieved text, along with the user's question, is then fed into Gemma to generate an answer. The prompt might be something like: "*Here are some relevant pieces of information:\n[INFO]\nUsing this, answer the question: [User Question]*". Gemma then produces a tailored answer that hopefully uses the info to be correct and detailed. This is the **augmentation** part of RAG—real data augments the model's generation.

- **Fine-tuning with LoRA:** The author also fine-tuned Gemma (specifically the 2B model, since that's feasible to fine-tune on Kaggle with LoRA) on a collection of Q&A pairs about Python. Using LoRA, he injected knowledge of Python directly into the model weights in a parameter-efficient way. LoRA essentially learns small rank update matrices for the model's layers, which can be trained quickly. The result is that Gemma becomes more proficient in *the style of Q&A and the content of Python.* For example, after fine-tuning, it might recognize that when asked about code, it should provide code snippets in the answer, or that "open a file" implies showing the usage of the open( ) function.

- **Interactive and educational touches:** The notebook likely includes not just the raw Q&A system, but also explanatory content. For instance, it might demonstrate the effect of retrieval by showing what info was retrieved for a given question. It might also discuss limitations. This educational aspect was important for judges.

To illustrate how PyGEM works, let's walk through a hypothetical interaction:

The user asks: *"How do I sort a list in Python?"*

1. **Retrieve:** The system takes this query and finds relevant entries in its knowledge base. Likely, it finds a Q&A about the sorted( ) function or the list.sort( ) method. For example, a stored Q&A might be:

   a.   Q: "How can I sort a list in Python?"

   b.   A: "Use the list.sort( ) method to sort a list in place, or the sorted(list) function to get a sorted copy."

   This could be a direct match. It might also fetch a snippet from Python's documentation on sorting.

2. **Generate with Gemma:** It then constructs a prompt for Gemma that includes the retrieved answer. Perhaps:

   a.   Q: How can I sort a list in Python?

   b.   A: You can sort a list in Python using either the list.sort( ) method or the sorted( ) function. The list.sort( ) method will sort the list in place and return None, whereas the sorted( )function will create a new sorted list from the iterable you pass it.

Gemma, being fine-tuned on Q&A style, will likely continue to provide answers in a well-structured format, possibly expanding on them (perhaps adding example code if the prompt or its training prompts it to do so).

3. **Result**: The assistant replies to the user: "You can sort a list in Python by using the `list.sort()` method or the `sorted()` function. For example:

```
numbers = [5, 2, 9]
numbers.sort()
# Now numbers is [2, 5, 9]
```

If you need a sorted copy without modifying the original list, you can use `sorted_list = sorted(numbers)`."

This answer is accurate and even provides a useful tip about `sorted()` vs `sort()`.

What if the user asks something the system doesn't have directly, like a particular question? The solution likely covered **limitations**: they mention that if the sole goal was best chatbot, they could use more advanced techniques or larger models (maybe hinting at GPT-4), but since the goal was to demonstrate Gemma, they stuck to techniques one can implement with Gemma. So an extremely novel or niche question might not be perfectly answered. However, the system could still try using Gemma's base knowledge in the absence of retrieved info.

The key achievement of PyGEM was showing that a relatively small LLM (Gemma 2B) could be turned into a specialist assistant with clever techniques:

- RAG mitigated the size limitation by providing factual grounding from external text. This counters Gemma 2B's possibly limited internal knowledge.

- **LoRA fine-tuning** gave the model a focused expertise without needing an exorbitant amount of compute or losing generality.

- The notebook explained these steps clearly, essentially becoming a how-to guide for others. It discussed concepts like using embedding models and evaluating LLM output quality, which the judges likely appreciated for its pedagogical value.

Several other top notebooks followed similar patterns, each tackling a different category:

- Luca Massaron's "*Data Science Assistant*" (https://www.kaggle.com/code/lucamassaron/data-science-ai-assistant-with-gemma-2b-it) focused on explaining basic data science concepts. His notebook was essentially a RAG 101 tutorial: it walked through building a knowledge base of data science definitions and then using Gemma to answer questions like "What is gradient descent?" with that knowledge. He utilized ScaNN (a Google vector search library) for retrieval and developed everything from scratch to demonstrate its functionality. His assistant could, for instance, answer "Explain random forest in simple terms" by retrieving a description of random forests and then summarizing it. This entry won recognition for its clarity.

- Another team built an assistant to summarize Kaggle solution write-ups (`https://www.kaggle.com/code/marianadeem755/kaggle-mastery-summarize-kaggle-solution-write-up`). They curated a set of past solution posts and fine-tuned or prompted Gemma to summarize long text. That involves prompt engineering for summarization (e.g., "Summarize the above in 5 bullet points.") and maybe some semantic splitting of long documents.

By the end of the competition, the top entries demonstrated the feasibility of tailoring an LLM to niche but practical tasks. The use of open models like Gemma means these solutions could, in principle, be deployed without relying on proprietary APIs, which is attractive for the community.

In conclusion, the AI Assistants with Gemma competition entries, such as PyGEM, highlighted a few key lessons:

- Even a relatively compact model can be very powerful when augmented with relevant data (the "open-book" approach vs "closed-book" model).

- Fine-tuning (even lightweight like LoRA) can significantly improve domain performance, making the difference between a generic model and an expert assistant.

- Clarity in explaining the solution is important—many Kagglers produced notebooks that not only solved the task but taught readers *how* it was solved, aligning with the spirit of knowledge sharing.

- These projects gave a template for building custom AI assistants—something many companies and teams are now doing in various fields (finance, law, customer support, etc.). Kagglers essentially prototyped those ideas in the competition.

To wrap up, let's reflect on what these three competitions tell us about the evolving role of generative AI on Kaggle and in data science.

Summary

In this chapter, you have seen how top data scientists approach fine-tuning, prompt design, and LLM integration in competitive environments. These case studies demonstrate that success often hinges not on raw model size, but on thoughtful data curation, clever metric-aware engineering, and robust, hybrid solutions. There are several key insights from these competitions that transcend Kaggle and are applicable in a wide variety of data science, machine learning, and AI tasks, such as:

- Data quality and strategy matter because those who engineered better data (be it synthetic instructions or prompt-response pairs) and leveraged domain knowledge outperformed those who might have just thrown brute-force compute at the problem

- Combining approaches leads to robust solutions, because generative AI doesn't eliminate traditional methods; instead, it often works best in tandem with them

- Importance of sharing and open models because all these competitions involve a strong open-source component, proving that open source accelerates progress in a way that closed models cannot

This knowledge is directly applicable to real-world problems, providing a blueprint for building specialized models, reverse-engineering system behavior, and creating helpful AI assistants. As Kaggle continues to incorporate generative AI, we anticipate even more diverse challenges, such as multimodal generative tasks and AI agent competitions. By experimenting with the techniques explored here, you will be well-equipped to tackle these future challenges and turn generative AI into a powerful ally in any data science project.

Get This Book's PDF Version and Exclusive Extras

UNLOCK NOW

Scan the QR code (or go to `packtpub.com/unlock`). Search for this book by name, confirm the edition, and then follow the steps on the page.

Note: Keep your invoice handy. Purchases made directly from Packt don't require an invoice.

Join our book's Discord space

Join our community's Discord space for discussions with the authors and other readers:

`https://packt.link/kaggle`

14

Simulation and Optimization Competitions

Reinforcement learning (RL) is an interesting case among the different branches of machine learning. On the one hand, it is quite demanding from a technical standpoint; various intuitions from supervised learning do not hold, and the associated mathematical apparatus is significantly more advanced. On the other hand, it is the easiest to explain to an outsider or layperson. A simple analogy is teaching your pet (I am very intentionally trying to steer clear of the dogs versus cats debate) to perform tricks: you provide a treat for a trick well done, and refuse it otherwise.

RL was a latecomer to the competition party on Kaggle, but the situation has changed in the last few years with the introduction of simulation competitions. In this chapter, we will describe this new and exciting part of the Kaggle universe. As of the time of writing, there have already been ten **Featured** competitions and five **Playground** ones. We expect this number to continue growing as time passes, as the interest generated by this type of competition is particularly strong.

In this chapter, we will demonstrate solutions to the problems presented in several simulation competitions:

- We start working with *Connect X*.
- We follow with a deep dive into *Rock, Paper, Scissors*, where a dual approach to building a competitive agent is shown.
- We finish by demonstrating a solution to the *Santa* competition based on multi-armed bandits.

We conclude with an overview of the remaining competitions, which are slightly outside the scope of this chapter.

Technical requirements

If RL is an entirely new concept for you, it is probably a good idea to get some basic understanding first. A great way to begin the RL adventure is the Kaggle Learn course, dedicated to this topic in the context of game AI (`https://www.kaggle.com/learn/intro-to-game-ai-and-reinforcement-learning`). The course introduces basic concepts such as agents and policies, also providing a (crash) introduction to deep RL. All the examples in the course use the data from the Playground competition *Connect X*, in which the objective is to train an agent capable of playing a game of connecting checkers in a line (`https://www.kaggle.com/c/connectx/overview`).

On a more general level, it is worth pointing out that an important aspect of simulation and optimization competitions is the **environment**: due to the very nature of the problem, your solution needs to exhibit more dynamic characteristics than just submitting a set of numbers (as would be the case for "regular" supervised learning contests). A very informative and detailed description of the environment used in simulation competitions can be found at `https://github.com/Kaggle/kaggle-environments/blob/master/README.md`.

Working with Connect X

In this section, we demonstrate how to approach the simple problem of playing checkers using heuristics. While not a deep learning solution, we believe that this bare-bones presentation of the concepts is more useful for individuals without significant prior exposure to RL.

The objective of Connect X is to get a number (X) of your checkers in a row – horizontally, vertically, or diagonally – on the game board before your opponent. Players take turns dropping their checkers into one of the columns at the top of the board. This means each move may have the purpose of trying to win for you or trying to stop your opponent from winning.

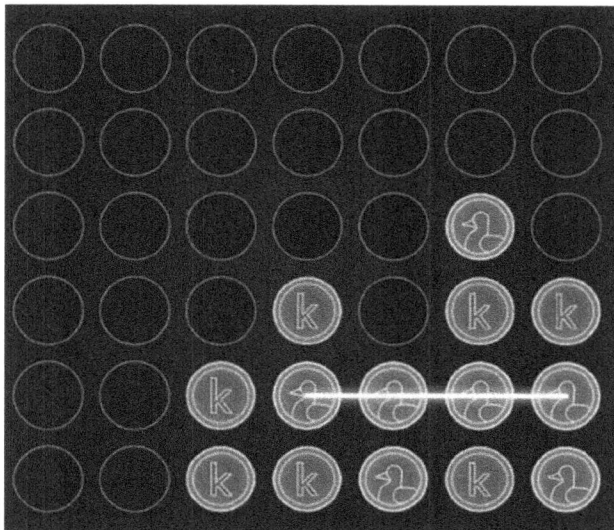

Figure 14.1: Connect X board

Connect X was the first competition that introduced **agents**: instead of a static submission (or a Notebook that was evaluated against an unseen dataset), participants had to submit agents capable of playing the game against others. The evaluation proceeded in steps:

1. Upon uploading, a submission plays against itself to ensure that it works properly.
2. If this validation episode is successful, a skill rating is assigned, and the submission joins the ranks of all competitors.
3. Each day, several episodes are played for each submission, and subsequently, rankings are adjusted.

With that setup in mind, let us proceed toward demonstrating how to build a submission for the *Connect X* competition. The code we present is for *X=4*, but it can be easily adapted for other values or *X* variables.

First, we install the Kaggle environments package:

```
!pip install kaggle-environments --upgrade
```

Then, we define an environment in which our agent will be evaluated:

```
from kaggle_environments import evaluate, make
env = make("connectx", debug=True)
env.render()
```

While a frequent impulse you might have is to try sophisticated optimization methods, it is useful to start simple – as we will do here, by using simple heuristics. These are combined into a single function in the accompanying code, but for the sake of presentation, we describe them one at a time here.

The first rule is to check whether either player has a chance to connect four checkers vertically and, if so, return the position at which it is possible. We can achieve this by using a simple variable as our input argument, which can take on two possible values, indicating which player opportunities are being analyzed:

```
def my_agent(observation, configuration):
    from random import choice
    # me:me_or_enemy=1, enemy:me_or_enemy=2
    def check_vertical_chance(me_or_enemy):
        for i in range(0, 7):
            if observation.board[i+7*5] == me_or_enemy \
            and observation.board[i+7*4] == me_or_enemy \
            and observation.board[i+7*3] == me_or_enemy \
            and observation.board[i+7*2] == 0:
                return i
            elif observation.board[i+7*4] == me_or_enemy \
            and observation.board[i+7*3] == me_or_enemy \
            and observation.board[i+7*2] == me_or_enemy \
            and observation.board[i+7*1] == 0:
                return i
            elif observation.board[i+7*3] == me_or_enemy \
            and observation.board[i+7*2] == me_or_enemy \
            and observation.board[i+7*1] == me_or_enemy \
            and observation.board[i+7*0] == 0:
                return i
        # no chance
        return -99
```

We can define an analogous method for horizontal chances:

```
def check_horizontal_chance(me_or_enemy):
    chance_cell_num = -99
    for i in [0,7,14,21,28,35]:
        for j in range(0, 4):
            val_1 = i+j+0
            val_2 = i+j+1
            val_3 = i+j+2
            val_4 = i+j+3
            if sum([observation.board[val_1] == me_or_enemy, \
                    observation.board[val_2] == me_or_enemy, \
                    observation.board[val_3] == me_or_enemy, \
                    observation.board[val_4] == me_or_enemy]) == 3:
                for k in [val_1,val_2,val_3,val_4]:
                    if observation.board[k] == 0:
                        chance_cell_num = k
                        # bottom line
                        for l in range(35, 42):
                            if chance_cell_num == l:
                                return l - 35
                        # others
                        if observation.board[chance_cell_num+7] != 0:
                            return chance_cell_num % 7
    # no chance
    return -99
```

We repeat the same approach for the diagonal combinations:

```
# me:me_or_enemy=1, enemy:me_or_enemy=2
def check_slanting_chance(me_or_enemy, lag, cell_list):
    chance_cell_num = -99
    for i in cell_list:
        val_1 = i+lag*0
        val_2 = i+lag*1
        val_3 = i+lag*2
        val_4 = i+lag*3
        if sum([observation.board[val_1] == me_or_enemy, \
                observation.board[val_2] == me_or_enemy, \
                observation.board[val_3] == me_or_enemy, \
```

```
                    observation.board[val_4] == me_or_enemy]) == 3:
            for j in [val_1,val_2,val_3,val_4]:
                if observation.board[j] == 0:
                    chance_cell_num = j
                    # bottom line
                    for k in range(35, 42):
                        if chance_cell_num == k:
                            return k - 35
                    # others
                    if chance_cell_num != -99 \
                    and observation.board[chance_cell_num+7] != 0:
                        return chance_cell_num % 7
    # no chance
    return -99
```

We can combine the logic into a single function checking the opportunities (playing the game against an opponent):

```
def check_my_chances():
    # check my vertical chance
    result = check_vertical_chance(my_num)
    if result != -99:
        return result
    # check my horizontal chance
    result = check_horizontal_chance(my_num)
    if result != -99:
        return result
    # check my slanting chance 1 (up-right to down-left)
    result = check_slanting_chance(
        my_num, 6, [3,4,5,6,10,11,12,13,17,18,19,20])
    if result != -99:
        return result
    # check my slanting chance 2 (up-left to down-right)
    result = check_slanting_chance(
        my_num, 8, [0,1,2,3,7,8,9,10,14,15,16,17])
    if result != -99:
        return result
    # no chance
    return -99
```

Those blocks constitute the basics of the logic. While a bit cumbersome to formulate, they are a useful exercise in converting an intuition into heuristics that can be used in an agent competing in a game.

Please see the accompanying code in the repository (https://github.com/PacktPublishing/The-Kaggle-Book-2nd-Edition) for a complete definition of the agent in this example.

The performance of our newly defined agent can be evaluated against a predefined agent, for example, a random one:

```
env.reset()
env.run([my_agent, "random"])
env.render(mode="ipython", width=500, height=450)
```

The code above demonstrates how to set up a solution from scratch for a relatively simple problem (there is a reason why *Connect X* is a Playground competition, not a Featured competition). Interestingly, this simple problem can be handled with (almost) state-of-the-art methods, such as AlphaZero: https://www.kaggle.com/connect4alphazero/alphazero-baseline-connectx.

With the introductory example behind us, you should be ready to dive into the more elaborate, or, in any case, not toy example-based contests.

Rock, Paper, Scissors

It is no coincidence that several problems in simulation competitions refer to playing games: at varying levels of complexity, games offer an environment with clearly defined rules, naturally lending themselves to the agent-action-reward framework. Aside from Tic-Tac-Toe, connecting checkers is one of the simplest examples of a competitive game. Moving up the difficulty ladder (of games), let's have a look at **rock-paper-scissors** and how a Kaggle contest centered around this game could be approached.

The idea of the *Rock, Paper, Scissors* competition (https://www.kaggle.com/competitions/rock-paper-scissors) was an extension of the basic rock-paper-scissors game (known as *roshambo* in some parts of the world): instead of the usual "best of 3" score, we use "best of 1,000."

We will describe two possible approaches to the problem: one rooted in the game-theoretic approach, and the other more focused on the algorithmic side.

We begin with the **Nash equilibrium**. Wikipedia defines this as the solution to a non-cooperative game involving two or more players, where each player is assumed to know the equilibrium strategies of the others, and no player can obtain an advantage by changing only their own strategy.

An excellent introduction to rock-paper-scissors in a game-theoretic framework can be found at `https://www.youtube.com/watch?v=-1GDMXoMdaY`.

Denoting our players as red and blue, each cell in the matrix of outcomes shows the result of a given combination of moves:

	Rock	Paper	Scissors
Rock	0, 0	-1, 1	1, -1
Paper	1, -1	0, 0	-1,-1
Scissors	-1,-1	1, -1	0, 0

Figure 14.2: Payoff matrix for rock-paper-scissors

As an example, if both play Rock (the top-left cell), both gain 0 points; if blue plays Rock and red plays Paper (the cell in the second column of the first row), red wins – so red gains 1 point and blue has -1 point as a result.

If we played each action with an equal probability of 1/3, then the opponent must do the same; otherwise, if they play Rock all the time, they will tie against Rock, lose against Paper, and win against Scissors – each with a probability of 1/3 (or one-third of the time). The expected reward, in this case, is 0, in which case we can change our strategy to Paper and win all the time. The same reasoning can be applied to the strategy of Paper versus Scissors and Scissors versus Rock, for which we will not present the matrix of outcomes due to redundancy.

The remaining option in order to be in equilibrium is that both players must play a random strategy, which is the Nash equilibrium. We can build a simple agent around this idea:

```
%%writefile submission.py
import random
def nash_equilibrium_agent(observation, configuration):
    return random.randint(0, 2)
```

The magic at the start (writing from a Notebook directly to a file) is necessary to satisfy the submission constraints of this particular competition.

How does our Nash agent perform against others? We can find out by evaluating the performance:

```
!pip install -q -U kaggle_environments
from kaggle_environments import make
```

At the time of writing, there is an error that pops up after this import (**Failure to load a module named 'gfootball'**); the official advice from Kaggle is to ignore it. In practice, it appears to have no impact on executing the code.

We start by creating the rock-paper-scissors environment and setting the limit to 1,000 episodes per simulation:

```
env = make("rps", configuration={"episodeSteps": 1000})
```

We will make use of a Notebook created in this competition that implemented numerous agents based on deterministic heuristics (`https://www.kaggle.com/ilialar/multi-armed-bandit-vs-deterministic-agents`) and import the code for the agents we compete against from there:

```
%%writefile submission_copy_opponent.py
def copy_opponent_agent(observation, configuration):
    if observation.step > 0:
        return observation.lastOpponentAction
    else:
        return 0
# nash_equilibrium_agent vs copy_opponent_agent
env.run(
    ["submission.py", "submission_copy_opponent.py"]
)
env.render(mode="ipython", width=500, height=400)
```

When we execute the preceding block and run the environment, we can watch an animated board for 1,000 epochs. A snapshot looks like this:

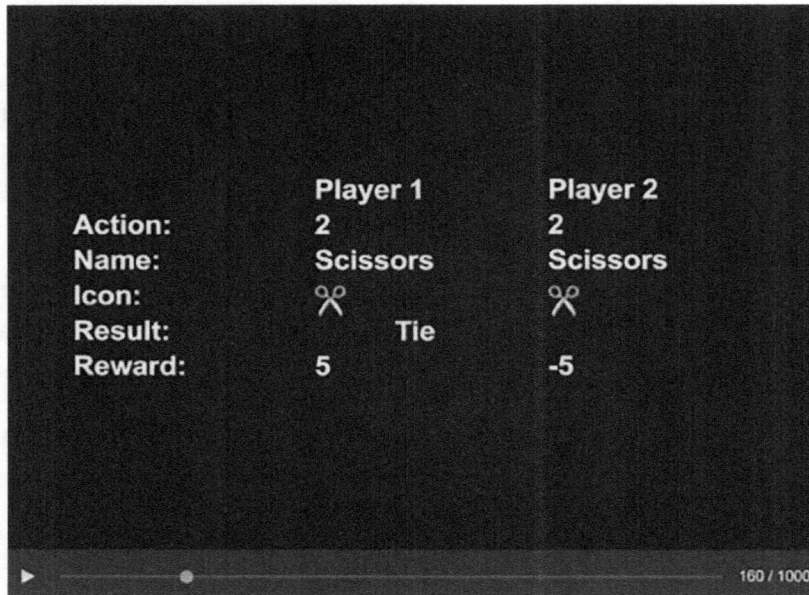

Figure 14.3: A snapshot from a rendered environment evaluating agent performance

In supervised learning – both classification and regression – it is frequently helpful to start approaching any problem with a simple benchmark, usually a linear model. Although not state-of-the-art, it can provide a valuable benchmark and a measure of performance. In RL, a similar idea holds; an approach worth trying in this capacity is the multi-armed bandit, the simplest algorithm we can honestly call RL (it is a simpler, stateless version). In the next section, we demonstrate how this approach can be used in a simulation competition.

Santa competition 2020

Over the last few years, a tradition has emerged on Kaggle: in early December, a Santa-themed competition is held. The actual algorithmic side varies from year to year, but for our purposes, the 2020 competition is an interesting case: https://www.kaggle.com/c/santa-2020.

The setup was a classical **multi-armed bandit (MAB)** trying to maximize reward by taking repeated action on a vending machine, but with two extras:

- **Reward decay**: At each step, the probability of obtaining a reward from a machine decreases by three percent.

- **Competition**: You are constrained not only by time (a limited number of attempts) but also by another player attempting to achieve the same objective. We mention this constraint primarily for completeness, as it is not essential to explicitly incorporate it into our demonstrated solution.

For a good explanation of the methods for approaching the general MAB problem, you can refer to https://lilianweng.github.io/lil-log/2018/01/23/the-multi-armed-bandit-problem-and-its-solutions.html.

The solution we demonstrate is adapted from https://www.kaggle.com/ilialar/simple-multi-armed-bandit, with code from *Ilia Larchenko* (https://www.kaggle.com/ilialar). Our approach is based on successive updates to the distribution of reward: at each step, we generate a random number from a Beta distribution with parameters *(a+1, b+1)*, where:

- *a* is the total reward from this arm (number of wins)
- *b* is the number of historical losses

When we need to decide which arm to pull, we select the arm with the highest generated number and use it to generate the next step; our posterior distribution becomes a prior for the next step.

The graph below shows the shape of a Beta distribution for different pairs of (a, *b*) values:

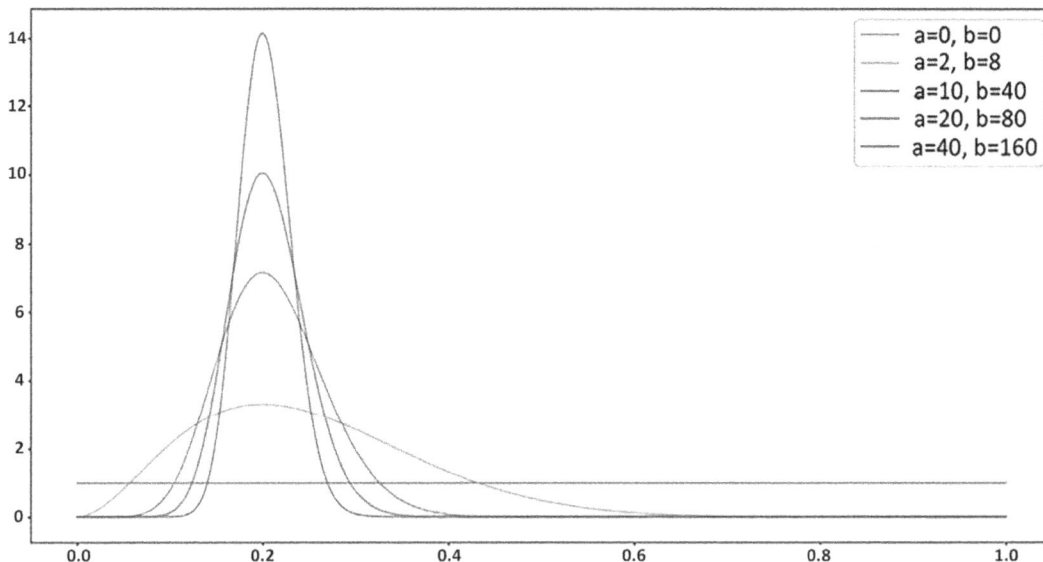

Figure 14.4: Shape of the Beta distribution density for different combinations of (a,b) parameters

As you can see, initially, the distribution is flat (Beta(0,0) is uniform), but as we gather more information, it concentrates the probability mass around the mode, indicating less uncertainty and greater confidence in our judgment. We can incorporate the competition-specific reward decay by decreasing the *a* parameter every time an arm is used.

We begin the creation of our agent by writing a submission file. First, the necessary imports and variable initialization:

```
%%writefile submission.py
import json
import numpy as np
import pandas as pd
bandit_state = None
total_reward = 0
last_step = None
```

We define a class that specifies an MAB agent. For the sake of reading coherence, we reproduce the entire code and include the explanations in comments within it:

```
def multi_armed_bandit_agent (observation, configuration):
    global history, history_bandit
    step = 1.0          # balance exploration / exploitation
    decay_rate = 0.97   # how much do we decay the win
                        # count after each call

    global bandit_state,total_reward,last_step

    if observation.step == 0:
        # initial bandit state
        bandit_state = [
            [1,1]
            for i in range(configuration["banditCount"])]
    else:
        # updating bandit_state using the result of the previous step
        last_reward = observation["reward"] - total_reward
        total_reward = observation["reward"]

        # we need to understand who we are Player 1 or 2
        player = int(last_step == observation.lastActions[1])
```

```
        if last_reward > 0:
            bandit_state[observation.lastActions[
                player]][0] += last_reward * step
        else:
            bandit_state[observation.lastActions[player]][1] += step

        bandit_state[observation.lastActions[0]][0] = (
            bandit_state[observation.lastActions[0]][0] - 1
        ) * decay_rate + 1
        bandit_state[observation.lastActions[1]][0] = (
            bandit_state[observation.lastActions[1]][0] - 1
        ) * decay_rate + 1
    # generate random number from Beta distribution for each agent
    # and select the most lucky one
    best_proba = -1
    best_agent = None
    for k in range(configuration["banditCount"]):
        proba = np.random.beta(bandit_state[k][0],bandit_state[k][1])
        if proba > best_proba:
            best_proba = proba
            best_agent = k

    last_step = best_agent
    return best_agent
```

As you can see, the core logic of the function is a straightforward implementation of the MAB algorithm. An adjustment specific to our contest occurs with the bandit_state variable, where we apply the decay multiplier.

Similar to the previous case, we are now ready to evaluate the performance of our agent in the contest environment. The code snippet below demonstrates how this can be implemented:

```
%%writefile random_agent.py
import random
def random_agent(observation, configuration):
    return random.randrange(configuration.banditCount)
from kaggle_environments import make
env = make("mab", debug=True)
```

```
env.reset()
env.run(["random_agent.py", "submission.py"])
env.render(mode="ipython", width=800, height=700)
```

We see something like this:

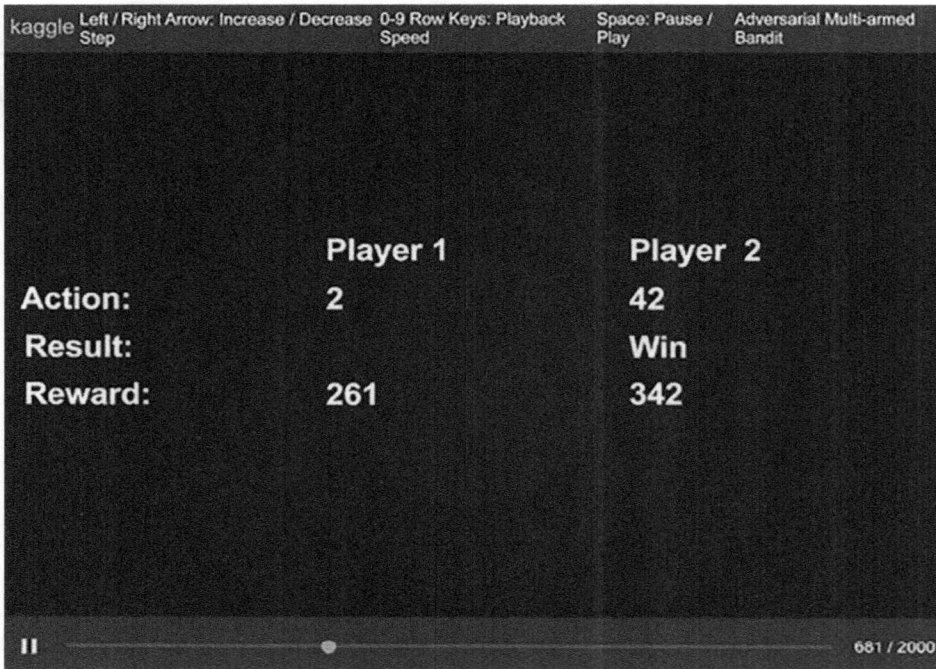

Figure 14.5: Snapshot from a rendered environment evaluating agent performance

In this section, we demonstrated how a vintage MAB algorithm can be utilized in a simulation competition on Kaggle. While useful as a starting point, this was not sufficient to qualify for the medal zone, where deep RL approaches were more popular.

We will follow up with a discussion of approaches based on other methods in a diverse range of competitions.

A few other Kaggle game agent competitions

Beyond the relatively elementary games discussed above, simulation competitions involve more elaborate setups. In this section, we will briefly discuss those. The first example is **Halite**, defined on the competition page at https://www.kaggle.com/c/halite.

This is what the game looks like:

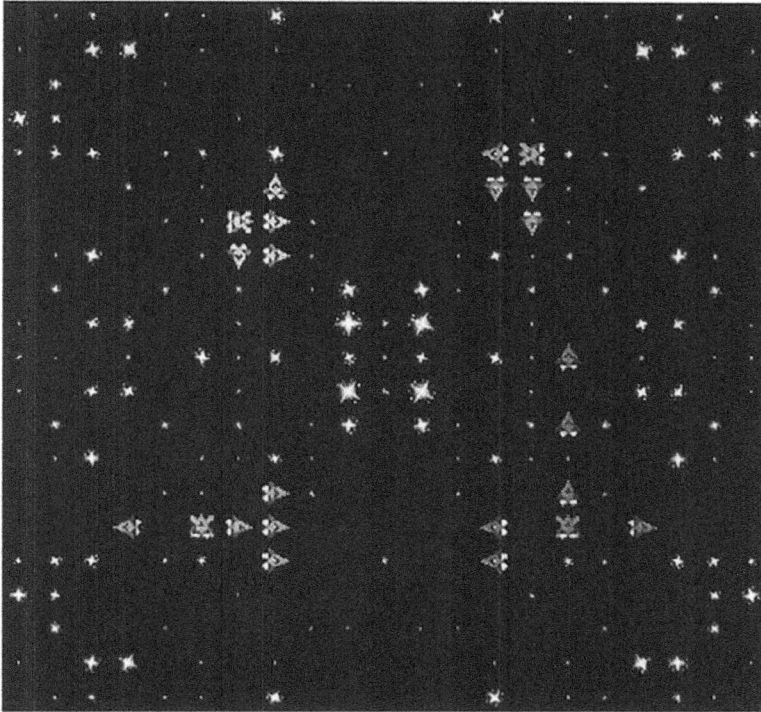

Figure 14.6: Halite game board

Kaggle organized two competitions around the game: a Playground edition (`https://www.kaggle.com/c/halite-iv-playground-edition`) and a regular Featured edition (`https://www.kaggle.com/c/halite`). The classic RL approach was less useful in this instance since, with an arbitrary number of units (ships/bases) and a dynamic opponent pool, the problem of credit assignment was becoming intractable for people with access to a "normal" level of computing resources.

Explaining the problem of credit assignment in full generality is beyond the scope of this book, but I encourage you to start with the Wikipedia entry (`https://en.wikipedia.org/wiki/Assignment_problem`) and follow up with this excellent introductory article by Mesnard et al.: `https://proceedings.mlr.press/v139/mesnard21a.html`.

A description of the winning solution by *Tom van de Wiele* (`https://www.kaggle.com/c/halite/discussion/183543`) provides an excellent overview of the modified approach that proved successful in this instance (deep RL with independent credit assignment per unit).

Another competition involving a relatively sophisticated game was *Lux AI* (`https://www.kaggle.com/c/lux-ai-2021`). In this competition, participants were tasked with designing agents to tackle a multi-variable optimization problem combining resource gathering and allocation, competing against other players. Additionally, successful agents had to analyze their opponents' moves and react accordingly. An interesting feature of this contest was the popularity of a "meta" approach: **imitation learning** (`https://paperswithcode.com/task/imitation-learning`). This is a fairly novel approach in RL, focused on learning a behavior policy from demonstration – without a specific model to describe the generation of state-action pairs. A competitive implementation of this idea (at the time of writing) is given by Kaggle user *Ironbar* (`https://www.kaggle.com/c/lux-ai-2021/discussion/293911`).

Finally, no discussion of simulation competitions in Kaggle would be complete without the *Google Research Football with Manchester City F.C.* competition (`https://www.kaggle.com/c/google-football/overview`). The motivation behind this contest was for researchers to explore AI agents' ability to play in complex settings, such as football. Unlike some examples given above, this competition was dominated by RL approaches:

- Team Raw Beast (3rd) followed a methodology inspired by *AlphaStar*: `https://www.kaggle.com/c/google-football/discussion/200709`

- Salty Fish (2nd) utilized a form of self-play: `https://www.kaggle.com/c/google-football/discussion/202977`

- The winners, WeKick, used a deep learning-based solution with creative feature engineering and reward structure adjustment: `https://www.kaggle.com/c/google-football/discussion/202232`

Studying the solutions listed above is an excellent starting point to learn how RL can be utilized to solve this class of problems.

These simulation competitions highlight diverse applications of RL, imitation learning, and innovative feature engineering in dynamic environments, offering valuable insights into AI's adaptability and creativity. Moving forward, we shift our focus to the Google FIDE Chess Optimization Challenge, exploring how strategic efficiency reshaped the classical battleground of chess AI under resource constraints.

FIDE and the Google Efficient Chess AI Challenge

Chess has long served as a cornerstone for AI research, a domain where the interplay of strategy, foresight, and computation has captivated scientists and enthusiasts alike.

As Claude Shannon noted in his seminal 1950 paper, "Thinking rigorously about the construction of a chess-playing computer might act as a wedge in attacking other problems of a similar nature and of greater significance." From Deep Blue's triumph over Garry Kasparov to AlphaZero's revolutionary self-learning approach, chess has been a proving ground for AI innovation. However, these advancements often rely on vast computational resources, such as pre-computed tables, extensive search trees, and powerful hardware that place them beyond the reach of many developers.

FIDE stands for **Fédération Internationale des Échecs**, which translates from French as **International Chess Federation**. It is the international organization responsible for governing chess competitions, setting standards and rules, maintaining player rankings, and awarding titles such as Grandmaster and International Master. FIDE was founded on **July 20, 1924**, and is currently headquartered in Lausanne, Switzerland.

The Google FIDE Chess Optimization Challenge (`https://www.kaggle.com/competitions/fide-google-efficiency-chess-ai-challenge/`), hosted on Kaggle, introduced a refreshing twist to this narrative by prioritizing efficiency over brute force. Launched in 2024 and closed on March 6, 2025, this simulation competition challenged participants to develop chess-playing agents under stringent CPU and memory constraints. Rather than rewarding sheer computational power, the competition emphasized elegant design and strategic optimization, leveling the playing field and encouraging innovative solutions. Participants were tasked with creating agents that could compete effectively against others within these limits, pushing the boundaries of AI efficiency in a resource-constrained environment.

The competition's structure mirrored other Kaggle simulation contests. Each team could submit up to five agents daily, which then played episodes against opponents of similar skill ratings. Skill ratings, modeled as a Gaussian distribution $N(\mu, \sigma^2)$, adjusted dynamically: wins increased μ, losses decreased it, and ties balanced it toward the mean. The uncertainty parameter σ decreased as more games provided data, refining the rating's accuracy. Agents underwent a validation episode upon submission—playing against themselves to ensure functionality—before entering the ongoing evaluation pool, initialized with a skill rating of $\mu_0 = 600$. Teams were limited to three active submissions, with older agents deactivated as new ones were added. The final leaderboard, determined after a post-deadline evaluation period from February 11, 2025, to approximately March 6, 2025, showcased each team's highest-scoring agent.

This setup created a unique challenge: how to craft a powerful chess engine when traditional crutches like massive memory or endless computation were unavailable? The answer lay in blending classical chess programming with modern optimization techniques, as exemplified by the 4th place solution, which we'll explore in depth.

The 4th place solution: enhancing Stockfish with a small neural network

Among the standout submissions was the 4th place solution, implemented by the Kaggle user nagiss (https://www.kaggle.com/nagiss), which cleverly modified Stockfish 16—the last version supporting **Hand-Crafted Evaluation** (**HCE**)—and augmented it with a small neural network. This hybrid approach boosted the engine's Elo rating by approximately 30 points, demonstrating the power of combining traditional heuristics with lightweight machine learning under tight constraints.

Let's break down this solution step by step, from engine selection to implementation and results.

> You can read the full solution description here: https://www.kaggle.com/
> competitions/fide-google-efficiency-chess-ai-challenge/writeups/
> nagiss-4th-place-solution.

Choosing the base engine

As stated in the preceding section, nagiss selected Stockfish 16 as the foundation for their agent, a choice driven by both practicality and prior experience. Stockfish, an open-source chess engine renowned for its strength and optimization, offered a robust starting point. Version 16 was particularly appealing because it retained HCE, unlike later versions that adopted **Neural Network Universal Evaluation** (**NNUE**), which demands more memory—potentially exceeding the competition's limits. Discussions on Kaggle and Discord suggested that HCE could still compete effectively, making Stockfish 16 a strategic fit.

The decision was also personal: the participant's interest in Stockfish stemmed from their earlier work in the Hungry Geese competition, where they explored shogi AI and noted its influence from Stockfish. Rather than experimenting with alternative engines, they doubled down on familiarity, leveraging Stockfish's mature code base to focus on targeted enhancements.

Optimizing memory usage

Memory constraints were a defining challenge of the competition, and the participant addressed this head-on by reducing Stockfish's resource footprint. Here's how they restructured the engine's memory usage:

- **Pawn Hash Table:** Scaled down from 131,072 entries to 8,192, thus reducing the memory usage to approximately 640 KiB. Some member variables were converted to smaller data types to further save space.

- **Continuation History:** Replaced with a hash table of 262,144 elements, using about 512 KiB of memory. Notably, collision detection was not considered, prioritizing speed and simplicity at the expense of perfect accuracy.

- **Transposition Table:** Stockfish's minimum configurable size is set to 1 MiB. This was a practical choice based on intuition, requiring no further tuning beyond.

- **Other Components:** Miscellaneous tables and features were trimmed, collectively using around 1 MiB. For instance, the participant swapped Stockfish's Rook Magic Bitboard for CFish's Classical Approach (sourced from the Chess Programming wiki), reducing memory demands.

Additional optimizations included removing Large Page support—irrelevant in Kaggle's environment—and rewriting C++ standard library-dependent code to eliminate reliance on libstdc++. While the engine linked against glibc (typically 3 MiB), Kaggle's pre-loaded environment excluded this from the 5 MiB limit. Linking libstdc++, however, increased usage and occasionally triggered errors, so its removal lowered failure rates in submissions.

These adjustments created a lean, efficient engine capable of operating within the competition's memory cap, setting the stage for evaluation improvements.

Enhancing the evaluation function with a neural network

With memory optimized, the participant turned to the evaluation function—Stockfish's HCE—as the key to boosting performance. While NNUE offered a tempting upgrade (as top teams likely used), the participant chose a novel path: extending HCE with a small, custom neural network. This decision was partly practical (fitting within constraints) and partly creative (more "fun" than mimicking NNUE).

With memory optimization paving the way, the participant seamlessly shifted focus to designing a neural network architecture tailored to enhance evaluation precision within the strict constraints of the competition.

Neural network architecture

The network was a three-layer **Multi-Layer Perceptron (MLP)**, designed for efficiency, which consisted of the following:

1. **Input Layer:** 99 features extracted from the board state.
2. **First Hidden Layer:** Produced 14 outputs, computed simultaneously using 256-bit registers. With biases, this layer had $99 \times 14 + 14 = 1,400$ parameters.

3. **Combination:** Outputs were calculated for white and black separately, then combined based on the turn, yielding a 32-dimensional vector.

4. **Subsequent Layers:** The vector passed through a clipped ReLU (borrowed from NNUE), a 32×32 fully connected layer, another clipped ReLU, and a 32×1 layer to output a single value. Total parameters: 2,489.

This compact design ensured that the network fit within memory limits while enhancing evaluation precision.

Training the neural network

The training focused on helping Stockfish's NNUE to understand and improve what the HCE evaluation was doing. It used data from about 70,000 games run via a `Kaggle-environments` setup, with positions sampled at a 1/8,192 rate. These games mixed in HCE, NNUE, and developmental evaluation functions to create a range of training scenarios. They used **Mean Squared Error (MSE)** as the loss function, and **Quantization-Aware Training (QAT)** was applied as well—though it's still a bit unclear how much it helped. All of this was run on Kaggle Notebooks to stay aligned with the competition environment.

Integration into Stockfish

Integrating the network posed challenges. Initially, adding its full output to HCE yielded a modest Elo gain (<10 points). After experimentation, scaling the output by 0.5 (i.e., adding half its value) boosted performance by ~30 Elo points—a breakthrough noticed only on the competition's final day, leaving no time for larger models. The pseudocode below illustrates this:

```
// Traditional HCE evaluation
int hce_eval = evaluate_hce(board);
// NN prediction
float nn_output = neural_network.predict(board_features);
// Scaled integration
int adjusted_eval = hce_eval + static_cast<int>(0.5 * nn_output);
// Final evaluation
return adjusted_eval;
```

This hybrid evaluation combined reliable heuristics with learned adjustments and worked well within the limits.

Submission strategy

Uncertain of the neural network's performance in Kaggle's environment, the participant hedged their bets by submitting two agents: one using pure HCE and another with the HCE-NN hybrid. This dual approach ensured competitiveness regardless of which excelled, a pragmatic tactic given the late-stage discovery of the scaling factor.

Performance and insights

The HCE-NN agent secured 4th place, a testament to its efficacy. The ~30 Elo improvement over Stockfish 16's baseline HCE placed it in the upper echelon, though it trailed teams likely using full NNUE or deeper optimizations. Locally, the optimized HCE alone achieved gold-medal contention, suggesting the base engine's strength, but the neural enhancement pushed it further in the dynamic leaderboard.

Key insights from this solution include:

- **Leveraging Legacy Code:** Starting with Stockfish 16 capitalized on decades of refinement, allowing focus on incremental gains.

- **Memory Discipline:** Strategic cuts (e.g., table resizing and library avoidance) demonstrated how to thrive under limits.

- **Hybrid Innovation:** Blending HCE with a neural network bridged classical and modern AI, offering a scalable model for constrained settings.

- **Tuning Matters:** The 0.5 scaling factor's impact underscored the importance of integration finesse.

- **Strategic Flexibility:** Dual submissions mitigated risk, a lesson in adapting to uncertainty.

Code example and repository

The GitHub repository (`https://github.com/Lgeu/kaggle-stockfish/tree/main`) contains the full implementation. Below is a simplified snippet of the evaluation adjustment:

```cpp
#include "stockfish.h"
#include "neural_network.h"

int evaluate_position(Board& board) {
    int hce_score = stockfish::evaluate_hce(board);
    float nn_score = nn::predict(board.get_features());
    return hce_score + static_cast<int>(0.5 * nn_score);
}
```

This code assumes that `stockfish.h` provides the HCE function and `neural_network.h` defines the MLP, reflecting the participant's approach.

Broader implications

The Google FIDE Chess Optimization Challenge reframed chess AI as a test of efficiency, not just strength. It challenged participants to rethink resource-intensive paradigms, aligning with real-world scenarios where computational power is limited, such as embedded systems or mobile devices. The 4th place solution exemplifies this shift, demonstrating how modest neural enhancements can enhance traditional methods without depleting resource banks.

For Kaggle learners, this competition offers a masterclass in optimization and hybrid design. It builds on this chapter's themes—simulation, reinforcement learning, and dynamic environments—by adding a layer of practical constraint management. Aspiring competitors can study the repository, experiment with Stockfish modifications, and explore how small networks amplify heuristic engines, all while honing skills transferable to diverse AI challenges.

Interview: Firat Gonen

`https://www.kaggle.com/frtgnn`

For this chapter's interview, we spoke to Firat Gonen, a Grandmaster in Datasets and Notebooks, and an HP Data Science Global Ambassador. He is also the deputy general manager and executive vice president of GTech, a company that crafts state-of-the-art software solutions for the financial sector. He shares his perspective on his Kaggle approach and how his attitude has evolved over time.

What's your favorite kind of competition and why? In terms of technique and solving approaches, what is your specialty on Kaggle?

My favorite kind evolved over time. I used to prefer very generic tabular competitions where a nice laptop and some patience would suffice to master the trends. I felt like I used to be able to see the outlying trends between training and test sets pretty good. Over time, with being awarded the ambassadorship by Z by HP and my workstation equipment, I kind of converted myself toward more computer vision competitions, though I still have a lot to learn.

How do you approach a Kaggle competition? How different is this approach from what you do in your day-to-day work?

I usually prefer to delay the modeling part for as long as I can. I like to use that time for **Exploratory Data Analysis (EDA)**, identifying outliers, reading the forum, and other tasks, all while trying to be patient. After I feel like I'm done with feature engineering, I try to form only benchmark models to get a grip on different architecture results. My technique is very similar when it comes to professional work as well. I find it useless to try to do the best in a huge amount of time; there has to be a balance between time and success.

Tell us about a particularly challenging competition you entered, and what insights you used to tackle the task.

The competition hosted by François Chollet was extremely challenging; the very first competition to force us into AGI. I recall feeling pretty powerless in that competition, where I learned several new techniques. I think everybody did that, while keeping in mind that data science is not just machine learning. Several other techniques, such as mixed-integer programming, resurfaced on Kaggle.

Has Kaggle helped you in your career? If so, how?

Of course, I learned a lot of new techniques and stayed up to date thanks to Kaggle. I'm in a place in my career where my main responsibility lies mostly in management. That's why Kaggle is very important to me for staying up to date on several things.

How have you built up your portfolio thanks to Kaggle?

I believe the advantage was in a more indirect way, where people saw both practical skills (thanks to Kaggle) and more theoretical skills in my more conventional education qualifications.

In your experience, what do inexperienced Kagglers often overlook? What do you know now that you wish you'd known when you first started?

I think there are two things newcomers do wrong. The first one is having a fear of entering a new competition, thinking that they will get bad scores, and it will be registered. This is nonsense. Everybody has bad scores; it's all about how much you devote to a new competition. The second one is that they want to get to the model-building stage ASAP, which is very wrong; they want to see their benchmark scores, and then they get frustrated. I advise them to take their time in feature generation and selection, and also in the EDA stages.

What mistakes have you made in competitions in the past?

My mistakes are, unfortunately, very similar to new rookies. I got impatient in several competitions where I didn't pay enough attention to the early stages, and after some time, I felt like I didn't have enough time to go back.

Are there any particular tools or libraries that you would recommend using for data analysis or machine learning?

I would recommend PyCaret for benchmarking to get you up to speed, and PyTorch for a model-building framework.

What's the most important thing someone should keep in mind or do when they're entering a competition?

Exploratory data analysis and previous similar competition discussions.

Do you use other competition platforms? How do they compare to Kaggle?

To be honest, I haven't rolled the dice outside Kaggle, but I have had my share of them from a tourist perspective. It takes time to adjust to other platforms.

Summary

In this chapter, we explored simulation competitions, a dynamic and increasingly popular contest format. Unlike vision or NLP-centered challenges, simulation competitions require a versatile approach that combines diverse methods with higher mathematical complexity, rooted in the principles of reinforcement learning rather than traditional supervised learning. This knowledge gave you a broader understanding of various types of competition and the skills necessary to tackle them effectively.

The information shared serves as a valuable tool for expanding your knowledge in data science competitions, highlighting how adapting methods to specific contest formats can significantly enhance performance. Understanding these distinctions helps you prepare for all the challenges you may encounter on Kaggle that require multidisciplinary approaches and deep analytical thinking.

Looking ahead, the next chapter shifts focus to professional development. You will learn how to transform your Kaggle Notebooks into a portfolio of projects, leveraging these to unlock career opportunities and build a stronger professional presence in the data science community.

Join our book's Discord space

Join our community's Discord space for discussions with the authors and other readers:

`https://packt.link/kaggle`

Part 3

Kaggle for Your Career: Building Your Profile and Finding Opportunities

In this final part, we shift our focus from competition mechanics to career impact, showing you how to translate your Kaggle achievements into tangible professional growth. We'll guide you through building a compelling portfolio that showcases your skills and innovative solutions developed through competitions. Furthermore, you'll discover strategies for leveraging your Kaggle profile and network to uncover exciting job opportunities and advance your data science career. By the end of this part, you'll understand how to strategically utilize your Kaggle experience to build a strong professional brand, attract potential employers, and open doors to new career paths in the data science field.

This part of the book includes the following chapters:

- *Chapter 15, Creating Your Portfolio of Projects and Ideas*
- *Chapter 16, Finding New Professional Opportunities*

15
Creating Your Portfolio of Projects and Ideas

Participation in Kaggle has its benefits. First, scoring well in the four areas and consequently ranking highly in the esteem of other Kagglers certainly brings satisfaction and a sense of accomplishment. In addition, your experience also has an impact beyond Kaggle and can help advance your career. This is partly due to the technical side: the experience you gain from participating in competitions, experimenting on data you have never worked on before, or repeating experiments with new techniques. Although Kaggle is not fully recognized as a qualification by many companies and recruiters, your work in competitions can demonstrate a lot about your capabilities and help you stand out from the crowd.

The Reddit discussion on Kaggle (which you can read at https://www.reddit.com/r/datascience/comments/s4p93s/an_honest_conversation_about_kaggle/) reveals a range of opinions on its utility in the data science community. Some participants perceive Kaggle as a valuable learning tool, providing a safe environment to learn from real-world datasets and observe how others approach similar problems. However, others express concerns about the artificiality of Kaggle competitions and questioning their relevance to real-world problems.

Despite these criticisms, some participants acknowledge that Kaggle has helped them in their careers, particularly in learning and showcasing their skills to companies they were interested in. However, we add that it is not just that. Your career is also boosted by the connections you create with other data scientists, such as in the Kaggle community, besides the attention you may get from companies and the role they may propose to you.

This chapter will explore effective ways to gain opportunities and stand out by effectively showcasing your work on Kaggle itself and other data science competition sites. We will cover the
following topics:

- Building your portfolio with Kaggle
- Arranging your online presence beyond Kaggle
- Monitoring competition updates and newsletters

In the next chapter, we will conclude the book by exploring how Kaggle can directly affect your
career by enhancing your professional network and providing you with career opportunities.

Building your portfolio with Kaggle

Kaggle's claim to be the *"home of data science"* has to be taken into perspective. As we have discussed at length, Kaggle is open to everyone willing to compete to figure out the best models in
predictive tasks according to a given evaluation metric.

There are no restrictions in participation based on where you are in the world (apart from a few
locations, see `https://www.kaggle.com/terms`), your education, or your proficiency in predictive modeling. Sometimes, some competitions are not even predictive in nature, for instance,
reinforcement learning competitions, algorithmic challenges, and analytical contests that accommodate a larger audience than just data scientists. However, despite the variety of offered
challenges, making the best predictions from data according to an evaluation metric is the core
purpose of Kaggle competitions.

Real-world data science, on the other hand, has many facets. First, with modeling, your priority
is to solve problems relevant to a business or other relevant purposes, and the metric for scoring
your model is simply a more or less exact measurement of how well it solves the problem. In
certain contexts, you may not only deal with a single metric but also consider multiple ones. In
addition, problems are open to being solved in different ways, and much depends on how you
initially formulate them.

Besides modeling, with data, you seldom get specifications about the data you have to use, and
you can modify any existing dataset to fit your needs. Sometimes, you have to create your own
dataset from scratch because it is necessary. Usually, there are no indications of how to assemble
or process data.

When solving a real-world problem, you also have to consider:

- Technical debt

- Extracting insights from data that can provide valuable information for improving business processes, such as identifying trends, understanding customer behavior

- Maintainability of the solution over time

- Time and computational costs for running the solution

- Explainability of the workings of the model

- Impact on the operating income (if the real-world project is a business one, increasing profits and/or reducing costs is the leitmotif)

- Operations and business optimizations

- Communication of the results at different levels of complexity and abstraction

- Proposing and making informed decisions

Often, all these aspects count more than raw performance against evaluation metrics, which is just a technical and unidimensional way to evaluate if a model is performing well in a problem.

Technical debt is the primary and most frequently cited criticism when discussing shortcomings in Kaggle's approach. It is a term more common in software development than data science, though it is relevant. For technical debt, you should consider whatever you have to do to deliver your project faster, but you will have to redo it later at a higher cost. Usually, in the heat of the competition, you assemble functions (some taken from other Kagglers' notebooks, too) in a rush to experiment more and score better without much consideration of the future implications of taking a solution.

> The classic paper *Hidden Technical Debt in Machine Learning Systems* by *David Sculley*, previously Kaggle CEO, and other Google researchers should enlighten you on the relevance of the problem for data science: `https://proceedings.neurips.cc/paper/2015/file/86df7dcfd896fcaf2674f757a2463eba-Paper.pdf`.

Not all this expertise can be supplemented by Kaggle competitions. Such hands-on knowledge should be gained by direct practice and experience-building in an enterprise environment. Yet, by saying so, we don't mean to imply that what you learn and do on Kaggle is irrelevant. The knowledge and skills attached to Kaggle competitions are not completely detached from many of the considerations we discussed above, and they are a good complement to many of the enterprise-level data science processes. Consider this: by competing on Kaggle, you are exposed to different types of data and problems; you must execute extensive feature engineering and fast iterations of model hypotheses; and you must also devise methods to assemble state-of-the-art solutions using common open-source packages. This is a set of valuable skills, and it should be promoted on your side as much as possible (just don't rely completely on it).

In our experience, the best way to highlight your knowledge and skills is to build a **portfolio**, i.e., a collection of your best solutions and work based on Kaggle competitions and other Kaggle resources. By a portfolio, we also mean a specifically engineered sample of your work on Kaggle, not all the notebooks you made public, and not in the form you presented to the Kaggle community.

When starting or progressing your career in tech, having a good portfolio of projects is the best way to demonstrate your abilities and make a lasting impression on potential employers. In particular, having a portfolio not only hints at your ability to use technologies and your enthusiasm for them but also reflects your capacity to gain ownership of your learning and growth, rather than constantly depending on directions provided by others.

You can take multiple approaches to build an effective portfolio from Kaggle competitions. The easiest is to leverage the facilities offered by Kaggle, especially **Datasets**, **Notebooks**, and **Discussions**. Before delving into that, we'll provide you with the golden rules for a good portfolio. But before we proceed, we present a discussion on career opportunities derived from Kaggle in our interview with *Gilberto Titericz*. Gilberto is a Grandmaster in Competitions and Discussions, the former number 1 in rankings, and the current number 1 in total gold medals from Kaggle competitions. He is also a senior data scientist at NVIDIA and was featured not long ago in an article on Wired on the topic of Kaggle (`https://www.wired.com/story/solve-these-tough-data-problems-and-watch-job-offers-roll-in/`).

Interview: Gilberto Titericz

`https://www.kaggle.com/titericz`

What's your favorite kind of competition and why? In terms of techniques and solving approaches, what is your specialty on Kaggle?

Since I started to compete on Kaggle in 2011, the types of competitions that I prefer are the ones with structured tabular data. The techniques that I use more in Kaggle are target encoding of categorical features (there are infinite ways to do it wrong) and stacking ensembles.

How do you approach a Kaggle competition? How different is this approach to what you do in your day-to-day work?

Kaggle is a great playground for **machine learning (ML)**. The main difference from real-life projects is that, in Kaggle, we already have the problem very well defined and formatted, the dataset created, the target variable built, and the metric chosen. So, I always start a Kaggle competition playing with **exploratory data analysis (EDA)**. Understanding the problem and knowing the dataset is one of the keys to gaining an advantage over other players. After that, I spend some time defining a proper validation strategy. It is very important to validate your model correctly and in line with the way that Kaggle scores the private test set. Besides the fact that using a stratified K-fold is something that works for most binary classification problems, we must evaluate if a grouped K-fold or a time-based split must be used in order to validate correctly, avoid overfitting, and mimic, as much as possible, the private test set. After that, it is important to spend some time running experiments on feature engineering and hyperparameter optimization. Also, I always end a competition with at least one gradient-boosted tree model and one deep learning-based approach. A blend of such diverse approaches is very important to increase diversity in the predictions and boost the competition metric.

Has Kaggle helped you in your career? If so, how?

Yes, Kaggle was the main reason I changed the direction of my career. Up to 2016, I worked as an electronic engineer, and due to everything that I learned competing since 2011, I was able to switch to the data science area. Kaggle helped me to understand the concepts of ML and apply everything I learned from the theory. In addition, Kaggle is an excellent place for experimentation, where you can download a dataset and play with it to extract the maximum information possible from the data. That, combined with the competitive environment, makes it perfect to learn coding and ML, and at the same time, it gets addictive and makes you want to learn more and more. Winning a few competitions puts your name at the top of the leaderboard and this is priceless for anyone's career. Headhunters all around the world look at Kaggle to find good matches for their positions, and the knowledge and experience gained from competitions can boost any career.

How have you built up your portfolio thanks to Kaggle?

Once I joined Kaggle, I spent some years learning all the techniques, algorithms, and tricks to extract more information from data and boost the metrics as much as possible. High accuracy is the main goal of most of the competitions, but to do that relying on luck alone is almost impossible; knowledge and experience play a big role when the goal is to win or at least finish in the gold medal zone. The number of medals I have in Kaggle competitions is my portfolio; up to now (December 2024), it's 63 gold and 56 silver, which summarizes well the ML experience I got from Kaggle. Taking into account that each competition runs for at least 1 month, this is more than 119 consecutive months of experience doing competitive ML.

In your experience, what do inexperienced Kagglers often overlook? What do you know now that you wish you'd known when you first started?

Novices often overlook a proper validation strategy. That doesn't just happen in Kaggle; I've seen data scientists all around the world building models and neglecting one of the most important things in the experimentation theory. There is no general rule when setting a proper validation strategy, but the data scientist must take into account how the model is going to be used in the future, and make the validation as close as possible to that.

What mistakes have you made in competitions in the past?

Several mistakes; it is impossible to list them all. I have probably made all the possible combinations of mistakes. The good thing about mistakes is that you can learn from them. Once you make a mistake and you detect it, it is very likely that you won't make it again.

The main mistake people make in Kaggle is to trust in the leaderboard score and not in their local validation score. Overfitting to the leaderboard is a constant in Kaggle and this is the main difference from the real world. In a real project, we must build a strong validation strategy that we can trust, because in the real world, the models will be tested on real data and you have only one chance to hit the mark, not multiple submissions per day.

Are there any particular tools or libraries that you would recommend using for data analysis and ML?

Some years ago I would have recommended R, but taking into account how fast Python is growing in the ML space and how generic and easy it is to use in production, I recommend to anyone starting ML that they learn it. In terms of libraries for tabular data, I recommend pandas for manipulation, and if you want speed, then go with cuDF (the RAPIDS.ai GPU version of pandas). For EDA, I recommend using DataFrame with the Seaborn or Matplotlib libraries, and for ML, scikit-learn, SciPy, cuML (GPU), XGBoost, LightGBM, CatBoost, and PyTorch. Keep in mind that building a simple XGBoost model using the raw features is fast and can give you a good benchmark to compare with further models.

What's the most important thing someone should keep in mind or do when they're entering a competition?

Entering a Kaggle competition and submitting a public notebook is easy, but finishing a competition in the gold zone can be extremely challenging. So, the most important thing, at least for me, is to keep in mind that independent of the final ranking, we should use Kaggle to have fun and learn as much as possible from the discussion forums, from the public notebooks, and even from the post-deadline winners' posts describing their ideas and what worked.

Also, keep in mind that what makes a competition winner is not just replicating what everyone else is doing but thinking out of the box and coming up with novel ideas, strategies, architectures, and approaches.

Do you use other competition platforms? How do they compare to Kaggle?

I have won a couple of competitions on other competition platforms, but the main difference compared to Kaggle is the number of users. Kaggle has 536k active users as of April 2023, which makes the forums, notebooks, and dataset interactions much richer in terms of content. Also, Kaggle offers something unique: notebooks where you can write and run code for free using Google servers, which can be priceless if you don't have access to good hardware.

The golden rules of a good portfolio

Here, the discussion is more general and can be applied to any project you want to showcase besides Kaggle. Given the purpose of showcasing to get an interview or to make a positive impression during the desk evaluation of candidates or a technical interview, you have to consider these three key points:

- Achieve something relevant applied to interesting data.
- Structure your project or code based on your target: will you impress the layperson, the technical expert, or the manager?
- Tell an interesting story you can reuse in an interview.

Deciding on the examples to present in your portfolio is important. Avoid presenting commonly found projects, such as the Titanic passenger survival classification and the MNIST hand-written digit classification, or, decisively worse, copying the selection of projects from another data scientist. Your choices must be unique to you and tailored to you. A good starting point may be your passion or data science interests: if you are keen on recommender systems, just build a fancy one. If you'd love to showcase forecasting time series, find an interesting series (or collect one) to show off the range of your abilities, from classical auto-regressive models to deep learning architectures.

In addition to your passion, having a range of examples related to the sectors you would like to work in will greatly help you. Having made up your mind on what to revolve the models around, you can use this checklist for your project:

- The data you work on is not so common and interesting in itself. You should use data that you collected and curated. Otherwise, you have plenty of choices from Kaggle competitions and datasets. Special data will automatically make your project special. Key questions you should ask yourself are "Are you using a special and relevant dataset? Why is it special and relevant?"

- Your project achieves some interesting results, and it is not just a mere exercise to show you can use Python. It could be business value or the ability to beat a reference benchmark. Consequently, you'll have a notable achievement to highlight when discussing the project. The key question here is "Does your project highlight a significant achievement?"

- Check that tables, charts, appealing graphics, or interactive visualizations can easily illustrate and summarize your project. A project that cannot be described briefly won't be helpful. Here, ask yourself "Can you easily illustrate and summarize your project with tables, charts, or visualizations?"

Once you have the data, execution, and results, all the materials should be properly arranged to reach your target audience. If your audience is non-technical, present results before the code, especially if the results can be shown as an amazing, possibly interactive, chart. If your audience is technical, keep your code correct, clean, and well-arranged. You can use various developer tools for this, such as the Black formatter (`https://github.com/psf/black`) or code readability tools (use Cyclomatic complexity, McGabe complexity, or the maintainability index, for instance). Find how to deploy your data as an API or web app. For instance, using Streamlit and its community cloud (`https://streamlit.io/cloud`) or using Gradio (`https://www.gradio.app`) are good and accessible choices. Don't forget to highlight the tricky parts of your code and any collaboration you had because it signals that you can be a team player. Besides this, always remember to document your project, upload it to a repository you can refer to by an internet address, and keep it updated (by adding fresh data occasionally, for instance).

Finally, you cannot underestimate the possibility that your portfolio may not be seen or examined superficially before your interview. Therefore, having a speech and a narrative related to the examples in the portfolio, how you made them, the things you learned, and what challenges you faced is quite handy because you can mention your work with a reason. For instance, in my personal experience, I once discussed a Kaggle competition on causal modeling, pointing out how I had previously used the same techniques in customer satisfaction analysis for various businesses.

Connecting your example with your previous experience or existing interests is up to you.

Now that you know about the golden rules of a good portfolio, let's get more practical and see how the tools already offered by Kaggle can help you in your portfolio building.

Leveraging notebooks

Besides rankings, notebooks are a way to get noticed on Kaggle because they demonstrate how you solve problems, present ideas, and code them. Conceived as a way to easily and openly share solutions and ideas among participants, notebooks are the most important tool (after rankings) for demonstrating abilities that employers appreciate.

In fact, one of the most important changes in the world of data science in recent years has been its transition from a game of outstanding talents (unicorn data scientists) to a team game, where data scientists have to collaborate with each other and with other departments to ensure the success of a project. Consequently, in their hiring processes, companies often care more about you being able to communicate ideas and results and coding in a clean and effective way.

In the previous section, we discussed how real-world projects require a wider range of skills, ranging from dealing with technical debt to designing cost-effective solutions. You can still demonstrate these skills on Kaggle, even if they are not the ones that will make you win a competition. Notebooks are the best tools for doing this.

> Refer to *Chapter 3, Working and Learning with Kaggle Notebooks*, for an introduction to Kaggle Notebooks.

You will find different types of notebooks on Kaggle. As a good approximation, we can group them into four categories:

- Solutions and ideas for ranking in a competition
- EDA on the data
- Tutorials explaining ML models or data science principles
- Fresh implementations of models derived from papers or other original solutions

Each of these can provide you with an edge by means of an interesting set of skills. Solutions and ideas for competitions are the classic way to demonstrate that you know how to tackle a complex problem in data science, while the other three can show the world that you can:

- Manipulate, represent, and extract visual and non-visual insights from data (through EDA), which is a skill deemed very important in every setting, from scientific research to business
- Educate users on data science, opening the door to roles in education, mentorship, and developer advocacy
- Translate research into practice, a key skill at a time when innovations in data science (especially in deep learning) appear daily and need to be translated into working solutions quickly

Even if you don't rank highly in Kaggle competitions or have astonishing solutions to present, these other kinds of notebooks (EDA, tutorials, and paper implementations) can provide you with opportunities in the real world if you can promote them best. To do so, you must understand how to code readable and interesting notebooks, which you learn from practice and experience. Since it is an art, we suggest learning from others, especially the Notebooks Grandmasters, who place high in the Notebooks user ranking (`https://www.kaggle.com/rankings?group=notebooks&page=1&pageSize=20`).

We recommend you look at what kind of notebooks they have developed, how they have arranged their work using figures, how they have structured their code, and then, finally, based on your skills and interests, try to imitate one of their notebooks. We also suggest that you do not bet your chances for success only on code and charts but also on the **narrative** that you present. Whether you are showing off a solution, teaching, or implementing a neural architecture in TensorFlow, explaining the code in the notebook through a story is very important in leaving a lasting positive impression, as it could be a great discourse to bring up during an interview.

Aside from browsing the notebooks of high-rankers, there is also a way to be notified about less mainstream – yet still finely crafted – notebooks that have recently appeared on Kaggle. The astrophysicist and passionate Kaggle user Heads or Tails, *Martin Henze* (`https://www.kaggle.com/headsortails`), published on the discussion forums a weekly *Notebooks of the Week: Hidden Gems* post, a collection of the most interesting notebooks around. There are 100 volumes whose references are collected in this notebook: `https://www.kaggle.com/code/headsortails/hidden-gems-a-collection-of-underrated-notebooks`. The publication of the Hidden Gems post seems to be currently suspended, but if you would like to be updated about new issues, just follow Martin Henze's profile on Kaggle or occasionally check if he has published something new under his discussions.

If you love digging through notebooks, looking for ideas, and learning from them, we never tire of stressing that you should not brainlessly copy other people's work. There are many notebooks on Kaggle, and often, someone copies one, makes some small changes, and re-presents the notebook to other Kagglers as if it were their own original idea. It is also customary to cherry-pick a function or part of the code from a notebook and insert it into your own. In both these cases, please always quote the source and the author. If you cannot retrace something to the original author, even referring to the last notebook where you found the code you used is enough. While the main purpose of a showcase is to display your own efforts and skills, it is very important to recognize that some parts of your code or some ideas are taken from elsewhere. Aside from being a sign of respect toward your fellow Kagglers, a source attribution highlights that you are knowledgeable enough to recognize other people's efforts and inventions and know how to employ them in your work.

Leveraging discussions

In a minor way, discussions on Kaggle's forums can help you get noticed for specific roles in data science and software development. Initially, discussions on Kaggle were just for communicating with organizers or asking pressing questions about the competition. Participants seldom felt compelled to present or discuss their solutions at the end of competitions. However, since discussions obtained their own user rankings and mastery grades, you have been able to find much more information on forums.

> Refer to *Chapter 4*, *Leveraging Discussion Forums*, for an introduction to discussions on Kaggle.

In our experience, discussions on Kaggle can be split into four categories:

- Competition solutions that explain in detail (sometimes with the help of an associated notebook) how a team managed to reach a certain position on the private leaderboard
- Help with and an explanation of requirements during a competition
- Thanks, compliments, and chit-chat
- Posts that help and tutor other competitors, explaining things to them

We have observed that excelling in the last type of post and being widely noticed for it can help you achieve the role of developer advocate, especially if you also have other active channels where you interact with your fellow data scientists (for instance, a Twitch or YouTube channel, an X (formerly Twitter) account, or a Medium blog).

With the growth of developer advocate roles in large companies and start-ups, there is an important demand for experts skilled at helping other data scientists and developers in their projects. If you want to learn more about this role, the following article on *draft.dev* is quite explanatory and exhaustive: `https://draft.dev/learn/what-is-a-developer-advocate`.

Leveraging datasets

Kaggle competitions are often criticized for presenting data that is already cleaned, well arranged, and far from representative of data found in the real world. Our point of view is slightly different; the data that Kaggle presents in competitions can also be messy or noisy. Sometimes, the data presented will not suffice in terms of quality and quantity to get a top score, and you will need to look around for additional data on the internet. Recent examples are LLM-based (LLM stands for *large language model*) competitions, such as *LLM – Detect AI Generated Text* and *LLM Prompt Recovery*, and more classical competitions involving NLP, such as the **named entity recognition** (**NER**) competition, *The Learning Agency Lab – PII Data Detection*. In all these competitions, producing a useful dataset was a determinant of victory.

What Kaggle does miss out about data in a data science project is the process of collecting and gathering data in organized repositories and files, a process that, in real-world settings, is not possible to standardize because it differs from company to company and problem to problem. Data handling in the real world should mostly be learned in the field.

Introducing datasets into Kaggle aimed to mitigate the idea that Kaggle was just focused on modeling problems. Kaggle Datasets are very helpful in this sense because they allow you to create and upload your own data and document the features and their values; they also require you to manage your data over time by planning the frequency with which you will update or completely replace it.

> Refer to *Chapter 2, Organizing Data with Datasets*, for an introduction to Kaggle Datasets.

More interestingly, in Kaggle Datasets, you are also allowed to attach different analyses and models built using Kaggle Notebooks, uploaded from your data or a competition. These models could be work you came up with during a competition or something you devised because you studied the uploaded data attentively and found a set of interesting problems you could solve with it.

In addition, Kaggle Datasets offer a template to check for the completeness of the meta-information accompanying your data. A description, tags, a license, sources, and the frequency of updates are only a few of the required pieces of information (used to calculate a usability score) that will help anyone using your data understand how to use it. You may even point out (in the description or discussions) tasks for the dataset related to pending work you would like to do with it. This is a good way to communicate your full understanding of the potential value of the data you have uploaded.

Previously, Tasks were part of the Kaggle Datasets functionality, but they have recently been removed: `https://www.kaggle.com/product-feedback/292674`. Nevertheless, you can use the data description and discussions to point out what you expect your data to be used for.

All these characteristics make Kaggle Datasets a perfect way to show off your experience with problems on Kaggle and, in general, your ability with data and ML algorithms because they allow you to:

- Publish and maintain a dataset
- Demonstrate that you have understood the value of the data with a task roadmap
- Show coded and fully working solutions (since Kaggle Notebooks can immediately work on the same data without any preparation), ranging from data preparation to EDA to predictive modeling

We strongly recommend using Kaggle Datasets to show off your work during Kaggle competitions or on any other project because they separate your work from others and integrate data and notebooks. In short, Kaggle Datasets can demonstrate to anyone a working solution you have implemented. There is a downside, though: you are mostly tied to a notebook environment (even when you use scripting), which is not perfectly transparent regarding the package and version requirements necessary for someone to know to run the code in other environments.

In fact, Kaggle Notebooks depend on a Docker environment (`https://www.docker.com/`) set by a configuration file, a **Dockerfile**, that determines which versions have been installed. When browsing a notebook, it is not immediately evident what package version is being used until you inspect this configuration file. For this purpose, as well as for replicating the settings, the Dockerfile can be found in the Kaggle repository on GitHub (`https://github.com/Kaggle/docker-python/blob/main/Dockerfile.tmpl`), though it changes over time, and you may need to keep track of the one used in your work.

Finally, in addition to this aspect, don't forget that getting even a glimpse of a dataset and its related notebooks requires access to the Kaggle community.

Next, we present an inspiring career-oriented talk that we had with Gabriel Preda, a Kaggle Grandmaster in Datasets, Notebooks, and Discussions, and principal data scientist at Endava. Gabriel has a PhD in computational electromagnetics and had a long career in software development before devoting himself completely to data science. When he discovered Kaggle, he felt at home on the platform and invested a lot of time and effort into it, which paid dividends for him professionally. He is also the author of another Packt book on Kaggle, *Developing Kaggle Notebooks*, and he is a Google Developer Expert on Kaggle.

Interview: Gabriel Preda

`https://www.kaggle.com/gpreda`

Has Kaggle helped you in your career? How?

Kaggle helped me to accelerate my learning curve in data science. Before Kaggle, I was looking all around for sources of information or problems to solve, but it was not very methodical or effective. On Kaggle, I found a community of people interested in the same things as me. I was able to see the work of top experts in the field, learn from their published notebooks with analyses or models, get insights from them, ask them questions, and even compete against them. I was mostly in data analysis at the time I joined Kaggle, but very quickly, I started to compete; that means learning how to build, validate, and iteratively improve models. After around two years on Kaggle, I switched my career; I went from managing software projects to a full-time data science job. Kaggle also gave me some visibility, and during interviews with candidates at my present company, they mentioned that they wanted to join because they saw that I worked there.

Have you ever used something you have done on Kaggle as part of your portfolio to show potential employers?

I use my Kaggle portfolio as the main source of information for potential employers; my LinkedIn profile points to my Kaggle profile. Also, in recent years, employers have become more aware of Kaggle, and some of them ask specifically about your Kaggle profile. There are also potential employers who make it very clear that they do not consider Kaggle relevant. I disagree with this view; personally, before interviewing candidates, I normally check their GitHub and Kaggle profiles. I find them extremely relevant. A good Kaggle profile will demonstrate not only technical skills and experience with certain languages, tools, techniques, or problem-solving skills but also how well someone is able to communicate through Discussions and Notebooks. This is a very important quality for a data scientist.

You reached Grandmaster in Notebooks (kernels) first, then in Discussions, and finally, in Datasets. Can you tell us about your journey?

I became the seventh Kernels Grandmaster and I got as high as the third rank. For maybe two years, I think I was in the top 10 in the kernels hierarchy as well. I started writing kernels primarily to improve my knowledge of the R language while analyzing datasets I found more interesting. I also experimented with all kinds of techniques, including polygon clips, building dual meshes of Voronoi polygons, and 2D Delaunay tessellation. I gradually started to focus on EDA, followed by building models for datasets and then for competitions. Also, once I started to compete more, I started to write kernels for competing in Python. Around the same time, I began to notice that some of my kernels attracted attention from Kagglers, primarily upvotes and forks but also favorable comments. Some of my kernels written for the exploration of data in active competitions reached a very wide audience and brought me many gold medals; therefore, I reached the Master and then Grandmaster tier. Currently, I do not publish many kernels related to competitions; mostly, I create starting kernels related to datasets that I publish.

Next, I also obtained the Discussions Grandmaster level. I never anticipated that I would reach this tier in Discussions. First, I started commenting on other people's kernels. Then, gradually, as I got more involved in competitions, most of my comments were in the discussion sections of active competitions, either asking questions about topics of interest in these competitions or starting new topics, for example, suggesting solutions for one problem in a competition or collections of resources to address various open issues related to the competition. I want to mention a special set of comments that I added. As a Kaggle Kernels Grandmaster (one of the first), I frequently upvoted new Kagglers' Notebooks when I discovered very good content.

In such cases, I try to find a few moments to also praise the achievement of the author (especially if the content is of good quality). Especially to beginners, giving not only the expression of your appreciation by upvoting their work but also the addition of some positive feedback about their contribution might give them a boost of confidence so that they will invest more in learning and contributing even more on Kaggle. I like to do this, and I hope it helps. I once also compiled a list of recommendations about how to comment on Kaggle. This is the list: be short (but not too short); be specific; provide information, not opinions; praise other people's work when you have the opportunity; keep calm and try to be helpful; and do not tag people in your comments unless it makes sense (for example, if it is a discussion, and you need to direct your comment to someone that addressed you in that thread).

The last Grandmaster tier I reached is in Datasets. This is also the tier where I reached the highest ranking, second. My progress through the ranks was slow. I started with something I liked. Getting a high profile in Datasets requires investment in curating, cleaning, and documenting the data. If it is not something that you really like, you most probably will not keep going. I pursued things that were important to me but also to a wider community: to my country, my continent, or the whole world. I published datasets about elections in my country and about various social, demographic, and economic topics in Europe. I focused on subjects of actuality that were both relevant and of high importance to the community. For example, during the pandemic, I published datasets on COVID-19 cases, about vaccinations, tests, and virus variants both from my country and worldwide. I captured data that went beyond simple numerical, tabular values. Text data, especially originating from direct contributions from people, provided important insights for many people. One of my most upvoted datasets consists of collections of Reddit posts and comments or X (formerly Twitter) posts (formerly tweets) on subjects as diverse as vaccine myths, cricket, pandemics, sports events, and political personalities. I invested significantly in automating data collection, data cleaning, and data processing scripts. This saved me precious time (especially for datasets updated frequently – some of them were collected continuously, with scripts triggered every hour) but also made it possible to have better control of the process. Every time I publish a new dataset, I also write one or more starting kernels. These kernels are not intended to reach a large audience. I create them as helper kernels for potential users of my datasets so that they find it easier to use the data. In many cases, I prefer to keep the original data (as I collected it, or downloaded from an alternative source) and include a kernel for data cleaning, transformation, and preliminary analysis of the data in a more accessible format, as well as the result of this process. In this way, in the dataset, I try to capture more than the data itself; I also provide information about techniques for data transformation.

Arranging your online presence beyond Kaggle

Since Kaggle Datasets and Notebooks require a Kaggle account, you have to consider that not everyone may already have one or want to create one just to look at your work. You also have to consider alternatives that are more accessible. More frequently, Kagglers choose to use a project on GitHub (https://github.com/), write an article on Medium (https://medium.com/) as well as other publishing platforms, or post on their own blog. There are other opportunities to promote your work and skills, however, such as:

- Publishing code relevant to Kaggle competitions that can be executed from the browser on https://deepnote.com/.

- Setting up a Discord community that gathers Kagglers, such as *Abhishek Thakur*'s MLSpace (https://discord.com/invite/4RMwz64gdH), or running a YouTube channel (also from *Abhishek Thakur*: https://www.youtube.com/channel/UCBPRJjIWfyNG4X-CRbnv78A).

- Setting up a Twitch channel like *Rob Mulla*'s, where he demonstrates coding relevant to Kaggle competitions: https://www.twitch.tv/medallionstallion (also on GitHub: https://github.com/RobMulla/twitch-stream-projects).

- Delivering a weekly newsletter on Kaggle news, like *Shotaro Ishihara*: https://substack.com/@weeklykagglenews.

- Interviewing Kagglers and other data science experts, as *Sanyam Bhutani* is doing, and broadcasting the interviews using videos, podcasts, and blog posts: https://www.youtube.com/@ChaiTimeDataScience (you can browse the dataset containing all the data about the interviews held so far, prepared by *Rohan Rao*: https://www.kaggle.com/rohanrao/chai-time-data-science).

- Writing papers about the insights you got from the competitions is often overlooked. To raise attention to this opportunity, there has been an analytical competition about writing comprehensive documents on the latest AI advancements and salient topics in modern ML based on Kaggle competitions (see https://www.kaggle.com/competitions/2023-kaggle-ai-report). Moreover, if you look for papers with a Kaggle citation in Google Scholar, it returns over 90,000 results (https://scholar.google.com/scholar?hl=en&as_sdt=0%2C5&q=Kaggle&btnG=).

There are many opportunities and media through which you can diffuse your work and skills on Kaggle, depending on your goal. In the following sections, our focus is on blogs and a GitHub presence (the most common and effective choices), but you are free to decide on any different approach you deem suitable for your purposes.

Blogs and publications

Writing can be a way to refine your knowledge – because you need to read up on a topic to write about it – and to let others know about you and your skills. Getting famous for your writing helps you in various ways, from being spotted by recruiters and companies to building your connections for both Kaggle competitions and your wider professional life.

Social media (LinkedIn, X, and Facebook) allows you to post ideas and short text, which we suggest you leverage. Since data science and Kaggle competition topics require discussion and reasoning at length, the best approach is to write **long articles** and publish them employing a blog or a website that publishes writing. Ideally, we suggest you coordinate your communication between social media and your articles in order to promote them, with dedicated posts announcing them or discussing key points in your writing.

Let's first discuss how and where you could publish your articles:

- **Medium:** An article on Medium can get much attention. Medium publications are shared spaces for stories written around a common theme or topic, usually by multiple authors. As a website, Medium can reach a wide audience of readers, and some publications have a very good reputation in the data science community for the quality of their articles. A publication can have one or more editors who select the pieces and ensure that their contents are consistent with the policies of the publication and its quality level. Medium publications where you could post your articles are:

  - **Towards Data Science** (`https://towardsdatascience.com/questions-96667b06af5`)

  - **Better Programming** (`https://betterprogramming.pub/write-for-us-5c4bcba59397`)

  - **Becoming Human** (`https://becominghuman.ai/write-for-us-48270209de63`)

  - **Towards AI** (`https://pub.towardsai.net/submit-your-medium-story-to-towards-ai-a4fa7e8b141d`)

Each of these publications has the great advantage of already having a large audience, probably larger than your following on social media. You will get more readers than expected, reaching people at companies and other professionals you can network with.

- **Hacker Noon** (`https://www.publish.hackernoon.com/`): This is quite popular among tech bloggers and contains anything tech-related (it is a generalist website). With a monthly audience of four million people, it is the right place to reach many tech lovers with anything tech-related. Being featured on the top pages is extremely difficult and a double-edged sword: you will get a lot of attention and many critics.

- **Dev.to** (`https://dev.to/`): This mainly has an audience of developers (almost 800,000) and features articles and tutorials on coding. Your posts should focus more on the quality and efficacy of your code (modeling is in the background).

- **FreeCodeCamp** (`https://www.freecodecamp.org/news/developer-news-style-guide/`): This is more focused on tutorials; people go there to learn how to code. It is ideal for promoting courses on ML and new packages.

- **Analytics Vidhya** (`https://www.analyticsvidhya.com/about/`): This gained great popularity in India; it centers more on articles explaining ML and deep learning building blocks.

- **KDnuggets** (`https://www.kdnuggets.com/news/submissions.html`: This is one of the oldest publications in data mining. It still has many followers (one million unique visitors in March 2021) among the old guard of data scientists and academics.

Each publication has strong and weak points and differs in the audience it reaches, so you have to decide which one better suits your content. Start by browsing the publications they offer to understand how your writing could fit in.

Of course, if you prefer, you can use your blog instead. Having your own blog has its advantages, such as no advertising or editorial scrutiny over what you write. On the other hand, you cannot leverage a pre-existing audience, and you will have to work to create one by promoting your articles on social media. You can set up your own website from scratch on a web domain of your choice or create your blog on GitHub, too.

If you decide to use GitHub (since it is free and you may already use it as a repository for your code), here is a simple and fast guide to creating GitHub blog posts: `http://jmcglone.com/guides/github-pages/`.

If you need something even more automated, using a platform such as *Jeremy Howard*'s **fastpages** (`https://github.com/fastai/fastpages`) can simplify the way you deal with writing content together with code examples because it automatically converts notebooks and Word documents into blog pages and publishes them for you.

If you prefer to be completely independent and set up your own website, this will require more effort and some expense; domain names and web space are not free. In this case, self-promotion of your content becomes critical.

The main advantage of writing about your solutions is the storytelling element because you have to accompany your code snippets with descriptions and explanations and write more verbosely than you could in a notebook. How you describe your work becomes as important as the code you write. By adjusting the tone of your writing, you can reach different audiences. Writing concepts in an accessible way means you will enlarge your audience and connect with more professionals. Writing in a highly technical way instead could impress more potential companies that may consider hiring you, though limiting the number of readers you get.

Since writing is a very personal act and our hints and suggestions won't apply to every scenario, our general suggestion is to decide beforehand the purpose of your writing and who you would like to reach with it.

Moreover, nowadays, you can leverage the power of LLMs to help you in your work's storytelling and refining process. Starting from the appearance of **ChatGPT** on November 30th, 2022, many proprietary or open LLMs have appeared, from Google's **Gemini** and **Gemma** to Meta's **Llama 3** and **Mistral** from Mistral AI. LLMs are widely recognized as having strong capabilities in text (and code) generation, extraction, and transformation, which can help you in your task to both prepare write-ups and clean your code.

Even if you are not an expert in writing, use prompts such as "Rephrase this text more clearly and appealingly while maintaining unaltered the meaning" or "Improve this text using the writing style of *STYLE* while maintaining the original meaning but altering the tone" (where *STYLE* has to be replaced with your favorite style). These prompts can produce an improved text starting from your existing writings. You can also correct or translate any text you wrote from mistakes using prompts such as "Correct this text only where necessary, maintaining unaltered the tone and style."

The idea is to have the LLM improve your work and not replace or substitute you, something that can be detected, for instance, if you use the word delve (see: `https://www.technollama.co.uk/to-delve-or-not-to-delve-ai-detection-made-easy`). In addition, quite a few LLM detectors on the market now can estimate the probability of a text being written by a machine, not a human. As a rule, remember that an improved text is generally accepted, but a completely generated one is not. You can also apply the same for your code by asking for help with refactoring, spotting errors, making the code clearer or performing, or simply adding comments or explanations to the code you wrote.

GitHub

Aside from writing articles and having a code repository to which you can direct readers, having your code on GitHub will also help you not reinvent the wheel in every competition you enter. You can store the code you want to reuse in a project or in **Gists** (`https://docs.github.com/en/ github/writing-on-github/editing-and-sharing-content-with-gists`), which are small snippets of code that can be accessed individually.

Even if it may appeal to you to leave all your code on Kaggle, with time, you will find it difficult to access, and you may even have trouble finding it altogether. This is because you cannot arrange your Kaggle Notebooks into separate projects; they will just be presented as a long list that you can order by a few attributes, such as the number of votes or when you last ran the notebook. GitHub makes it much easier to find what you need and reuse it. For instance, you can create scripts containing all your code and then download and import them into a Kaggle Notebook without needing to copy anything. You can also import your own project repository as a dataset if needed.

In the following example, we download and reuse helper functions for a `tabular` neural network:

```
!wget https://raw.githubusercontent.com/lmassaron/deep_learning_for_
tabular_data/master/tabular.py
# Importing from Tabular
from tabular import (gelu, Mish, mish, TabularTransformer, DataGenerator)
```

A `wget` command will directly access the code on GitHub and download it onto the notebook storage space; afterward, you can just import the functions and classes you need. To obtain the link providing direct access to your code, you just need to look for the file containing it in the GitHub repository and then click on the **Raw** button on the header of the page:

Figure 15.1: The header of a visualized file on GitHub; notice the Raw button on the upper-right part of the header bar

After clicking the **Raw** button, you will be taken to the web address where the file is stored on GitHub. You can use that web address to refer to the file outside of GitHub.

GitHub is also useful for storing images you can use on Kaggle discussions (since you can no longer upload images on the Kaggle forums). In the case of images, you won't have a **Raw** button to click, but you can right-click on the image and open the file in another tab; this will have the same effect.

GitHub is another great way to showcase your work. Still, given the nature of the website (it is targeted at developers) and the content you can put on it (files containing code), you should expect a very technical audience. In companies, human resources probably won't look too deeply at your GitHub account, instead stopping at README.md, which should be well-written and visually appealing. Recruiting managers, on the other hand, will be more interested in the code in your projects. You should put effort into having well-structured code in your files, procedures, and classes, including the instructions to install and replicate your results.

Making an online demo

You have a few options if you want to put in live demonstrations of your models. The easiest is to run the code on the original notebooks (just by putting a link to your Kaggle Notebook in the README.md file of your GitHub project) or on **Google Colab**. To have the notebook you stored on GitHub run automatically in Google Colab, just post its link with the domain changed from github.com to githubtocolab.com. The link will open your notebook in Colab.

However, the most impressive showcase you can prepare is using **Hugging Face Spaces** (https://huggingface.co/spaces) to demonstrate how your Kaggle model could be used in an online application. Spaces are a simple way to host ML demonstrations and create online portfolios of your work, as explained in the documentation (https://huggingface.co/docs/hub/spaces). They are limited to 16 GB of RAM and 8 CPU cores, but they are free and sufficient for demonstrating how your model can run in a dedicated application. You can install your dependencies on the Hugging Face remote machine, sync code and models with GitHub, or build an app using **Streamlit** (https://streamlit.io/) or **Gradio** (https://gradio.app/).

As an example, *Rashmi Banthia*, a Kaggle Master and a Teaching Fellow at Harvard University (https://www.kaggle.com/rashmibanthia), has posted a demonstration of a model capable of detecting which, out of 7 possible LLMs, produced a particular response. The model achieved the third place solution in the H2O.ai hackathon **Predict the LLM**. Presenting your model with a few examples in a real-time demonstration lets you immediately convey its effectiveness even to a non-ML audience.

Writing a paper on arXiv

This idea is not just about publishing a detailed write-up outlining your solution or sharing your code publicly, but it instead involves preparing a detailed technical report or paper illustrating the numerous ideas, multiple trials (including unsuccessful ones), and occasionally achieved **state-of-the-art (SOTA)** solutions.

This will result in a comprehensive summary of these experiences, which will be of great help to other researchers and students when selecting approaches, model architectures, loss functions, or augmentation techniques in similar problems and it will make you surely noticed.

The benefits of writing a paper range from polishing up your code, making it readable and fully reproducible, to writing down your reasoning more scientifically. After writing the paper, you can publish it on arXiv (`https://arxiv.org/`), which does not require a peer-review process, or in any scientific journal, which instead requires a full review. The reviewing process may require you to prepare additional analysis and tests, perform ablation studies (i.e., removing components from your solution to assess its performance), or even rewrite a large portion of your paper. However, if it leads to a publication in a renowned journal, the result will be even more significant and impressive for your CV than the participation and results you may get in a Kaggle competition.

Monitoring ongoing competitions

By now, you can see that it is important to showcase your work on Kaggle so you can communicate your interest in certain types of models and data problems to the world. From this perspective, you must always know the opportunities competitions offer.

The main way to do this is to visit the Kaggle website frequently and agree to receive emails from them. You can set this option from your profile on the **Notification and Email Settings** page, where you can agree to receive notifications both on the site and by email. You can also choose to receive emails containing tips on new features and initiatives on Kaggle, along with news about recently launched competitions:

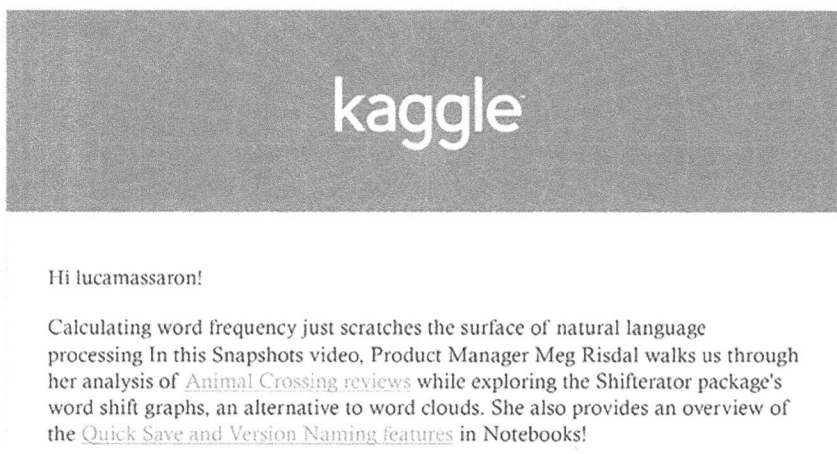

Hi lucamassaron!

Calculating word frequency just scratches the surface of natural language processing In this Snapshots video, Product Manager Meg Risdal walks us through her analysis of Animal Crossing reviews while exploring the Shifterator package's word shift graphs, an alternative to word clouds. She also provides an overview of the Quick Save and Version Naming features in Notebooks!

Figure 15.2: A Kaggle email announcing a series of videos from the Kaggle Team

As we know from *Chapter 1*, Kaggle is not the only organization that holds data science competitions. To keep better track of what is actually happening both on Kaggle and other data science competition websites, we suggest using websites such as `https://mlcontests.com/` or `https://ods.ai/competitions`, which monitor all ongoing competitions on Kaggle, as well as on other platforms such as AICrowd and DrivenData. For instance, `mlcontests.com` provides information on prizes, deadlines, and useful links for each competition. It also gives you cloud GPU performance, machines, and price comparisons. You can register your email address and receive much of this information directly to your inbox.

Summary

In this chapter, we discussed how to showcase your work and how this can be valuable for progressing your career. It helps you to demonstrate capabilities that, while (of course) not covering the entire span of your data science knowledge and experience, still represent a great asset.

To display your work, you can either use Kaggle resources or external resources. Kaggle resources offer an integrated environment and, provided you have everything at hand, are quite accessible and quick to set up. External resources (Medium publications, GitHub, Hugging Face Spaces, and so on) are more widely known and accessible for most recruiters, human resource officers, and hiring managers because they use them routinely.

In the next chapter, we will complete our discussion of the opportunities that Kaggle competitions offer you by talking about network building and how to use your Kaggle efforts to get an interview.

Join our book's Discord space

Join our community's Discord space for discussions with the authors and other readers:

`https://packt.link/kaggle`

16

Finding New Professional Opportunities

After introducing how to highlight your work and achievements in competitions in the previous chapter, we will conclude our overview of how Kaggle can positively affect your career. This last chapter discusses the best ways to leverage all your efforts to find new professional opportunities. We expect you now have all the previously described instruments (your Kaggle Discussions, Notebooks, and Datasets, and a GitHub account with quite a few projects derived from Kaggle), so this chapter will move on to softer aspects: how to network and how to present your Kaggle experience to recruiters and companies.

It is common knowledge that **networking** opens up many possibilities, from being contacted about new job opportunities that do not appear on public boards to having someone to rely on for data science problems in which you are not an expert. Networking on Kaggle is principally related to team collaboration during competitions and connections built during meetups and other events organized by Kagglers.

Regarding job opportunities, as we have often repeated previously, Kaggle is not a widely recognized source used by human resources and hiring managers for selecting candidates. Some companies do take your Kaggle rankings and achievements into reasonable consideration, but those are particular cases, not the general rule. Typically, you should expect your Kaggle experience to be ignored or sometimes even criticized by employers, especially those outside of data-driven fields, and by fellow data scientists interviewing you because of concerns about the real-world application of the skills demonstrated in competitions.

Our experience tells us, however, that what you learn and practice on Kaggle is highly valuable, and it can be promoted by showcasing your coding and modeling efforts and also by being able to talk about your experiences working alone or in a team.

Here, we will cover:

- Building connections with other competition data scientists
- Participating in Kaggle Days and other Kaggle meetups
- Getting spotted and other job opportunities

Building connections with other competition data scientists

Connections are essential for finding a job, as they help you get into contact with people who may know about an opportunity before it becomes public and the search for potential candidates begins. In recent years, Kaggle has increasingly become a place to connect with other data scientists, collaborate, and make friends. In the past, competitions did not give rise to many exchanges on forums, and teams were heavily penalized in the global rankings because competition points were split equally among the team members. Improved rankings (see https://www.kaggle.com/general/14196) helped many Kagglers see teaming up in a more favorable light.

Teaming up in a Kaggle competition works fine if you already know the other team members and have an established approach to assigning tasks and collaborating remotely. In these situations, each team member already knows how to collaborate by:

- Taking on part of the experimentation agreed by the team members
- Collaborating with another team member to build a solution
- Exploring new solutions based on their skills and experience
- Preparing models and submissions so they are easily stacked or blended

However, if you are new to teaming, you will find it challenging to enter a team or organize one yourself. Unless you have contacts, getting in touch with other people on the leaderboard will be hard. First, not all of them will want to team up because they prefer to compete alone. Furthermore, some competitors might be interested in teaming but will be too wary to accept your proposal. When forming a team with Kagglers you don't know, there are a few general hindering concerns regarding new persons joining because they may:

- Not bring any value to the team
- Not actually collaborate but just be a freeloader

- Have infringed (or will infringe) Kaggle rules, which will lead to the disqualification of the entire team
- Be actually interested in spying and leaking information to other teams

Most of these situations are pathological in a competition, and you should be aware that these are common considerations that many make when evaluating whether or not to team up with another Kaggler for the first time. You can only dispel these perceived potential problems by presenting yourself as someone with a strong background in Kaggle; that is, someone who has taken part in some competitions alone and, in particular, published Notebooks and participated in discussions. Such premises will add significant credibility to your proposal and will likely bring you acceptance into a team. However, remember that you also have to do your due diligence and verify that the team you are going to join is made up of equally trustworthy Kagglers.

When you have finally joined a team, it is essential to establish **efficient and dedicated forms of communication** between the team members (for instance, by creating a channel on Slack or Discord). It is also essential to agree on daily operations that involve deciding the following:

- How to divide your experimentation efforts and generally the workload of the competition. Different team members will display different time availability and different commitments.
- How to use the daily submissions, which are limited in number, and their use is often a cause of conflict in the team. In the end, only the team leader chooses the final two submissions, but the process of getting there naturally involves discussion and disagreement. Be prepared to demonstrate to your teammates why you have decided on certain submissions as final by showing them your local cross-validation strategy and results.

After you have experienced working together in a team in a positive manner, you will surely have gained the respect and trust of other team members. In future competitions, you will probably find it easier to team up again with the same people or join a different team that they are part of with their help.

The people you will meet and get to work with on Kaggle include data scientists, data enthusiasts, students, domain specialists, and more. Next, we speak to a diverse cross-section of Kagglers, who describe their day jobs and how Kaggle fits into their lives.

We first talked to *Yirun Zhang*, who completed his PhD at King's College London and presently works at Jane Street. A Notebooks and Discussion Grandmaster, he was a member of the winning team in the *Jane Street Market Prediction* competition (https://www.kaggle.com/c/jane-street-market-prediction).

Interview: Yirun Zhang

`https://www.kaggle.com/gogo827jz`

Can you tell us about yourself?

My research area lies in the field of applying machine learning algorithms to solving challenging problems in modern wireless communication networks such as time series forecasting, resource allocation, and optimization. I have also been involved in projects that study AI privacy, federated learning, and data compression and transmission.

Apart from daily PhD research, I have been active on Kaggle for almost two years, since the second year of my PhD. The first competition I took part in on Kaggle was Instant Gratification, in which I utilized a diversity of machine learning and statistical methods from the `sklearn` library. This competition helped me develop a general sense of what a machine learning modeling pipeline is for Kaggle competitions.

I have been actively sharing my knowledge with the community in terms of Notebooks and discussion posts on Kaggle, and am now a Kaggle Notebooks and Discussion Grandmaster. Through sharing and discussing with others on the forum, I have gained precious feedback and new knowledge, which has also helped me to finally become the winner of a Kaggle competition.

Tell us a bit about the competition you won.

Jane Street Market Prediction was a really tough one. The reason is that it was hard to build a robust **cross-validation (CV)** strategy and lots of people were just using the public leaderboard as the validation set. They were training a neural network for hundreds of epochs without using a validation strategy to overfit the public leaderboard. Our team tried hard to maintain our own CV strategy and survived in the shake-up.

How different is your approach to Kaggle from what you do in your day-to-day work?

Kaggle competitions are very different from my daily PhD research. The former is very tense and involves instant feedback, while the latter is a long-term process. However, I have found that the new knowledge and methodology I learned from Kaggle competitions is also very useful in my PhD research.

Next, we spoke to *Osamu Akiyama*, aka OsciiArt, a Kaggler whose day job does not involve data science. He's a medical doctor at Osaka University Hospital and a Competitions Master.

Interview: Osamu Akiyama

```
https://www.kaggle.com/osciiart
```

Can you tell us about yourself?

I'm a medical doctor working at Osaka University Hospital. I received my master's degree in Life Science from Kyoto University. After I worked in an R&D job for a pharmaceutical company, I transferred to the Faculty of Medicine of Osaka University and obtained a medical license for Japan.

I started to learn data science and AI on my own because I was shocked by AlphaGo. I started participating on Kaggle in order to learn and test my skills in data science and AI. My first competition was NOAA Fisheries Steller Sea Lion Population Count in 2017. I participate in Kaggle competitions constantly and I've got three gold medals.

Has Kaggle helped you in your career?

Because I'm not educated in information science, I used my results in Kaggle to demonstrate my skills when I applied for an internship at an AI company and when I applied to be a short-term student in an AI laboratory. As I'm just a medical doctor, I've never used my data science skills in my main job. However, thanks to my Kaggle results, I sometimes have the opportunity to participate in medical data research.

What is your favorite type of competition and why?

My favorite kind of competition is medical data competitions. I love to try to find some insight from the medical data using my knowledge of medicine.

How do you approach a Kaggle competition?

I love to find a secret characteristic of competition data that most other competitors are not aware of or to try a unique approach customized to the characteristics of competition data. Actually, such an approach is not successful in most cases, but still, it's fun to try.

Tell us about a particularly challenging competition you entered, and what insights you used to tackle the task.

I'd like to mention Freesound Audio Tagging 2019, which was a multi-label classification task for sound data. The training data was composed of a small amount of reliably labeled data (clean data) and a larger amount of data with unreliable labels (noisy data). Additionally, there was a difference between the data distribution in the curated data and the noisy data. To tackle this difficulty, we used two strategies. The first was multitask learning, in which training on noisy data was treated as a different task from clean data. The second was pseudo-labeling (a kind of semi-supervised learning), in which noisy data was relabeled by predicted labels from a model trained with the clean data.

Our next interview was with *Mikel Bober-Irizar*, aka Anokas, a Competitions Grandmaster (the youngest ever Grandmaster on Kaggle), a Master in Notebooks and Discussions, having completed his Master's degree at Cambridge and started his own company on AI gaming, Iconic.

Interview: Mikel Bober-Irizar

`https://www.kaggle.com/anokas`

Can you tell us about yourself?

I joined Kaggle in 2016, back when I was 14 and I had no idea what I was doing – I had just read about machine learning online and it seemed cool. I started in my first few competitions by copying other people's public code from the forums and making small tweaks to it. Throughout a few competitions, I slowly gained an understanding of how things worked, motivated by trying to climb the leaderboard – until I started making good progress, which culminated in coming second in the Avito Duplicate Ads Competition later that year.

Since then, I have participated in 75 competitions, in 2018 becoming the youngest competition Grandmaster and the first ever Triple Master.

What's your favorite kind of competition and why? In terms of techniques and solving approaches, what is your speciality on Kaggle?

I really enjoy competitions with lots of opportunities for feature engineering, and those with lots of different types of data, which allow you to be really creative in the approach you take to solving them – it's a lot more fun than a competition where everyone has to take the same approach and you're fighting over the last decimal place.

I wouldn't say I have a specialty in terms of approach, but enjoy trying different things.

Tell us about a particularly challenging competition you entered, and what insights you used to tackle the task.

A few years ago, Google ran a competition for detecting objects within images and the relationships between them (e.g., "chair at table" – see: `https://www.kaggle.com/competitions/google-ai-open-images-visual-relationship-track`). Other teams spent ages taking a conventional approach and training large neural networks to tackle the tasks, which I didn't have the knowledge or compute to compete with. I chose to attack the problem from a different angle, and using some neat heuristics and tree models I ended up in seventh place with just a few hours of work.

Has Kaggle helped you in your career?

Kaggle has led to lots of opportunities for me and has been a really great community to get to know. I've met lots of people and learned a lot throughout all the competitions I've participated in. But Kaggle is also how I got into machine learning in the first place – and I don't think I would be in this field otherwise. So yes, it's helped a lot.

What mistakes have you made in competitions in the past?

It's quite easy to end up with a complicated solution that you can't replicate from scratch, since chances are you'll be using various versions of code and intermediate datasets in your final solution. Then, if you're lucky enough to win, it can be very stressful to deliver working code to the host! If you're doing well, it's a good idea to pin down what your solution is and clean up your code.

It's also easy to get into a situation where you use different validation sets for different models, or don't retain validation predictions, which can make it hard to compare them or do meta-learning later on in the competition.

Are there any particular tools or libraries that you would recommend using for data analysis or machine learning?

I really like XGBoost, which still tends to beat neural networks on tabular data (as well as its newer cousin, LightGBM). SHAP is really nice for explaining models (even complex ones), which can give you more insights into what to try next.

What's the most important thing someone should keep in mind or do when they're entering a competition?

I think it's important to try not to get bogged down in implementing ultra-complicated solutions, and instead try to make incremental solutions.

Competitions now are a lot harder than when I first started out, so it's a good idea to look at other people's code (lots of people make this public during the competition) and try to learn from them. You might want to consider joining a team with some other Kagglers: competitions in teams have been the most fun competitions for me, and have always been a fantastic learning experience.

And finally: most ideas tend to not work – if you want to win a competition, you need to persevere and keep experimenting!

Kaggle has certainly been influential in the previous three interviewees' rich lives and careers, and they are only just getting started. Next, we speak to two Kagglers who now hold senior roles in their respective companies and have also had long and fruitful journeys thanks to Kaggle.

First, we have *Dan Becker*, a Notebooks Grandmaster and co-founder and CTO at LightTable AI. Kaggle has played a significant part in Dan's career.

Interview: Dan Becker

https://www.kaggle.com/dansbecker

Can you tell us about yourself?

I first tried using machine learning at a 3-person start-up in 2000 where we tried to use neural networks to help retailers optimize the reserve prices they set for items on eBay. We had no clue what we were doing, and we failed miserably.

By 2002, I was confident that machine learning could never work. I got a PhD in economics and took a job as an economist for the US government. I wanted to move to Colorado, but there weren't many jobs there looking for economics PhDs. So I was looking for a less academic credential.

In 2010, I saw a newspaper article about the Heritage Health Prize. It was an early Kaggle competition with a $3 million prize. I still believed that simpler models like what I used as an economist would give better predictions than fancy machine learning models. So I started competing, thinking that a good score in this competition would be the credential I needed to find an interesting job in Colorado. My first submission to that competition was not last place, but it was pretty close. My heart sank when I watched my model get scored, and then saw everyone else was so far ahead of me. I briefly gave up any hope of doing well in the competition, but I was frustrated to not even be average.

I spent all my nights and weekends working on the competition to climb up the leaderboard. I relearned machine learning, which had progressed a lot in the 10 years since I'd first tried it. I'd learn more and upload a new model each day. It took a lot of time, but it was rewarding to march up the leaderboard each day. By the time my score was in the middle of the leaderboard, I thought continued work might get me in the top 10%. So I kept working. Soon I was in the top 10%, thinking I might get in the top 10 competitors.

When I was in the top 10, an analytics consulting company reached out to me to ask if I wanted to be hired and compete under their company name, which they would use for marketing. I told them I would do it if I could work from Colorado. So the Kaggle competition helped me achieve my original goal.

We finished in 2nd place. There was no prize for 2nd place, but everything I've done in my career since then has been enabled by this one Kaggle competition. It was a bigger success than I ever could have imagined.

How else has Kaggle helped you in your career?

Kaggle has almost entirely made my career. My first job as a data scientist came when someone recruited me off the leaderboard. My next job after that was working for Kaggle. Then I worked at DataRobot, whose recruiting strategy at the time was to hire people who had done well in Kaggle competitions. Then I went back to Kaggle to start Kaggle Learn, which is Kaggle's data science education platform. The list goes on. Every job I've had in the last decade is clearly attributable to my initial Kaggle success.

As I switched from economics to data science, my Kaggle achievements were at the heart of why I was hired. Being further in my career now, I don't think in terms of portfolios... and I'm fortunate that I'm recruited more than I look for work.

What's your favorite kind of competition and why? In terms of techniques and solving approaches, what is your speciality in Kaggle?

I've been around the community for a long time, but I haven't intensely dedicated myself to a competition in 7 or 8 years. I enjoy new types of competitions. For example, I was first exposed to deep learning in 2013 as part of Kaggle's first competitions where deep learning was competitive. This was before Keras, TensorFlow, PyTorch, or any of the deep learning frameworks that exist today. No one in the community really knew how to do deep learning, so everyone was learning something new for the first time.

Kaggle also ran an adversarial modeling competition, where some people built models that tried to manipulate images slightly to fool other models. That was very experimental, and I don't know if they'll ever run anything like that again. But I really like the experimental stuff, when everyone in the community is figuring things out together in the forums.

How do you approach a Kaggle competition? How different is this approach to what you do in your day-to-day work?

The last few times I've done competitions, I focused on "What tooling can I build for this competition that would automate my work across projects?". That hasn't been especially successful, but it's an interesting challenge. It's very different from how I approach everything else professionally.

Outside of competitions, I LOVE analytics and just looking at data on interesting topics. I sometimes say that my strength as a data scientist is that I just look at the data (in ways that aren't filtered by ML models).

I also spend a lot of time thinking about how we go from an ML model's prediction to what decision we make. For example, if a ML model predicts that a grocery store will sell 1,000 mangos before the next shipment comes, how many should that grocery store hold in stock? Some people assume it's 1,000... exactly what you forecast you can sell. That's wrong.

You need to think about trade-offs between the cost of spoiling mangos if you buy too many vs the cost of running out. And what's their shelf life? Can you carry extra stock until after your next shipment comes? There's a lot of optimization to be done there that's part of my day-to-day work, and it's stuff that doesn't show up in Kaggle competitions.

Tell us about a particularly challenging competition you entered, and what insights you used to tackle the task.

I tried to build an automated system that did joins and feature engineering for the Practice Fusion Diabetes Classification challenge. The main thing I learned was that if you have more than a few files, you still need a person to look at the data and understand what feature engineering makes sense.

In your experience, what do inexperienced Kagglers often overlook? What do you know now that you wish you'd known when you first started?

New participants don't realize how high the bar is to do well in Kaggle competitions. They think they can jump in and score in the top 50% with a pretty generic approach... and that's usually not true. The thing I was most surprised by was the value of using leaderboard scores for different models in assigning weights when ensembling previous submissions.

What mistakes have you made in competitions in the past?

I've screwed up last-minute details of submissions in multi-stage competitions several times (and ended up in last place or near last place as a result).

Are there any particular tools or libraries that you would recommend using for data analysis or machine learning?

Mostly the standard stuff.

Outside of Kaggle competitions, I personally like Altair for visualization... and I write a lot of SQL. SQL is designed for looking at simple aggregations or trends rather than building complex models, but I think that's a feature rather than a bug.

Finally, we have *Jeong-Yoon Lee*, a multiple-medal-winning Competitions Master and Senior Manager, Applied Science at Uber. Jeong is also leading the CausalML project, an open-source Python package that provides a suite of causal machine-learning algorithms based on recent research. He co-organized the causal machine learning tutorial and workshops at KDD in 2021, 2023, and 2024.

Interview: Jeong-Yoon Lee

`https://www.kaggle.com/jeongyoonlee`

Can you tell us about yourself?

My name is Jeong, and I'm a Senior Manager at Uber. I started Kaggle back in 2011 when I finished my PhD and joined an analytic consulting start-up, Opera Solutions. There, I met avid Kaggle competitors including Michael Jahrer, and we participated in KDD Cups and Kaggle competitions together. Since then, even after leaving the company, I continue working on competitions both as a competitor and an organizer. Lately, I don't spend as much time as I did before on Kaggle, but still check it out from time to time to learn the latest tools and approaches in ML.

Has Kaggle helped you in your career?

Tremendously. First, it provides credentials in ML. Many hiring managers (when I was an interviewee) as well as candidates (when I was an interviewer) mentioned that my Kaggle track records had caught their attention. Second, it provides learning in state-of-the-art approaches in ML. By working on over 100 competitions across different domains, I'm familiar with more approaches to almost any ML problem than my peers. Third, it provides a network of top-class data scientists across the world. I've met so many talented data scientists at Kaggle and enjoy working with them. I translated Abhishek Thakur's book, organized a panel at KDD with Mario, Giba, and Abhishek, and am interviewing for Luca's book. ;)

In 2012, I used Factorization Machine, which was introduced by Steffen Rendle at KDD Cup 2012, and improved prediction performance by 30% over an existing SVM model in a month after I joined a new company. At a start-up I co-founded, our main pitch was the ensemble algorithm to beat the market-standard linear regression. At Uber, I introduced adversarial validation to address covariate shifts in features in the machine learning pipelines.

What's your favorite kind of competition and why? In terms of techniques and solving approaches, what is your specialty on Kaggle?

I like competitions with small to medium-sized datasets, which are mostly tabular data competitions, because I can quickly iterate different approaches even on my laptop anytime anywhere. During my peak time at Kaggle in 2015, I often built my solutions on the airplane or in between my babysitting shifts. My triplets were born in late 2014 and I was working at a new start-up I'd co-founded.

I don't think I have any special modeling techniques, but my specialty is more around competition management, which includes recruiting team members, setting up a collaboration framework (e.g., Git, S3, Messenger, Wiki, internal leaderboard, cross-validation splits), helping the team work effectively throughout the competition, etc. So I'm not a competition Grandmaster myself, but was able to reach the top 10 because other Grandmasters liked to work with me.

How do you approach a Kaggle competition? How different is this approach to what you do in your day-to-day work?

I try to build a pipeline that enables fast iterations and incremental improvements. The more ideas you try, the better chance you have to do well in a competition. The principle applies to my day-to-day work as well. The scope is different, though. At work, we start by defining problems and identifying the data, while at Kaggle, both are given, and we start from EDA.

In your experience, what do inexperienced Kagglers often overlook? What do you know now that you wish you'd known when you first started?

Recently, I noticed that many users simply fork a Notebook shared by other users and fine-tune it to get better scores. Eventually what matters is learning, not the Kaggle ranking or points. I recommend that new Kagglers spend more time building their own solutions.

What's the most important thing someone should keep in mind or do when they're entering a competition?

It's about learning, not about winning.

Do you use other competition platforms? How do they compare to Kaggle?

I'm advising Dacon AI, a Korean ML competition platform company. It started in 2018 and has hosted 96 competitions so far. It's still in an early stage compared to Kaggle but provides similar experiences to Korean users.

The stories from all these Kagglers prove how diverse and successful career paths can be undertaken, starting from your passion for data science competitions. Whether you are at the beginning of your data science journey or aiming for senior leadership roles, Kaggle offers valuable tools and opportunities to help you progress. Next, we will explore how effective Kaggle is as a networking tool.

Participating in Kaggle Days and other Kaggle meetups

A good way to build connections with other Kagglers (and be more easily accepted into a team) is simply to meet them. Meetups and conferences have always been a good way to do so, even if they do not specifically deal with Kaggle competitions, because the speakers talk about their experiences on Kaggle or because the topics have been dealt with in Kaggle competitions. For instance, many Research competitions require successful competitors to write papers on their experience, and the paper could be presented or quoted during a conference speech.

There were no special events directly connected with Kaggle until 2018, when LogicAI, a company created by Maria Parysz and Paweł Jankiewicz, arranged the first Kaggle Days event in Warsaw, Poland, in collaboration with Kaggle. They gathered over 100 participants and 8 Kaggle Grandmasters as speakers. More Kaggle Days events followed. At the moment, such large events are suspended but there are still smaller events going on in the form of local meetups held in various cities. Participating in a major event or a meetup is a perfect opportunity to meet other Kagglers and make friends, and could be helpful both for career purposes or for teaming up for future Kaggle competitions.

Getting spotted and other job opportunities

For some time, Kaggle was a hotspot where employers could find rare data analysis and machine learning modeling skills. Kaggle itself offered a job board among the discussion forums, and many recruiters roamed the leaderboard looking for profiles to contact. Companies themselves held contests explicitly to find candidates (Facebook, Intel, and Yelp arranged recruiting competitions for this purpose) or conveniently pick up the best competitors after seeing them perform excellently on certain kinds of problems (such as the insurance company AXA did after its telematics competitions). The peak of all this was marked by a Wired interview with *Gilberto Titericz*, where it was stated that *"highly ranked solvers are flooded with job offers"* (https://www.wired.com/story/solve-these-tough-data-problems-and-watch-job-offers-roll-in/).

Recently, things have changed somewhat, and many Kagglers report that the best you can expect when you win or score well in a competition is some contact from recruiters for a couple of months. Let's look at how things have changed and why.

How Kaggle can help

Nowadays, you seldom find job offers requiring Kaggle experience since companies most often need previous experience in the field (even better, in the same industry or knowledge domain), an academic background in math-heavy disciplines, or certifications from Google, Amazon, or Microsoft. Your presence on Kaggle will still have some effect because it will allow you to:

- Be spotted by recruiters who monitor Kaggle rankings and competitions
- Be spotted by companies themselves since many managers and human resource departments keep an eye on Kaggle profiles
- Have some proof of your coding and machine learning ability that could help companies select you, perhaps not requiring you to take any further tests
- Have specific experience with problems highly relevant to certain companies that you cannot acquire otherwise because data is not easily accessible to everyone (for instance, telematics, fraud detection, or deepfakes, which have all been topics of Kaggle competitions)

Seldom will your results and rankings be taken into account at face value, though, because it is difficult to distinguish the parts that are actually due to your skill from other factors affecting the results that are of less interest to a company thinking of hiring you (for instance, the time you have available to devote to competitions, hardware availability, or some luck).

Your Kaggle rankings and results will more likely be noticed in the following cases:

- You have scored well in a competition whose problem is particularly important for the company.
- You have systematically scored well in multiple competitions around topics of interest for the company. This is a sign of real competency, meaning you are not simply labeling yourself a "data scientist" or a "machine learning engineer" without a solid basis.
- Through your Kaggle participation, you are showing a true passion for data analysis to the point where you are investing your free time for free. This is a positive, but it may also turn into a double-edged sword and bring lower monetary offers unless you show that you recognize your value.

While they might not make the difference alone, your Kaggle rankings and results can act as **differentiators**. Recruiters and companies may use Kaggle rankings to make lists of potential candidates. The two most noticed rankings are in Competitions and Notebooks (hence, they also have the more intense competition and larger numbers of Grandmasters out of the four ranked areas). Still, sometimes, they also watch the rankings for a *specific* competition. When certain skills (for instance, in NLP or computer vision) are sought after, it is easier to find them in competitions that require you to use them skillfully to be successful.

Another great differentiator comes at interview time. You can quote your competitions to show how you solved problems, coded solutions, and interacted and collaborated with teammates. On these occasions, more than the ranking or medal you got from Kaggle, it is important to talk about the specifics of the Kaggle competition, such as the industry it referred to, the type of data you had to deal with and why it interested you, and also to present your actions during the competition using the **STAR approach**, often used in job interviews.

The STAR approach

The **STAR** (short for **Situation, Task, Action, Result**) method is a narrative framework frequently employed in interview readiness, particularly when you anticipate behavioral questions. Behavioral interviews involve inquiries regarding your prior experiences in various situations, aimed at showcasing your values and approaches when dealing with challenges and opportunities.

Behavioral questions range from the classic and general "Tell me about yourself" to "Tell me a time when you…" and the phrase is completed by a situation. The interviewer wants to hear about your past behavior in order to check if there is a match to their expectations for soft skills, position fit, and culture fit. For instance, during Amazon interviews, recruiters routinely ask behavioral questions to check your fit for their leadership principles (https://www.amazon.jobs/content/en/our-workplace/leadership-principles), while at Google, they will interview you about your Googliness (https://www.thinkwithgoogle.com/future-of-marketing/emerging-technology/missions-that-matter/).

Each company, regardless of its size, has specific expectations regarding a candidate's behavior, and in one way or another, you will encounter a behavioral question during your interview. Indeed, regarding technical positions in data science, the lion's share should be related to your technical skills. Still, there is always a minor part of the interview dedicated to assessing your past behavior, serving as an indicator of your expected future conduct. Acing the behavioral questions also means gaining an important hedge against other candidates, especially if you are new to data science and can fare well technically but not excel.

In the STAR approach, you should structure what you did in a competition based on the framework **Situation**, **Task**, **Action**, and **Result**. This method aims to have you talk more about behaviors than techniques, thus putting more emphasis on your capacities than the capabilities of the algorithm you chose; anyone else could have used the same algorithm, but you managed to use it so successfully.

The method shines when dealing with success stories. Still, you can also apply it to unsuccessful cases, especially for situations where you gained important insights about the reasons for your failure that stopped you from failing in the same way again in the future.

To apply the method, you break down your story into four components:

- **Situation**: Describe the context and the details of the case, which could be a challenge, a problem, or an opportunity so that the interviewer can understand, at a glance, your story.

- **Task**: Describe your specific role and responsibilities in the situation, helping your interviewer to frame your individual contribution in terms of skills and behaviors

- **Action**: Explain in first person (even if working in a team) what specific actions you took to handle your role and objectives

- **Result**: Illustrate the results of your efforts as well as the overall result. If there was a positive outcome, bring examples of measurable accomplishments. If it was negative, explain what you learned and what you would do differently in future similar situations.

Some companies explicitly ask for the STAR approach (or its corollary, the **Goal-Impact-Challenges-Finding** method, where more emphasis is placed on the results); others implicitly expect something similar. The best answers are those that suit the values and objectives of the company you are interviewing for.

Since just reporting the rankings and medals you got in a competition may not be enough to impress your interviewer, reformulating your successful experience in a Kaggle competition is paramount. The approach can work when you have competed solo or in a team; in the latter case, an important aspect to describe is how you interacted with and positively influenced the other teammates. Let's discuss some ways you could do that:

- First, you describe the situation that arose in the competition. This could be in the initial phases, in the experimentation phases, or the final wrap-up. You must provide a clear context for the listener to evaluate whether your behavior was correct for the situation. Be detailed and explain the situation and why it required your attention and action.

- Then, you should explain the task that you took on. For instance, it could be cleaning your data, doing explorative analysis, creating a benchmark model, or continuously improving your solution.

- Next, you describe how you executed the task. Here, it would be handy to present a Medium article or a GitHub project supporting your description (as discussed in the previous chapter). Systematically showcasing your experience and competence through well-written documentation and good coding will enforce your value proposition in front of the interviewer.

- Finally, you have to explain the result obtained, which could be qualitative (for instance, how you coordinated the work of a team competing on Kaggle) or quantitative (for instance, how much your contribution affected the final result).

Summary (and some parting words)

In this chapter, we have discussed how competing on Kaggle can help improve your career prospects. We have touched on building connections by teaming up on competitions, participating in events related to past competitions, and using your Kaggle experience to find a new job. Based on our experience and those of other Kagglers, we have discussed how you need more than competition results on Kaggle to ensure you get a position. However, they can help you get attention from recruiters and human resource departments and reinforce how you present skills in data science (if a carefully built portfolio supports them, as we described in the previous chapter).

This chapter also marks the conclusion of the book. Through all these chapters, we have talked about Kaggle competitions, Datasets, Notebooks, and Discussions. We covered technical topics in machine learning and deep learning (from evaluation metrics to simulation competitions), intending to help you achieve more both on Kaggle and after Kaggle.

Having been involved in Kaggle competitions for ten years, we know very well that you can find everything you need on Kaggle – but everything is dispersed across hundreds of competitions and thousands of Notebooks, discussions, and Datasets. Finding what you need, when you need it, can prove daunting for anyone starting on Kaggle. We compiled what we think is essential knowledge to guide you through all the competitions you may want to participate in. That is why this has not been a book on data science in a strict sense but a book specifically on data science on Kaggle.

Aside from technical and practical hints, we also wanted to convey that, in over ten years, we have always found a way to turn our experiences on Kaggle into positive ones. You can re-read this work as a book describing our endless journey through data science competitions. A journey on Kaggle does not end when you get all the Grandmaster titles and rank first worldwide.

It actually never ends because you can re-invent how you participate and leverage your experience in competitions in endless ways. As this book ends, your journey on Kaggle starts, and we wish you a long, rich, and fruitful experience – as it has been for us. Have an incredible journey!

Join our book's Discord space

Join our community's Discord space for discussions with the authors and other readers:

`https://packt.link/kaggle`

17
Unlock Your Exclusive Benefits

Your copy of this book includes the following exclusive benefits:

- ⌲ Next-gen Packt Reader
- 📄 DRM-free PDF/ePub downloads

Follow the guide below to unlock them. The process takes only a few minutes and needs to be completed once.

Unlock this Book's Free Benefits in 3 Easy Steps

Step 1

Keep your purchase invoice ready for *Step 3*. If you have a physical copy, scan it using your phone and save it as a PDF, JPG, or PNG.

For more help on finding your invoice, visit `https://www.packtpub.com/unlock-benefits/help`.

> **Note:** If you bought this book directly from Packt, no invoice is required. After *Step 2*, you can access your exclusive content right away.

Step 2

Scan the QR code or go to `packtpub.com/unlock`.

On the page that opens (similar to *Figure 17.1* on desktop), search for this book by name and select the correct edition.

Figure 17.1: Packt unlock landing page on desktop

Step 3

After selecting your book, sign in to your Packt account or create one for free. Then upload your invoice (PDF, PNG, or JPG, up to 10 MB). Follow the on-screen instructions to finish the process.

Need help?

If you get stuck and need help, visit `https://www.packtpub.com/unlock-benefits/help` for a detailed FAQ on how to find your invoices and more. This QR code will take you to the help page.

Note: If you are still facing issues, reach out to `customercare@packt.com`.

<packt>

packtpub.com

Subscribe to our online digital library for full access to over 7,000 books and videos, as well as industry leading tools to help you plan your personal development and advance your career. For more information, please visit our website.

Why subscribe?

- Spend less time learning and more time coding with practical eBooks and Videos from over 4,000 industry professionals
- Improve your learning with Skill Plans built especially for you
- Get a free eBook or video every month
- Fully searchable for easy access to vital information
- Copy and paste, print, and bookmark content

At www.packtpub.com, you can also read a collection of free technical articles, sign up for a range of free newsletters, and receive exclusive discounts and offers on Packt books and eBooks.

Other Books You May Enjoy

If you enjoyed this book, you may be interested in these other books by Packt:

EXPERT INSIGHT   kaggle

The

Kaggle Workbook

Self-learning exercises and valuable insights for
Kaggle data science competitions

Konrad Banachewicz
Luca Massaron   <packt>

The Kaggle Workbook

Konrad Banachewicz, Luca Massaron

ISBN: 978-1-80461-121-0

- Take your modeling to the next level by analyzing different case studies
- Boost your data science skillset with a curated selection of exercises
- Combine different methods to create better solutions
- Get a deeper insight into NLP and how it can help you solve unlikely challenges
- Sharpen your knowledge of time-series forecasting
- Challenge yourself to become a better data scientist

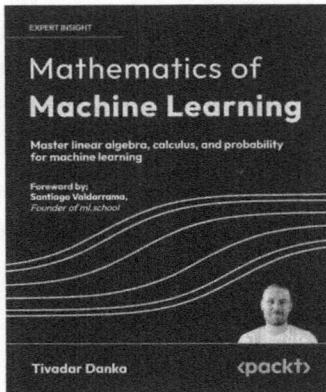

Mathematics of Machine Learning

Tivadar Danka

ISBN: 978-1-83702-787-3

- Understand core concepts of linear algebra, including matrices, eigenvalues, and decompositions
- Grasp fundamental principles of calculus, including differentiation and integration
- Explore advanced topics in multivariable calculus for optimization in high dimensions
- Master essential probability concepts like distributions, Bayes' theorem, and entropy
- Bring mathematical ideas to life through Python-based implementations

AI by Hand

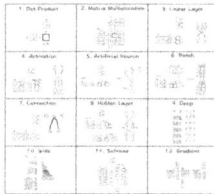

Deep Learning
Math Workbook

Prof. Tom Yeh

Deep Learning Math Workbook

Tom Yeh

ISBN: 978-1-80667-477-0

- Connect hand calculations to the behavior of modern deep neural networks
- Break down deep learning math into small, solvable puzzles
- Discover the meaning behind core operations, such as dot product, matrix multiplication, and normalization
- See how linear layers, activations, and loss functions fit together
- Link the math of simple neurons to modern deep networks
- Develop true intuition for AI, not by memorizing formulas, but by reasoning step by step

Packt is searching for authors like you

If you're interested in becoming an author for Packt, please visit `authors.packt.com` and apply today. We have worked with thousands of developers and tech professionals, just like you, to help them share their insight with the global tech community. You can make a general application, apply for a specific hot topic that we are recruiting an author for, or submit your own idea.

Share your thoughts

Now you've finished *The Kaggle Book, Second Edition*, we'd love to hear your thoughts! Scan the QR code below to go straight to the Amazon review page for this book and share your feedback or leave a review on the site that you purchased it from.

`https://packt.link/r/183508320X`

Your review is important to us and the tech community and will help us make sure we're delivering excellent quality content.

Index

Symbols

2.5D U-Net 469

3D object detection task 453

3D semantic segmentation 461
baseline model 461

3D U-Net 465

4th place solution
base engine, choosing 598
broader implications 602
code example and repository 601, 602
evaluation function, enhancing with neural network 599
memory usage, optimizing 598

β -amylase 453

A

acquisition function 310

adaptive overfitting 187

adversarial testing 27

adversarial validation 198
example implementation 223, 224
training and test data distributions, handling 225-227
using 221-223

AI Ethics
reference link 90

Akiyama, Osamu
approach, to Kaggle 639, 640

Analytics competitions 20

Analytics Vidhya
URL 627

API token 56

architecture 325

artifacts 356

arXiv
URL 631

attention 335

augmentations
applying, offline and online methods 398

augmentation strategies 396
brightness 396
cropping 396
flipping 396
rotation 396
scaling 396

auto-correlation 205

autoencoders 271
denoising with 271

AutoGluon
URL 276

Automated Machine Learning (AutoML) 276, 528
for tabular competitions 276, 277

AutoViz
reference link 248

average precision 153

average precision at k (AP@K) 170

averaging ensembling technique 366-368
cross-validation strategy, averaging 375

majority voting 369-371

model predictions, averaging 372, 373

ROC-AUC evaluation averaging, correcting 376

weighted averages 373, 374

averaging techniques 363

B

Badges

reference link 38

bagging technique 363

Bayesian optimization (BO) 309-310

extending, to neural architecture search 325-333

scikit-optimize, using 311-317

search customization 317-324

Becker, Dan

approach, to Kaggle 644-646

BERT

URL 103

bias 193-197

BitsAndByte 563

bivariate analysis 248

Black formatter

URL 616

blending solutions

creating 389-391

Bober-Irizar, Mikel

approach, to Kaggle 641-643

bootstrap

using 213-215

bootstrapping 298-300

Boruta

reference link 269

BorutaShap

reference link 269

bounding box 419

C

calibration function 179

capstone case study 452

CatBoost 304, 305

reference link 304, 305

URL 259

ChatGPT 628

Chesler, Ryan

insights, on competition 216, 217

classification tasks 132, 133

binary problems 133

multi-class prediction problem 133

multi-label problem 133

class weighting 455

cluster analysis 204

code competitions 24, 186

Cohen Kappa score 159

Colab Secrets 58

column block 303

Common Objects in Context (COCO) 434

Common Task Framework (CTF) paradigm 25, 26

competition considerations 26-28

Community competitions 20

competitions 37

validation, importance 189-191

complex stacking solutions

creating 389-391

computational resources 28-30

Kaggle Notebooks 30, 31

concept drift 226

conditional tabular generative adversarial network (CTGAN) 238

confusion matrix 149

Connect X
   working with 582-587

Conort, Xavier 392

ConvNeXt-based 3D CNN 469

correlated 198

correlation matrix 374

cost function 131

cropping 396

cross-entropy 154

cross-entropy loss 466

cross-validation 180

Cryo-electron tomography (CryoET) 452

CZII - CryoET Object Identification
      competition 452
   3D U-Net segmentation approach 461
   challenges 454, 455
   data exploration overview 456-461
   data format 453
   dataset characteristics 453
   evaluation strategy and challenges 454
   matching criteria 454
   modeling implications 453

D

Danese, Alberto
   insights, on Kaggle 307, 309

Darragh
   reference link 80

data
   gathering 49-54

data augmentation 396

data cleaning techniques
   missing values treatment 259
   outlier capping or removal 259

data drift 226

data leakage
   feature leakage 231, 232
   handling 230
   handling, in Kaggle 233, 234
   in, training examples 232, 233

data processing
   speeding up 254-256

DataRobot
   URL 276

data science competition platforms
   AIcrowd 12
   Alibaba Cloud 12
   Analytics Vidhya 12
   CodaLab 12
   CrowdANALYTIX 11
   DrivenData 11
   EvalAI 11
   Kaggle competition platform 7-10
   Numerai 11
   rise 4, 5, 7
   Signate 11
   Zindi 12

data science competitions 3

data science portfolio 83-86

data scientists
   connections, building with 636, 637

dataset 612
   legal caveats 60
   leveraging 620, 621
   setting up 43-49
   working with 55, 56

data size
   reducing 252, 254

Data Version Control
   reference link 352

decision trees 178

deep neural networks (DNNs) 325

denoising autoencoder (DAE) 238, 274, 275
  bottleneck DAE 272
  decoder part 272
  deep stack DAE 273, 274
  encoding part 272
  using 272

DenseVNet 465

Deotte, Chris
  insights, on Kaggle 408, 409

Detectron2
  reference link 433

Dev.to
  URL 627

Dice coefficient 168, 169

Dice loss 466

Dice score 463

differentiators 651

Direct Preference
    Optimization (DPO) 561, 562

discussions 612
  leveraging 619, 620

DistilBERT 477

Docker
  URL 621

Dockerfile 621

E

early stopping 416

Efficiency Tracks 28

EfficientNet-B0 (EffNetB0) 415

EfficientNet networks
  reference link 411

embargo 207

embarrassingly parallel algorithm 289

embeddings 489

empirical Bayesian approach 265

ensemble learning 468

ensemble probability volume 467

ensemble selection technique 380

ensembling algorithms 362, 363

error function 132
  post-processing tuning 175-177
  probabilistic adjustments,
      of predictions 178-180

evaluation metrics 131
  custom metrics 172-175
  custom objective functions 172-175
  optimizing 171, 172

exploitation 310

exploration 310

exploratory data analysis (EDA) 237
  dimensionality reduction, with t-SNE and
      UMAP 250, 252
  importance 247
  performing, in Kaggle 248, 249

Exponential Moving Average (EMA) 469

eXtreme gradient boosting (XGBoost) 302
  reference link 303

extremely randomized trees
    (extra trees) 298, 299

F

F1 score 153

false negatives (FNs) 454

false positives (FPs) 454

fastpages
  reference link 627

F-beta score 153, 454

Featured and Research competitions 24

feature engineering 237, 486
  applying 256, 257
  categorical feature, encoding 258
  categorical features, splitting and
      aggregating 258
  meta-features, based on
      rows and columns 260, 261
  numeric features, binning 258
  numeric feature transformations 258
  polynomial features, creating 258
  target encoding 262-267
  time feature processing 258

feature importance
  using, for evaluating work 267-269

feature leakage 231, 232

Fink, Laura
  insights, on Kaggle 448-451

flipping 396

focal loss 173

FreeCodeCamp
  URL 627

Freedman-Diaconis rule 204

G

Game AI
  reference link 90

Gaussian process (GP) 310

Gemini 628

Gemma 628
  AI assistants for data tasks 573
  fine-tuning 561
  fine-tuning, techniques 562

Gemma 2 560

general leaderboard

reference link 36

generative AI 554, 555

geometric mean 372

Gists 629

GitHub 629, 630
  URL 625
  used, for saving Kaggle Notebook 76-79

Goal-Impact-Challenges-Finding
    method 652

golden features 230

Gonen, Firat
  insights, on Kaggle 602-604

Google-AI Assistants for Data Tasks with
    Gemma competition
  overview 573-578

Google Cloud AI notebooks 81

Google Cloud Platform (GCP)
  upgrading 81-83

Google Colab 55, 630
  Kaggle datasets, using in 56-59
  URL 56

Google Dataset Search engine 60

Google Efficient Chess AI Challenge 596, 597
  reference link 598

Googliness
  reference link 651

gradient boosting 363

gradient boosting decision
    trees (GBDT) 255, 300

gradient tree boosting
  CatBoost 304, 305
  HistGradientBoosting 306, 307
  LightGBM 300-302
  XGBoost 302-304

Gradio 630
  URL 616
grid search 288-291

H

H2O Driverless AI
  URL 276
Hacker Noon
  URL 627
Halite
  reference link 594
halving algorithm 288
Hand-Crafted Evaluation (HCE) 598
harmonic mean 372
Henze, Martin
  insights, on Kaggle 86, 88
high-cardinality feature 262
HistGradientBoosting 306
holdout 197
Hugging Face Spaces 630
hyperband optimization 333
Hyperband tuner
  reference link 341
hyperopt 343

I

image brightness 396
image classification 409-419
ImageDataGenerator approach 401-407
imitation learning
  reference link 596
independent and identically
    distributed (i.i.d.) 188

instance normalization
    (InstanceNorm3d) 466
instruction tuning 557
inter-annotation agreement 159
Intermediate ML
  reference link 89
International Collegiate Programming
    Contest (ICPS) 4
interquartile range (IQR) 204, 259
intersection over union (IoU) 167, 168, 440
Ishihara, Shotaro
  insights, on Kaggle 496-498
isotonic regression 180

J

Jaccard index 167
Janson, Giuliano
  insights, on competitions 227-229

K

Kaggle 48
  portfolio, building 610-612
  URL 43
  uses 650, 651
Kaggle baseline
  ensemble of lightweight 3D models 465
  ensembling and inference, performing 467
  overall trends emerged 470, 471
  starting from 463, 465
  top solutions and insights 468, 469
Kaggle competition 13
  examples 17-20
  leaderboard 186-188
  limitations 39, 40
  stages 13-17

types 17-20

Kaggle datasets
using, in Google Colab 56-59

Kaggle Days
reference link 33

Kaggle Days event 649

Kaggle Days Milan
reference link 33

Kaggle environment 30

Kaggle forums
discussion approaches 120
working 114-120

Kaggle forums, discussion approaches
common challenges 120, 121
information leakage 122
overfitting 122

Kaggle game agent competitions 594-596

kagglehub package 59

Kaggle Learn
courses 89, 90
reference link 89

Kaggle meetups 649

Kaggle Models 95
Data Type selection 97
fine tuneable 101
Framework selection 98
Language selection 99
License type 100
model, uploading to 108-110
selecting 96
size 102
task 96
using 103-106

KaggleNoobs 33

Kaggle Notebook 55, 56
reference link 334
resources 80, 81
running 74, 76
saving, to GitHub 76-79
setting, as utility script 79, 80
setting up 65-73

Kaggle Notebook, methods
from a dataset 66
from the frontpage 66
from the homepage 65

Kaggler-ja
reference link 33

KDD Cup 5

KDnuggets
URL 627

Keras built-in augmentations 401
albumentations package 406, 407
ImageDataGenerator approach 401-403
preprocessing layers 404

KerasTuner 333
lighter and faster models,
creating with 333-342

kernel function 290

key parameters 296

k-fold cross-validation 199
in financial time series 207-210
nested cross-validation 210, 211
OOF predictions, producing 212
sequential cross-validation,
in time series 205, 206
working 200-205

K-fold variations 202, 203

Kyara 561

L

LabelEncoder 262

Large Language Models (LLMs) 237, 554
  fine-tuning for languages 560
  use cases 557, 558
  working 555-559

Larko, Dmitry
  insights, on competition 192, 193

law of large numbers 199

leadership principles
  reference link 651

leaf-wise tree growth strategy 300

leaky validation strategy 230

Lee, Jeong-Yoon
  approach, to Kaggle 647, 648

level-wise tree growth strategy 300

lexical diversity 486

LightGBM 300-302
  URL 259

linear models 296

linear transformations 146

Llama 3 628

**LLM Nerd-Off Sharpened
    Cosine Similarity** 565

LLM prompt recovery 564
  challenge evaluation 565, 566
  competition overview 564
  data examination 564, 565
  third-place solution 567, 568
  top solution approaches 566, 567

logarithmic mean 372

logarithmic transformation 145

log loss 154

LoRA fine-tuning 577

loss function 131

Low-Rank Adaptation (LoRA) 570

Lukyanenko, Andrey
  insights 160-164

M

Machine Learning Explainability
  reference link 90

Machine Learning (ML)
  reference link 89

MAE evaluation metric 146-148

magic features 230

Maranhão, Andrew
  approaches, to Kaggle 50-54

**Matthews correlation coefficient
    (MCC)** 156, 157

MCRMSLE 145

**mean average precision at k
    (MAP@K)** 169, 170

mean of powers 372

mean prompt
  attribute isolation, with tag model 571
  cluster spectrum 571
  end-to-end inference flow 571
  full-prompt predictor, building 570
  gateclassifier, use for safeguarding 570
  origins 569
  quantitative outcome and
      qualitative lessons 572

Mean Squared Error (MSE) 143, 600

Medium
  URL 625

Meta Kaggle dataset 135-138
  URL 135

meta-model 377
  models, blending with 377, 378

metrics
  for multi-label classification and
      recommendation problems 169, 170

metrics for classification 148
  accuracy 148-150
  F1 score 153
  log loss 154
  Matthews correlation
      coefficient (MCC) 156, 157
  precision 151-153
  recall 151-153
  ROC-AUC 154, 155

metrics for multi-class classification
  macro averaging 157
  Macro-F1 158, 159
  Mean-F1 159
  micro averaging 158
  multiclass accuracy 158
  multiclass log loss 158
  weighting 158

metrics, for object detection and
      segmentation problems 164-166
  Dice coefficient 168, 169
  intersection over union (IoU) 167, 168

metrics for regression 142
  MAE 146-148
  mean squared error (MSE) 143
  RMSE 144, 145
  RMSLE 145, 146
  R squared 143, 144

Mistral 628

mixup 274

MLflow
  reference link 352

model blending
  best practices 378-383
  with meta-model 377

model ensembling 470

model validation system
  tuning 218-220

ModernNCA 280

Monigatti, Leonie
  approaches, to Kaggle 106-108

Mulla, Rob 364

multi-armed bandit (MAB) 590

multi-head attention 336

Multi-Layer Perceptron (MLP) 599

multi-stage training 455

multivariate analysis 248

N

nagiss
  reference link 598

Nash equilibrium 588

natural language processing (NLP) 335, 473

Neighborhood Component
      Analysis (NCA) 280

Neptune
  reference link 352

nested cross-validation 210-212

networking 32, 635

neural architecture search (NAS) 296

Neural Information Processing Systems
      (NIPS) 188

neural network
  architecture 599
  for tabular competitions 277, 278
  evaluation function, enhancing with 599
  integration, into Stockfish 600
  performance and insights 601
  submission strategy 601
  training 600

Neural Network Universal
     Evaluation (NNUE) 598
Neural Oblivious Decision
     Ensembles (NODE) 279
never-before-seen metrics
     handling 138, 140
nlpaug 548
     URL 548
noise 194
noise swapping 274
     column-wise noise swapping 274
     random noise swapping 274
     row-wise noise swapping 274
non-linear transformations 146
non-parametric model 340
Notebook-only competitions 29
notebooks 612
     leveraging 617-619
NVIDIA KGMoN team 254

O

object detection 164, 419-433
objective function 131
OCR error 549
Olteanu, Andrada
     insights, on Kaggle 91, 92
OneHotEncoder 262
one-hot encoding 258, 304
ongoing competitions
     monitoring 631, 632
online presence, beyond Kaggle
     blogs and publications 626, 628
     GitHub 629, 630
     online demo, making 630
     references 625

Onodera, Kazuki
     insights, on competitions 294, 295
OOF predictions 212
     generating 213
Open Data Science Network
     reference link 33
open domain Q&A 483-496
optimization techniques 288
     grid search 289-291
     halving search 293, 294
     random search 291, 292
Optuna
     reference link 311
     TPE approach 342-346
OrdinalEncoder 262
ordinal tasks 132, 134
out-of-fold (OOF) prediction strategy 383
out-of-fold prediction 180
out-of-memory errors 252
overfitting 194
oversampling 455

P

pandas
     reference link 89
pandas profiling
     reference link 249
patch-based training 455
performance tiers 36, 37
pixel accuracy 167
portfolio 612
     building, with Kaggle 610-612
     good portfolio, golden rules 615, 616
precision 151

precision at k (P@K) 170

precision/recall curve 152

precision/recall trade-off 152

Preda, Gabriel 622-624

PReLU activations 466

principal component analysis (PCA) 204

prioritized recall over precision 454

private repositories 76

private test set 17

probabilistic approach 196

proxy function 309

pruning 344

pseudo-labeling 269-271

public repositories 76

public test set 16

Puget, Jean-François
  insights on Kaggle 282-284

PyTabKit
  reference link 279

PyTorch Tabular
  reference link 279

Q

Quadratic Weighted Kappa 138, 159

Quantization-Aware Training (QAT) 600

R

radial basis function (RBF) kernel 290

Rajkumar, Sudalai
  insights 181, 182

random forest algorithms 363

random forests 298, 299

random search 288-292

random state
  setting for reproducibility 245, 246

rankings 36, 37

Rao, Rohan
  insights, on Kaggle 141, 142

RAPIDS
  reference link 251

RAPIDS libraries 255

RealMLP 280

receiver operating
  characteristic (ROC)-AUC 153

Rectified Adam (RAdam) 336

recurrent neural networks (RNNs) 335
  text classification with 514

Reddit discussion, on Kaggle
  reference link 609

regression tasks 132

regularization 542

Reinforcement Learning from Human
  Feedback (RLHF) 557

reinforcement learning (RL) tasks 132

remote URL 45

replicable 383

Retrieval-Augmented
  Generation (RAG) 5, 553, 574

RMSE 144, 145

ROC curve 154

rock-paper-scissors 587-590

root mean squared log error (RMSLE) 145

rotation 396

R squared 143, 144

run-length encoding (RLE) 434

S

sampling function 329

sampling strategies 197

Santa competition 2020 590-594

scaling 396

scikit-multilearn package
  reference link 203

scikit-optimize
  using 311-317

scoring function 132

Scott's rule 204

segmentation 164, 166, 433
  instance segmentation 166
  semantic segmentation 166

segmentation-to-centroid approach 465

SegResNet 465

semantic segmentation 433-448

sentiment analysis 473-480

shake-up 187, 193

shared words 487

sigmoid method 180

simple format 23

simulated typo 548

Spearman correlation 483

special loss functions 455

splitting strategies 196
  options 196

stability selection procedure 268

stacking 363, 383
  models 383, 384
  performing 385-388
  variations 388, 389

STAR approach 651, 652, 653

state of the art (SOTA) solutions 395, 630

Stochastic Weighted Averaging (SWA) 336

Stockfish
  neural network, integrating to 600

stratification 198

stratified k-fold 203

Streamlit 630

SOLT - Streaming Over Lightweight data
  Transformations 398-400
  URL 616

study 343

subsampling 213

successive halving 293

sum of squared errors (SSE) 143

sum of squares total (SST) 143

Supervised Fine-Tuning (SFT) 562

support vector classifier (SVC) 289

support vector machines (SVM) 297, 298

surrogate function 309, 342

swapping 542

Sweeps optimization 356-359

Sweetviz
  reference link 249

synonym replacement 540

synthetic data generation 560

systematic experimentation 189

T

TabArena 277

TabICL 281

TabM 280

TabNet 279

TabPFNv2 281

TabR 280

tabular competitions
AutoML, using 276, 277
neural networks, using 277-279

Tabular Playground Series 238, 240, 243

Takuoko
reference link 497

target encoding 263, 304

teaming 32

technical debt 611

tensor processing units (TPUs) 398

test phase 453

test set
private part 186
public part 186

Test-Time Augmentation (TTA) 465, 470

text augmentation strategies 540
basic techniques 540-545
nlpaug 548-550
with back-and-forth translation 546, 548

text classification, with AutoTrain 528
AutoTrain parameters, setting up 529
Autotrain project, creating 530
Autotrain project, initializing 530
custom Dataset class, creating 531
environment and dependencies,
setting up 528, 529
predictions, running with trainer 532, 533
pretrained model and tokenizer, loading 531
test data, preparing for inference 531

text classification, with DistilBERT 519
custom Dataset class, creating for multi-
label classification 521, 522
data loaders, creating 523

data splitting, into training and
validation sets 522
environment and dependencies,
setting up 520
model and optimizer,
preparing for training 524, 525
model architecture, defining 524
predictions, formatting 527, 528
predictions, saving 527, 528
test data, inferencing 526, 527
test data, preparing 526
test data, processing 526
tokenizer, initializing 523
training data, loading 520
training data, preparing 520
training loop 525, 526

text classification, with LLM embeddings
and logistic regression 533
cross-validation and training 538
data, cleaning 535
data, loading 535
embeddings, converting to
NumPy arrays 536
embeddings, generating for data 535
embeddings, saving for later use 536
function, defining 537
helper function, defining for
embeddings 534
iterating, over each target 538, 539
NVIDIA embeddings 536
OpenAI client, initializing 534
OpenAI embeddings 533
performance, recording 539
predictions, making 539
specific data anomalies, handling 535
stage, setting 537, 538
submission, preparing 539, 540

text classification, with RNNs 514
  datasets, splitting 516
  embeddings, loading 516
  imports and environment setup 514
  Keras model, building 516
  multiple seeds, averaging 517
  multiple seeds, training 517
  preprocessed data, loading 515
  submission file, creating 519

TF-IDF representations 491

Thakur, Abhishek
  insights, on Kaggle 481, 482

time_ids 209

Titericz, Gilberto
  insights, on Kaggle 613-615

toxic comments classification 499
  character-level TF-IDF vectorization 502
  contraction patterns, defining 508
  data normalization 512, 513
  data, reading 512, 513
  dataset, loading 501
  files, loading 505, 506
  global variables, setting up 505, 506
  libraries, importing 505, 506
  logistic regression model and cross-
      validation, training 503
  main normalization function 511
  necessary libraries, importing 500
  normalize_by_dictionary() function 510
  predictions, saving to CSV file 504
  preprocessing parameters, setting 507
  pretrained dictionaries, loading 507
  spaCy model, loading 511
  target labels, defining 501
  text classification, with TF-IDF and logistic
      regression 500

  text preprocessing and cleanup 504
  total cross-validation score, calculating 504
  toxic words, splitting 509
  TweetTokenizer, tokenizing with 509
  word and character features, combining 503
  word-level TF-IDF vectorization 501, 502

train-test split 197, 198
  k-fold cross-validation 199
  probabilistic evaluation methods 199

transformations
  working 396, 397

Tree Parzen Estimators 310

**Tree-structured Parzen
      Estimators (TPEs)** 310, 343

true positives (TPs) 440, 454

t-SNE 250

Tversky loss 463, 466

two-stage competition 23

U

UMAP 250

UNet2E3D 466

univariate analysis 248

Universal Sentence Encoder 489

**Unlocking Global Communication with
      Gemma competition**
  format and data 560
  LLM prompt recovery 564
  mean prompt, origins 569
  top solutions overview 560, 561

usability index 48
  factors 48

V

validation 190
  importance, in competitions 189-191
validation loss 194
validation set 198
validation strategy 190
variance 193-195
version control 76
volumetric data 452
voxel count statistics 467
VoxHRNet 465, 466
VoxResNet 465, 466

W

walk-forward method 207
weighting process (attention) 336
Weights & Biases
  experiments, tracking 352-355
  reference link 352
  Sweeps optimization,
    implementing 356-359
  using 352
  versioning, with artifacts 356
Worksheets
  URL 12

X

X 473
XGBoost
  reference link 302
  URL 259
Xie, Yifan
  insights, on Kaggle techniques and
    approaches 123, 124

Y

YOLOv5 419

Z

Zhang, Yirun
  approach, to Kaggle 639

www.ingramcontent.com/pod-product-compliance
Lightning Source LLC
Chambersburg PA
CBHW081209220326
41598CB00037B/6729